Biochemistry: Concepts, Techniques and Applications

Biochemistry: Concepts, Techniques and Applications

Edited by **Oliver Stone**

Callisto Reference
New York

Published by Callisto Reference,
106 Park Avenue, Suite 200,
New York, NY 10016, USA
www.callistoreference.com

Biochemistry: Concepts, Techniques and Applications
Edited by Oliver Stone

International Standard Book Number: 978-1-63239-680-8 (Hardback)

Printed in China.

Contents

Preface

Biochemistry is the combination of two disciplines; biology and chemistry. It studies chemical reactions occurring within the living organisms or related to them. Applied biochemistry, biochemistry of cell metabolism and the endocrine system, bioenergetics, biochemical mechanisms, proteins, carbohydrates, lipids, nucleic acids and other biologically active molecules, protein synthesis, cell membrane transport and signal transduction, industrial applications of biochemistry like food and chemical toxicology and pharmacology are some of the crucial topics discussed herein, in the light of new findings and researches. Different approaches, evaluations, methodologies and advanced studies in the field of biochemistry have been included in this book. It will help the readers in keeping pace with the rapid changes in this field. This book collaborates researches of international experts to provide a truly global outlook on the subject.

Significant researches are present in this book. Intensive efforts have been employed by authors to make this book an outstanding discourse. This book contains the enlightening chapters which have been written on the basis of significant researches done by the experts.

Finally, I would also like to thank all the members involved in this book for being a team and meeting all the deadlines for the submission of their respective works. I would also like to thank my friends and family for being supportive in my efforts.

Editor

1

Vascular Endothelial Growth Factor Gene Polymorphisms in Type 2 Diabetes Mellitus; Insertion/Deletion at -2549 is Associated with Retinopathy but not + 405 G/C Polymorphism

Hanaa M. Abdeen[1*], Nadia El-Menshawy[2], Ziad Tawheed[2], Nader El-Malky[3] and Abeer M. Khattab[4]

[1]Department of Medical Biochemistry and Molecular Biology, Faculty of Medicine, Mansoura University, Mansoura, Egypt.
[2]Department of Clinical Pathology, Faculty of Medicine, Mansoura University, Mansoura, Egypt.
[3]Department of Internal Medicine, Faculty of Medicine, Mansoura University, Mansoura, Egypt.
[4]Department of Ophthalmology, Faculty of Medicine, Mansoura University, Mansoura, Egypt.

Authors' contributions

This work was carried out in collaboration between all authors. Author HMA designed the study, wrote the protocol. Authors Nader El-Malky and AMK supplied the patients' samples. Authors ZT, Nadia El-Menshawy and HMA carried out all laboratories work and performed the statistical analysis. Author HMA wrote the first draft of the manuscript. All authors read and approved the final manuscript.

Editor(s):
(1) Chunying Li, Department of Biochemistry and Molecular Biology Wayne State University School of Medicine, Detroit, USA.
Reviewers:
(1) Ge Wu, The Fourth Hospital of Jilin University, China.
(2) Samah Mohamed Ahmed Elattar, Cairo University, Egypt.

ABSTRACT

Background: Retinopathy is a serious ocular complication of diabetes. Vascular endothelial growth factor (VEGF) is massively unregulated in diabetic retinopathy. The objective of this study was to investigate the G + 405 C polymorphism in the 5'-untranslated region and the I/D polymorphism at the – 2549 position of the promoter region of the VEGF gene in patients with type 2 diabetes mellitus.

**Corresponding author: E-mail: hanaamaher1@yahoo.com*

Methodology: In this study, 103 unrelated type 2 diabetic patients and 40 control subjects were involved in a case-control design. Genotyping of the I/D polymorphism at -2549 was analyzed by PCR and the G/C +405 polymorphism by PCR-RFLP.
Results: For +405 G/C polymorphism, there is no statistical significant difference in genotype distribution and allele frequencies in both diabetes groups when compared to control group or to each other. While I/D polymorphism at -2549 shows significant increase in ID genotype distribution in diabetes without and with retinopathy groups versus control group [OR (CI) =3.67 (1.27-10.8) & 5.5 (1.98-15.73) and P =0.01 & <0.001, respectively]. I allele is significantly increased in diabetes without and with retinopathy groups versus control group [OR (CI) = 3.99 (1.8-8.97) & 3.98 (1.84-8.71), respectively and P <0.001 for both]. There is no significant difference in genotype and allele frequencies in both diabetes groups when compared to each other (P =0.59 & 0.99, respectively).
Conclusions: Heterozygous form of I/D polymorphism and I allele at -2549 of VEGF gene might possibly be associated with a higher susceptibility to retinopathy as a complication of type 2 diabetes mellitus. While, no association was found with VEGF +405 G/C gene polymorphism.

Keywords: VEGF gene; diabetic retinopathy; -2549 I/D polymorphism; +405 G/C gene polymorphism; type 2 diabetes.

1. INTRODUCTION

Diabetic retinopathy (DR) is one of the most prominent vascular complications in diabetes and is the most common cause of blindness among diabetic adults [1]. It is characterized by increased vascular permeability, tissue ischemia and angiogenesis [2]. DR progresses from non-proliferative (background diabetic retinopathy) into proliferative (high-risk proliferative retinopathy) and finally advanced diabetic eye disease which is accompanied by recurrent hemorrhage and fibrosis, this results in tractional retinal detachment and neovascular glaucoma [1].

The pathogenesis of DR is influenced by environmental and genetic factors. Ethnic differences in the prevalence of DR may offer understandings into the relative importance of genetic or environmental risk factors. Therefore, it is important to identify molecular markers that may help in the diagnosis of DR in multiple populations [3].

Vascular endothelial growth factor (VEGF) is one of the most potent endothelial cell mitogens and plays a critical role in angiogenesis [4]. It is a highly conserved homodimeric glycoprotein [5], which specifically binds to two transmembrane VEGF tyrosine kinase receptors on endothelial cells to initiate intracellular signal transduction pathways that mediate angiogenesis and vascular permeability [4]. VEGF is massively unregulated in DR and not only promotes neovascularization but also increases vascular permeability and leakage [6]. It is produced from many cell types within the eye [5] and plays a key role in the pathogenesis of diabetic microvascular complications [6].

Several studies have shown that VEGF expression is increased in patients with DR [7,8] and others documented that VEGF levels are markedly elevated in vitreous of the eyes of individuals with proliferative DR [9]. The genetic variations in the VEGF gene can influence levels of VEGF protein expression [2].

The human VEGF gene is located on chromosome 6 (6p21.3) and is highly polymorphic. It includes eight exons and seven introns. Of particular interest is an insertion/deletion (I/D) polymorphism of the 18 base pair (bp) fragment at – 2549 position of the promoter region that has been implicated in a number of diseases, especially those with angiogenic basis [10,11]. Also, at least 30 single-nucleotide polymorphisms (SNP) in this gene have been described. The G + 405 C (rs2010963) polymorphism in the 5′-untranslated region is a common SNP and is related to VEGF protein production [12,13]. The +405 G/C polymorphism alters the activity of the internal ribosomal entry site B enhancing initiation of translation at the AUG start codon. Also, regulates the production of the large VEGF isoform translated at an alternative CUG codon. Therefore, it is believed that the +405 G/C polymorphism likely affects expression at the post-transcriptional level [14,15].

The purpose of this study was to investigate the G + 405 C polymorphism in the 5′-untranslated region and the I/D polymorphism of the 18 bp fragment at the – 2549 position of the promoter region of the VEGF gene in patients with type 2 diabetes mellitus and to assess their possible role in the risk of developing diabetic retinopathy.

2. METHODS

A case-control study enrolled 143 subject; 103 unrelated type 2 diabetic patients and 40 controls was conducted. Diagnosis of type 2 diabetes based on the current American Diabetes Association Criteria for the Diagnosis and Classification of Diabetes [16]. Diabetic patients were classified into two groups. Diabetes without retinopathy group (46 patients) recruited from Diabetic & Endocrinology Unit at Specialized Medical Hospital, Mansoura University, Egypt. After carful fundus examination to exclude any retinopathy. Diabetes with retinopathy group (57 patients), that are proved to have retinopathy and recruited from Fluorescein Angiography Unit and Argon Laser Clinic, Mansoura Ophthalmic Centre, Mansoura University, Egypt.

Control subjects were unrelated healthy volunteers with matched age and sex with no history of diabetes or any major clinical disorders and had normal glycosylated hemoglobin (HbA1c). Also, carful fundus examination was done to exclude any retinopathy. Also, HbA1c was measured for all diabetic patients. Fully informed and written consent was obtained from all participants. The protocol of the study was approved by the Institutional Ethics Committee and in accordance with the guidelines of the Declaration of Helsinki.

2.1 Samples Collection and DNA Extraction

Venous blood samples (6 ml) were obtained on EDTA containing tubes from all patients and control subjects; 1ml of the sample was used for HbA1c estimation (COBAS, INTEGR, Roche Diagnostics, USA). Genomic DNA was isolated from peripheral blood leukocyte using the QiaAmp kit (Qiagen, GmbH, USA). The extracted DNA was stored at -20℃ until analysis.

2.2 VEGF Gene I/D Polymorphism at -2549 Analysis

Genotyping of the I/D polymorphism was analyzed by polymerase chain reaction (PCR) using the following primers: forward 5'-GCTGAGAGTGGGGCTGACTAGGTA-3' and reverse 5'-GTTTCTGACCTGGCTATTTCCAGG-3'. PCR amplification was performed using Dream Tag Green PCR Master Mix (Fermentas, USA). Genomic DNA (300 ng) was amplified in a final volume of 30 µl using the following conditions: denaturation at 95℃ for 6 min followed by 35 cycles at 94℃ for 1 min, 57℃ for 1.5 min and 72℃ for 2 min. A final extension was at 72℃ for 10 min. The amplification products were separated by electrophoresis on 2.5% agarose gel stained with ethidium bromide; GeneRuler 50 bp DNA ladder (Germentas, USA) was used as a marker. For the VEGF I/D polymorphism two bands were observed, 211 bp for D allele and 229 bp for I allele (Fig. 1).

2.3 VEGF Gene G + 405 C Polymorphism Analysis

VEGF G + 405 C polymorphism was analyzed by polymerase chain reaction and restriction fragment length polymorphism (PCR-RFLP). PCR amplification performed as described above using 5′-ATTTATTTTTGCTTGCCATT-3′ and 5′-GTCTGTCTGTCTGTCCGTCA-3′ as the forward and reverse primer pairs, respectively. Then the PCR product was digested with FastDigest BsmFI (Thermo Scientific, USA) restriction nuclease at 37℃ _overnight. The digestion products were separated by electrophoresis through 2.5% agarose gel stained with ethidium bromide; GeneRuler 50 bp DNA ladder (Germentas, USA) was used as a marker. The uncut fragment was 304 bp (C allele) and digestion products were 193 and 111 bp (G allele) (Fig. 2).

2.4 Statistical Analysis

Statistical analyses of data were done by Statistical Package for Social Science version 22 (SPSS, Inc., Chicago, IL). Qualitative variables expressed as number and percentage. Quantitative variables expressed as mean ± standard deviation. Statistical significance was defined as *P* value ≤0.05. For quantitative date, student t-test was used to compare between two groups. Repeated measure ANOVA to compare same group in different times followed by post Hoc test for intra-time comparison. Chi square test was used for qualitative data and odds ratio & 95% confidence interval (OR & 95% CI) for risk assessment.

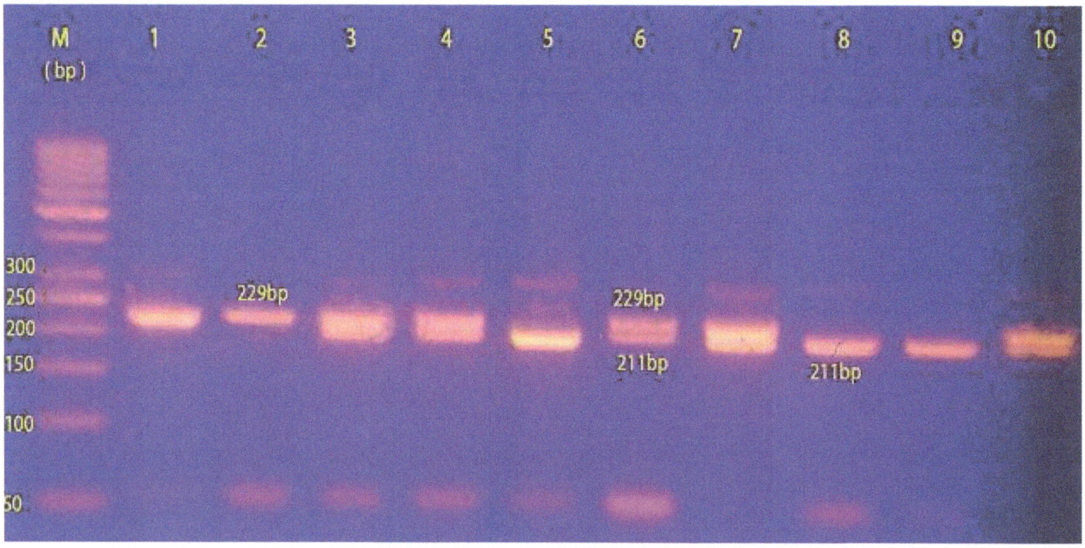

Fig. 1. VEGF gene I/D polymorphism at -2549. marker ladder, lane M; II genotype (229 bp), for example lane 2; DD genotype (211 bp), for example lanes 5, 8 and ID genotype (211 and 229 bp), for example lanes 3, 4, 6 & 10

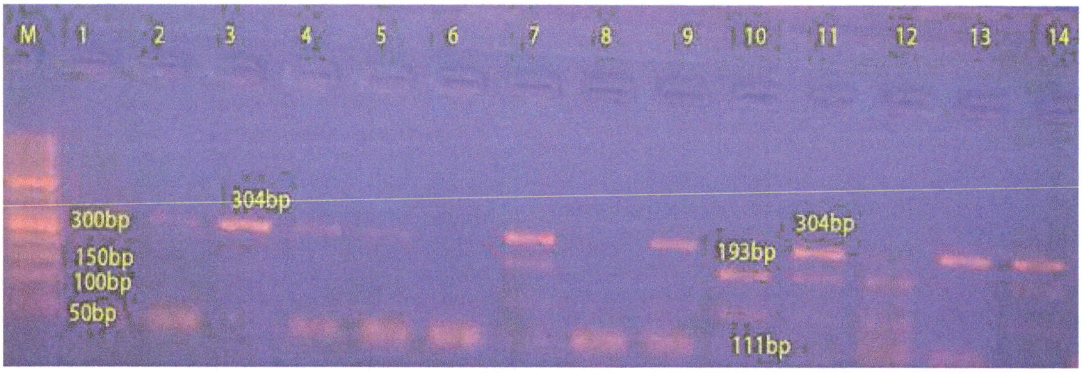

Fig. 2. VEGF gene G + 405 C polymorphism. Marker ladder, lane M; CC genotype (304 bp), for example lane 3; GG genotype (193 and 111 bp), for example lane 10; and GC genotype (304, 193 and 111 bp), for example lanes 7 & 11

3. RESULTS

The characteristics of patients and control subjects are listed in Table 1. There is significant difference in HbA1c in the three groups (P <0.001). Pairwise comparison revealed significant increase in HbA1c in both diabetes groups versus control group (P <0.001, for both) and in diabetes with retinopathy versus without retinopathy (P = 0.002) (not shown in the table).

3.1 VEGF Gene G + 405 C Polymorphism

There is no significant difference in genotype distribution and allele frequency in the diabetes groups (without and with retinopathy) when compared to the control group or to each other (Table 2).

3.2 VEGF Gene I/D Polymorphism at -2549

There is significant decrease in DD genotype distribution in both diabetes groups versus control group [OR (CI) =0.18 (0.06-0.5) & 0.14 (0.05-0.39) for diabetes without and with retinopathy groups, respectively and P <0.001, for both]. While, there is significant increase in ID genotype distribution in both diabetes without and with retinopathy groups versus control [OR (CI) =3.67 (1.27-10.8) & 5.5 (1.98-15.73) and P =0.01 & <0.001, respectively]. However, there is no significant difference for II genotype distribution in both diabetes groups versus control group. I allele is significantly increased in diabetes without and with retinopathy groups versus control group [OR (CI) =3.99 (1.8-8.97) & 3.98 (1.84-8.71), respectively and *P* <0.001 for both]. Meanwhile, D allele is significantly increased in controls. There is no significant difference in genotype distribution and allele frequency in both diabetes groups (without and with retinopathy) when compared to each other (*P* =0.59 & 0.99, respectively) (Table 3).

4. DISCUSSION

Retinopathy is a serious ocular complication of diabetes that causes loss of vision in adults [17]. DR is characterized by increased vascular permeability, tissue ischemia and neovascularization. VEGF can stimulate angiogenesis and increases the permeability of the microvasculature [18] and has been implicated in the development of diabetic retinopathy [11]. It has been suggested that VEGF protein expression is affected by genetic variant in the VEGF gene [18].

In the present study, no significant association was found between VEGF G + 405 C gene polymorphism and DR. This result is in accordance with Watson et al. [12], Ray et al. [19] and Buraczynska et al. [11] who found no significant association between retinopathy status and the VEGF G + 405C polymorphism. Uthra et al. [20] and Petrovič et al. [21] reported no significant association between the rs2010963 (G + 405 C in this study) and DR. In addition, Bleda et al. [22] analyze the VEGF gene +405 polymorphism and found no association with DR but when analyze allele frequency; they observed that the presence of the G allele is a risk marker for developing retinopathy. Lu et al. [18] in their meta-analysis found that VEGF rs2010963 (G + 405 C) polymorphism was not associated with diabetic retinopathy in both Europeans and East Asians ethnity. Also, Han et al. [3] in their met-analysis confirm the non-significant association between VEGF rs2010963 polymorphism and DR.

The result is contradictory with Awata et al. [2] where +405 polymorphism was strongly associated with DR in Japanese subjects. Suganthalakshmi et al. [23] found the heterozygous form is significantly higher in patients with diabetic retinopathy compared with diabetic without retinopathy. While, Szaflik et al. [24] found that the C allele of VEGF +405 polymorphism was associated with an increased risk of DR. This discrepancy might be due to different ethnicity of the patients or may be related to other contributing factors among those populations that need further exploration.

Also, the same patients and controls have been screened for VEGF gene I/D polymorphism of the 18 bp fragment at -2549 position of the promoter region. The current study showed a significant association of ID genotype and I allele with retinopathy in diabetic patients when compared to healthy control group, while no significant difference was found when compared to diabetes without retinopathy group. The results in accordance with that obtained by Bleda et al. [22] where the ID genotype was identified as a potential risk marker of diabetic retinopathy. Moreover, the result is partly in accordance with Foad et al. [25] who found no significant difference in genotype distribution and allele frequency in diabetic patients with retinopathy, diabetics without retinopathy and control subjects.

Table 1. Characteristics of the control, diabetes with and without retinopathy groups

	Control (n= 40)	**Diabetes without retinopathy (n=46)**	**Diabetes with retinopathy (n=57)**	***P***
Age (years)	47.88±7.1	46.72±4.58	48.75±6.13	0.23
Sex:				
Male	18 (45)	20 (43.5)	28 (49.1)	
Female	22 (55)	26 (56.5)	29 (50.9)	0.84
Diabetes duration (years)	-	10.13±3.92	11.63±4.7	0.2
HbA_{1c} (%)	5.24±0.86	7.69±0.65	8.24±1.0	0.000*

Data are number (%) & Mean ± SD.
**: Significant (P = 0.05)*

Table 2. Genotype and allele frequencies of the G + 405C polymorphism in the control, diabetes without and with retinopathy groups

	Control (n= 40)	Diabetes without retinopathy (n= 46)	Diabetes with retinopathy (n= 57)	Diabetes without retinopathy versus control			Diabetes with retinopathy versus control		
				OR (CI)	Pearson x^2 (*df.=2*)	*P*	OR (CI)	Pearson x^2 (*df.=2*)	*P*
Genotype									
CC	32 (80)	28 (60.9)	39 (68.4)	0.39 [0.13-1.14]	2.86	0.09	0.46 [0.16-1.3]	1.94	0.16
GC	6 (15)	14 (30.4)	14 (24.6)	2.48 [0.76-8.33]	2.06	0.15	1.72 [0.54-5.66]	0.59	0.44
GG	2 (5)	4 (8.7)	4 (7.0)	1.81[0.26-15.21]	0.06	0.8	2.51 [0.44-18.59]	0.62	0.43
		P' = 0.73							
Allele: C	70 (87.5)	70 (76.1)	92 (80.7)	0.45 [0.18-1.1]	2.97	0.08	0.6 [0.025-1.43]	1.12	.28
G	10 (12.5)	22 (23.9)	22 (19.3)	2.2 [0.91-5.41]			1.67 [0.7-4.07]		
		P' = 0.42							

*Data are number (%); OR, odd ratio; CI, confidence interval. *: Significant (P = 0.05).*
P': For diabetes without versus with retinopathy groups

Table 3. Distribution of I/D polymorphism at – 2549 position of the promoter reg ion in the control, diabetes without and with retinopathy groups

	Control (n= 40)	Diabetes without retinopathy (n= 46)	Diabetes with retinopathy (n= 57)	Diabetes without retinopathy versus control			Diabetes with retinopathy versus control		
				OR (CI)	Pearson x^2 (*df.=2*)	*P*	OR (CI)	Pearson x^2 (*df.=2*)	*P*
Genotype									
DD	30 (75)	16 (34.8)	17 (29.8)	0.18 [0.06-0.5]	12.34	<0.001*	0.14 [0.05-0.39]	17.4	<0.001*
ID	8 (20)	22 (47.8)	33 (57.9)	3.67 [1.27-10.8]	6.12	0.01*	5.5 [1.98-15.73]	12.3	<0.001*
II	2 (5)	8 (17.4)	7 (12.3)	4 [0.71-29.3]	2.1	0.14	2.66[0.46-19.74]	0.74	0.38
		P' = 0.59							
Allele: D	68 (85)	54 (58.7)	67 (58.8)	0.25 [0.11-0.56]	13.11	<0.001*	0.25 [0.11-0.54]	14.07	<0.001*
I	12 (15)	38 (41.3)	47 (41.2%)	3.99 [1.8-8.97]			3.98 [1.84-8.71]		
		P' = 0.99							

Data are number (%); OR, odd ratio; CI, confidence interval, Significant (P<0.05);
P': For diabetes without versus with retinopathy groups

In contrast, Awata et al. [2] and Buraczynska et al. [11] found an association between the DD genotype and increase in the risk of DR. The discrepant findings might be due to differences in the statistical power, the recruitment of studied population, the type of retinopathy, and the genetic and environmental backgrounds.

It is well established that DR is determined by both genetic and environmental factors. It would be useful to identify molecular markers that may help to predict the development of retinopathy at earlier stages of diabetes [18]. In this study, we have focused on the 5'-untranslated region and promoter region of VEGF gene for genetic analysis as they have been shown to be highly polymorphic. In addition, most of the hypoxia responsive elements are present in these regions and retinal hypoxia has been observed in DR. During hypoxia, hypoxia-inducible factors bind to the hypoxia-response elements and induce the expression of VEGF, which leads to the stimulation of angiogenesis [23]. Moreover, VEGF could induce the earliest changes in retinopathy including leukostasis, blood-retinal barrier breakdown, and macular edema & neovascularization in progression of DR. In addition, VEGF inhibition has been indicated to cause a marked reduction in retinal neovascularization and prevention of the blood–retinal barrier breakdown [18].

The current study also showed that diabetes with retinopathy group is associated with higher level of HbA_{1C} than in controls and diabetes without retinopathy group. This result is in harmony with Stratton et al. [26] who confirmed the steep association between HbA1c exposures and both incidence and progression of DR in type 2 diabetes. Longo-Mbenza et al. [27] found significant association of poor control of HbA1c (poor glycemic control) and long-term microvascular complications of diabetes as DR in African patients. Moreover, Foad et al. [25] found that Glycosylated hemoglobin was significantly higher in diabetic patients (with and without retinopathy) than in controls.

Hyperglycemia has been recognized as the primary pathogenic factor in the development and progression of diabetic retinopathy [28]. Several biochemical pathways are supposed to be activated secondary to hyperglycemia. These abnormal pathways in turn, may influence several vasoactive factors that are likely to create functional and morphological changes in the retina of patients with diabetes [17]. Furthermore, improving control of blood glucose reduced development of retinopathy [29].

5. CONCLUSION

The present study suggests that, the heterozygous form of I/D polymorphism and I allele at -2549 of VEGF gene might possibly be associated with a higher susceptibility to retinopathy as a complication of type 2 diabetes mellitus. While, no association is found with VEGF G + 405 C gene polymorphism. Further well-designed large-scale studies with the consideration of gene-gene and gene-environment interactions should be conducted to investigate the association is recommended.

ACKNOWLEDGEMENT

The authors gratefully acknowledge the helpful comments of the anonymous reviewers.

COMPETING INTERESTS

Authors have declared that no competing interests exist.

REFERENCES

1. Early Treatment Diabetic Retinopathy Study Research Group. Risk factors for proliferative diabetic retinopathy and severe visual loss. Ivest Ophthalmol Vis Sci. 1998;39:233-52.
2. Awata T, Inoue K, Kurihara S, Ohkubo T, Watanabe M, Inukai K, Inoue I, Katayama S. A common polymorphism in the 5'-untranslated region of the VEGF gene is associated with diabetic retinopathy in type 2 diabetes. Diabetes. 2002;51:1635–1639.
3. Han L, Zhang L, Xing W, Zhuo R, Lin XL, Hao Y, Wu Q, Zhao J. The associations between VEGF gene polymorphisms and diabetic retinopathy susceptibility: A meta-analysis of 11 case-control studies. J Diabetes Res; 2014. Article ID 805801: 10 pages.
4. Ferrara N, Gerber HP, Le Couter J. The biology of VEGF and its receptors. Nat Med. 2003;9:669–76.
5. Zhou L, Sun H, Xu J, Kang J. Level of vascular endothelial growth factor and interleukin-6 in aqueous humor in diabetic retinopathy patients. Yan Ke Xue Bao. 2010;25(1):26-30.
6. Ferrara N, Davis-Smyth T. The biology of vascular endothelial growth factor. Endocr Rev. 1997;18:4–25.

7. Cavusoglu AC, Bilgili S, Alaluf A, Doğan A, Yilmaz F, Aslanca D, Karaca B, Yüksel B, Topaloglu E. Vascular endothelial growth factor level in the serum of diabetic patients with retinopathy. Ann Ophthalmol (Skokie). 2007;39(3):205-208.
8. Yan H, Cui J, Yu JG, Han JD, Chen S, Zhang JK, Xu YH. The expression of vascular endothelial growth factor of vitreous in patients with proliferative diabetic retinopathy. Zhonghua Yan Ke Za Zhi. 2009;45(3):206-209.
9. Kant S, Seth G, Anthony K. Vascular endothelial growth factor-A (VEGF-A) in vitreous fluid of patients with proliferative diabetic retinopathy. Ann Ophthalmol (Skokie). 2009;41(3-4):170-173.
10. Vincenti V, Cassano C, Rocchi M, Persico G. Assignment of the vascular endothelial growth factor gene to human chromosome 6p21.3. Circulation. 1996;93:1493–1495.
11. Buraczynska M, Ksiazek P, Baranowicz-Gaszczyk I, Jozwiak I. Association of the VEGF gene polymorphism with diabetic retinopathy in type 2 diabetes patients. Nephrol Dial Transplant. 2007;22:827–832.
12. Watson CJ, Webb NJA, Bottomley MJ, Brenchley PEC. Identification of polymorphisms within the vascular endothelial growth factor gene: Correlation with variation in VEGF protein production. Cytokine. 2000;12:1232–5.
13. Stevens A, Soden J, Brenchley PE, Ralph S, Ray DW. Haplotype analysis of the polymorphic human vascular endothelial growth factor gene promoter. Cancer Res. 2003;63:812–6.
14. Huez I, Bornes S, Bresson D, Créancier L, Prats H. New vascular endothelial growth factor isoform generated by internal ribosome entry site-driven CUG translation initiation. Mol Endocrinol. 2001;15(12): 2197-2210.
15. Farzadi Haghighi S, Salehi Z, Mohamad RS. Analysis of MnSOD and VEGF genes polymorphisms in diabetic retinopathy molecular and biochemical diagnosis. MBD. 2014;1(1):13-20.
16. Report of the expert committee on the diagnosis and classification of diabetes mellitus. Diabetes Care. 2003;26: 3160–3167.
17. Nakamura S, Iwasaki N, Funatsu H, Kinato S, Iwamoto Y. Impact of variants in the VEGF gene on progression of proliferative diabetic retinopathy. Graefes Arch Clin Exp Ophthalmol. 2009;247:21–26.
18. Lu Y, Shi Y, Ge Y, Yin J, Huang Z. Two Polymorphisms (Rs 699947, Rs 2010963) in the VEGFA gene and diabetic retinopathy: An updated meta-analysis. J Diabetes Metab. 2012;S3:006.
DOI: 10.4172/2155-6156
19. Ray D, Mishra M, Ralph S, Read I, Davies R, Brenchley P. Association of the VEGF gene with proliferative diabetic retinopathy but not proteinuria in diabetes. Diabetes. 2004;53:861–864.
20. Uthra S, Raman R, Mukesh BN, Rajkumar SA, Padmaja KR, Paul PG, Lakshmipathy P, Gnanamoorthy P, Sharma T, McCarty CA, Kumaramanickavel G. Association of VEGF gene polymorphisms with diabetic retinopathy in a south Indian cohort. Ophthalmic Genet. 2008;29(1):11-15.
21. Petrovič MG, Korošec P, Košnik M, Osredkar J, Hawlina M, Peterlin B, Petrovič D. Local and genetic determinants of vascular endothelial growth factor expression in advanced proliferative diabetic retinopathy. Mol Vis. 2008;14(1): 1382–1387.
22. Bleda S, De-Haro J,Varela C, Esparza L, Ferruelo A, Acin F. Vascular endothelial growth factor polymorphisms are involved in the late vascular complications in Type II diabetic patients. Diabetes & Vascular Disease Research. 2011;9(1):68–74.
23. Suganthalakshmi B, Anand R, Kim R, Mahalakshmi R, Karthikprakash S, Namperumalsamy P, Sundaresan P. Association of VEGF and eNOS gene polymorphisms in type 2 diabetic retinopathy. Mol Vis. 2006;12(1):336-341.
24. Szaflik JP, Wysocki T, Kowalski M, Majsterek I, Borucka AI, Blasiak J, Szaflik J. An association between vascular endothelial growth factor gene promoter polymorphisms and diabetic retinopathy. Graefes Arch Clin Exp Ophthalmol. 2008; 246:39–43.
25. Fouad H, Abdel Hamid MA, Abdel Azeem AA, Labib HM, Khalaf NA. Vascular endothelial growth factor (VEGF) gene insertion/deletion polymorphism and diabetic retinopathy in patients with type 2 diabetes. Journal of American Science. 2011;7(3):199-205.
26. Stratton IM, Kohner EM, Aldington SJ, Turner RC, Holman RR, Manley SE, Matthews DR. UKPDS 50: Risk factors for incidence and progression of retinopathy in

type II diabetes over 6 years from diagnosis. Diabetologia. 2001;44:156-163.
27. Longo-Mbenza B, Muaka MM, Mbenza G, Mbungu-Fuele S, Mabwa-Mbalanda L, Nzuzi-Babeki V, Mbadi-A-Sungu J. Risk factors of poor control of HBA1c and diabetic retinopathy: Paradox with insulin therapy and high values of HDL in African diabetic patients. Int J Diabetes & Metabolism. 2008;16:69-78.
28. Cai J, Boulton M. The pathogenesis of diabetic retinopathy: Old concepts and new questions. Eye. 2002;16:242–260.
29. UK Prospective Diabetes Study Group. Intensive blood- glucose control with sulphonylureas or insulin compared with conventional treatment risk of complications in patients with type 2 diabetes (UKPDS 33). Lancet. 1998; 352:837–853.

2

Inhibitory Activity of Ethanol Extract of *Manihot esculenta* on Mitochondrial Membrane Permeability Transition Pore and Caspase 3 in Type 2 Diabetes Mellitus

Ebenezer Idowu O. Ajayi[1,2,3*], Emmanuel U. Modo[1], Adetayo O. Adebamowo[1], Uttam Chand Banerjee[3], Olumide O. Tewe[4] and Olufunso Olabode Olorunsogo[1]

[1]Department of Biochemistry, Laboratories for Biomembrane Research and Biotechnology, University of Ibadan, Ibadan, Nigeria.
[2]Department of Chemical Science, Biochemistry Unit, Faculty of Basic and Applied Sciences, College of Science, Engineering and Technology, Osun State University, Osogbo, Nigeria.
[3]Department of Pharmaceutical Technology/Biotechnology, National Institute of Pharmaceutical Education and Research, SAS Nagar, Punjab, 160 062, India.
[4]Department of Animal Science, Agricultural Biochemistry and Nutrition Division, University of Ibadan, Nigeria.

Authors' contributions

This work was carried out in collaboration between all authors. Authors EIOA, OOT and OOO designed the study. Authors EIOA and OOO wrote the protocol. Authors UCB and OOO supervised the work. Authors EIOA, EUM and AOA carried out all laboratory work and performed the statistical analysis. Authors EUM and AOA managed the literature searches. Author EIOA managed the analyses of the study and wrote the first draft of the manuscript. Authors OOT, UCB and OOO edited the manuscript. All authors read and approved the final manuscript.

Editor(s):
(1) Richard A. Manderville, Departments of Chemistry and Toxicology University of Guelph, Canada.
Reviewers:
(1) N. De Kimpe, Ghent University, Belgium.
(2) James Adams, University of Southern California, USA.

**Corresponding author: E-mail: ebenezer.ajayi@uniosun.edu.ng*

ABSTRACT

The modulatory effect of different concentrations of ethanol extract of matured leaves of *M. esculenta*, Crantz at 200, 600, 1000 and 1400 µg/ml was investigated *in vitro* on mitochondrial membrane permeability transition pore in liver, kidney and heart of diabetic animals in the absence and presence of 20 µM exogenous Ca^{2+}. The extract at these concentrations had no significant ($p< 0.05$) effect on mitochondrial membrane permeability transition (MPT) pore of the three organs in the absence of Ca^{2+}. However, in the presence of Ca^{2+}, the extract exhibited significant ($p< 0.05$) inhibition of mitochondrial membrane transition pore opening: liver by 62.66%, 42.47%, 22.44% and 17.63% at 1400, 1000, 600, and 200 µg/ml, respectively. In the heart, inhibition of MMPT pore opening was by 92.86%, 71.43%, 64.29% and 57.14% at 200, 600, 1400, and 1000 µg/ml, respectively. In the kidney, the extract also inhibited mitochondrial membrane transition pore opening in a concentration-dependent manner by 92.65%, 91.18%, 89.71% and 72.06% at 1400, 1000, 600, and 200 µg/ml, respectively. Caspase-3 activity *in vitro* was also reduced with increasing extract concentrations; thus confirming that the extract inhibits MMPT pore opening in the presence of Ca^{2+}. The extract may be able to protect these organs against damage, resulting from Ca^{2+} overload that may trigger cell death, and as such it may be useful in the management of diseases related to tissue wastage such as cardiomyopathy and nephropathy which are associated with Type 2 diabetes mellitus.

Keywords: Type 2 diabetes; high-energy diet; low-dose streptozotocin; MMPT; Caspase-3.

1. INTRODUCTION

Multicellular organisms ensure optimum tissue homeostasis to maintain harmony among the processes of cell division and growth, differentiation, aging and death by activating a cell suicide process known as apoptosis. This program is an evolutionary conserved cell suicide program whose execution falls under the regulation of finely controlled signalling pathways through the interplay of a group of putative proteins known as the Bcl-2 family. For apoptosis to occur, these proteins modulate mitochondrial outer membrane permeabilization thus controlling the inevitable release of cytochrome c which is sequestered in the intermembrane space of mitochondria [1]. Upon receipt of an apoptotic stimulus, Bax or Bak oligomerize leading to the formation of a high-conductance, non-specific pore across the inner and outer mitochondrial membrane thus allowing the release of the key players of apoptosis initiation into the cytosol [2]. These pro-apoptotic factors in turn may activate the cysteine-dependent aspartate-directed proteases known as caspases or act in a caspase-independent fashion to bring about cell suicide [3].

Major and minor phytonutrients have found usefulness in herbal medicine. They are usually secondary plant metabolites that have the potential to mediate biological processes in such a manner that the antioxidant systems of the body can respond to their challenge to ameliorate or offset the risk of causation of chronic diseases in humans [4,5]. However, there is paucity of scientific evidences to support the hypoglycemic and antioxidative effects of leaf extracts of *Manihot esculenta* Crantz. Several food crops, vegetables, spices, marine and tropical medicinal herbs have been tested in an effort to identify new chemical entities that can serve as potent antioxidants [6]. Traditional medicines, which originate from natural products, are useful anti-diabetic drugs that improve the immunity of the body. They prove to be very effective for diabetes treatment when they are single used and they show measurable effects such as reducing blood sugar, as with western medicines. Thus they are promising for finding active compounds which can be developed into new anti-diabetic drugs [7].

Interestingly, ethno-botanical medicine often offers vast, relatively untapped sources of such drug candidates from natural products. Reactive oxygen species (ROS) formation is an unavoidable process that occurs naturally [8], and capable of leading to amplified increases in oxidative stress and enhanced ROS-mediated cell damage. ROS and free radicals have been implicated in a variety of diabetic complications such as nephropathy [9], retinopathy [10], cardiomyopathy, ischemia reperfusion injury and chronic tubulointerstitial injury [11].

Type 2 diabetes mellitus is a multifactorial disease which is characterised by hyperglycaemia and abnormal lipoprotein levels [12]. These traits are indicated to cause cell membrane damage and high preponderance of ROS. Generation of free radicals accompanying cellular metabolism in obesity, dyslipidemia, hypertension, and insulin resistance with their underlying biochemical mechanisms, the involvement of certain environmental factors, including lifestyles are risk factors for the pathogenesis of T2DM [13]. It is estimated that > 100, 000 people die annually in India, and between 40, 000–99, 999 people die in Nigeria from diabetes-related complications. The 5th edition of the Diabetes Atlas's new figures revealed that the number of people living with diabetes in the world would likely rise from 366 million in 2011 to 552 million by 2030 [14]. India continues to be at the fore-front of diabetes incidence in the world with Gujarat state having high prevalence of the dreaded disease, while more than 5 million diabetics have been statistically identified in Africa and they are expected to increase to 15 million by 2025.

This lifestyle disease potentiates continuous and increased propagation of free radical-mediated oxidative stress thus elongating the chain of damage [15,16].

Common names for *Manihot esculenta* Crantz include cassava, yuca, tapioca, manioc, Brazilian arrowroot, kappa, gbaguda, paki, ege. The leaves serve as vegetable in many African cultures and are a good source of plant nutrients [17]. The levels of cyanogenic glycosides – linamarin (95%) and lotaustralin (5%) – in the leaves and roots of cassava make the food crop potentially toxic [18, 19]. Poorly processed cassava has been shown to cause goitre and tropical ataxic neuropathy as a result of exposure to chronic, low cyanide levels [1,20,21]. A paralytic disorder, konzo [22], as well as death [23] are popular outcomes of acute cyanide poisoning. In order to render cassava- based foods safe, it is advisable that the roots and leaves be thoroughly processed. Linamarin-free roots are now available in transgenic cassavas [24].

This study was conducted determine the *in vitro* effect of varying concentrations of the ethanol extract of *Manihot esculenta* on mitochondrial membrane permeabilization in diabetic rat kidney, heart and liver both in the absence and presence of exogenous Ca^{2+}; and the effect on Caspase 3 activation, *in vitro*.

2. MATERIALS AND METHODS

2.1 Plant Material

Dried matured *Manihot esculenta* leaves were powdered (500 g) and soaked overnight in 95 % ethanol (1: 3 w/v) in a Soxhlet extractor. The suspension filtered and residue treated same manner again for 48 hours. Continuous hot extraction was done until the final drop of the extract became colorless and the extract was then concentrated *in vacuo* at 60°C using a rotary evaporator. The yield was 14.5 g of extract, which was kept on the bench until further use.

2.2 Experimental Animals

Twenty-four Sprague-Dawley rats (180–250 g) were used for in vivo experiment. All animals were maintained under standard conditions and control groups were rodent normal pellet diet *ad libitum*, while test groups manipulated for T2DM were fed on high fat diet. All experimental procedures were approved by the Institutional Animal Care and Use Committee at the National Institute of Pharmaceutical Education and Research, Punjab, India. The animals were fasted for 12 hours before sacrifice by decapitation but were allowed access to water. Experiments were conducted thrice.

High fat diet – HFD (58% fat, 25% protein and 17% carbohydrate, as a percentage of total kcal) was given to animal groups 2–6 *ad libitum*, for the initial period of 2 weeks, and continuously throughout the treatment period that lasted 4 weeks (1 week during which low-dose STZ was given i.p. —35 mg/kg) [25] and 21 days of extract intubation.

2.3 Preparation of Mitochondria

2.3.1 Isolation of rat liver mitochondria

The principle is based on isolation of low ionic strength mitochondria [26,27]. Rat liver was excised and trimmed after the animal was sacrificed by cervical dislocation under the influence of diethyl ether. The liver was weighed and washed several times in isolation buffer containing 210 mM mannitol, 70 mM sucrose and 1 mM EGTA in 5 mM HEPES -HCl, pH 7. 4 until a clear washing was obtained. A 50% suspension was prepared by homogenising the liver in a glass-Teflon homogeniser. The whole process was carried out on ice to preserve the integrity of the mitochondria. The homogenised liver was centrifuged at 2300x g for 5 minutes in a SIGMA-6K15 refrigerated centrifuge so as to sediment the nuclear fraction and cell debris 45by low speed centrifugation. The supernatant obtained was centrifuged at the same speed to remove unbroken cells. The supernatant was centrifuged at 13000x g for 10 minutes to pellet the mitochondria. The brown mitochondria pellet was washed by re-suspending in isolation buffer containing 0.5% BSA and centrifuged at 12000x g for 10 minutes. The washing was done twice. The mitochondria were then re-introduced in a buffer containing 210 mM mannitol, 70 mM sucrose, and 5 mM HEPES -HCl, pH 7.4; and dispensed in Eppendorf tubes kept on ice in aliquots.

2.3.2 Isolation of rat kidney mitochondria

Rat kidney mitochondria were isolated according to the method described by Schneider & Hogeboom [28] with slight modifications. Briefly, the kidneys were excised and placed in the ice-cold medium made up of 210 mM mannitol, 70 mM sucrose and 1 mM EGTA in 5 mM HEPES-HCl, pH 7.4. The kidneys were minced in homogenizing buffer (1:10 w/v) and subjected to 3 Dounce strokes with a glass-Teflon homogeniser. The homogenates was diluted to 10 ml per kidney and centrifuged at 900x g for 5 minutes and kept at 4°C. The supernatant was decanted and further centrifuged at 4500x g for 5 minutes the pellet was suspended in 20 ml of the ice-cold isolation buffer containing 0.5% BSA and centrifuged at 4500x g for 10 minutes. The brown mitochondrial pellet was suspended in 1 ml of medium containing 210 mM mannitol, 70 mM sucrose, and 5 mM HEPES- HCl, pH 7.4.

2.3.3 Isolation of rat heart mitochondria

Rat heart mitochondria were isolated by mincing and homogenising it in isolation medium (210 mM mannitol, 70 mM sucrose, 1 mM EGTA in 5 mM HEPES- HCl, pH 7.4) (1: 10 w/v) using glass-Teflon homogeniser [29] and centrifuged at 700x g for 10 minutes. The resulting supernatant was centrifuged at 8000x g for 10 minutes and the ensuing mitochondria were washed in isolation buffer containing 0.5% BSA at 7000x g for 10 minutes, and the pellets were suspended in 350 µl of isolation buffer without BSA.

Mitochondrial protein estimation was done by the method of Lowry et al. [30] expressed as bovine serum albumin equivalent (BSA equiv.) and the fractions were analysed immediately or within 4 hours of isolation, while the cytosolic fraction was kept frozen at -80°C until further use.

2.4 Treatment of Mitochondria

Mitochondria that have accumulated calcium can be induced to undergo permeabilization by which pores non-selectively open on the inner membrane to small (1500 Daltons) solutes [31]. Isolated mitochondria undergoing permeability transition (MPT) show colloidosmotic i.e. large amplitude swelling which results in a decreased photometric absorption at 540 nm. Several experimental parameters have been used to study the MPT in isolated mitochondria but in this study, the permeability transition was assessed by measuring the swelling of mitochondria by monitoring the associated decrease in light scattering. The principle behind this method is that, when mitochondria swell, their refractive index changes and thus less light is passed across the cuvette which results in a decrease in the light absorbance measured at the isobestic point from the cytochromes (520 nm or 540 nm). 1 mg/ml organ (liver, kidney and heart) mitochondria were incubated in the assay medium (210 mM mannitol, 70 mM sucrose in 5 mM 2HEPES- HCl, pH 7.4) at 30°C in the presence of 0.8 µM rotenone [in order to inhibit Complex I (NADH dehydrogenase) thus to experimentally eliminate confounding factors of convergence of electrons in the electron transport system as the system is fed only through Complex II (Succinate dehydrogenas)] [32]. Various concentrations of the extract (200, 600, 1000 and 1400 µg/ml) were added to the mitochondria 3 minutes before 20 µM $CaCl_2$ was added and 30 seconds afterwards, which the assay was energised by 25 mM Sodium Succinate. Mitochondria swelling was measured by continuous time scan at 540 nm over 12 minutes with Hitachi U–3010 Spectrophotometer.

2.5 Statistical Analysis

This was done with the aid of SigmaStat 3.5 using Analysis of Variance (ANOVA). Data were reported as Mean ± Standard Deviation.

3. RESULTS

Exogenous calcium induced the opening of MPT in the kidney of the diabetic control whereas in the absence of Ca^{2+}, there was no significant swelling. There was significant inhibition of MPT in the absence of exogenous calcium in kidney mitochondria of diabetic rat treated with varying concentration of ethanol extract of *Manihot esculenta*; with 200 µg/ml showing the least inhibition effect (Fig. 1). The varying concentrations of ethanol extract of matured leaves of *M. esculenta* significantly inhibited the kidney MPT of the diabetic control in the presence of calcium. The inhibition was found to be concentration dependent with the high concentrations (1000 and 1400 µg/ml) giving the greatest inhibition (Fig. 2). In the heart, there was a significant decrease in the absorbance at 540 nm in the presence of exogenous calcium ($p<0.05$), although a low amplitude swelling of the mitochondria with mean difference (-0.0136±0.009) compared with intact mitochondria (-0.033±0.017) i.e. in the absence of exogenous calcium. The varying concentrations of the extract inhibited mitochondria permeabilization in the heart (Figs. 3 and 4). The results obtained from the assessment of permeability transition in diabetic rat liver mitochondria showed a significant decrease in the absorbance at 540 nm in the presence of exogenous calcium ($p<0.05$) i.e. large amplitude swelling of the mitochondria with mean difference (-0.307±0.126) compared with intact mitochondria (-0.024±0.007). The varying concentrations of the extract of significantly inhibited the MPT in the liver of diabetic rats in the absence of calcium. The inhibition was found to be concentration-dependent (1400 µg/ml< 200 µg/ml < 600 µg/ml< 1000 µg/ml), while 1400 µg/ml gave the least inhibition (Fig. 5). When 200, 600 and 1000 µg/ml extract were incubated with exogenous calcium in the diabetic group, they had high induction effects in a time-dependent manner, while 1400 µg/ml had inhibitory effect compared to spermine in a concentration-dependent manner (Fig. 6). Table 1 summarises the effect of the extract on Caspase 3 activity, *in vitro*. Increases in the activities of caspase 3 were observed in organs of diabetic rats. Howbeit, the extract significantly reduced caspase 3 activities in a concentration-dependent manner, with the least activity observed at 1400 µg/ml. Across the selected organs, 1400 µg/ml was found to be most potent in the heart, followed by the kidney and least potent in the liver.

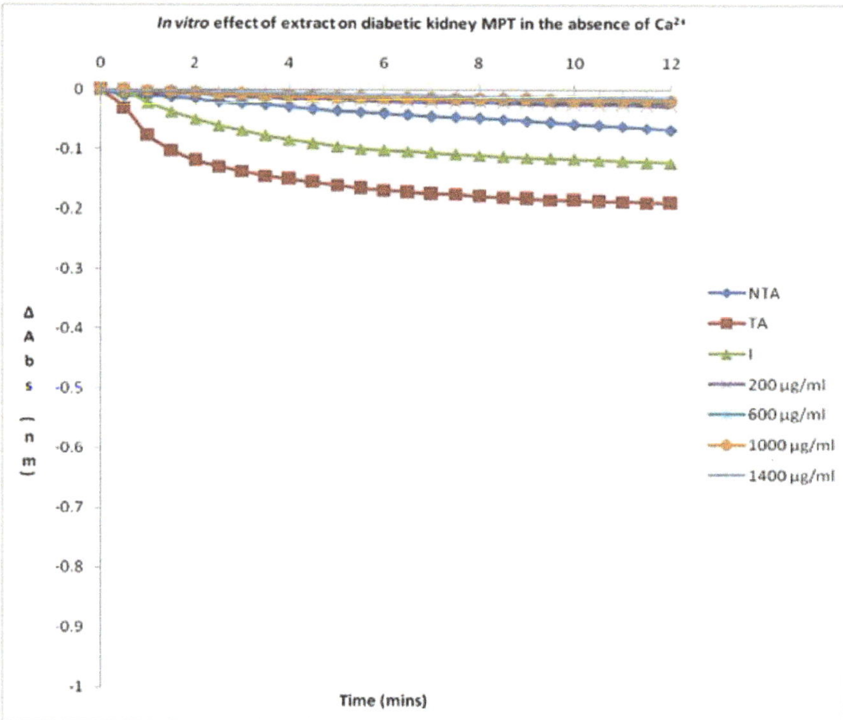

Fig. 1. *In vitro* effect of extract on diabetic kidney MPT in the absence of calcium

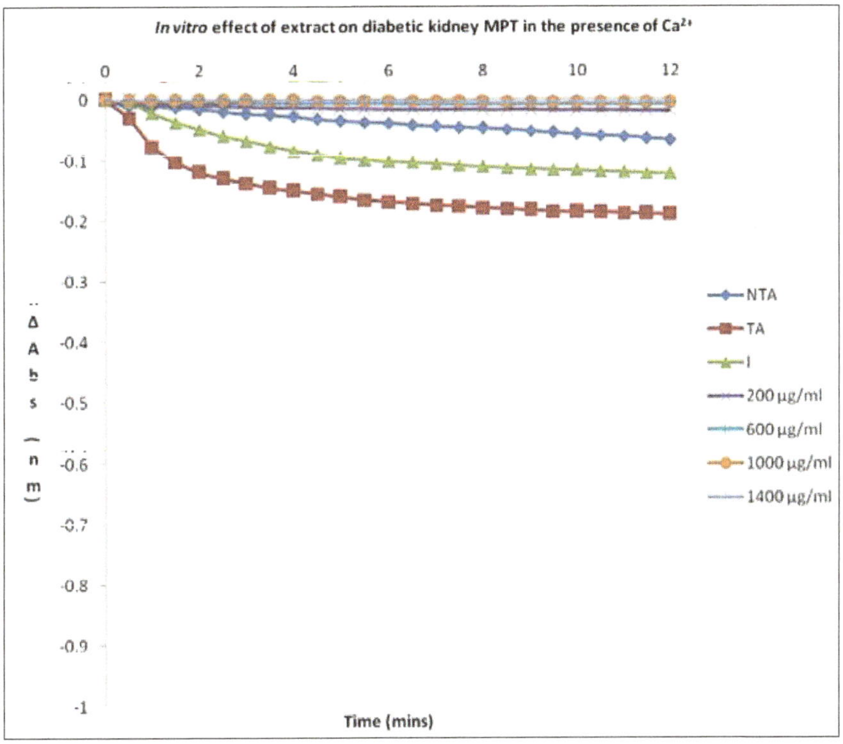

Fig. 2. ***In vitro*** **effect of extract on diabetic kidney MPT in the presence of calcium**

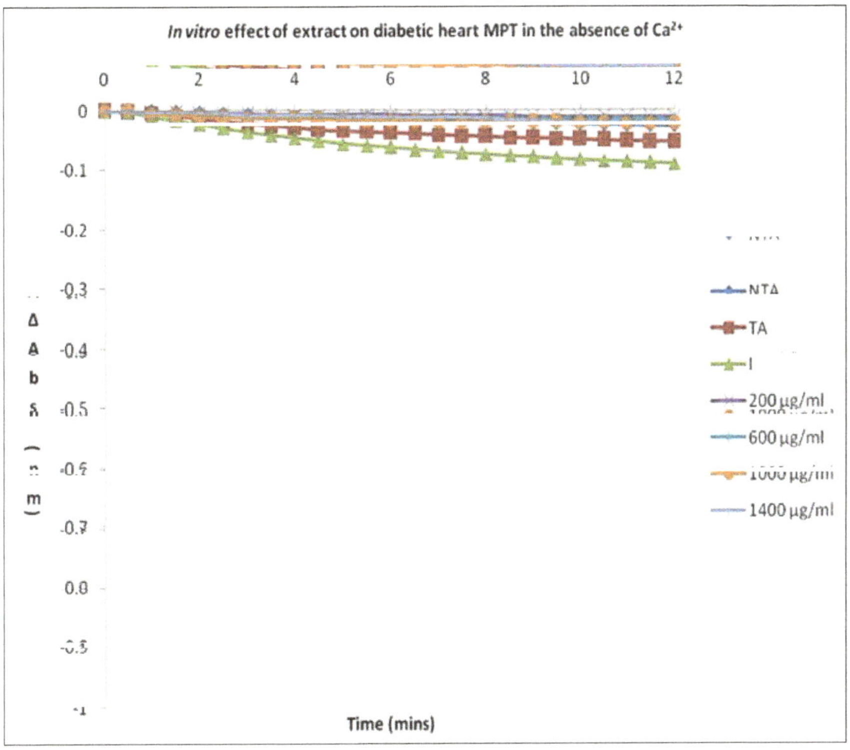

Fig. 3. ***In vitro*** **effect of extract on diabetic heart MPT in the absence of calcium**

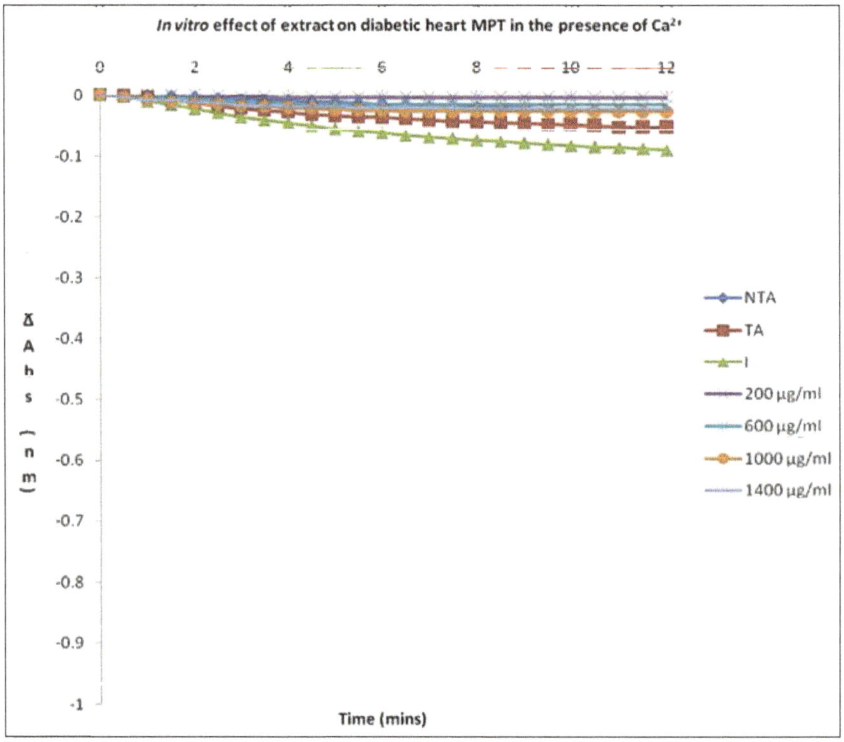

Fig. 4. ***In vitro*** **effect of extract on diabetic heart MPT in the presence of calcium**

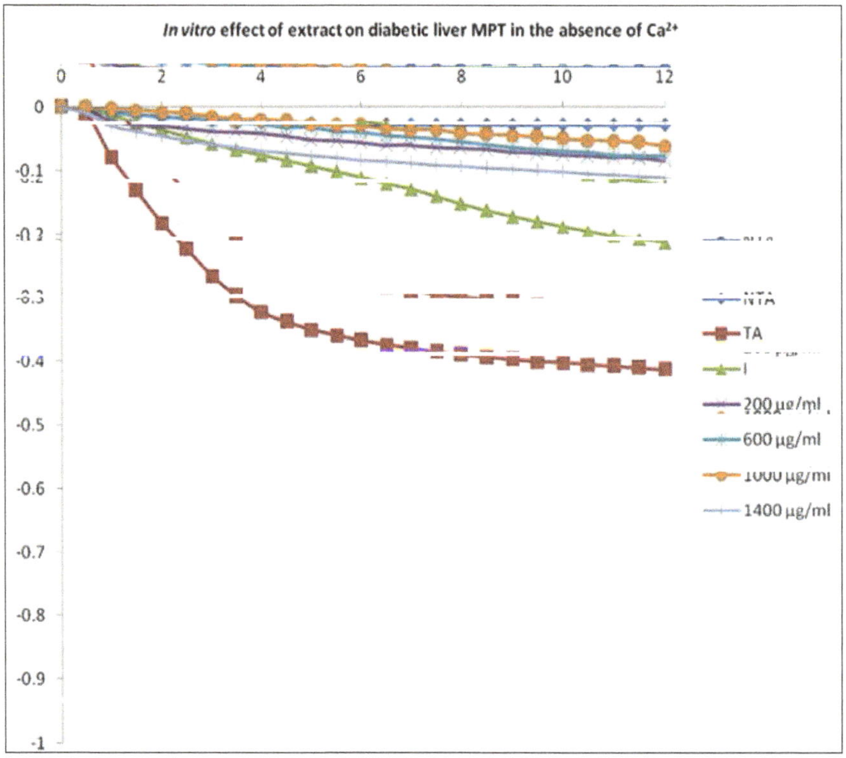

Fig. 5. ***In vitro*** **effect of extract on diabetic liver MPT in the absence of calcium**

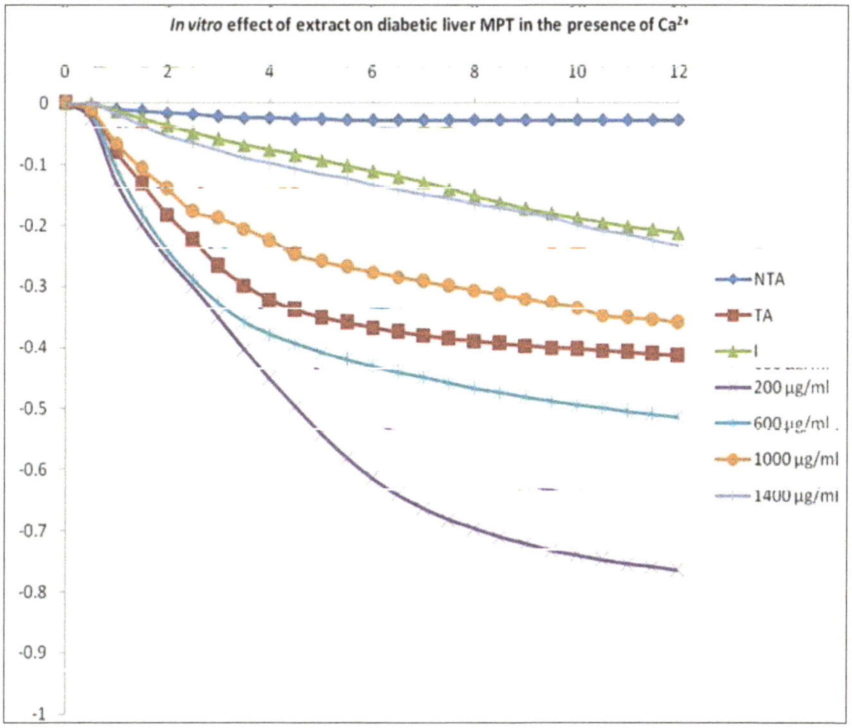

Fig. 6. *In vitro* effect of extract on diabetic liver MPT in the presence of calcium

Table 1. Effect of extract on caspase 3 activity, *in vitro* (U/ mg protein)

Extract (µg/ml)	0	200	600	1000	1400
Kidney normal	2.24±0.22	1.11±0.34*[c]	0.89±0.02* [c]	0.67±0.01* [c]	0.67±0.02* [c]
Kidney diabetic	3.10±0.74*	2.25±1.53[a]	1.55±1.06[a]	0.97±0.76* [a]	0.71±0.14* [a]
Heart normal	1.40±0.44	0.83±0.26*	0.54±0.23*	0.43±0.08*	0.54±0.37*
Heart diabetic	2.19±0.20*	1.43±0.84[a]	0.83±0.47*[a b]	0.83±0.37*[a b]	0.58±0.16*[a b]
Liver normal	2.80±0.31*	2.61±0.84*	2.49±0.42*	2.22±0.24*	1.73±0.25*
Liver diabetic	4.40±2.17	3.79±1.11	2.89±1.93*	2.10±0.46*	2.02±1.02*

4. DISCUSSION

Both in the presence of calcium, the extract significantly inhibited mitochondrial membrane permeability transition pore opening (MPT) in the selected organs of diabetic rats as compared to results obtained in the absence of Ca^{2+}. This indicates that the extract was able to inhibit large amplitude swelling caused by exogenous calcium alone. It should be pointed out that spermine is not a good candidate inhibitor of kidney mitochondrial MPT due to inconsistencies in inhibitory effect. Cyclosporine A is considered a better candidate for kidney mitochondrial MPT [33]. The extract reduced caspase-3 activity with increasing extract concentrations, *in vitro*. Increases in the activities of caspase 3 were observed in organs of untreated diabetic rats compared to those pre-incubated with various concentrations of the extract. Howbeit, the extract significantly reduced caspase 3 activities in a concentration-dependent manner, with the least activity observed at 1400 µg/ml. Across the selected organs, 1400 µg/ml was found to be most potent in the heart, followed by the kidney and least potent in the liver.

5. CONCLUSION

It is clear that apoptosis occurs under both physiological and pathological conditions, such as diabetes [34]. Elevated caspase activity in the kidney, heart and liver of intact mitochondria of diabetic rats was thus predictable. The extract, with increasing concentration showed potency in attenuating the processes leading to the activation of caspase 3 in the kidney and heart. The extent to which the extract inhibited caspase

3 activation in the liver was not as high as observed in the kidney and heart. The results confirmed that the extract inhibits MMPT pore opening in the presence of Ca^{2+}. The inhibition of MPT by the extract invariably led to significant decreases in the activity of caspase 3, which is a downstream factor necessary for apoptotic cell death to occur. Thus, the ethanol extract of *Manihot esculenta* Crantz may be able to protect these organs against damage, resulting from Ca^{2+} overload that may trigger cell death, and as such it may be useful in the management of diseases related to tissue wastage such as cardiomyopathy and nephropathy which are associated with Type 2 diabetes mellitus.

ACKNOWLEDGEMENT

The authors thank the Third World Academy of Science and the Department of Biotechnology, India for the 2009 DBT-TWAS Sandwich Postgraduate Fellowship awarded to AEIO for this project.

COMPETING INTERESTS

Authors have declared that no competing interests exist.

REFERENCES

1. Gottlieb RA. Mitochondria: Execution central. FEBS Lett. 2000;482:6-12.
2. Korsmeyer SJ, Wei MC, Saito M, Weiler S, Oh KJ. Proapoptotic cascade activates BID, which oligomerizes BAK or BAX into pores that result in the release of cytochrome c. Cell Death Differ. 2000;7:1166-1173.
3. Kuwana T, Newmeyer DD. Bcl-2-family proteins and the role of mitochondria in apoptosis. Curr. Opin. Cell Biol. 2003;15:691-699.
4. Halliwell B. Antioxidants in human health and disease. Annual Review of Nutrition. 1996;16:33–50.
5. Chen J, Mangelinckx S, Adams A, Wang Z-T, Li W-L, De Kimpe N. Natural flavonoids as potential herbal medication for the treatment of diabetes mellitus and its complications. Natural Product Communications. 2015;10:187-200.
6. Zheng W, Wang SY. Antioxidant activity and phenolic compounds in selected herbs. J. Agric. Food Chem. 2001; 49:5165-5170.
7. Li WL, Zheng HC, Bukuru J, De Kimpe N. Natural medicines used in the traditional Chinese medical system for therapy of diabetes mellitus. J. Ethnopharm. 2004;92: 1-21.
8. Djamali A. Oxidative stress as a common pathway to chronic tubulointerstitial injury in kidney allografts. Am J Physiol Renal Physiol. 2007;293(2):F445-55.
9. Weight SC, Bell PR, Nicholson ML. Renal ischaemia-reperfusion injury. Br. J. Surg. 1996;83:162-170.
10. Suganthalakshmi B, Anand R, Kim R, Mahalakshmi R, Karthikprakash S, Namperumalsamy P. et al. Association of VEGF and eNOS gene polymorphisms in type 2 diabetic retinopathy. Mol. Vis. 2006;12:336-341.
11. Verzola D, Bertolotto MB, Villaggio B, Ottonello L, Dallegri F, Salvatore F, et al. Oxidative stress mediates apoptotic changes induced by hyperglycemia in human tubular kidney cells. J. Am. Soc. Nephrol. 2004;15(Suppl.1):S85-S87.
12. Scoppola AJ, Montecchi FR, Mezinger G, Lala A. Urinary mevalonate excretion rate in type 2 diabetes: Role of metabolic control. Atherosclerosis. 2001;156:357-361.
13. Hartnett EM, Stratton RD, Browne RW, Rosner BA, Lanham RJ, Arnstrong D. Serum markers of oxidative stress and severity of diabetic retinopathy. Diabet. Care. 2000;23:234-240.
14. International Diabetes Federation. IDF Diabetes Atlas, 5th ed. Brussels, Belgium: International Diabetes Federation; 2011.
15. Baynes JW. Role of oxidative stress in development of complications in diabetes. Diabetes. 1991;40:405-12.
16. Varvarovská J, Racek J, Stetina R. Aspects of oxidative stress in children with type 1 diabetes mellitus. Biomed. Pharmacother. 2004;58:539-45.
17. Bokanga M. Processing of cassava leaves for human consumption. Acta Hort. 1994;375:203-207.
18. Balagopalan C, Padmaja G, Nanda SK, Moorthy SN. Cassava in food, feed and industry. CRC Press, Boca Raton, Florida; 1988.
19. Conn EE. Cyanogenesis–a personal perspective. Acta Horticulturae. 1994;375: 31-43.
20. Osuntokun B. Cassava diet, chronic cyanide intoxification and neuropathy in

Nigerian Africans. World Rev. Nutr. Diet. 1981;36:141-173.
21. Tylleskar T, Cooke R, Banea M, Poulter N, Bikangi N, Rosling H. Cassava cyanogens and konzo, an upper motor neuron disease found in Africa. Lancet. 1992;339:208-211.
22. Sreeja V, Nagahara N, Li Q, Minami M. New aspects in pathogenesis of konzo: Neural cell damage directly caused by linamarin contained in cassava (*Manihot esculenta* Crantz). Brit. J. Nutr. 2003;90:467-472.
23. Ernesto M, Cardoso AP, Nicala D, Mirione E, Massaza F, Cliff J, et al. Persistent konzo and cyanogen toxicity from cassava in Northern Mozambique. Acta Tropica, 2002;82:357-362.
24. Siritunga D, Richard S. Engineering cyanogen synthesis and turnover in cassava (*Manihot esculenta*). Plant Mol. Biol. 2004;56:661-669.
25. Srinivasan K, Viswanad B, Asrat L, Kaul CL, Ramarao P. Combination of high-fat diet and low-dose streptozotocin- treated rat: A model for type 2 diabetes milletus and pharmacological screening. Pharmacological Research. 2005;52:313-320.
26. Schneider WC, Hogeboom GH. Intracellular distribution of enzyrnes. J. Biol. Chem. 1948;176:259.
27. Johnson D, Lardy H. Isolation of liver or kidney mitochondria. Methods Enzymol. 1967;10:94-96.
28. Schneider WC, Hogeboom GH. Intracellular distribution of enzymes. V. Further studies on the distribution of cytochrome c in rat liver homogenates. J. Biol. Chem. 1950;183:123-128.
29. Mela L, Seitz S. Isolation of mitochondria with emphasis on heart mitochondria from small amounts of tissue. Methods Enzymol. 1969;55:39-46.
30. Lowry OH, Rosebrough NJ, Farr AL, Randall RJ. Protein measurement with the Folin phenol reagent. J Biol Chem. 1951;193:265-275.
31. Lapidus RG, Sokolove PM. Spermine inhibition of the permeability transition of isolated rat liver mitochondria: An investigation mechanism. Arch. Biochem. Biophys. 1993;3061:246-253.
32. Li B, Chauvin C, De Paulis D, De Oliveira F, Gharib A, Vial G, et al. Inhibition of complex I regulates the mitochondrial permeability transition through a phosphate-sensitive inhibitory site masked by cyclophilin D. Bioenergetics. 2012;1817(9):1628-1634.
33. Xinming Q, Yan C, Likun G, Linlin L, Fangping C, Ying X, et al. Role of mitochondrial permeability transition in human renal tubular epithelial cell death induced by aristolochic acid. Toxicol. and Appl. Pharmacol. 2007;222:105-110.
34. Merkle CJ. Cellular adaptation, injury, and death. In Pathophysiology: Concepts of altered health states. 8th edition. Edited by Porth CM, Matfin G. Philadelphia: Wolters Kluwer/ Lippincott Williams and Wilkins. 2009;94-111.

3

Protective Effect of *Dorema glabrum* on Diazinon-induced Oxidative Stress in Rat Kidney

Parisa Sistani[1], Mina Adampourezare[1], Gholamreza Dehghan[1*], Amin Adami[1] and Leila Mehdizadeh Fanid[1]

[1]*Department of Biology, Faculty of Natural Science, University of Tabriz, Tabriz, Iran.*

Authors' contributions

This work was carried out in collaboration between all authors. All authors read and approved the final manuscript.

Editor(s):
(1) Richard A. Manderville, Departments of Chemistry and Toxicology University of Guelph, Canada.
Reviewers:
(1) Anu Rahal, Indian Veterinary Research Institute, UP, India.
(2) Bhaskar Sharma, Suresh Gyan Vihar University, Rajasthan, India.

ABSTRACT

Introduction: A wide number of pesticides such as diazinon (DZN) have deteriorating effect on fauna and flora by inducing oxidative stress and cell damage. The aim of current study was to investigate the protective effect of *D. glabrum* against the subchronic effect of DZN on serum urea and creatinine, oxidative damage markers such as lipid peroxidation (LPO) and the antioxidant defense system (catalase (CAT), glutathione peroxidase (GPx), superoxide dismutase (SOD)) in the kidney of male Wistar rats.
Methods: Twenty-four adult male Wistar rats randomly divided into four groups including a control group, and three experimental groups. Two of three experimental groups received different doses of *D. glabrum* (40 and 80 mg/kg) as pre-treatment for 21 days along with DZN (100 mg/kg) that injected intraperitoneally in the last day of *D. glabrum* usage, and one group received only DZN (100 mg/kg).
Results: The levels of LPO, urea, creatinine, and the decreased antioxidant defenses, like free radical scavenging enzymes viz., catalase (CAT), superoxide dismutase (SOD) and glutathione peroxidase (GPx) significantly increased in DZN-treated group as compared to control group. Distinctly increased LPO levels, with alterations in endogenous antioxidant enzyme were evident in

**Corresponding author: E-mail: lfanid@yahoo.co.uk*

nephrotoxicity of DZN. Specific marker enzymes were restored to normalcy in rats supplemented with *D. glabrum* following treatment with DZN which otherwise was decreased in the DZN-treated rats.

Conclusion: The obtained results revealed that the oxidative stress of rats treated via DZN can be decreased if it is co-treated by *D. glabrum* with DZN. In addition, this co-treatment might serve a putative protective agent against kidney damage generated by DZN.

Keywords: Diazinon; oxidative stress; Dorema glabrum; kidney.

1. INTRODUCTION

One of the chemical compounds widely utilized in agriculture is Organophosphorous (OP) pesticides which are believed to harness the agricultural plagues. Among the most frequently used OP insecticides, diazinon is used worldwide to eliminate crop and cattle plagues, as well as in household pest control [1]. Considering the significant role of DZN and its potential impacts on the biochemical and hematological parameters of rats, rabbits, and mice, relevant findings have been reported in the field [2-4]. During inhalation, digestive system, the skin, or the respiratory tract are more likely to absorb DZN. It is largely argued that the kidney is able to remove the absorbed DZN; however, DZN tends to create such potent acetylcholinesterase inhibitors as diazoxon, hydroxydiazoxon, and hydroxydiazinon when it is oxidized by microsomal enzymes in the liver [5]. Mammals can also experience DZN when lipid metabolism is negatively affected by the presence of DZN. Proposed insights from in vivo animal studies, additionally, revealed that oxidative stress is likely to be induced through organophosphorus insecticides (OPIs). The resulted oxidative stress is more likely to play a pivotal role in its induced poisoning [6], observations in humans [7] and in vitro studies [8] over the past several years. Acute effects of DZN as inducers of oxidative stress on certain biomarkers in various tissues have been studied [9]. One of the biomarkers for oxidative stress is lipid peroxidation which is claimed to possess predictive significance in quite a number of investigations [10,11]. Oxidative stress, moreover, is triggered as an absence of balance between antioxidant system and pro-oxidant condition which pesticide toxicity brings about. To convert reactive oxygen species (ROS), Endogenous enzymatic and non-enzymatic antioxidants are considered indispensable since they are capable of making the metabolites harmless. In addition, they have the potential to protect and revitalize the metabolism and functions of normal cellular [12]. To detoxify ROS, an enzyme defense system including superoxide dismutase (SOD), catalase (CAT), and selenium-dependent glutathione peroxidase (GPx), or non-enzymatic systems via reduced glutathione is required. To detoxify organic peroxides, however, the activity of glutathione S-transferase (GST) is required. (GST) [13].

Native to Nakhchevan and Armenia zone (Transcaucasia) and Northwest of Iran is a species called *D. glabrum*. Regarding Iranian flora, there exist seven species which are believed to represent the genus Dorema from Apiaceae family [14]. It is useful as an food additive or herbal remedy in mentioned regions [15]. Members of this genus possess expectorant, antispasmodic, carminative, mild diuretic, stimulant, vasodilator [16], and hepatoprotective [17] properties and are intensively used as a folk medicine for treatment of many diseases [18]. According to the common folk beliefs of Armenian and Azeri people, *D. glabrum* can cure many diseases especially different kinds of cancer. In their seminal work, Dehghan et al. (2009) demonstrated that the crude extract of the plant can bring about antioxidant activity and anti-lipidemic effects. Therefore, the present study was aimed at evaluating the toxic effect of diazinon in rats with special reference to its possible reactive oxygen species (ROS) generating potential in kidney, which caused biochemical, hematological and histopathological alterations.

2. MATERIALS AND METHODS

2.1 Animals

Twenty-four adult male Wistar rats (weighting approximately 250-350 g) used in the present study. Animals were obtained from the animal house of the veterinary department of Tabriz University. During the experiment, every attempt was made to continuously maintain the rats under temperature 20-22°C and 12/12 h cycle of light and darkness. The rats, throughout the experiments, were provided with sufficient nutrition and water as well. Ethics of working with laboratory animals have been followed during all procedures.

2.2 Animal Treatment Schedule

The rats randomly divided into four groups including group I (C): normal control rats, group II (DZN): received DZN in single dose (100 mg/kg), group III (DGE 40): received (40 mg/kg) of *D.glabrum* and (100 mg/kg) of DZN, group IV (DGE 80): Received *D. glabrum* (80 mg/kg) and DZN (100 mg/kg). This two experimental groups received different doses of *D. glabrum* (40 and 80 mg/kg) as pre-treatment for 21 days with DZN (100 mg/kg) that injected intraperitonealy on the last day of *D. glabrum* usage.

2.3 Plant Materials

Seeds of *D. glabrum* were collected during the fruiting stage from the slopes of Aras river; Jolfa, Eastern Azerbaijan, Iran. Air dried and then powdered seeds were subjected to extraction by refluxing ethanol using a soxhlet. Then the extract was dried by a rotary evaporator (Heidolph, Germany).

2.4 Chemicals

DZN was applied from Aria chemistry Co. (Iran) containing 96% active ingredies. It was diluted with corn oil as DZN solvent. Thiobarbituric acid (TBA), trichloroacetic acid (TCA), H_2O_2 (30%), ethylenediaminetetraacetic acid (EDTA), Tris–HCl, 2,2′-dinitro-5,5′dithio-dibenzoic acid (DTNB), ethanol of technical grade and the other chemicals used in this study were procured from Merck Co. (Germany).

2.5 Sample Collection and Processing

All animals were anesthetized under ether after 21 days of treatment; then blood samples were taken out to measure the level of serum urea via cardiac puncture. Following this, we removed creatinine and kidneys from the animals and washed with cold saline buffer, then were stored at -80°C. To obtain the enzyme extract, tissues were homogenized in ice-cold KCl 1.15% to yield 10% (W/V) homogenate. Then the homogenates were centrifuged at 1000 rpm for 10 min at 4°C [19]. Employing global units per mg protein (IU/mg protein), an attempt was made to discriminate and utilize the supernatants for enzyme activity of SOD, CAT and GPx. Biomarkers for tissue damage were measured using the UV kinetics methodology and total protein was determined using bovine serum albumin (BSA) as standard and the values were expressed as mg/dl.

2.6 Analytical Procedures

Pars Azmun kit and the method of Satho were utilized to determine and measure the concentration of urea and creatinine and thiobarbituric acid reactive substances (TBARS) as an index of lipid peroxidation respectively [20]. SOD was determined according to the method described by Ukeda [21]. CAT was measured by monitoring the decomposition of hydrogen peroxide, as described by Aebi [22]. GPx was evaluated by the method of Paglia and Valentine [23]. To measure protein and total thiol content (TSH), Lowry [24] and Hu [25] methods were employed respectively along with bovine serum albumin as standard.

2.7 Statistical Analysis

To statistically analyze the collected data, one-way analysis of variance (ANOVA) followed by Tukey's post hoc test (level of significance $P<0.05$) was run using SPSS version 16, statistical program. The results are expressed as mean ± SEM. Statistical analysis was based on comparing the values between the DZN and control group, while DZN-treated groups concomitantly with extract of *D. glabrum* (DGE) were compared with their corresponding group of DZN-treated rats.

3. RESULTS

3.1 Lipid Peroxidation

In the kidney increased lipid peroxidation was observed as measured by a significant increase in MDA (malondialdehyd) that expressed as nano moles of TBARS/g of protein in the rats treated with DZN when compared to control rats (p=0.001,48.9%) (Fig. 1). The group which received DZN and DGE 40 as cotreatment showed decreased levels of MDA when compared with DZN treated group (P=0.003,22.5%) and the group, which received DZN and DGE 80 showed decreased levels of MDA when compared with DZN treated group (p=0.007,20.44%) (Fig. 1).

3.2 Antioxidant Enzymes

The activities of CAT (U/mg of protein) in kidney were reduced significantly in DZN treated animals compared to controls (P=0.008, 21.14%). Whereas in DGE 40 group a significant increase in the activity of this enzyme was observed compared to rats treated with DZN alone (P>0.05, 9.7%). In the DGE 80 group a significant increase in the activity of CAT was observed compared to rats treated with DZN alone (P>0.05, 12.5%) (Fig. 2). As presented in Fig 3, SOD activities were diminished by 20.53% by treatment with DZN (p=0.006).Treatment with DGE 40 and DGE 80 showed a significant increase of SOD activities by 4.7%, 15.9% respectively (p>0.05). In GPx activities, there was no significant difference between the groups (Fig. 4).

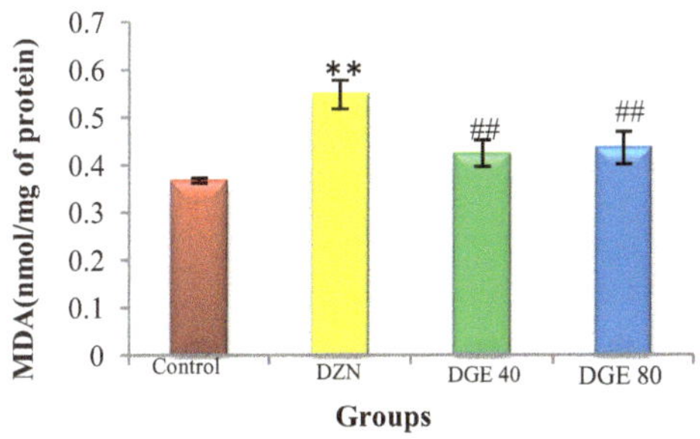

Fig. 1. Effect of DGE 40 and DGE 80 on MDA level of DZN treated rats and effect of DZN (100 mg/Kg) on MDA level of normal rats in the kidney

*Values are mean±SEM (n=6). ** P<0.01, compared to control group, ##P<0.01, compared to DZN group*

3.3 Antioxidant Enzymes

The activities of CAT (U/mg of protein) in kidney were reduced significantly in DZN treated animals compared to controls (P=0.008, 21.14%). Whereas in DGE 40 group a significant increase in the activity of this enzyme was observed compared to rats treated with DZN alone (P>0.05, 9.7%). In the DGE 80 group a significant increase in the activity of CAT was observed compared to rats treated with DZN alone (P>0.05, 12.5%) (Fig. 2). As presented in Fig. 3, SOD activities were diminished by 20.53% by treatment with DZN (p=0.006). Treatment with DGE 40 and DGE 80 showed a significant increase of SOD activities by 4.7%, 15.9% respectively (p>0.05). In GPx activities, there was no significant difference between the groups (Fig. 4).

3.4 Total Thiol Content

There was not significant difference in values of total thiol in rats which received DZN, DGE 40 and DGE 80 when compared with control rats (Fig. 5).

3.5 Concentration of Serum Creatinine

Data showed that DZN significantly increased level of creatinine (p < 0.0001) compare to control group. DGE 40 and DGE 80 significantly decreased level of creatinine (p < 0.001) compare to DZN group (Fig. 6).

3.6 Concentration of Serum Urea

Data showed that DZN significantly increased level of urea (p< 0.01) compare to control group.

DGE 40 and DGE 80 significantly decreased level of urea (p<0.05, p=0.001) compare to DZN group (Fig. 7).

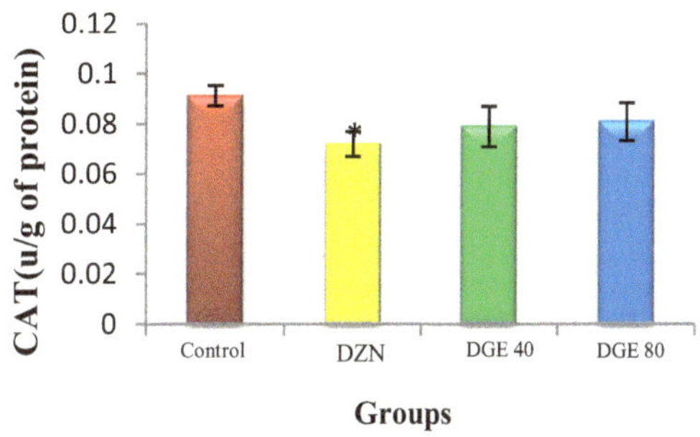

Fig. 2. Effect of DZN (100 mg/Kg) on CAT activity of normal rats (p<0.01) and the effect of DGE 40 and DGE 80 on CAT activity of DZN treated rats (p>0.05)**

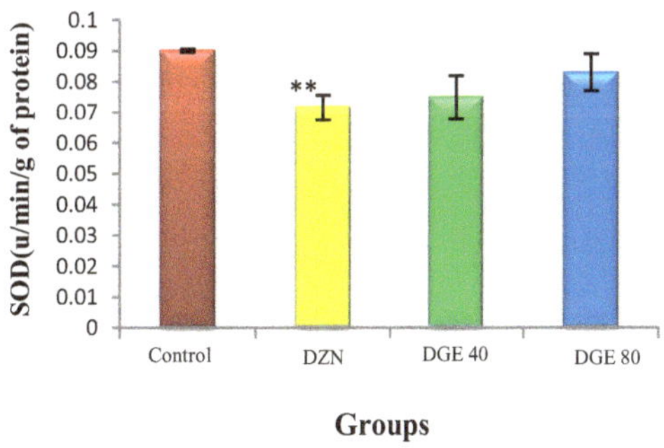

Fig. 3. Effect of DGE 40 and DGE 80 on SOD activity of DZN treated rats and effect of DZN (100 mg/Kg) on SOD activity of normal rats

*Values are mean±SEM, (n=6) ** P<0.01, compared to control group, P>0.05, compared to DZN group*

4. DISCUSSION

Recently, an increasing amount of attention has been concentrated towards free radical mediated damage triggered by pesticide exposure in biological systems [26]. It has been recognized that oxidative stress plays a role in the pathogenesis of kidney toxicity by many toxins and drugs [27]. With respect to antioxidant activity, a substantial body of research has been devoted to herbal drugs since the antioxidants generated by natural sources tend to be often secure and free of adverse side effects [28]. *D. glabrom* has extensive biological properties but most of the biological action of *D. glabrum* seems to be associated with its antioxidant potential. In the present study, the protective effect of *D. glabrum* has been evaluated against DZN-induced toxicity in the kidney of rats. This study indicated that DZN has induced oxidative stress and kidney dysfunction, also acute toxicity of DZN increases serum creatinine and urea. Creatinine concentration, as a marker of kidney function, increases when the kidney gets injured. Serum creatinine displays kidney clearance as well. This implies that the associated elevation in creatinine appears to be the consequence of DZN-induced kidney damage and clearance reduction. According to our findings, DZN treatment promotes MDA content in the kidney of rats. The increase of MDA observed in the kidney following DZN exposure, was probably ascribed to the excessive production of ROS, which could be related to nephrocyte enzyme leakage. Mitochondria are the main source of ROS production [29,30]. Moussa and Hafez [31] demonstrated that the activity of mitochondrial enzyme is inhibited by the organophosphate and dimethoate. Significant cellular macromolecules like lipids, proteins, and nucleic are resulted when ROS is excessively generated [32]. Our research revealed that oxidative stress is induced by DZN treatment. The enhanced LPO in the kidney and relevant antioxidant defenses demonstrated the presence of the oxidative stress.

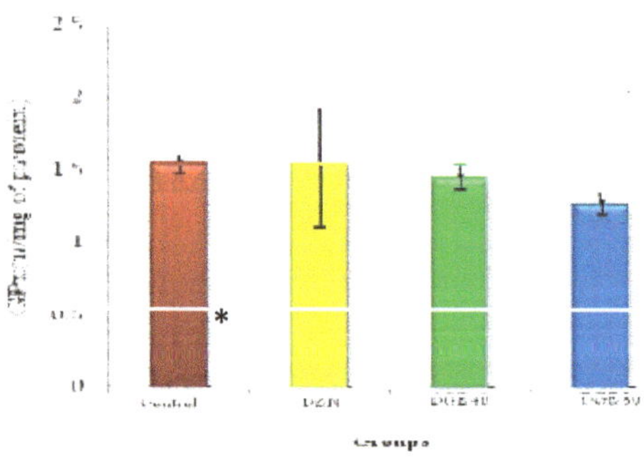

Fig. 4. Effect of DZN (100 mg/Kg) on GPx activity of normal rats (*p<0.05) and effect of DGE 40 and DGE 80 on GPx activity of DZN treated rats (p>0.05)

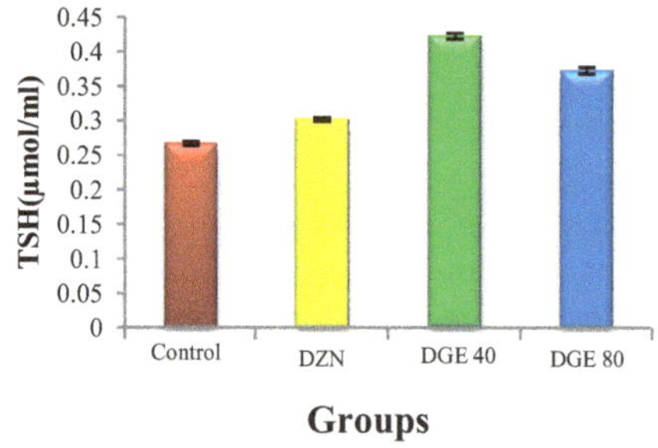

Fig. 5. Values of total thiol in control group and treated rats with DZN, DGE 40 and DGE 80. There was not significant difference between the groups

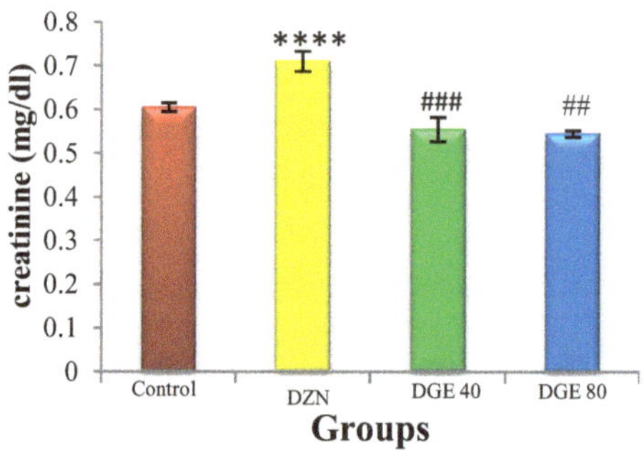

Fig 6. Effect of DZN (100 mg/Kg) on creatinine activity of normal rats (**p<0.0001) and effect of DGE 40 and DGE 80 on GPx activity of DZN treated rats ($^{###}$p<0.001, $^{###}$p<0.001 respectively)**

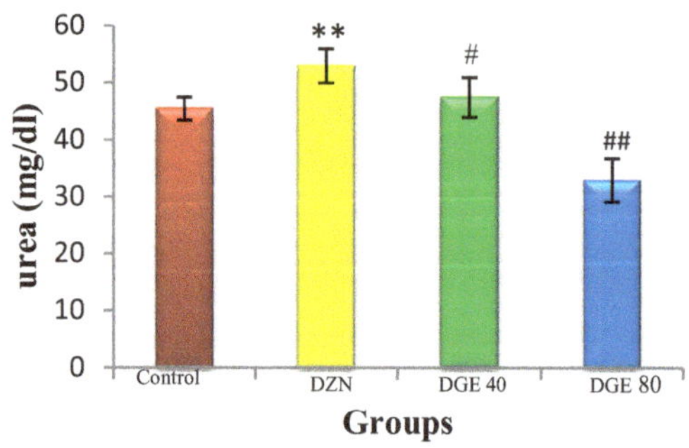

Fig. 7. Effect of DZN (100 mg/Kg) on urea activity of normal rats (p<0.01) and effect of DGE 40 and DGE 80 on urea activity of DZN treated rats ($^{#}$p<0.05, $^{##}$p<0.001 respectively)**

These findings are in accordance with the studies of [33,34] who demonstrated a significant increase in LPO level in kidney tissue, serum urea and creatinine, and a significant decrease in

SOD,CAT and GPx activities in kidney of rats exposed to DZN.

Furthermore, GSH redox cycles, as a crucial component of cellular antioxidant defenses, are essential for the tissues to resist against the ROS damage. The cycles, serving both as a non-enzymatic oxygen radical scavenger and as a substrate for diverse enzymes like GPX, play a significant role in removing ROS [35]. As an important enzyme, GPX contains selenocysteine which has the potential to defend cellular antioxidant [36]. In this study, a significant decrease in GPx activity was observed in DZN treated animals. This may be due to enhanced free radical production, as evident in increased level of MDA. Moreover, the decrease of GPx activity induced by DZN may be attributable to a direct inhibitory oxidative effect on the enzyme. To prevent biological macromolecules resulted from oxidative damage, SOD and CAT as antioxidant enzymes are regarded as the primary defense systems. That is to say, superoxide anion ($O_2^{\cdot -}$) is swiftly dismutased to a less reactive molecule (H_2O_2) via SOD. The reactive molecule (H_2O_2) is then degraded by CAT and GPX enzymes to water and oxygen [37]. Following the DZN treatment, the results of the current study indicated that SOD activity decreased significantly.

Obviously, mitochondria superoxide radical production is induced by xenobiotics [38]. In addition, if SOD is inhibited the amount of ($O_2^{\cdot -}$) formed in a cell can reach dangerous levels. Superoxide radical is a potent inhibitor of CAT [39]. The activity of antioxidant enzyme is likely to be exhausted when DZN-induced ROS is generated, the enzyme substrates are depleted, and the transcription and translation processes are down-regulated. Consequently, this process tends to impose a negatively direct impact on the antioxidant enzyme activity. DZN can affect both animals and human, although by a single exposure [40]. Whereas the toxicological effects incurred by DZN are mainly because acetylcholinesterase is inhibited. DZN toxicity is also argued to induce oxidative stress. From our findings it can be assumed that nephrotoxic effects of DZN are mainly attributed to oxidative stress increase since the effects were prevented by *D. glabrum* supplementation. The antioxidant activity promoted by *D. glabrum* is essentially performed in the lipid membrane environment. *D. glabrum* allows free radicals to abstract a hydrogen atom from the antioxidant molecule, thus breaking the chain of free radical reactions [41].

5. CONCLUSION

We have concluded that DZN toxicity in the kidney could be attributed to the oxidative stress on the nephrotic cells. An increase in DZN toxicity results in the depletion of the antioxidant enzymes which are capable of scavenging the toxic superoxide and hydrogen peroxide radicals, resulting in an elevation of LPO. Nonetheless, *D. glabrum* treatment could have a protective effect against DZN-induced nephrotoxicity in rat and could be useful to decrease DZN toxicity by quenching oxidative stress imposed by DZN.

Moreover, similar studies indicated that diazinon treatment enhances renal damage and also demonstrated a fall in the activities of antioxidant enzymes including the enzyme involved in glutathione metabolism and excessive production of oxidants with consequent renal damage. The occurrence of these events brings about diazinon-mediated renal oxidative stress and toxicity as the ultimate consequence [42].

COMPETING INTERESTS

Authors have declared that no competing interests exist.

REFERENCES

1. Sarabia L, Maurer I, Bustos-Obrego E. Melatonin prevents damage elicited by the organophosphorous pesticide diazinon on the mouse testis. Ecotoxicol. Environ. Safety. 2009;72:663–668.
2. Kalender S, Ogutcu A, Uzunhisarcikli M, Acikgoz F, Durak D, Ulusoy Y, et al. Diazinon-induced hepatotoxicity and protective effect of vitamin E on some biochemical indices and ultrastructural changes. Toxicol. 2005;211:197–206.
3. Kalender Y, Uzunhisarcikli M, Ogutcu A, Acikgoz F, Kalender S. Effects of diazinon on pseudocholinesteras activity and haematological indices in rats: The protective role of vitamin E. Environ. Toxicol. Pharmacol. 2006;22(1):46– 51.
4. Yehia MAH, El-Banna SG, Okab AB. Diazinon toxicity affects histophysiological and biochemical parameters in rabbits. Exper. Toxicol. Pathol. 2007;59:215–225.
5. Quistad GB, Sparksand SE, Casida GE. Fatty acid amide hydrolase inhibition by neurotoxic organophosphorus pesticides.

Toxicol. Appl. Pharmacol. 2001;173:48–55.

6. Monteiro DA, de Almeida JA, Rantin FT, Kalinin AL. Oxidative stress biomarkers in the freshwater characid fish, *Brycon cephalus*, exposed to organophosphorus insecticide Folisuper 600 (methyl parathion). Comp. Biochem. Physiol. C Toxicol. Pharmacol. 2006;143:141–149.
7. Zhou JF, Xu GB, Fang WJ. Relationship between acute organophosphorus pesticide poisoning and damages induced by free radicals. Biomed. Environ. Sci. 2002;15:177–186.
8. Gultekin F, Ozturk M, Akdogan M. The effect of organophosphate insecticide chlorpyrifos-ethyl on lipid peroxidation and antioxidant enzymes (*in vitro*). Arch. Toxicol. 2000;74:533–538.
9. Isika I, Celik I. Acute effects of methyl parathion and diazinon as inducers for oxidative stress on certain biomarkers in various tissues of rainbow trout (Oncorhynchus mykiss). Pest. Biochem. Physiol. 2008;92(1):38–42.
10. Ogutcu A, Uzunhisarcikli M, Kalender S, Durak D, Bayrakdar F, Kalender Y. The effects of organophosphate insecticide diazinon on malondialdehyde levels and myocardial cells in rat heart tissue and protective role of vitamin E. Pest. Biochem. Physiol. 2006;86(2):93–98.
11. Amirkabirian N, Teimouri F, Esmaily H. Protection by pentoxifylline of diazinon-induced toxic stress in rat liver and muscle. Toxicol. Lett. 2007;172(1):S61–S62.
12. Bebe FN, Panemangalore M. Exposure to low doses of endosulfan and chlorpyrifos modifies endogenous antioxidants in tissue of rats. J. Environ. Sci. Health B. 2003;38: 349–363.
13. Halliwell B, Gutteridge JMC. Free radicals in biology and medicine. Oxford University Press. London. 1999;936.
14. Mozaffarian VA. Dictionary of iranian plant names. Tehran: Farhang Moaser; 1996.
15. Asnaashari S, Dadizadeh E, Talebpour AH, Eskandani M, Nazemiyeh H. Free radical scavenging potential and essential oil composition of the dorema glabrum fisch. C.A. mey roots from Iran. BioImpacts. 2001;1(4):241-4.
16. Mozaffarian V. Tehran, Research Institute of Forests and Rangelands. Flora of Iran, Umbelliferae. 2007;54:368- 374.
17. Govind P. Medicinal plants against liver diseases. International Research Journal of Pharmacy. 2011;2(5):115-121.
18. Ibadullayeva S, Movsumova N, Gasymov H, Mamedli T. Protection of some rare and endangered vegetable plants in the flora of the Nakhichevan AR. International Journal of Biodiversity and Conservation. 2011; 3(6):224-229.
19. Salehi M, Jafari M, Asgari A, Saleh R, Moghaddam M, Salimian MS, Abbasnejad M, et al. Study of Diazinon effect on antioxidant enzymes and lipid peroxidation in rat's brain. Journal of Iran University of Medical Science. 2010;70(23-28).
20. Satho K. Serum lipid peroxidation in cerebrovascular disorders determined by a new colorimetric method. Clin Chem Acta. 1978;90:37–43.
21. Ukeda H, Maeda S, Ishii T, Sawamura M. Spectrophotometric assay for superoxide dismutase based on tetrazolium salt 3′–1–(phenylamino)-carbonyl–3, 4-tetrazolium]-bis(4-methoxy-6-nitro)benzenesulfonic acid hydrate reduction by xanthine-xanthine oxidase. Anal Biochem. 1997;251:206–9.
22. Aebi H. Catalase *in vitro* assay. Meth Enzymol. 1984;105:121–6.
23. Paglia DE, Valentine WN. Studies on the quantitative and qualitative characterization of glutathione peroxidase. J Lab Med. 1967;70:158–65.
24. Lowry OH, Roserbrough NJ, Farr AL, Randal RJ. Protein measurement with the folin phenol reagent. J Biol Chem. 1951; 193:265–75.
25. Hu ML, Dillared CJ. Plasma SH and GSH measurement. Methods Enzymol. 1994; 233:385–7.
26. Banerjee BD, Seth V, Bhattacharya A, Pasha ST, Chakraborty AK. Biochemical effects of some pesticides on lipid peroxidation and free-radical scavengers. Toxicol Lett. 1999;107(1–3):33–47.
27. Sabolic I. Common mechanisms in nephropathy induced by toxic metals. Nephron Physiol. 2006;104:107-14.
28. Dehghan G, Khoshkam Z. Tin(II)-quercetin complex: Synthesis, spectral characterization and antioxidant activity. Food Chemistry. 2012;131:422-427.
29. Wang YM, Peng SQ, Zhou Q, Wang MW, Yan CH, Yang HY, et al. Depletion of intracellular glutathione mediates butenolide-induced cytotoxicity in HepG2 cells. Toxicol. Lett. 2006a;164:231–238.

30. Wang YM, Peng SQ, Zhou Q, Wang MW, Han G, Yan CH, et al. Role of mitochondria in cellular oxidative injuries induced by butenolide. Chin. J. Pharmacol. Toxicol. 2006b;20:484–489.
31. Moussa TA, Hafez MM. The effect of dimethoate on the mitochondria of the guinea pig. Egypt. J. Histol. 1995;6:101–106.
32. Bergamini CM, Gambetti SS, Dondi A, Cervellati C. Oxygen, reactive oxygen species and tissue damage. Curr. Pharm. Des. 2004;10:1611–1626.
33. Shah MD, Iqbal M. Diazinon-induced oxidative stress and renal dysfunction in rats. Food Chem. Toxicol. 2010;48:3345-53.
34. Boroushaki MT, Sadeghnia HR, Banihasan M. Protective effect of pomegranate seed oil on Hexacholorobutadiene -induced nephrotoxicity in rat. J. Ren. Fail. 2013;32: 612-617.
35. Tsukamoto M, Tampo Y, Sawada M, Yonaha M. Paraquatinduced oxidative stress and dysfunction of the glutathione redox cycle in pulmonary microvascular endothelial cells. Toxicol. Appl. Pharm. 2002;178:82–87.
36. Arthur JR. The glutathione peroxidises. Cell. Mol. Life Sci. 2000;57:1825–1835.
37. Wetscher GJ, Bagchi M, Bagchi D, Perdikis G, Hinder PR, Glaser K, et al. Free radical production in nicotine treated pancreatic tissue. Free Radic Biol Med. 1995;18(5):877-82.
38. Peixoto F, Vicente J, Madeira VMC. A comparative study of plant and animal mitochondria exposed to paraquat reveals that hydrogen peroxide is not related to the observed toxicity. Toxicol. *in Vitro*. 2004; 18:733–739.
39. Ashakumari L, Vijayammal PL. Addictive effect of alcohol and nicotine on lipid peroxidation and antioxidant defense mechanism in rats. J. Appl. Toxicol. 1994; 16:305–308.
40. Sarabia L, Maurer I, Bustos-Obrego E. Melatonin prevents damage elicited by the organophosphorous pesticide diazinon on the mouse testis. Ecotoxicol. Environ. Safety. 2009;72:663–668.
41. Pascoe G, Olafs F, Read D. Vitamin E protection against chemical induced cell injury. Maintenance of cellular protein thiols as a cytoprotective mechanism Arch. Biochem. Biophys. 1987;256:150–158.
42. Dawood Shah M, Iqbal M. Diazinon-induced oxidative stress and renal dysfunction in rats. Food and Toxicology. 2010;48:3345-3353.

4

Antifungal Activity of Roots Barks Extract of *Securinega virosa* (Roxb. ex Willd.) Baill and *Anogeissus leiocarpa* (DC.) Guill. & Perr, Two Plants Used in the Traditional Treatment of Candidiasis in Northern Côte d'Ivoire

Kouangbé Mani Adrien[1,2*], Bahi Calixte[1], Tia Honoré[2], Boga Gogo Lucien[1], Edoh Vincent[2], Djaman Allico Joseph[1,3] and N'Guessan Jean David[1]

[1]Biosciences Department, Biochemical Pharmacodynamy Laboratory, University of Félix Houphouët-Boigny, Côte d'Ivoire.
[2]Department of Bacteriology-Virology, Central Laboratory of University Hospital of Treichville, Abidjan, Côte d'Ivoire.
[3]Department of Clinical and Basic Biochemistry, Pasteur Institute of Côte d'Ivoire, Côte d'Ivoire.

Authors' contributions

This work was done as a team with all authors. Author KMA designed the study, performed the different tests and wrote this article and performed statistical analysis. Authors BC and BGL wrote the first draft of the manuscript. Authors TH and EV wrote the protocol, managed the analyses of the study analysis and managed literature. Authors DAJ and NJD provided the data included to the manuscript and managed the literature search. All authors read and approved the final version of the manuscript.

Editor(s):
(1) Tomio Yabe, Department of Applied Life Scinece, Gifu University, Japan.
Reviewers:
(1) Anonymous, Mahidol University, Thailand.
(2) Patricia García Palencia, Department of Microbiology and Infection Control, Autonomous University of Nuevo León, Mexico.
(3) Dorota Wojnicz, Department of Biology and Medical Parasitology, Wroclaw Medical University, Poland.

Corresponding author: E-mail: kmania1@yahoo.fr

ABSTRACT

Aims: To evaluate the anticandidal activity of some solvent extracts of *Securinega virosa* and *Anogeissus leiocarpa* from northern Côte d'Ivoire.
Study Design: *In vitro* assay of antifungal activity.
Place and Duration of Study: Biochemical Pharmacodynamy Laboratory, Biosciences Department, University Félix Houphouët-Boigny and Bacteriology-Virology Laboratory University Hospital of Treichville, Côte d'Ivoire between January and December 2013.
Methods: The herbs studied were examined for diameter of inhibition zone using agar well diffusion method; minimal inhibitory concentration (MIC) and minimal fungicidal concentration (MFC) using microdilution method.
Results: All tested plants extracts, except the aqueous extracts, showed varying zones of inhibition against fungi tested. The diameters of inhibition zones for all organic extracts are greater than 10 mm for a sample concentration of 500 mg/ml and were significantly higher than for nystatine ($p < 0.05$; $p < 0.01$). The ethanol extract of *Anogeissus leiocarpa* revealed the strongest anticandidal activity against all tested strains with MICs ranging from 0.195 to 12.5 mg/ml, and MFCs from 0.390 to 50 mg/ml. The phytochemical screening of extracts shows the presence of polyterpenes and sterols, polyphenols, flavonoids, catechin tannins and alkaloids.
Conclusion: *S. virosa* and *A. leiocarpa* possesse compounds with good anticandidal properties. This results support their traditional use in treatment of infectious diseases caused by certain Candida species.

Keywords: Anogeissus leiocarpa; Securinega virosa; anticandidal activity; northern Côte d'Ivoire.

1 INTRODUCTION

Whatever its origins and its civilization, man has sought in his plant or animal environment remedies to overcome his health problems. Research and knowledge of the properties of drugs have spawned real pharmacopoeia [1]. Medicinal plants and their knowledge are for the African continent a rich heritage hitherto insufficiently unexplored. Adjanohoun identified nearly 50.000 species [2]. In Côte d'Ivoire are nearly 800 species of medicinal plants and more than 1.400 drug recipes that have been identified [3]. The study of biological and chemical properties showed that the Ivorian flora has a real therapeutic and nutritional potential and can be used to treat or prevent many diseases [4]. These plants have a real interest in human health for their antibacterial [5-7], antifungal [8,9], antiplasmodial [10] and antioxidant [11] activities. Infectious diseases are a serious public health problem in developing countries where they are the main cause of high mortality rates, and in industrialized countries where resistance to existing antibiotics are growing alarmingly. This creates a growing need to find new antimicrobial compounds and/or inhibitory mechanisms of antibiotic resistance. Therefore, some 250-300.000 plants species inventoried, only 5 to 15% have been investigated for bioactive molecules, represent a huge pool of potential new bioactive compounds [12]. According to some authors, the naturally occurring compounds have the advantage of a great diversity of chemical structures and they also have a wide range of biological activities [13]. In this context, the present work was focused on, two plants *Securinega virosa* and *Anogeissus leiocarpa* whose root barks are used in traditional treatment of microbial infections in northern Côte d'Ivoire. Various researches have focused on the study of the biological activities of leaves and stem barks of these plants. However, there are very few studies on the evaluation of the antifungal activity of the root bark of plants.

This work intends to evaluate the *in vitro* antifungal properties of different extracts of both plants of the Ivorian flora on three species of *Candida* namely *Candida albicans, Candida tropicalis* and *Candida glabrata.*

2. MATERIALS AND METHODS

2.1 Plant Material

The plant material consisted of the root bark of *A. leiocarpa* and *S. virosa*. These organs were harvested in January 2013 to Kouto, town located to 725 km north of Abidjan after an ethnobotanical survey of traditional healers in the community. Plants used in this study were identified by Professor AKé-Assi Laurent of the National Floristic Center, University of Félix Houphouet Boigny Cocody Abidjan.

These barks were dried out of the sun for two weeks before being ground into a fine powder by grinding. From the powder obtained after spraying, the different extracts to be tested were prepared.

2.2 Fungal Strains

The fungal strains tested were *C. albicans*, *C. glabrata* and *C. tropicalis*, yeasts isolated to pathological products from consulted patients with candidiasis in the Laboratory of Bacteriology-Virology University Hospital Center of Treichville (Abidjan, Côte d'Ivoire). These strains were maintained in a refrigerator at 4°C for the duration of the experiment.

2.3 Preparation of Plant Extracts

2.3.1 Preparation of aqueous total extract

The total aqueous extract was obtained by dissolving 50 g of the plant powder in 2 l of distilled water. The macerate was mixed for 48 h using a magnetic stirrer type IKAMAG RCT at a temperature of the laboraty. The homogenate was filtered two times successively on cotton wool and then once on Whatman filter paper N°3. The filtrate obtained was reduced using a rotary evaporator type IKA Labortechnick before being evaporated at 50°C using an oven type Med Center Venticell. The powder obtained for each plant was the total aqueous extract and were stored at 5°C.

2.3.2 Preparation of the organic extracts

The extraction method used was that described by Manga et al. [14]. The plant material was subjected to a first out with solvents of increasing polarity of dichloromethane, ethyl acetate and ethanol.

The liquid-liquid chromatography also called partition chromatography was based on the sharing of the solute in the two immiscible liquid phases. This chromatography required the use of three different polarity solvents namely dichloromethane, ethyl acetate, ethanol. To achieve this chromatography, 20 g of aqueous extract plant were added to 100 ml of water. This solution was transferred into a separatory funnel of 500 ml in which 100 ml of dichloromethane were then added. The vial was stirred for 2 min and then allowed to sedimentation. The lower aqueous phase was then collected. This operation was repeated four times and the fractions collected dichloromethane (400 ml) were evaporated to dryness. The aqueous phase was then extracted with 100 ml of ethyl acetate four times. The ethyl acetate fractions (400 ml) were combined and then evaporated to dryness. The aqueous phase is then extracted with 100 ml of ethanol 4 times then ethanol fractions (400 ml) were combined and then evaporated to dryness. The aqueous phase is evaporated to dryness. These different fractions then were kept in refrigerator at 5°C before use.

2.4 Preparation of Inoculum

Fungal strains tested were seeded on Petri dishes containing Sabouraud agar and incubated for 48 to 72 h to obtain young cultures and isolated colonies. From these boxes, using a platinum loop two well-isolated colonies and identical were perfectly picked and put in 10 ml of Sabouraud broth and incubated for 3-5 h at 37°C for pre-culture. A volume of 1 ml of this broth was taken and added to 10 ml of Sabouraud's broth. This fungal suspension produced was valued approximately at 10^5cells/ml and was the pure inoculum.

2.5 Preparation of Concentration Ranges

The range of extracts concentrations was prepared by double dilution method according to a geometric progression of ratio 1/2 with concentrations ranging from 500 mg/ml to 7.81 mg/ml. These tubes were then autoclaved at 121°C for 15 min.

2.6 Antifungal Activity

2.6.1Sensitivity test

Antifungal activity was screened by agar well diffusion method described by Bakkiyaraj and Pandiyaraj [15], Irshad et al. [16] and Adekunle et al. [17]. The principle of this method was based on the spread of antimicrobial in solid media into a Petri dish by creating a concentration of gradient after a certain time of contact between the drug and the tested microorganisms. The antimicrobial effect of the drug on the tested microorganisms was assessed by the measurement of a zone of inhibition; and depending on the diameter of inhibition.

Sabouraud medium was poured into sterilized Petri dish to a thickness of 8 mm. After inoculation by flooding adequate dilution (approximately 10^5cells/ml) of the tested strains made, wells of 6 mm diameter were formed concentrically in the agar. Each well was then filled with 20 µl of a given concentration of the extract. After a pre-diffusion of 45 min at room temperature in the hood, the strains were incubated at 37°C for 24 h after which the diameters of the zones of inhibition were determined. Alongside Nystatin a standard antifungal was used.

2.6.2 Determination of antifungal parameters: minimum inhibitory concentration (MIC) and minimum fungicide concentration (MFC)

The incorporation of the different extracts was made in Sabouraud in inclined tubes according to the double dilution method previously described by Yayé et al. [18] and Ouattara et al. [19].

A serie of 7 tubes were prepared according to the method of double dilution with concentrations ranging respectively from 50 to 0.390 mg/ml.

The tubes containing were then inclined with a small base at the temperature of the room so that the agar solidified. After solidification, agar seeding or not containing the extract (control) is performed on the surface with 1 µl of pure inoculum. The range of both inoculated and control tubes were incubated at 37°C for 24 h and the MIC was determined. It corresponded to the concentration of the first tube where there was no observable growth visible to the naked eye of the tested germ. Then we inoculated the surface of new Sabouraud agar with 0.1 ml of pipe content with concentration greater than or equal to the MIC to determine the MFC. To appreciate the power of each extract, the ratio MFC/MIC was calculated. Indeed, if the ratio MFC/MIC equals to 1 or 2, the effect is fungicidal and if it was between 4 to 16 the effect was fungistatic [20,21].

2.7 Phytochemical Screening

Vegetable extracts obtained were subjected to a phytochemical screening to reveal the major chemical groups it contains. The methods used are those described by Bekro et al. [22], Yusuf et al. [23], Mehta et al. [24] and Shittu et al. [25].

2.8 Statistical Analysis

Analysis and graphical representations of data were performed using GraphPad Prism version 5.00 for Windows (GraphPad Software, San Diego California USA, www.graphpad.com) software. The values expressed were the means of three experiments with the standard error of the mean (mean ± SEM).

3. RESULTS

3.1 Extractions Yields

Table 1 shows the yield and characteristics of each plant extract. All the extracts were friable except the ethyl acetate extract of *A. leiocarpa* which was powdery and dichloromethane extract of *S. virosa* which looked pasty. All extracts were brown.

Aqueous extracts haved the highest yields with 4.5% for aqueous extract of *S. virosa* and 7.8% for the aqueous extract of *A. leiocarpa*. The lowest yields were 0.19% and 0.91% respectively for dichloromethane extracts of both plants.

3.2 Phytochemical Screening

Table 2 lists the major chemical groups encountered in different extracts of root bark of *A. leiocarpa*. The ethanol extract contained polyterpenes, polyphenols, flavonoids, catechin tannins and alkaloids. In the ethyl acetate extract, we noted the presence of polyterpenes, flavonoids and alkaloids. Polyterpenes, polyphenols and alkaloids were also present in dichloromethane extract. As to the total aqueous extract, it contained polyterpenes, polyphenols, flavonoids, catechin tannins, quinone substances, alkaloids and saponins.

The results of the phytochemical screening of extracts of roots bark of *S. virosa* were summarized in Table 3. This table shows that ethanol extract contained polyterpenes, polyphenols, catechin, tannins, alkaloids and quinone substances. Ethyl acetate extract contained only polyterpenes and alkaloids while dichloromethane extract contained only polyterpenes. In the total aqueous extract, polyterpenes, polyphenols, flavonoids, catechin tannins, quinone substances and alkaloids haved been highlighted.

3.3 Sensitivity Tests

The results of susceptibility testing of plant extracts against the strains investigated were shown in Figs. 1 and 2.

Table 1. Yield and features excerpts

Extract	*A. leiocarpa*			*S. virosa*		
	Appearance	Color	Y (%)	Appearance	Color	Y(%)
Ethanol	Friable	Brown	6,10	Friable	Brown	2,60
Aqueous	Friable	Dark brown	7,80	Friable	Brown	4,50
Dichloromethane	Friable	Brown	0,19	Pasty	Brown	0,91
Ethyl acetate	Powdery	Brown	0,80	Friable	Brown	1,11

Y: yield

Table 2. Phytochemical analysis of different solvents extracts of root bark of *A. leiocarpa*

Chemical groups		Extracts			
		ETHA	EAA	EDMA	ETAA
Sterols (polyterpenes)		+	+	+	+
Polyphenols		+	-	+	+
Flavonoids		+	+	-	+
Tannins	Catechin	+	-	-	+
	Gallic	-	-	-	-
Quinone substances		-	-	-	+
Alkaloids		+	+	+	+
Saponins		-	-	-	+

ETHA: ethanol extract of A. leiocarpa; ETAA: total aqueous extract of A. leiocarpa; EDMA: dichloromethane extract of A leiocarpa; EAA: ethyl acetate extract of A. leiocarpa
- : absence; +: presence

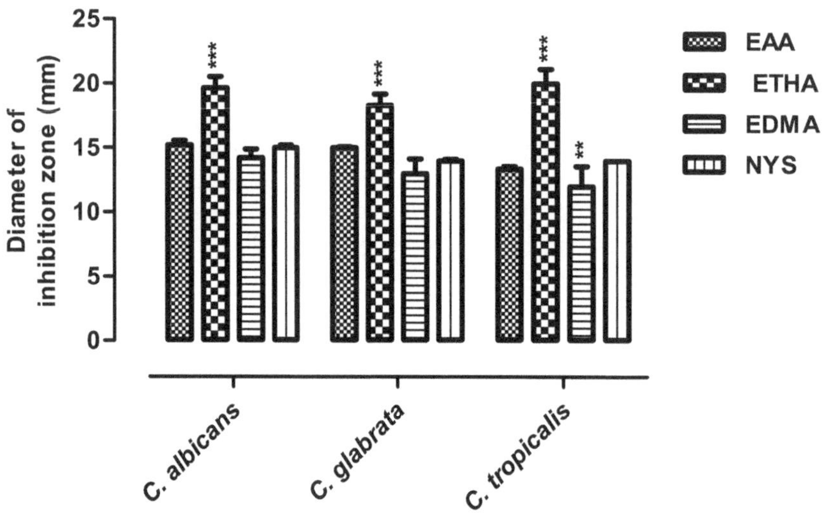

Fig. 1. Activity of *A. leiocarpa* extracts on growth of *Candida*

The diameters of the inhibition zones are the arithmetic mean (mm) of 3 experiments (n = 3) affected the standard error (mm ± SEM).
*** There is a significant difference at p < 0.05 compared to the average of NYS;*
**** There is a significant difference at p < 0.01 compared to the average of NYS*

Comparing the diameters of inhibition zones, extracts of *A. leiocarpa* (Fig. 1) showed that the effect of the extracts at the same concentration was more marked with ETHA (ethanol extract of *A. leiocarpa*). It induced inhibition zone diameters of 19.67±0.88 mm; 18.33±0. and 20.00±1.15 mm respectively on *C. albicans, C. glabrata* and *C. tropicalis*. EDMA (dichloromethane extract of *A. leiocarpa*) was the less active extract with diameters of inhibition of 14.23±0.66 mm;

13.00±1.15 mm and 12±1.58 mm respectively on *C. albicans, C. glabrata* and *C. tropicalis.* Furthermore the average diameters of inhibition induced by ETHA was significantly greater than those induced by NYS ($p < 0.05$; $p < 0.01$).

As for *S. virosa* (Fig. 2), the diameters of inhibition zones induced by its various extracts were lower than those of *A. leiocarpa*. At the same concentration, all extracts of *S. virosa* induced any inhibition zones except ETAS (total aqueous extract of *S. virosa*), which had no effect on all the microorganisms tested. For this plant, ETHS (ethanol extract of *S. virosa*) induced inhibition diameters larger on *C. tropicalis* (14.23±1.53 mm) followed by *C. albicans* (14±0.38 mm) and finally *C. glabrata* (12.05±0.56 mm). These diameters were significantly higher ($p < 0.05$; $p < 0.01$) compared to NYS. Germs were resistant overlook aqueous extracts of both plants.

Tables 4, 5 and 6 show the values of various antifungal parameters of extracts tested.

To *A. leiocarpa* MICs determined varied between 0.195 to 12.5 mg/ml and MFCs 0.390 to 50 mg/ml.

Table 3. Phytochemical analysis of different solvents extracts of root bark of *S. virosa*

Chemical groups		Extracts			
		ETHS	EAS	EDMS	ETAS
Stérols (polyterpenes)		+	+	+	+
Polyphenols		+	-	-	+
Flavonoids		-	-	-	+
Tanins	Catechin	+	-	-	+
	Gallic	-	-	-	-
Quinone substances		+	-	-	+
Alkaloids		+	+	-	+
Saponins		-	-	-	-

ETHS: ethanol extract of S. virosa; ETAS: total aqueous extract of S. virosa;
EDMS: dichloromethane extract of S. virosa; EAS: ethyl acetate extract of S. virosa
- : absence; +: presence

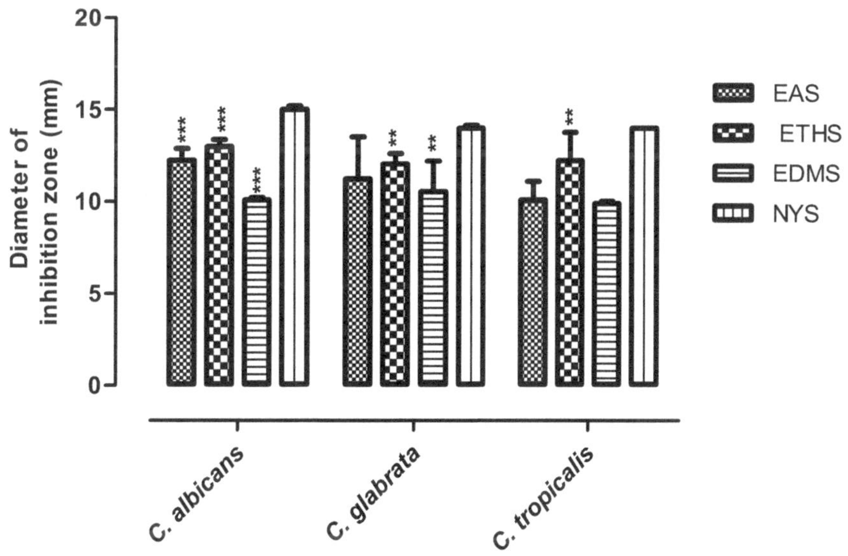

Fig. 2. Activity of *S. virosa* on growth of *Candida*

The diameters of the inhibition zones are the arithmetic mean (mm) of 3 experiments (n = 3) affected the ndard error (mm ± SEM).
*** There is a significant difference at $p < 0.05$ compared to the average of NYS;*
**** There is a significant difference at $p < 0.01$ compared to the average of NYS*

MICs for *S. virosa* ranged 1.562 to 25 mg/ml and MFCs 3.125 to 50 mg/ml. The fungicidal effect appreciated by the ratio MCF/CMI showed that ETHA, ETHS, EAS (ethyl acetate extract of *S. virosa*) and EDMS (dichloromethane extract of *S. virosa*) were fungicidal because this ratio was between 1 to 2 and EAA was funcicidal on *C. albicans* and *C. tropicalis* but fungistatic on *C. glabrata*. EDMA was fungistatic on *C. albicans* and *C. glabrata*.

Figs. 3 and 4 show that *A. leiocarpa* presents the lowest MIC among tested plants. Ethanol extracts have the lowest MIC.

Table 4. Values of antifungal parameters and ratios MFC/MIC of ethanol extracts of *A. leiocarpa* and *S. virosa*

	ETHA			ETHS		
	MIC (mg/ml)	**MFC (mg/ml)**	**MFC/CMI**	**MIC (mg/ml)**	**MFC (mg/ml)**	**MFC/MIC**
C. albicans	0.195	0.390	2	3.125	6.250	2
C. glabrata	0.781	0.781	1	1.562	3.125	2
C. tropicalis	3.125	6.250	2	12.50	25.00	2

Table 5. Values of antifungal parameters and ratios MFC/MIC of ethyl acetate extracts of *A. leiocarpa* and *S. virosa*

	EAA			EAS		
	MIC (mg/ml)	**MFC (mg/ml)**	**MFC/CMI**	**MIC (mg/ml)**	**MFC (mg/ml)**	**MFC/CMI**
C. albicans	6.25	6.25	1	12.5	12.5	1
C. glabrata	12.5	50	4	25	50	2
C. tropicalis	6.25	12.5	2	25	25	1

Table 6. Values of antifungal parameters and ratios MFC/MIC of dichloromethane extracts of *A. leiocarpa* and *S. virosa*

	EDMA			EDMS		
	MIC (mg/ml)	**MFC (mg/ml)**	**MFC/MIC**	**MIC (mg/ml)**	**MFC (mg/ml)**	**MFC/MIC**
C. albicans	12.5	50	4	25	50	2
C. glabrata	6.25	31.25	5	6.25	6.25	1
C. tropicalis	12.5	12.5	1	12.5	12.5	1

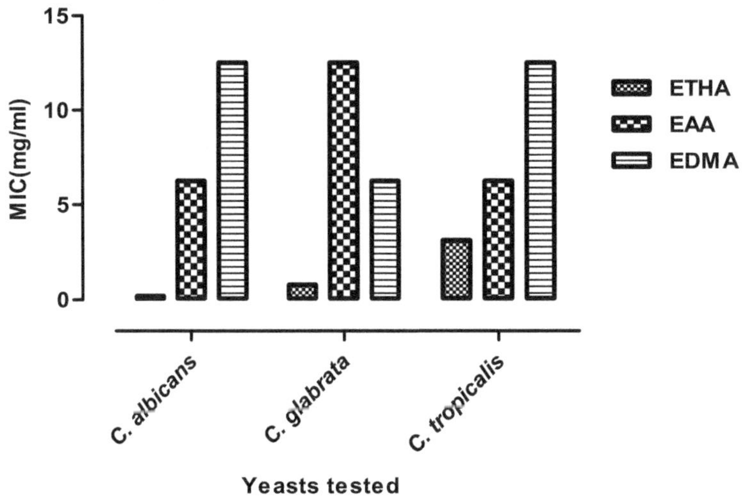

Fig. 3. MIC of *A. leiocarpa* extracts of fungi tested

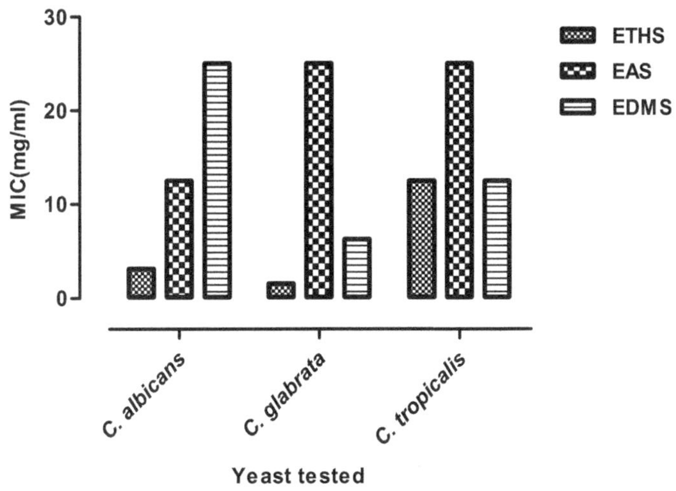

Fig. 4. MIC of *S. virosa* on fungi tested

4. DISCUSSION

We performed the extraction of the plant powder with solvents of increasing polarity. The polar extracts have the highest yields with 7.80% for aqueous extract of *A. leiocarpa* and 4.50% for aqueous extract of *S. virosa*. The extraction yields reveal that root barks of these plants are richer in polar compounds than unpolar because yields increase with polarity of the solvent used.

In this study the antifungal activity of the extracts of root bark of *A. leiocarpa* and *S. virosa* was evaluated on three species of *Candida* which are: *C. albicans, C. tropicalis* and *C. glabrata*. These saprophytes and opportunistic yeast are present in the mucosa where they take advantage of an imbalance of existing flora or immune deficiency to multiply and cause infections. Resistance to azoles, including fluconazole, may explain clinical failures [26]. Determining the diameter of the inhibition zones and determination of antifungal parameters (MIC and MFC) were used to assess their sensitivity. In view of the results, we note that the best extraction yields are obtained with the aqueous extracts, but the best antifungal activity is obtained with ethanol extracts.

About *A. leiocarpa* the diameters of inhibition zones are 18 mm for ETHA the most active extract and 12 mm for EDMA at a concentration of 500 mg/ml. Kubmarawa et al. [27] showed that the ethanol extract of the stem bark of *A. leiocarpa* has significant inhibitory activity against *C. albicans*. Similarly Adigun et al. [28] showed that the acid 3,3,4-Tri-o-methylflavellagic glucoside isolated in the stem bark of *A. leiocarpa* has an antimicrobial effect on *Staphylococcus aureus*, *Escherichia coli*, *Pseudomonas aeruginosa* and *Candida albicans*. In Togo, studies conducted by Batawila [29] show that the MIC of *A leiocarpa* for twenty fungal germs vary between 0.25 and 4 mg/ml. Our results show that the MIC range from 0.195 to 12.5 mg/ml and are in agreement with these authors.

As *A leiocarpa*, *S. virosa* induces diameters of inhibition zone greater than 10 mm and the MIC range from 1.562 to 25 mg/ml. These results suggest that the plant has an antifungal activity. This is corroborated by the results of Agassounon et al. [30] which have shown that this plant has fungicidal properties.

The both plants have antifungal activity but *A. leiocarpa* seems to be more active than *S. virosa*. The roots of *A leiocarpa* concentrate more active compounds than roots of *S. virosa*.

Therefore, the roots concentrate polar active substances. These results are supported by the work of Moroh [31] which showed that the roots of *Morinda morindoides* concentrate better the antimicrobial active compounds than other organs of the plant. It should be noted that the antimicrobial activity of medicinal plants varies from one geographic region to another. Then, leaf extracts of *A. leiocarpa* from Ghana are deemed to have a better antibacterial activity than extracts of leaves of the same plant from Nigeria [32].

The antifungal activity of both plants would be related to their phytomolecules composition. Indeed phytochemical screening of extracts of *A. leiocarpa* shows the presence of polyterpenes and sterols, polyphenols, flavonoids, catechin tannins and alkaloids. These results are consistent with those of Arbab et al. [33] showing the same chemical constituents in all organs of the plant.

The phytochemical screening of *S. virosa* extracts shows also polyterpenes and sterols, polyphenols, flavonoids alkaloids and catechin tannins. These compounds have also put in evidence by Ezeonwumelu et al. [34] in extracts of this plant. These extracts are less rich in bioactive compounds than *A. leiocarapa*. The ethanol extracts were the most active; indicating that they concentrate most of the compounds. Ethanol would be the best extraction solvent of active substances from these plants.

The presence of large chemical groups such as polyterpenes and sterols, polyphenols, flavonoids, catechin tannins and alkaloids in ethanol extracts explains their best activity because these compounds are recognized for their antimicrobial activities [35-37]. It may also be explained that the activity of antibiotics in plant extracts against fungi or growth may be due to their mechanism of action, chemical structure or spectrum of activity [38].

Flavonoids are phenolic in nature and they act as cytoplasmic poisons, which inhibit the activity of cytoplasmic enzymes such as aldose reductase, xanthine oxidase, phosphodiesterase and ATPase [39,40]. Tannins coagulate the cell wall proteins resulting in bactericidal activity at higher concentrations, while saponins are surface active agents and they alter the permeability of the cell wall thus facilitating the entry of toxic materials or leakage of vital constituents from the cell [41].

5. CONCLUSION

The extraction of bioactive substances with solvents of increasing polarity showed that the inhibitory activity of the extracts is due to better polar compounds concentrated in ethanol. The extracts are either fungistatic or fungicidal. These results validate the traditional use of these plants in the treatment of candidiasis.

ACKNOWLEGEMENTS

The authors thank Blégui Douh Blaise, Anoh Bodouin, Yoboué Sylvestre, Dakrom Richard, Ahoussi Corneille and Mrs Honorine Coulibaly for their technical assistance and as well as the entire staff of the department of Bacteriology-Virology of the Central laboratory of University Hospital of Treichville, Abidjan Côte d'Ivoire.

COMPETING INTERESTS

Authors have declared that no competing interests exist.

REFERENCES

1. Videau JY. Access for all to quality medicines. Med Trop. 2002;62:396-400. French
2. Adjanahoun EJ. State of the African ethno-pharmacopoeia. Bull Med Trad Pharm. 1990;4(1):59-63. French
3. Ake-Assi L. Report on the International Symposium on African traditional medicine and pharmacopoeia in Abidjan, Côte d'Ivoire. Bull Med Trad Pharm Afri. 1991; 4(2):203. French
4. Dro B, Soro D, Kone MW, Bakayoko A, Kamanzi K. Evaluation of the abundance of medicinal plants used in traditional medicine in Northern Côte d'Ivoire. J An &Plant Sci. 2013;17(3):2631-2646. French
5. Thenmozhi K, Saradha M, Manian S, Paulsamy S. *In vitro* antimicrobial potential of root extracts of the medicinal plant species, *Emilia sonchifolia* (linn.) Dc. Asian J Pharm Clin Res. 2013;6(3):149-151.
6. Konan Kouadio F, Guessennd NK, Karamoko O, Bahi C, Adama C, Dosso M. Antibacterial action of the ethanol extract 70% of *Clerodendrum splendens* (G. Don) (Verbenacae) on bacterial strains isolated from diarrheal stools in children. Int. J. Biol. Chem. Sci. 2013;7(3):1332-1337. French
7. Sapna P, Vaibhav S, Kakasaheb M. Antimicrobial properties of different Ocimum species (Lamiaceae). J. Adv Pharm Res Biosci. 2013;1(1):16-19.
8. Toure A, Bahi C, Ouattara K, Djaman AJ, Coulibaly A. Phytochimical screening and *in vitro* antifungical activities of extracts of leaves of *Morinda morindoides* (Morinda, Rubiaceae). J of Med Plants Res. 2011; 5(31):6780-86.
9. Sithapa O, Kra KAM, Kporou KEL, Zihiri GN, Yapi HF, N'Guessan JD, Djaman AJ. Comparative study chromatographic fractions activities from *Terminalia ivorensis* and ketoconazole as standard antifungal on *in vitro* growth of

Trichophyton mentagrophytes var interdigitale. J. Drug Delivery Ther. 2013; 3(5):18-21.

10. Shuaibu MN, Wuyep PA, Yanagi T, Hirayama K, Tanaka T, Kouno I. The use of microfluorometric method for activity guided isolation of antiplasmodial compound from plant extracts. Parasitol Res. 2008;102:1119-1127.
11. Bidie AP, N'Guessan BB, Yapo AF, N'Guessan JD, Djaman AJ. Antioxidant activities of ten medicinal plants of the Ivorian Pharmacopoeia. Sci Nat. 2011; 8(1):1-11. French
12. Larsonn J, Gottfries J, Bohiin L, Backlun D. Epanding the chemGPS chemical space with natural product. A. J. Nat. Prod. 2005; 68:985-991.
13. Rollinger JM, Haupt S, Stuppner H, Langer TJ. Combining ethnopharmacology and virtual screening for lead structure discovery: COX-Inhibitors as application example. Chem. Inf. Comput. Sci. 2004; 44:480-488.
14. Manga A, Gassama A, Sy YG, Bassene E, Lavaud C. Structural determination of news flavones C-glycosides and trans (S, E) - (-) clovamide isolated *Icacina senegalensis Juss* leaves (Icacinaceae). J. Soc. Ouest-Afr. Chim. 2013;035:15-27.
15. Bakkiyaraj S, Pandiyaraj S. Evaluation of potential antimicrobial activity of some medicinal plants against common food-borne pathogenic microorganism. Int J Pharm Bio Sci. 2011;2(2):484-491.
16. Irshad S, Mahmood M, Perveen F. *In vitro* anti-bacterial activities of three medicinal plants using agar well diffusion method. Res. J. Bio. 2012;2(1):1-8.
17. Adekunle O. Phytochemical Screening and Antimicrobial Efficacy of Aqueous and Methanolic Extract of *Mangifera indica* (Mango Stem Bark). World J Life Sci. and Medical Research. 2012;2(2):81-85.
18. Yaye YG, Kra AKM, Ackah JAAB, Djaman AJ. Evaluation of the antifungal activity assay and purification of extracts from *Terminalia Mantaly* (H. Perrier) a Combretaceae on the growth *in vitro* of Candida albicans. Bull Soc R Sci Liège. 2011;80:953-964. French
19. Ouattara S, Kporou KE, Kra KAM, Yapi HF, Zirihi GN, N'Guessan JD, Bidié ADP, Djaman AJ. Optimization of the in vitro antifungal activity of hydroalcoholic extract of *Terminalia ivorensis*. A. Chev. Nat. Prod. Plant Resour. 2013;3(4):29-33.
20. Berche P, Gaillard JL, Simonet M. The bacteria of human infections. Editor: Flammarion, Medecine-Sciences. 1991; 660. French
21. Loubaki BC, Ouattara AS, Ouattara AS, Ouedraogo-Traore R, Traore AS. Antimicrobial activities of total aqueous extracts of *Detarium microcarpum* (Cesalpinaceae) on eight bacterial species involved in infectious diseases in Burkina Faso. Rev. CAMES-Serie A. 1999;1:6673. French
22. Bekro Y, Bekro J, Boua BB, Tra BF, Ehile E. Ethnobotanical study and phytochemical screening of *Caesalpinia benthamiana* (Baill.) Herend and Zarrucchi (Caesalpiniacea). Sci Nat. 2007;4(2):217-225. French
23. Mehta K, Patel BN, Jain BK, Phytochemical analysis of leaf extract of *Phyllantus fraternus*. Res. J. Recent. Sci 2013;2:12-15.
24. Yusuf AZ, Zakir A, Shemau ZI, Abdullahi M, Halima SA. Phytochemical analysis of methanol leaves extract of *Paullinia pinnata* linn J. Pharmacognosy Phytother 2014;6(2):10-16.
25. Shittu OB, Olabode OO, Omemu AM, Oluwalana SA, Adeniran S, Akpan I.. Phytochemical and antimicrobial screening of *Spondias mombin*, *Senna occidentalis* and *Musa sapientum* against *Vibrio cholerae* O1. Int. J. Curr. Microbiol. App. Sci. 2014;3(5):948-961.
26. Vanessa C. Vaginal candidiasis recurrent to *Candida albicans*: Diagnosis, pathophysiology, treatment. University Henri Poincaré -Nancy 1, France. Thesis Pharmacy. 2001;131. French
27. Kubmarawa D. Preliminary phytochemical and antimicrobial screening of 50 medicinal plants from Nigeria. Afr J Biotechnol. 2007;6(14):1690-1696.
28. Adigun JO, Amupitan JO, Kelly DR. Isolation and investigation of antimicrobial effect of 3,4,3 '-tri-Omethylflavellagic acid and its glucoside from *Anogeissus leocarpus*. B Chem Soc Ethiopia. 2000; 14(2):169-174.
29. Batawila K. Antifungal activities of five Combretaceae used in Togolese traditional medicine. Fitoterapia. 2005;76(2):264-8.
30. Agassounon DTM, De Souza C, Anani KT, Koumaglo K, Toukourou F, Gbeassor M, Evaluation of the cytotoxic, antiviral, antibacterial and antifungal activities of six

medicinal plants. Pharm. Méd. Trad. AF JOOl. 2001;11:93-l05. French
31. Moroh AJL. Bacterial resistance and antimicrobial phytomolecules from *Morinda morindoides*. University of Bretagne Occidentale, France, University Félix Houphouet Boigny, Côte d'Ivoire. Thesis 2013;205. French
32. Mann A, Yahaya Y, Banso A, Ajayi GO. Phytochemical and Antibacterial screening of *Anogeissus leiocarpus* against some microorganisms associated with infectious wounds. African J. of Microbiol Res. 2008; 2:060-062.
33. Arbab AH. Review on *Anogeissus leiocarpus* a potent african traditional drug. Int. J. Res. Pharm. Chem. 2014;4(3):496-500.
34. Ezeonwumelu COJ, Matuki KE, Ajayi MA, Okoruwa GA, Tanayen KJ, Adiukwu PC, Goji TDA, Dare S, Okonkwo OC, Byarugaba F. Phytochemical screening, acute toxicity and analgesic properties of aqueous extract of *Flueggea virosa*'s root in rats. Ibnosina J Med BS. 2013;15-21.
35. Mada SB, Garba A, Muhammad A, Mohammed A, Adekunle DO. Phytochemical screening and antimicrobial efficacy of aqueous and methanolic extract of *Mangifera indica* (mango stem bark). World J Life Sci. and Medical Research. 2012;2(2):81-85.
36. Mbengui DR, Guessennd KN, M'Boh GM, Golly KJ, Okou OC, N'Guessan JD, Dosso M, Djaman AJ. Phytochemical screening and study of comparative antibacterial activity of aqueous and alcoholic extracts of the leaves and barks of *Terminalia catappa* on multiresistant strains. J. Appl. Biosci. 2013;66:5040-5048.
37. Gbadamosi ET, Ogunsuyi A. An appraisal of the potency of roots of *Anogeissus leiocarpus* (DC.) Guill. & Perr. and *Terminalia glaucescens* Benth. in the management of *E. coli* related infections J. Appl. Biosci. 2014;78:6646-6653
38. Calderon CB, Sabundayo BP. Antimicrobial Classifications: Drugs for Bugs. In Schwalbe R, Steele-Moore L, Goodwin AC. Antimicrobial Susceptibility Testing Protocols. CRC Press. Taylor & Frances Group; 2007.
39. Rathee P, Chaudhary H, Rathee S, Rathee D, Kumar V, Kohli K. Mechanism of action of flavonoids as anti-inflammatory agents. Inflamm Allergy Drug Targets. 2009;8(3): 229-35.
40. Arullappan S, Rajamanickam P, Thevar N, Kodimani CC. In vitro screening of cytotoxic, antimicrobial and antioxidant activities of *Clinacanthus nutans* (Acanthaceae) leaf extracts. Trop J Pharm Res, September. 2014;13(9):1455-1461.
41. Iwu MM, Igboko OA, Okeniyi CD, Tempesta MS. Inhibition of the enzyme activity of aldosoreductase of some flavonoids by some flavonoids. J Pharm Pharmacol. 1990;42:290-292.

5

LH/FSH, BMI and Clinical Profile in Polycystic Ovarian Syndrome: A Correlative Study

Hifsa Mobeen[1*], Hamdan Hamid[1], Almina Shafiq[1], Muhammad Adil[1], Muhammad Kashif[1] and Noshin Wasim Yousuf[2]

[1]Department of Immunology, University of Health Sciences Lahore, Pakistan.
[2]Department of Pathology, Allama Iqbal Medical College Lahore, Pakistan.

Authors' contributions

This work was carried out in collaboration between all authors. Authors HM and NWY designed the study. Authors HH, MK and HM, wrote the protocol and wrote the first draft of the manuscript. Authors HH, HM, MA and AS managed the literature searches, analyses of the study. All authors read and approved the final manuscript.

Editor(s):
(1) Chunying Li, Department of Biochemistry and Molecular Biology Wayne State University School of Medicine, Detroit, USA.
(2) Luisa Di Paola, Chemical and Biochemical Engineering Fundamentals, Faculty of Engineering Università Campus Biomedico, Via Alvaro del Portillo, Roma, Italy.
Reviewers:
(1) Samira Yarak, Federal University of Sao Paulo, Brazil.
(2) Mra Aye, Melaka Manipal Medical College, Malaysia.
(3) Anonymous , Erzincan University, Turkey.

ABSTRACT

Background and Objectives: Polycystic ovarian disease (PCOS) is probably the most prevalent endocrinological disorder affecting females and is the most common cause of menstrual disturbance during the reproductive age. It is characterized by polycystic ovaries on ultrasound and/or clinical and biochemical signs and symptoms of hyperandrogenism and/or oligo- anovulation. Therefore, this study was designed to determine relationship among LH/FSH ratio, BMI and the clinical profile of females suffering from PCOS.

Materials and Methods: Blood samples from 50 study subjects were taken after getting informed consent for hormone profile (FSH and LH) by ELISA kit of Kamiya Biomedical company. Body Mass index (BMI) and Ultra-sonogram related findings of polycystic ovarian syndrome patients were recorded. Statistical analysis was done by using SPSS-20.

*Corresponding author: E mail: Hifsamobeen035@gmail.com

Results: LH/FSH values were significantly higher among infertile and fertile females (P=0.05). BMI was also significantly correlated statistically in female patients with and without hirsutism (P=0.005). BMI (P=0.00) and LH/FSH (P=0.004) were also found to be associated significantly with obesity in patients with PCOS.

Conclusion: Since FSH was significantly correlated with obesity and infertility while BMI was associated with obesity and hirsutism among the clinical profiles, lifestyle interventions to decrease the overweight and obesity would be able to reduce the unwanted clinical symptoms of infertility and hirsutism in females with polycystic ovary syndrome.

Keywords: Polycystic ovarian syndrome; FSH; LH; BMI; polycystic ovaries; obesity; hirsutism; oligomenorrhea; amenorrhea.

1. INTRODUCTION

Polycystic Ovarian Syndrome (PCOS) is probably the most prevalent endocrinological disorder affecting females and it is the most common cause of menstrual disturbance during the reproductive age. It is characterized by presence of polycystic ovaries on ultrasound and/or clinical & biochemical signs of hyperandrogenism and/or oli-anovulation [1].

PCOS was first described by Stein and Leventhal in 1935 in women with of PCOS is multifactorial and complex. The disease has genetic basis affected by environmental factors [2,3].

Obesity is an important risk factor for PCOS. Literature review shows that 30-70% of PCOS females are obese, though disease is also seen in females of normal weight but with less frequency. The presence of obesity markedly modifies the clinical and biochemical expression of disease and makes the management of the syndrome very difficult [4].

World Health Organization (WHO) and National Institute of Health (NIH) criteria defined obesity; underweight as body mass index (BMI) < 18.5, normal weight as BMI between 18.5-24.9, overweight as BMI between 25-29.9 and obesity as a BMI of 30 or greater. Obesity is further divided into class 1 (BMI 30-34.9), class II (BMI 35-39.9) and class III (BMI > 40) [5].

When defined by the presence of oligomenorrhea and hyperandrogenism, 75% of women with PCOS have an LH level that is above the normal range for women in the early follicular phase and 94% have an increased LH to FSH ratio [6].

However, further studies have demonstrated an inverse relationship between LH and body weight that is continuous across a wide spectrum of body weights in PCOS patients, supporting the model of an intrinsic neuroendocrine abnormality in all oligomenorrheic PCOS patients that is modified by obesity [7,8].

Moreover, manifestations of PCOS may include menstrual irregularities, obesity and elevated serum LH levels [9].

The characteristic increase in LH relative to FSH release, have long been appreciated in PCOS. Because of the pulsatile nature of their release, a single test fails to detect an increased LH/FSH ratio. This, as well as its lack of specificity, has led to the recommendation that LH/FSH ratios not be included in the diagnostic criteria for PCOS [10,11].

Approximately half of PCOS women are obese or overweight and obesity has important role in the development of the Hyperandrogenic state [12] [13]. Even those with normal BMI, PCOS sufferers tend to have android body type with waist to hip ratio greater than 0.8 [14].

The study is designed to demonstrate a correlation/ association among LH/FSH ratio, BMI, and the clinical profile of females suffering from PCOS.

2. MATERIALS AND METHODS

2.1 Subjects

It was a comparative study, carried out in the Department of Pathology, Allama Iqbal Medical College Lahore after approval of Ethical Review Committee. The samples were taken from Infertility clinic of Obstetrics and Gynaecology Department of Jinnah Hospital, Lahore. Purposive sampling technique was used and a total of 50 subjects were recruited for this study after an informed consent. The study included all diagnosed cases of females of reproductive age, married or unmarried presenting with clinical features of PCOS confirmed on Ultrasound. Patients with hyperandrogenism and anovulation other than PCOS like Cushing syndrome, untreated hypo or hyperthyroidism, ovarian tumors and adrenal hyperplasia/tumors were excluded from the study.

At the time of sample collection, relevant details of each participant such as age, sex, and clinical features of disease were noted. The detail about their ultrasonic findings and BMI was also noted. Samples were processed for fertility hormone profile by sandwich ELISA technique of Kamiya Biomedical Company with sensitivity of 0.48 miu/ml and high specificity without cross reactivity.

Statistical analysis: The data was entered and analyzed using IBM SPSS 20.0. Mean±SD was used for quantitative variables. Qualitative variables were expressed as frequencies, percentages and graphs. For normally distributed data one way ANOVA and for not normally distributed data Kruskal-Walis test was applied. Post Hoc Tukey test was applied to observe which group means differ in various groups. A p-value of ≤ 0.05 was considered as statistically significant.

3. RESULTS

When BMI, FSH and LH were compared between fertile and infertile femles, FSH was significantly correlated with infertility shown in Table 1.

Table 1. Comparison of BMI, FSH, LH b/w fertile and infertile females

		N	Mean±SD	P- value
BMI	Infertility	25	27.52±5.39	0.55
	Fertility	5	26.48±6.70	
FSH	Infertility	25	8.49±3.98	0.05
	Fertility	25	6.60±2.63	
LH	Infertility	25	10.40±7.98	0.09
	Fertility	25	14.24±7.80	

When BMI, FSH and LH were compared between menstrually disturbed and undisturbed females, there was no significant difference between two groups shown in Table 2.

When BMI, FSH and LH were compared between females with hirsutism and without hirsutism, BMI is significantly associated with hirsutism shown in Table 3.

When BMI, FSH and LH were compared between females with oligomenorrhea and without oligomenorrhea, there was no significant difference between two groups shown in Table 4.

Table 2. Comparison of BMI, FSH, LH b/w menstrually disturbed and undisturbed females

	History of menstrual disturbance	N	Mean±SD	P-value
BMI	No disturbance	1	23.9±6.09	0.609
	Disturbance	49	27.06±6.09	
FSH	No disturbance	1	10.50±3.48	0.397
	Disturbance	49	7.48±3.48	
LH	No disturbance	1	21.4±8.02	0.259
	Disturbance	49	12.13±8.02	

Table 3. Comparison of BMI, FSH, LH b/w females with hirsuitism and without hirsutism

	Hirsutism	N	Mean±SD	P- value
BMI	No hirsutism	19	23.98±4.60	0.005
	Hirsutism	31	28.85±6.13	
FSH	No hirsutism	19	6.93±3.30	0.331
	Hirsutism	31	7.92±3.58	
LH	No hirsutism	19	9.78±7.79	0.081
	Hirsutism	31	13.87±7.43	

When BMI, FSH and LH were compared between females with amenorrhea and without amenorrhea, there was no statistically significant difference between two groups shown in Table 5.

When BMI, FSH and LH were compared between females with menopause and without menopause, there was no statistically significant difference between two groups shown in Table 6.

When BMI, FSH and LH were compared between females with acne and without acne, there was no statistically significant difference between two groups shown in Table 7.

When BMI, FSH and LH were compared between females with and without obesity, BMI and FSH were significantly correlated with obesity shown in Table 8.

Table 4. Comparison of BMI, FSH, LH and oligomenorrhea

	Oligomenorrhea	N	Mean±SD	P- value
BMI	No	20	27.58±6.29	0.586
	oligomenorrhea	30	26.62±5.95	
FSH	No	20	8.18±3.10	0.298
	oligomenorrhea	30	7.12±3.69	
LH	No	20	13.77±8.42	0.302
	oligomenorrhea	30	11.35±7.78	

Table 5. Comparison of BMI, FSH, LH and amenorrhea

	Amenorrhea	N	Mean±SD	P-value
BMI	NO	47	26.9±5.99	0.810
	amenorrhea	3	27.8±8.31	
FSH	NO	47	7.70±3.51	0.200
	amenorrhea	3	5.03±1.33	
LH	No	47	12.45±8.16	0.643
	amenorrhea	3	10.20±6.97	

When BMI, FSH and LH were compared between patients with and without regular menstrual periods, there was no statistically significant difference between two groups shown in Table 9.

Table 6. Comparison of BMI, FSH, LH and menopause

	Menopause	N	Mean±SD	P-value
BMI	No	49	26.98±6.10	0.870
	Yes	1	28.00±6.10	
FSH	No	49	7.58±3.50	0.617
	Yes	1	5.80±3.50	
LH	No	49	12.41±8.10	0.559
	Yes	1	7.60±8.10	

Table 7. Comparison of BMI, FSH, LH b/w patients of acne and without acne

	Acne	N	Mean ±SD	P- value
BMI	No	20	27.34±6.63	0.75
	Acne	30	26.78±5.73	
FSH	No	20	8.50±4.09	0.112
	Acne	30	6.90±2.90	
LH	No	20	11.05±7.92	0.368
	Acne	30	13.16± 8.15	

4. DISCUSSION

In the current study, FSH, LH and BMI were compared among different groups of PCOS categorized on the basis of fertility, menstrual disturbance, hirsutism, oligomenorrhea, amenorrhea, menopause, acne, obesity and regular menstrual periods. FSH was significantly correlated with obesity and infertility while BMI was associated with obesity and hirsutism.

Table 8. Comparison of BMI, FSH, LH b/w patients with obesity and without obesity

	Obesity	N	Mean±SD	P- value
BMI	No	15	20.6±3.85	0.00
	Obesity	35	29.7±4.63	
FSH	No	15	6.06±2.96	0.04
	Obesity	35	8.18±3.52	
LH	NO	15	11.82±7.88	0.77
	Obesity	35	12.53±8.22	

Table 9. Comparison of BMI, FSH, LH b/w patients with regular menstrual periods and without regular menstrual periods

	Regular menstrual period	N	Mean±SD	P-value
BMI	No	49	26.9±6.10	0.870
	Yes	1	28.00±6.10	
FSH	No	49	7.58 ±3.50	0.617
	Yes	1	5.80±3.50	
LH	No	49	12.41±8.10	0.559
	Yes	1	7.600±8.10	

The current study is partially in agreement with the studies of Olooto et al. (2012) [15] who find statistical significant difference in FSH (P=0.00) and LH (P=0.00) but not in BMI (P=0.554) between fertile and infertile females. In our study only FSH is statistically significant between infertile and fertile females (P=0.05).

When FSH, LH and BMI were compared between females with and without menstrual disturbance, there was no statistically significant difference between two groups. The current study is in accordance with the studies of Abdulrazak et al. [16] and Dimitrios Panidis et al. (2015) but is partially in agreement with the study conducted by Manal Ibrahim Mahmoud et al. (2015) who find statistical significant difference in BMI (P=0.006) but not in FSH and LH.

BMI is significantly different between females with and without hirsutism (P=0.005) that is not in agreement with the studies of Farid-ur-rehman et al. [17] (P>0.05). FSH and LH values are insignificant between these groups that is also not in accordance with the studies of Shazia Rasool et al. [18] (P<0.05).

FSH, LH and BMI were compared between females with and without oligomenorrhea, no statistically significant difference is found that is partially in accordance with the studies of Singer et al. [19] whose studies FSH and LH were significantly higher in patients with oligomenorrhea (P<0.001) but no significant difference in BMI that is similar to current study. The current study is also partially in agreement with the study conducted by Manal Ibrahim Mahmoud et al. (2015) who find significant difference in BMI (P=0.006) but not in FSH and LH.

The current study is partially in agreement with the study of Shazia Rasool et al. [18]. FSH and LH were significantly different between females with and without amenorrhea that is not similar to the current study. BMI was not significantly different between these groups that is similar to the current study.

According to Ewenighi Chinwe et al. [20] BMI, FSH and LH were statistically significant (0.000) between these groups with and without menopause that is not in accordance with current study.

The current study is in agreement with the study of Sayera Begum et al. [21] who also found no difference in LH, FSH and BMI (P>0.05).

BMI (P=0.00) and FSH (P=0.04) are statistically significant between females with and without obesity but no difference is found in LH that is partially in agreement with the study of Abdulrazak et al. [16] who could not found any statistically significant difference in FSH, LH and BMI and study of Manal Ibrahim Mahmoud et al. (2015) who find significant difference in BMI (P=0.006), FSH (P=0.048) and LH (P=0.036).

According to Abdulrazak et al. [16] there was no statistically significant difference in LH, FSH and BMI in females with and without regular menstrual periods that is in accordance with the current study.

5. CONCLUSIONS

Since FSH was significantly correlated with obesity and infertility while BMI was associated with obesity and hirsutism among the clinical profiles, lifestyle interventions to decrease the overweight and obesity would be able to reduce the unwanted clinical symptoms of infertility and hirsutism in females with polycystic ovary syndrome.

COMPETING INTERESTS

Authors have declared that no competing interests exist.

REFERENCES

1. Ovesen PG, Moller N, Greisen S, et al. Polycystic ovary Syndrome. Clinical Presentation and Treatment Ugesker Laeger. 1998;160(3):260-4.
2. Stein IF, Leventhal ML. Amenorrhea associated with bilateral polycystic ovaries. American J Obstet Gynecol. 1935;29: 181-191.
3. Asuncion M, Clavo RM, San Millan, JL, et al. A prospective study of the prevalence of the polycystic ovary syndrome in unselected Caucasian women from Spain. Journal of Clinical Endocrinology and Metabolism. 2000;85:2434-2438.
4. Venkatesan AM, Dunaif A, Corbould A. Insulin resistance in Polycystic Ovarian Syndrome: Progress and Paradoxes: Recent Progress in Hormone Research 2001;56:295-308.
5. World Health Organization. Obesity: Preventing and Managing the Global Epidemic, Report of the WHO Consultation on Obesity. Geneva: World Health Organization; 1997.
6. National Institute of Health. Clinical guidelines on the identification, evaluation and treatment of overweight and obesity in adults: The evidence based report. Bethesda, MD: National Institute of Health, US Department of Health and Human Services; 1998.
7. Taylor AE, McCourt B, Martin KA, et al. Determinants of abnormal gonadotropin secretion in clinically defined women with

polycystic ovary syndrome. J Clin Endocrinol Metab. 1997;82:2248–2256.

8. Arroyo A, Laughlin GA, Morales AJ, et al. Inappropriate gonadotropin secretion in polycystic ovary syndrome: Influence of adiposity. J Clin Endocrinol Metab. 1997; 82:3728–3733.
9. Rotterdam ESHRE/ASRM-Sponsored PCOS Consensus Workshop Group Revised 2003 consensus on diagnostic criteria and long term health risks related to polycystic ovary syndrome. Fertil Steril. 2004;81:19-25.
10. Dunaif A, Givens JR, Hasltine F, et al. The polycystic ovary syndrome. Blackwell Scientific, Cambridge, MA; 1992.
11. Dunaif A. Insulin resistance and the polycystic ovary syndrome: Mechanism and implications for pathogenesis. Endocrine Reviews 1997;18(6):774.
12. Pasquali R, Vicennate V. Influence of weight and distribution of adipose tissue in functional hyperandrogenism. Contracept Fertil Sex. 1998;26(5):372-5.
13. Yen SS. The polycystic ovary syndrome. Clin.Endocrinol. 1980;12:177-183.
14. Hull L, Dunaif A. Menstrual disorders in insulin resistant states. Diabetes Spectr. 1997;10:185-190.
15. Olooto WE, Adeleye OA, Amballi AA, et al. Pattern of reproductive hormones: Follicle stimulating hormone, luteinizing hormone, estradiol, progesterone, and prolactin levels in infertile women in South Western Nigeria. Der Pharmacia Letter. 2012;4(2): 549-553.
16. Alnakash AH, Al-Tae'e NK. Polycystic ovarian syndrome: The correlation between the LH/FSH ratio and disease manifestations. Middle East Fertility Society Journal. 2007;12(1).
17. Rehman F, Sohail I, Hayat Z, Niazi NA. Etiology of hirsutism. Is there a correlation between menstrual regularity, body mass index and severity of hirsutism with the cause? Journal of Pakistan Association of Dermatologists. 2010;20:4-9.
18. Rasool S, Dar LR, Hameed A. Prevalence of polycystic ovaries among patients with hirsutism and menstrual abnormalities. Journal of Pakistan Association of Dermatologists. 2011;21(3):174-178.
19. Singer K, Rosenthal A, Kasa-Vubu J. Elevated Testosterone and Hypergonadotropism in Active Adolescents of Normal Weight with Oligomenorrhea Journal of Pediatric and Adolescent Gynecology 2009; 22(5):323-327.
20. Ewenighi CO, Uchechukwu D, Adejumo BI et al. Enzymatic activities of glucose 6 phosphate dehydrogenase and some liver enzymes among quarry workers in Umuoghara, Ebonyi state-Nigeria. Int J Health Sci Res. 2014;4(9):129-134.
21. Begum S, Hossain MZ, Rahman F, et al. Polycystic ovarian syndrome in women with acne. Journal of Pakistan Association of Dermatologists. 2012;22:24-29.

6

Anti-cancerous/Anti-bacterial Activities of Allicin Generated *In situ* from Diastereo Pure Alliins by Alliinase

Shalini Gupta[1], Lasanthi Jayathilaka[1], Jin-Sheng Huang[1], Jenny Lee[2] and Bao-Shiang Lee[1*]

[1]Protein Research Laboratory, Research Resources Center, University of Illinois at Chicago, 835 S. Wolcott Avenue, Chicago, IL 60612, USA.
[2]Department of Chemical Engineering, Columbia University, 500 W. 120th Street, New York, NY 10027, USA.

Authors' contributions

This work was jointly carried out in collaboration between all authors. All authors read and approved the final manuscript.

Editor(s):
(1) Chunying Li, Department of Biochemistry and Molecular Biology Wayne State University, School of Medicine, Detroit, USA.
Reviewers:
(1) Daniela Hanganu, Department of Pharmacognosy, "Iuliu Hatieganu" University of Medicine and Pharmacy, Romania.
(2) Tülay Aşkin Çelik, Department of Biology, Adnan Menderes University, Aydin, Turkey.

ABSTRACT

Aims: To facilitate allicin generation *in-situ* from pure diastereomers of alliin by enzymatic reaction of alliinase and assess its anti-cancerous/anti-bacterial activities.
Study Design: Chemical synthesis and *in-vitro* assay of anti-cancerous/anti-bacterial activities.
Place and Duration of Study: Protein Research Laboratory, Research Resources Center, University of Illinois at Chicago, between February 2014 and February 2015.
Methodology: Cancer cell viability assay MTT assay, bacterial plate-diffusion growth inhibition assay, and flow cytometry cell cycle analysis have been used to demonstrate the anti-cancerous/anti-pathogen activities of the *in-situ* allicin. Diastereomers of alliin are produced by H_2O_2 oxidation of deoxyalliin, which is prepared by mixing L-cysteine and allyl bromide. Deoxyalliin and diastereomers of alliin are purified to high purity with repeated fractional crystallization. In

**Corresponding author: E-mail: boblee@uic.edu*

addition, fluorenylmethyloxycarbonyl (Fmoc) protected alliin and alliin methyl ester are synthesized and purified with RP-HPLC to test the importance of amino and carboxyl groups of alliin in alliinase enzymatic reaction. Alliinase is produced by a simple and effective method from an aqueous garlic extract

Results: Results from spectrophotometric alliinase activity assay indicate that (+)-L-alliin is more reactive toward alliinase than (-)-L-alliin, and both amino and carboxyl groups of the cysteine portion of alliin are critical in alliinase enzymatic reaction. Results from cancer cell viability assay MTT assay, bacterial plate-diffusion growth inhibition, and flow cytometry cell cycle analysis confirm that the *in-situ* allicin is as active as allicin purified from aqueous garlic extract or allicin synthesized chemically in a dose-dependent manner.

Conclusion: We describe here facile pathways to synthesize diastereomerically pure alliins and isolate allinase. The *in-situ* allicin conversed from alliin by allinase is very active. The data obtained here provide useful information on the design of the *in-situ* allicin strategy.

Keywords: Allium sativum; alliin; allicin; alliinase; anti-cancerous; anti-bacterial.

ABBREVIATIONS

AGE, Aqueous Garlic Extract; MTT, 3-(4,5-Dimethylthiazol-2-yl)-2,5-diphenyltetrazolium bromide; SDS-PAGE, sodium dodecyl sulfate polyacrylamide gel electrophoresis.

1. INTRODUCTION

Allicin protects garlic from pests, insects, and microorganisms [1-3]. It has lipid-lowering, anti-blood coagulation, anti-hypertension, anti-thrombotic, anti-inflammatory, anti-oxidant, anti-cancerous, anti-viral, anti-microbial, and anti-parasitic activities with low toxicity towards human [2-4]. Not surprisingly, garlic is the best-selling herbal supplements in the United States and allicin has been explored as a potential lead compound for drug design [5,6]. Allicin is formed by the action of the enzyme alliinase on alliin. Purified allicin is unstable with a characteristic smell of garlic [7-9]. It decomposes into biologically less active oil-soluble organosulfur compounds (OSCs) such as diallyl sulfide, diallyl disulfide, diallyl trisulfide (allitridi), dithiins, and ajoene and some water-soluble OSCs such as S-allyl cysteine and S-allyl mercaptocysteine [10-12]. Allicin reacts rapidly with free thiol groups [13,14]. As soon as allicin is taken up by the body or is in the circulatory system, it reacts with accessible thiol groups, penetrating biological membranes and decomposing to other compounds. Therefore it disappears from the circulation within a few minutes after injection [13-15] This is why the majority of the potent anti-cancerous/anti-bacterial activities effects of allicin were demonstrated in vitro and makes a pharmaceutical application of allicin implausible [4,8,16].

However, the usage of allicin in medical areas mentioned above remains an attractive endeavor. In recent years there has been a growing interest of circumventing the instability of allicin by use a binary system consisting of the plant enzyme alliinase and its substrate alliin to generate allicin *in situ* [4,17-19]. In this binary system, the two inactive stable components generate an active compound that exhibit biological activity *in situ*. In one proof-of-concept study, alliinase is conjugated to an anti-CD20 antibody rituximab which successfully inhibits the proliferation of CD20+ B chronic lymphocytic leukemia via allicin produced from alliin *in situ* [18]. It is therefore possible to initiate biological activity in a controlled, effective manner. Apart from circumventing the instability of allicin, this binary system has additional advantages, such as low toxicity of its individual components and site-directed selective activation. Furthermore, alliinase is able to react besides alliin other S-alk(en)yl-L-cystene solfoxides such as dihydroalliin, S-isopropyl-L-cysteine sulfoxide, and S-butyl-L-cysteine sulfoxide [20]. In general, substrates of allinase reaction contain the following structural features: (1) They must be derivatives of cysteine. (2) The sulfur atom of the cysteine derivative must link to alk(en)yl group. (3) The amino group of the cysteine portion of the compound must not be blocked. (4) The sulfur atom of the cysteine derivative must be a sulfoxide. These observations pave the way to a range of novel binary systems with specific chemical and biochemical properties.

In situ allicin has shown promising applications *in vivo*. However, more information on reagent preparation and reaction conditions are needed to realize the biomedical potential of *in situ* allicin strategy. For example, there is lack of information on the efficacy of *in situ* allicin systems produced from pure diastereomers or derivatives of alliin. Here we developed a facial synthesis for mass production of diastereo pure alliins to alleviate the need of purchasing expensive alliin, an effective method of purifying alliinase, and simple methods of assessing the efficacy of the *in situ* allicin. In addition, the inactivity of fluorenylmethyloxycarbonyl (Fmoc) protected alliin and alliin methyl ester in alliinase enzymatic reaction enable us to conclude that the amino and carboxyl moieties of alliin are critical for alliinase reaction. Results indicate that the length of alliin in cancer cell culture before addition of alliinase is most effective within 3 days and ineffective after 7 days as a substrate of the enzymatic reaction of alliinase. In addition, allinase reacts less with (-)-L- alliin than with (+)-L- alliin. Cancer cell viability assay MTT assay, alliinase assay, bacterial plate-diffusion growth inhibition assay, and flow cytometry cell cycle analysis are effective in validating the anti-cancerous/anti-pathogen activities of the *in-situ* allicin.

2. MATERIALS AND METHODS

2.1 Materials

All reagents and solvents were obtained from Sigma–Aldrich Chemical (St. Louis, MO, USA) with the highest purity available and used without further purification. Allicin was prepared from either aqueous garlic extract or by H_2O_2 oxidation of allyl disulfide with HPLC purification as previously described [8,9]. Double distilled water was used throughout the experiments. *Staphylococcus aureus subsp. aureus Rosenbach* (ATCC® 6538™) and *Escherichia coli (Migula) Castellani and Chalmers* (ATCC® 8739™) were from ATCC (Manassas, VA, USA). PD10 desalting column packed with Sephadex G-25 medium (GE Healthcare Biosciences, Piscataway, NJ, USA) was used to separate low-molecular weight (< 1000 Da) from high-molecular weight (> 5000 Da) biomolecules. ^{1}H and ^{13}C nuclear magnetic resonance (NMR) spectra were recorded on a Bruker FT-NMR spectrometer (Billerica, MA, USA) at 900 and 225 MHz, respectively. The nmr spectra were measured in D_2O or CD_3CN at 25°C and the ^{1}H and ^{13}C chemical shift values (δ) were reported in parts per million (ppm) relative to the TSP (δ 0.0 ppm) and DMSO (δ 39.5 ppm), respectively. Coupling constants (J values) were given in Hz. Multiplicities are reported as follows: s = singlet, d = doublet, dd = doublet of doublets, t = triplet, td = triplet of doublets, tt = triplet of triplets, q = quartet, dq = doublet of quartet. Accurate mass measurements (ESI-HRMS) were recorded on a LC-MS IT-TOF (Ion Trap-Time of Flight) mass spectrometry (Shimadzu Scientific Instruments, Columbia, MD, USA). Resolution was set between 5000 and 10000. It delivered high mass accuracy and high mass resolution independent of MS mode with a dual source and a suitable external calibrant. Thin-layer chromatography was carried out on aluminum-backed 0.25-mm silica gel plates with fluorescence indicator (Dynamic Adsorbents, Norcross, GA, USA) and were developed with n-butanol: glacial acetic acid:water (2:1:1, v/v/v). Substance spots on the chromatographic plate were detected by illumination with UV light (254 nm), KMnO4 staining, or 0.5% ninhydrin solution spraying. Reaction products of Fmoc protected alliin and alliin methyl ester were purified using either a Perseptive Biosystems (Life Technologies, Carlsbad, CA, USA) or an Agilent (Agilent Technologies, Santa Clara, CA, USA) HPLC system with a Phenomenex (Torrance, CA, USA) Kinetex Reversed-Phase C18 150 x 2.1 mm or 250 x 4.6 mm column. Gradient elution was run using solvents A and B (solvent A: 0.1% TFA in water; solvent B: 0.1% TFA in acetonitrile) from 100% A for 5 min and then to 100% B for another 20 min by linearly increasing solvent B. UV absorbance at 220 nm or 260 nm was used as detection. The progress of the reaction was monitored by RP-HPLC. ^{1}H NMR spectrum in an appropriate deuterium solvent of the resultant reaction solution confirmed completion of the reaction. The products were confirmed by TLC, melting point, and spectroscopic (^{1}H,^{13}C 1-D or 2-D NMR and HRMS-ESI) analysis. Yields refer to chromatographic and spectroscopic pure compounds, second level heading.

2.2 Preparation of Aqueous Garlic Extract

Garlic extract was prepared from garlic cloves as in previously described protocol with modifications [8,9]. Briefly, garlic (The Garlic Company, Bakersfield, CA, USA) was purchased from the Costco Wholesale Corporation at Naperville (Naperville, IL, USA) and the aqueous garlic extract (AGE) was prepared from peeled garlic cloves in hepes buffer (50 mM, pH 7.2) containing 10% glycerol, 1 mM CaCl2, 1 mM MnCl2, 1 mM MgCl2, 20 uM pyridoxal 5'-phosphate, and 0.25 M NaCl at 1g/mL. A 900 W high-torque motor blender (NutriBullet 900, Los Angeles, CA) was used to crush and pulverize the garlic cloves using three 5-second pulses with 2-minute breaks between pulses at 4°C. The garlic puree was strained through 3 layers of cheesecloth. The filtrate was kept overnight at -20°C. The following day, AGE was spun at 9878 x g for 4 minutes at 4°C. Finally, the supernatant was filtered through a sterilized 0.22µm syringe-driven filter (Becton Drive, Franklin Lakes, NJ, USA) to remove any residual precipitate and to sterilize the AGE. The clear supernatant was aliquoted and stored at -80°C until use.

2.3 Alliinase Purification and Characterization

The enzyme alliinase was purified to at least 95% homogeneity using PD10 and gel-filtration columns. Briefly, 2.5 mL of AGE was loaded onto a PD10 column. High-molecular weight (>5000 Da) biomolecules eluted into the first 3.5 mL fraction and low-molecular weight compounds into next 5 mL fraction with hepes buffer (50 mM, pH 7.2) containing 10% glycerol, 1 mM CaCl2, 1 mM MnCl2, 1 mM MgCl2, 20 uM pyridoxal 5'-phosphate, and 0.25 M NaCl. Subsequently, 2 mL of the first 3.5 mL PD10 protein fraction was injected onto Phenomenex (Torrance, CA, USA) Yarra 3 µm gel-filtration 300 x 7.8 mm HPLC column and the column was developed using hepes buffer at a flow rate of 1 mL/min and 280 nm detection. Pure alliinase fraction was collected at 8 min which was verified by alliinase enzyme activity assay and mass spectrometry. Concentration of alliinase was measured by 1-D sodium dodecyl sulfate polyacrylamide gel electrophoresis (SDS-PAGE) band intensity using BSA as standard. The alliinase solution displayed a single protein band (~52 kDa) on 1-D SDS-PAGE containing 4-12% (w/v) polyacrylamide (Invitrogen, Carlsbad, CA) with MES running buffer and stained with Coomassie blue.

The enzyme activity assay of alliinase was determined by the spectrophotometric assay using 4-mercaptopyridine (4-MP) method in a cell free system according to Miron et al. [21] with modifications. The assay is based on the assessment of the production rate of allicin using alliin as subtract. One enzyme unit is defined as the amount of enzyme necessary to convert alliin into 1 µmol pyruvate per minute. The molar consumption of 4-MP in its reaction with allicin is equivalent to the production of pyruvic acid. One mL of the reaction solution contained 4-MP (10^{-4} M) in 50 mM sodium phosphate, 2 mM EDTA, 0.02 mM pyridoxal 5'-phosphate, 200 µg alliin (1.12 mM), and 20 µg alliinase (0.36 µM), pH 7.2. Enzymatic activity started upon addition of 20 µL alliinase (1mg/mL solution in hepes buffer, 18 µM). Enzymatic activity was monitored at room temperature by the initial rate of the decrease in UV light absorption at 324 nm over a period of 5 min using quartz cuvettes of 1 cm in path length and a Spectronic Genesys UV/Vis spectrometer (Thermo Spectronic, Rochester, NY, USA). One unit of enzyme activity was defined as releasing 1 µmol pyruvate per minute and calculated using molar absorptivity ε_{324} = 19,800 M^{-1} cm^{-1}.

2.4 Synthesis of (R_C)-(-)-S-2-Propenyl-L-Cysteine (3) (L-Deoxyalliin)

The synthesis of compound 3 is a modified literature procedure which is based on earlier work by Iberl et al. with precise temperature control throughout the reaction [22,23]. L-cysteine hydrochloride monohydrate (20 g, 0.11 mol) was dissolved in 20 mL of water and stirred in an ice bath; then, 20 mL of an aqueous solution of sodium hydroxide (16g, 0.4 mol) was added drop wise. Next allyl bromide (10.8 mL, 0.12 mol) was gradually added to the solution and the reaction mixture was stirred for 14 hours at room temperature, cooled in an ice bath, and glacial acetic acid was slowly added to neutralize the reaction solution to pH 5. The solution was filtered and the precipitate was washed three times with absolute ethanol. The white solid was crystalized using water to produce purified compound **3**. This was done by dissolving the crude L-deoxyalliin in a minimum volume of warm 60°C water. After complete dissolution, the solution was cooled to room temperature and was kept at 4°C overnight. The L-deoxyalliin crystals were separated, washed with cold ethanol and then dried under vacuum. The yield was 76% (14.2 g) with purity of higher than 97%. The characterization data of the product were consistent with the literature results [22,24-26]. TLC: R_f 0.68; $[\alpha]^{25}_D$ = - 10.4° (c=1, H_2O); HRMS: (ESI, MH^+) calcd for $C_6H_{11}NO_2S$, 162.0589, found 162.0584; ^{1}H NMR (900 MHz, D_2O): δ – 2.97 (dd, J = 7.6, 14.9 Hz, 1 H, $=CHCH_2SC\underline{H}_2$), 3.07 (dd, J = 4.2, 14.9 Hz, 1 H, $=CHCH_2SC\underline{H}_2$), 3.22 (dd, J = 7.3, 13.8 Hz, 1 H, $=CHC\underline{H}_2S$), 3.24 (dd, J = 7.1, 13.8 Hz, 1 H, $=CHC\underline{H}_2S$), 3.91 (dd, J = 4.3, 7.7 Hz, 1 H, $SCH_2C\underline{H}N$), 5.21 (ddd, 1 H, J = 1.3, 2.2, 9.9 Hz, $C\underline{H}_2=CH$), 5.23 (ddd, 1 H, J = 1.3, 2.8, 17.1 Hz, $C\underline{H}_2=CH$), 5.84 (dddd, 1 H, J = 7.1, 7.3, 9.9, 17.1

Hz, CH2=CH-S); ^{13}C NMR (DEPTQ, 225 MHz, D_2O): δ = 30.05 (1 C, SCH_2CHN), 33.30 (1 C, $=CHCH_2S$), 52.88 (1 C, $CH_2=CH$), 117.81 (1 C, $CH_2=CH$), 132.91 (1 C, SCH_2CHN), 172.32 (1 C, OC=O).

2.5 Synthesis of (2R)-2-Amino-3-[(S)-Prop-2-Enylsulfinyl] Propanoic acid (4) [(+)-L-Alliin]

S-2-propenyl-L-cysteine sulfoxides were prepared by oxidation of the corresponding S-2-propenyl-L-cysteines with hydrogen peroxide with quantitative yield following a procedure described previously which was based on earlier work by Yu et al. [22,27] with modifications. To a 200 mL aqueous solution of compound **3** (7.67 g, 0.047 mol) at 0°C, 5.4 mL of 30% H_2O_2 (0.047 mol) was added drop wise. The mixture was stirred at 20°C for 14 h and lyophilized. The crude diastereo mixture was determined to be 42 % (+)-L-alliin by NMR. A cascade fractional crystallization was used to produce diastereomericaly pure alliins. Crystallization was done using warm aqueous ethanol. This was achieved by adding minimum amount of 50°C water to dissolve alliin, followed by addition of ethanol until solution turned slight turbid. The mixture was then heated to yield a clear solution. This solution was kept overnight at 4°C to complete the crystallization. The crystals and the supernatant were separated and analyzed by NMR. Each crystallization supernatant contained more (-)-L-alliin while crystals were enriched in (+)-L-alliin. The crystals were subjected to fractional crystallization multiple times to isolate (+)-L-alliin (4) (1.0g, 13% yield, > 97% pure). The characterization data of the product were in accordance with published data [23,26,28,29]. TLC: R_f 0.48; $[\alpha]^{25}_D$ = 59° (*c*=1, H_2O); HRMS: (ESI, MH^+) calcd for $C_6H_{11}NO_3S$, 178.0532, found: 178.0534; ^{1}H NMR (900 MHz, D_2O): δ = 3.13 (dd, J = 7.7, 14.0 Hz, 1 H, $=CHCH_2SOCH_2$), 3.34 (dd, J = 6.2, 14.0 Hz, 1 H, $=CHCH_2SOCH_2$), 3.55 (dd, J = 8.1, 13.3 Hz, 1 H, $=CHCH_2SO$), 3.75 (dd, J = 7.0, 13.3 Hz, 1 H, $=CHCH_2SO$), 4.11 (dd, J = 6.2, 7.7 Hz, 1 H, $SOCH_2CHN$), 5.39 (d, 1 H, J = 17.2 Hz, $CH_2=CH$), 5.46 (d, 1 H, J = 10.2 Hz, $CH_2=CH$), 5.82 (dddd, 1 H, J = 7.0, 8.1, 10.2, 17.2 Hz, CH2=CH-SO); ^{13}C NMR (DEPTQ, 225 MHz, D_2O): δ = 51.59 (1 C, $SOCH_2CHN$), 52.73 (1 C, $SOCH_2CHN$), 57.08 (1 C, $=CHCH_2SO$), 126.31 (1 C, $CH_2=CH$), 127.33 (1 C, $CH_2=CH$), 173.34 (1 C, OC=O). In addition, 2-D ^{1}H, ^{1}H correlation spectrometry (COSY) has confirmed the ^{1}H NMR assignments and 2-D ^{1}H, ^{13}C heteronuclear multiple quantum coherence (HMQC) spectrum, which contains a peak for each unique proton attached to the carbon being considered, has confirmed the ^{13}C NMR assignments (δ 51.59 and δ 57.08 have been wrongfully assigned to $=CHCH_2SO$ and $SOCH_2CHN$ in previous publication [29]).

2.6 Synthesis of (2R)-2-Amino-3-[(R)-Prop-2-Enylsulfinyl] Propanoic Acid (5) [(-)-L-Alliin]

Compound **5** was synthesized using the same procedure as for the synthesis of compound **4**. After completion, the supernatant was subject to fractional crystallization in warm aqueous ethanol and the process was repeated a number of times to produce pure compound **5** (0.9 g, 12% yield, > 95% pure). The characterization data of the product were consistent with the literature results [23,26,28, 29]. TLC: R_f 0.48; $[\alpha]^{25}_D$ = - 61° (*c*=1, H_2O); HRMS (ESI, MH^+): m/z calcd for $C_6H_{11}NO_3S$: 178.0532; found: 178.0534; ^{1}H NMR (900 MHz, D_2O): δ = 3.28 (dd, J = 8.5, 14.5 Hz, 1 H, $=CHCH_2SOCH_2$), 3.32 (dd, J = 3.3, 14.5 Hz, 1 H, $=CHCH_2SOCH_2$), 3.59 (dd, J = 7.9, 13.3 Hz, 1 H, $=CHCH_2SO$), 3.74 (dd, J = 7.1, 13.3 Hz, 1 H, $=CHCH_2SO$), 4.15 (dd, J = 3.3, 8.5 Hz, 1 H, $SOCH_2CHN$), 5.41 (d, 1 H, J = 17.2 Hz, $CH_2=CH$), 5.46 (d, 1 H, J = 10.2 Hz, $CH_2=CH$), 5.82 (dddd, 1 H, J = 7.1, 7.9, 10.2, 17.2 Hz, CH2=CH-SO); ^{13}C NMR (DEPTQ, 225 MHz, D_2O): δ = 51.34 (1 C, $SOCH_2CHN$), 52.10 (1 C, $SOCH_2CHN$), 56.55 (1 C, $=CHCH_2SO$), 126.24 (1 C, $CH_2=CH$), 127.40 (1 C, $CH_2=CH$), 173.21 (1 C, OC=O). In addition, 2-D ^{1}H, ^{1}H COSY has confirmed the ^{1}H NMR assignments and 2-D ^{1}H, ^{13}C HMQC spectrum has confirmed the ^{13}C NMR assignments.

2.7 Synthesis of Fmoc-(2R)-2-Amino-3-[(S)-Prop-2-Enylsulfinyl] Propanoic Acid (6) ([Fmoc-(+)-Alliin])

Compound **6** was synthesized following a procedure with modifications [30-33]. Alliin (3.3 mg, 1 equiv) was dissolved in 660 μL of water and $NaHCO_3$ (6.2 mg, 4 equiv), Na_2CO_3 (3.9 mg, 2 equiv), and 50 μL of 1,4-dioxane was added in sequence. Subsequently, *N*-(9-fluorenylmethoxycarbonyloxy) succinimide (Fmoc-OSu) was added drop wise to this clear alliin solution, which became turbid at first but cleared after stirring overnight. Fmoc-OSu solution was prepared by dissolving Fmoc-OSu powder (7.5 mg, 1.2 equiv) in 165 μL of 1,4-dioxane. Fmoc-alliin was purified by RP-C18 HPLC and dried as a white powder (6.7 mg, 90

% yield). The purity of the product was higher than 97% determined by NMR. The characterization data of the product were as follows. TLC: R_f 0.75; HRMS (ESI, MH$^+$): m/z calcd for $C_{21}H_{21}NO_5S$: 400.1213; found: 400.1210; RP-HPLC (t_R, min): 18; ^{1}H NMR (900 MHz, CD$_3$CN): δ = 3.55 (dd, J = 7.2, 13.7 Hz, 1 H, =CHCH$_2$SOCH$_2$), 3.81 (dd, J = 5.8, 13.7 Hz, 1 H, =CHCH$_2$SOCH$_2$), 3.93 (dd, J = 7.8, 13.2 Hz, 1 H, =CHCH$_2$SO), 4.08 (dd, J = 7.1, 13.2 Hz, 1 H, =CHCH$_2$SO), 4.67 (dd, J = 6.7, 7.0 Hz, 1 H, SOCH$_2$CHN), 4.79 (dd, J = 6.5, 16.3 Hz, 1 H, CHCH$_2$OCONH), 4.81 (dd, J = 7.0, 16.3 Hz, 1 H, CHCH$_2$OCONH), 4.97 (dd, J = 6.5, 7.0 Hz, 1 H, CHCH$_2$OCONH), 5.83 (d, 1 H, J = 17.1 Hz, CH$_2$=CH), 5.87 (d, 1 H, J = 10.0 Hz, CH$_2$=CH), 6.29 (dddd, 1 H, J = 6.7, 7.1, 10.0, 17.1 , 7.2 Hz, CH2=CH-SO), 7.77 (dd, J = 7.2, 7.4 Hz, 2 H, CHCH=CCHCH$_2$OCONH), 7.85 (dd, J = 6.7, 6.9 Hz, 2 H, CH=CHCH=CCHCH$_2$OCONH), 8.08 (dd, J = 6.3, 6.7 Hz, 1 H, CH=CCHCH$_2$OCONH), 8.25 (dd, J = 6.0, 7.2 Hz, 1 H, CHCH=CHCH=CCHCH$_2$OCONH); ^{13}C NMR (DEPTQ, 225 MHz, D$_2$O): δ = 47.27 (1 C, SOCH$_2$CHN), 49.47 (1 C, CHCH$_2$OCONH), 51.98 (1 C, SOCH$_2$CHN), 55.06 (1 C, =CHCH$_2$SO), 67.26 (1 C, CHCH$_2$OCONH), 120.52 (1 C, =CHCH$_2$SO), 124.78 (1 C, CH$_2$=CH), 125.79 (2 C, CHCH=CHCH=CCHCH$_2$OCONH), 125.67 (2 C, CH=CCHCH$_2$OCONH), 127.80 (2 C, CHCH=CCHCH$_2$OCONH), 128.37 (2 C, CH=CHCH=CCHCH$_2$OCONH), 141.50 (2 C, CH=CHCH=CHCH=CCHCH$_2$OCONH), 144.28 (2 C, CH=CCHCH$_2$OCONH), 157.02 (1 C, OCONH), 172.53 (1 C, HOC=O).

2.8 Synthesis of Methyl (2R)-2-Amino-3-[(S)-Prop-2-Enylsulfinyl] Propanoate (7) ([(+)-L-Alliin Methyl Ester]) and Methyl (2S)-2-Amino-3-[(S)-Prop-2-Enylsulfinyl] Propanoate (8) ([(+)-D-Alliin Methyl Ester])

Compound **7** and **8** were synthesized as a mixture following a procedure described with modifications [34,35]. Five mg of alliin was dissolved in 1.2 M HCl in anhydrous methanol (700 µL) and mixed overnight. Seventy mg of NH_4HCO_3 was slowly added to neutralize the solution and the solution was lyophilized. The white powder containing 59% of (+)-L-alliin and 41% (+)-D-alliin methyl ester was dissolved in aqueous 50% acetonitrile and HPLC was used to isolate the (+)-DL-alliin methyl ester. Alliin methyl ester was eluted at 4.6 min and dried as a white powder with 91% yield. The purity of the product was higher than 97% determined by NMR. The characterization data of the product were as follows. TLC: R_f 0.46; HRMS (ESI, MH$^+$): m/z calcd for $C_7H_{13}NO_3S$: 192.0689; found: 192.0682; RP-HPLC (t_R, min): 4.7; ^{1}H NMR (900 MHz, D$_2$O): δ [(+)-L-alliin methyl ester] = 3.33 (dd, J = 5.9, 14.1 Hz, 1 H, =CHCH$_2$SOCH$_2$), 3.42 (dd, J = 6.4, 14.1 Hz, 1 H, =CHCH$_2$SOCH$_2$), 3.57 (dd, J = 8.0, 13.3 Hz, 1 H, =CHCH$_2$SO), 3.75 (dd, J = 8.2, 13.3 Hz, 1 H, =CHCH$_2$SO), 3.77 (s, 3 H, COOCH$_3$), 4.61 (dd, J = 6.2, 6.2 Hz, 1 H, SOCH$_2$CHN), 5.40 (d, 1 H, J = 17.2 Hz, CH$_2$=CH), 5.47 (d, 1 H, J = 10.1 Hz, CH$_2$=CH), 5.82 (dddd, 1 H, J = 8.0, 8.2, 10.1, 17.2 Hz, CH2=CH-SO); ^{13}C NMR (DEPTQ, 225 MHz, D$_2$O): δ = 50.74 (1 C, SOCH$_2$CHN), 50.80 (1 C, SOCH$_2$CHN), 55.96 (1 C, COOCH$_3$), 57.02 (1 C, =CHCH$_2$SO), 125.96 (1 C, CH$_2$=CH), 127.63 (1 C, CH$_2$=CH), 170.26 (1 C, OC=O).

^{1}H NMR (900 MHz, D$_2$O): δ [(+)-D-alliin methyl ester] = 3.36 (dd, J = 8.3, 14.6 Hz, 1 H, =CHCH$_2$SOCH$_2$), 3.40 (dd, J = 4.1, 14.6 Hz, 1 H, =CHCH$_2$SOCH$_2$), 3.60 (dd, J = 7.8, 13.4 Hz, 1 H, =CHCH$_2$SO), 3.76 (dd, J = 8.0, 13.3 Hz, 1 H, =CHCH$_2$SO), 3.79 (s, 3 H, COOCH$_3$), 4.58 (dd, J = 4.2, 8.3 Hz, 1 H, SOCH$_2$CHN), 5.41 (d, 1 H, J = 17.2 Hz, CH$_2$=CH), 5.47 (d, 1 H, J = 10.1 Hz, CH$_2$=CH), 5.82 (dddd, 1 H, J = 7.8, 8.0, 10.1, 17.2 Hz, CH2=CH-SO); ^{13}C NMR (DEPTQ, 225 MHz, D$_2$O): δ = 50.30 (1 C, SOCH$_2$CHN), 50.44 (1 C, SOCH$_2$CHN), 56.14 (1 C, COOCH$_3$), 56.69 (1 C, =CHCH$_2$SO), 125.96 (1 C, CH$_2$=CH), 127.66 (1 C, CH$_2$=CH), 170.45 (1 C, OC=O).

2.9 MTT Cell Viability Assay of *in situ* Allicin

The viability of the *Mus musculus* colon carcinoma CT26.WT cells after treatment with *in situ* allicin was evaluated by 3-(4,5-Dimethylthiazol-2-yl)-2, 5-diphenyltetrazolium bromide (MTT) cell viability assay (Wallert and Provost Laboratory, Minnesota State University, Moorhead, MN) following a procedure described previously with modifications [8]. This assay is based on the conversion of MTT to MTT-formazon by mitochondrial enzymes. MTT assay was performed as follows: first, cells were plated at a density of 7,500 cells/well (144 µL/well) in 96-well plates and incubated at 37°C in a humid environment containing 5% CO_2 overnight. Then, 8 µL of alliin at varying concentrations was added to each well. After 1-7 days of incubation, 8 µL of alliinase at varying concentrations was mixed into each well. The following day, 30 µL of MTT solution (5 mg/mL in PBS) was added to each well. After incubation for 3.5 h, the medium was aspirated and 150 µL of MTT solvent (4 mM HCl and 0.1% NP-40 in isopropanol) was added to each well. After mixing well on an orbital shaker for 20 min, absorbance of the maximum at 545 nm with a reference filter of 630 nm was read by a multimode microplate reader SpectraMax M2 (Molecular Devices, Sunnyvale, CA). The absorbance readings revealed the percentage of the cell viability. The half maximal inhibitory concentration (IC_{50}) was calculated using a four-parameter logistic curve of SigmaPlot (Systat Software, Richmond, VA, USA). A control group (PBS, alliin, or alliinase treated cells) or blank group (water) was also included and the absorbance was considered to be 100% cell viability. Experiments were performed in triplicates. Data are presented as means±standard deviations. Where appropriate, one-way analysis of variance (ANOVA) or student's paired t test was performed using SigmaPlot. P values less than 0.05 were considered as statistically significant. The CT26.WT cell line was used because it can be easily inoculated into BALB/c mice for later in vivo studies and there were reports that garlic consumption is beneficial for the suppression of colon cancer.

2.10 Antimicrobial Activity of *in situ* Allicin

A bacterial plate-diffusion growth inhibition assay was used to assess the efficacy of *in situ* allicin on eliminating human pathogens [36-39]. Gram-positive *Staphylococcus aureus subsp. aureus Rosenbach* (ATCC® 6538™) and gram-negative *Escherichia coli (Migula) Castellani and Chalmers* (ATCC® 8739™) were used as test microorganisms. Bacteria in log phase growth in LB medium (Difco laboratories, Detroit, MI, USA) were grown at 37°C oven to an OD_{560} = 0.18 and diluted 10 times with LB medium. 0.5 mL of the bacterium solution was spread on a 10 cm diameter petri dish with agar, which was prepared with 15 mL of a sterilized LB agar aqueous solution (40 g/liter). Ten mm diameter holes (wells) were punched out and filled with 100 µL PBS solution containing 20 µL of various concentrations of alliin, alliinase, allicin, AGE, a mixture of alliin and alliinase, or PBS. Plates were then incubated overnight at 37°C and the diameter of the inhibition zone was measured. A dilution series of the *in situ* allicin was used to establish the proportionality of the relationship between the amount of active allicin and diameter of inhibition zone.

2.11 Flow Cytometric Analysis of Cell Cycle Using Propidium Iodide DNA Staining

Mus musculus colon carcinoma CT26.WT cells were plated at a density of 1×10^5 cells/well in 6-well plates, permitted to adhere and grown for 48 h in RPMI-1640 with 10% FBS. Wells were treated separately with 100 µL PBS solution containing 20 µL of alliin, alliinase, allicin, AGE, a mixture of alliin and alliinase, or PBS at a range of concentrations for 12h. The cells (~ 3.6×10^5) were then harvested by trypsin-EDTA treatment into sterilized 15 mL of Falcon tubes. After washing twice with 2 mL PBS, centrifuging at 300 x g for 5 minutes, and decanting the buffer from the pelleted cells, cells were fixed in 70% cold ethanol at 4°C overnight. The 70% cold ethanol was added drop wise to the pellet while vortexing to ensure fixation of all cells and minimize clumping. Before analysis, cells were washed twice with PBS, suspended in 500 µL of PBS containing 0.5% BSA, and treated with 200 µg/mL RNase A (Roche Diagnostics, Indianapolis, IN) and 40 µg/mL propidium iodide (PI; Sigma Corp., St. Louis, MO). Propidium iodide binds to double stranded DNA by intercalating between base pairs. After incubation for 30 min at 37°C and single cell suspension with Falcon cell-strainer (Corning Inc., Durham, NC, USA), the cells were subjected to DNA content analysis. The PI fluorescence was analyzed by using a Cyan ADP flow cytometer with 488 nm line of argon-ion laser excitation (Beckman Coulter, Pasadena, CA). Flow cytometer was cleaned with bleach once and with water twice before it was used. In all the experiments, cells were gated on forward and side scatter to eliminate dead cells and debris. Data from at least 10,000 cells were analyzed with Summit software (Beckman Coulter, Pasadena, CA). Each DNA histogram was analyzed for peak position and for the percentage of cells in different histogram regions. Cell cycle distributions were obtained by deconvolution of histograms using Peakfit software (Systat Software, Richmond, VA, USA). This technique measures DNA content of 5,000 cells per second and trace the dynamic changes in cell proliferation cycle. Data show percentages of cell in different phases like sub-G_1, G_1, S, and G_2/M [40].

3. RESULTS AND DISCUSSION

Aqueous garlic extract (AGE) was prepared from garlic clove according to with modification in hepes buffer (50 mM, pH 7.2) containing 10% glycerol, 1 mM CaCl2, 1 mM MnCl2, 1 mM MgCl2, 20 µM pyridoxal 5'-phosphate (PLP), and 0.25 M NaCl at a concentration of 1g/mL [8,9]. Glycosylated dimeric alliinase, which is used to convert bioinactive alliin to bioactive allicin, is purified to homogeneity from AGE. Alliinase is the major protein in AGE, and NaCl is required to preserve the enzymatic activity of alliinase because it can stabilize the alliinase dimer [41]. The inclusion of PLP and glycerol in hepes buffers is necessary to stabilize the enzyme during purification [42,43]. Our results from cancer cell viability assay MTT assay, bacterial plate-diffusion growth inhibition assay, and flow cytometry cell cycle analysis confirm the importance of buffer on the activity of alliinase (data not shown). Alliinase is completely inactive if prepared in water, semi-active if prepared in PBS, and most active if prepared in the hepes buffer. However, there is no detrimental effect if AGE is prepared in water because garlic itself contains the necessary compounds to stabilize the alliinase. It is the subsequent purification steps which require the hepes buffer.

Many protocols have been used to prepare alliinase from AGE using gel-filtration, ion-exchange, hydroxyapatite, and/or ConA-Sepharose 4B affinity columns with varying enzymatic activities and protein purities obtained at low yield [41,44-46]. Reports on recombinant expression of alliinase are scarce [47-49]. Here PD10 and high-resolution gel-filtration columns were used to produce highly purified alliinase with high yield. Many different kinds of columns such as ion-exchange, hydroxyapatite, hydrophobic, and ConA-Sepharose columns do not perform well in this study. A PD10 column is used to separate high-molecular weight (> 5000 Da) biomolecules from low-molecular weight compounds. All purification steps are done at 4°C in this facile procedure. Analysis of samples in every steps of the alliinase preparation by gel electrophoresis is shown in Fig. 1. Data reveal that the AGE contains two major protein bands at 52 and 12 kDa together with several minor bands. LC/MS/MS data on trypsin in-gel digestion of these two protein bands together with MassMatrix and X! Tandem database searches have confirmed that the 52 kDa protein is the enzyme alliinase, which is a mannose rich glycoprotein, and the 12 kDa protein is lectin, which is specific for mannose (data not shown). This observation points to possible interaction between these two proteins [44]. Concentration of alliinase was determined to be 1.5 mg/mL by SDS-PAGE band intensity using BSA as standard. SDS-PAGE after purification on PD10 (Fig. 1, Lane A) shows a similar protein profile to AGE and purification on high-resolution gel-filtration (Fig. 1, Lane B) reveals that alliinase is the only protein stained with Coomassie blue. Eighty percent of the alliinase in AGE is recovered with this protocol.

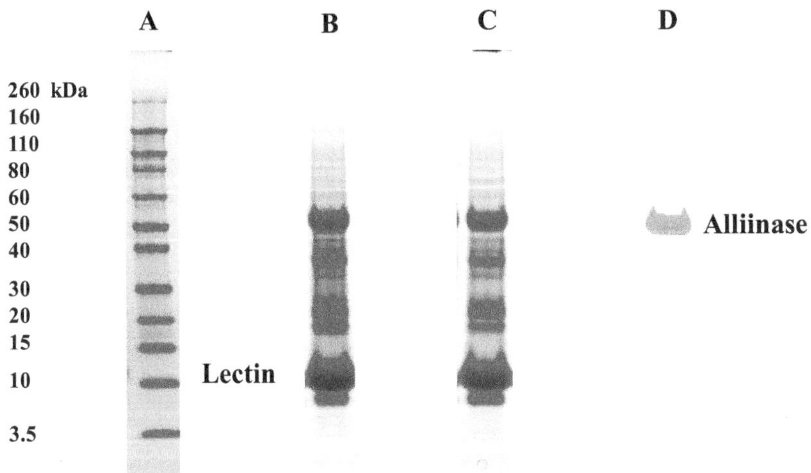

Lane A, protein molecular weight markers; Lane B, aqueous garlic extract (AGE);
Lane C, protein fraction of PD10 purification of AGE;
Lane D, alliinase purified on high-resolution HPLC size exclusion chromatography

Fig. 1. SDS-PAGE analyses of protein samples

The enzyme alliinase contains one PLP cofactor per subunit and belongs to the class I of PLP-dependent enzymes. It is able to cleave the C-S bond of sulfoxide derivatives of cysteine to produce allicin. The enzyme activity assay of alliinase was determined by the spectrophotometric assay using 4-mercaptopyridine (4-MP) method using pure (+)-L-alliin in a cell free system according to Miron et al. with modifications [21]. One unit of enzyme activity was defined as releasing 1 µmol pyruvate per minute. The final pure alliinase fraction prepared in this study shows an activity of 12 units/mg which is similar to the allinase purified in Miron et al. [21] and much higher than the allinase purified in Mallika et al. [46]. It is interesting to note that some of the partially purified allinase fractions containing lectin show higher enzyme activity than the pure allinase fraction (data not shown), suggesting that lectin might enhance the activity of allinase.

The course of the synthesis of (+)-L-alliin, (-)-L-alliin, Fmoc-(+)-alliin or (+)-DL-alliin methyl ester is illustrated in Schematic 1. The synthesis started with the formation of L-deoxyalliin (3) from the commercially available allyl bromide (1) and hydrochloride of L-cysteine (2) under alkaline conditions. A white precipitate was formed after the reaction solution was neutralized with acetic acid. Compound 3 was purified from the precipitate by crystalization in warm water and it does not show NMR signal for methyl group. It is interesting that proton coupling patterns of compound 3 between 2.97 and 3.07 and between 3.22 and 3.24 require non-first-order correction just as in the case of deoxyisoalliin, isoalliin, or allicin [8,9,22,50].

(+)-L-alliin (4) and (-)-L-alliin (5) were prepared by oxidation of the corresponding L-deoxyalliin (3) with hydrogen peroxide in H_2O. The reaction is very efficient with a high yield. By careful control of temperature and concentration, further oxidation to sulfate is avoided. This oxidation reaction converts the divalent sulfur of compounds 4 and 5 to trivalent sulfur, which becomes a chiral center. Because the oxygen can be introduced from either side of the sulfur, a mixture of separable diastereomers is produced.

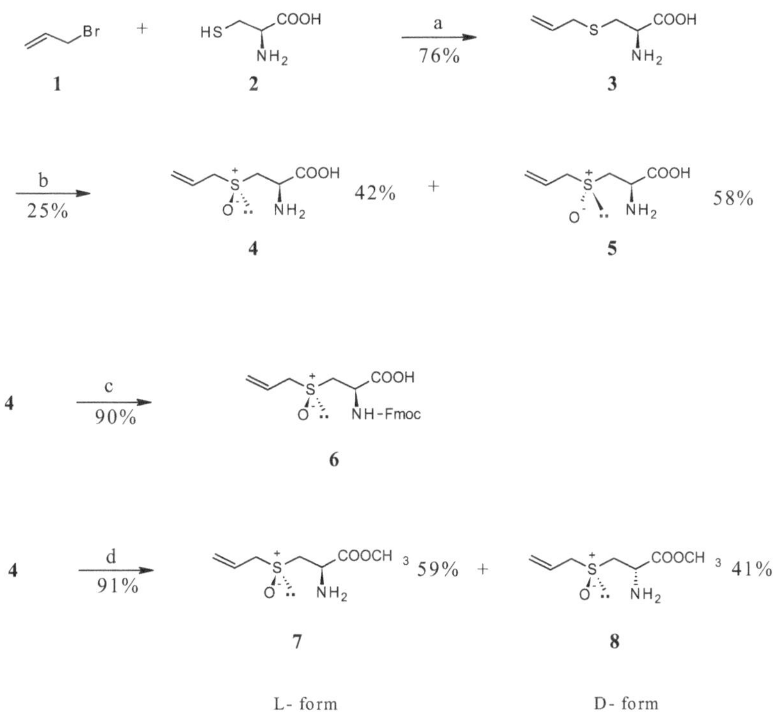

a) NaOH, H_2O, 20 °C, 14h, RP-HPLC; b) 30% H_2O_2, H_2O, 20 °C, 14 h, repeated fractional crystallization;
c) $NaHCO_3$, Na_2CO_3, 1,4 dioxane, Fmoc-OSu; d) $NaHCO_3$, Na_2CO_3, 1,4-dioxane, Fmoc-OSu;
d) HCl, anhydrous MeOH, neutralize with NH_4HCO_3 and lyophilized

Scheme 1. Chemical syntheses of (+)-L-alliin, (-)-L-alliin, (+)-Fmoc-L-alliin, and (+)-L,D-methyl alliin ester

Although many papers have reported synthesis of compounds 4 and 5, these methods produce either low yield or low purity [20,28,29]. Solvent extractions and multiple fractional crystallization are used to separate and purify compounds 4 and 5, which are dried as white powders with 13% and 12% total yield of > 97% purity, respectively. ^1H, ^{13}C NMR chemical shifts are quite similar between (+) and (-) isomers except for ^1H NMR chemical shift of $=CHCH_2SOC\underline{H}_2$. 2-D ^1H, ^1H correlation spectrometry (COSY) has confirmed the ^1H NMR assignments and 2-D ^1H, ^{13}C Heteronuclear Multiple Quantum Coherence (HMQC) spectrum, which contains a peak for each unique proton attached to the carbon being considered, has confirmed the ^{13}C NMR (δ 51.59 and δ 57.08 have been wrongfully assigned to $=CH\underline{C}H_2SO$ and $SO\underline{C}H_2CHN$ for Compound 4 in previous publication[29]). The spin coupling pattern between δ 3.13 and δ 3.34 and between δ 3.54 and δ 3.75 of compound 4 can be understood by non-first-order pattern resulting from the ratio between the chemical shift difference in Hz and the spin coupling of 14.0 (Δ v / J = 195.7/14.0) and 13.6 (Δ v / J = 182.7/13.4), respectively. The spin coupling pattern between δ 3.28 and δ 3.32 and between δ 3.59 and δ 3.74 of compound 5 can be understood by non-first-order pattern resulting from the ratio between the chemical shift difference in Hz and the spin coupling of 2.5 (Δ v / J = 36.6/14.5) and 10.1 (Δ v / J = 133.8/13.3), respectively. Alliinase activity measurement using 5 as the substrate reveal that the enzyme possess an activity of 4.6 units/mg compare to 12 units/mg by using 4 as the substrate.

Fmoc-(+)-alliin (6) and (+)-DL-alliin methyl ester have also been synthesized to test the roles of amino and carboxyl groups in alliinase reaction. Compound 6 was synthesized by mixing (+) alliin with FmocOSu. (+)-DL-alliin methyl ester was synthesized by mixing (+) alliin with anhydrous methanol in acidic conditions. The (+)-alliin methyl ester produced contains both (+)-L-alliin methyl ester (7) and (+)-D-alliin methyl ester (8) due to racemization [51,52]. Instead of fractional crystallization in 95% ethanol, RP-HPLC is used to purify compounds 6 and (+)-DL-alliin methyl ester. Results show that compounds 6 and (+)-DL-alliin methyl ester are eluted at 18 and 4.6 min, respectively. Compound 6 and (+)-DL-alliin methyl ester are dried as a white powder with 76% yield of > 97 % purity. Non-first-order correction is not needed to explain the pattern of the couplings for compounds 6 and (+)-DL-alliin methyl ester. The alliinase activity assay results of compounds 6 and (+)-DL-alliin methyl ester show that both compounds are inactive in alliinase enzymatic reaction. Alliinase belongs to the class of carbon-sulfur lyases named S-alk(en)yl-L-cystene solfoxide lyase and catalyzes the cleavage of carbon-sulfur bond of S-allyl-L-cysteine sulfoxide in a beta-elimination/ deamination reaction involving an aminoacryl bound PLP-amino acrylate (Schiff's base) reaction intermediate [45,47,53,54]. Alliinase provides acidic and basic groups in the active site of the enzyme, inducing polarization of the S=O group of the substrate, which initiates the carbon-sulfur bond breakage and produces an acrylamide bound PLP and allyl sulfenic acid. This catalytic mechanism is similar to that of cystathionine β-lyase [53]. Subsequently, the schiff's base of acrylamide bound PLP is spontaneously hydrolyzed and decomposed to pyruvate and ammonia. Simultaneously, condensation of two allyl sulfenic acids yields allicin. In agreement with this mechanism, it is plausible that the Fmoc group in Fmoc-alliin is blocking PLP cofactor binding and the inability to form carboxylate anion in (+)-alliin methyl ester inhibits the alliinase reaction. It is interesting to know that optimal pH value for alliinase catalytic activity is 6.5, while the optimal temperature is 33°C.

The *in vitro* cytotoxicity and anti-cancer effect of *in situ* allicin against *Mus musculus* colon carcinoma cells CT26.WT was determined by MTT assay [8]. Fig. 2 shows the IC_{50} dose-response curve of *in situ* allcin on viability of *Mus musculus* colon carcinoma cells CT26.WT. Results show clear evidence of cytotoxicity of the *in situ* allicin against the CT26.WT in a dose dependent manner, compared to the controls (PBS, alliin, or alliinase). Shrinkage, granulation of cytoplasm, and detachment were visible in allicin treated cells. The half maximal inhibitory concentration (IC_{50}) is the dilution of either the alliin or allinase solution to produce fifty percent inhibition of cells *in vitro*. The IC_{50} of 11.24 mM (+)-L-alliin solution is calculated to be 1:60 dilution measured by mixing diluted alliin with 0.58 μM alliinase. The IC_{50} of 11.54 μM alliinase solution is calculated to be 1:630 dilution measured by mixing diluted alliinase with 0.56 mM (+)-L-alliin. Since 2 mol of alliin produces 1 mol of allicin in the allinase enzymatic reaction, the IC_{50} in term of allicin is 93.67 μM. Comparable results were obtained with AGE, allicin, and *in situ* allicin [8,9]. Time course results have revealed that (+)-L-alliin in culture medium is fully active for 3 days, but is then gradually declines until it is inactive after 7 days. Data also reveal that with high (+)-L-alliin concentrations, less alliinase is needed to reach the 50% viable cells and vice versa. The IC_{50} of 11.24 mM (-)-L-alliin solution is calculated to be 1:19 dilution. In addition, anti-cancer effect has been tested with Fmoc-(+)-alliin and (+)-DL-alliin methyl ester with negative results, which confirm that both compounds are inactive in alliinase enzymatic reaction.

Experimental data have indicated that allicin is responsible for the antimicrobial activity of AGE against a broader spectrum of microorganisms [2,4,55,56]. Allicin is also reported to be active against methicillin resistant human pathogen *Staphylococcus aureus* (MRSA) [38]. The fact that diallyl disulfide has a significantly lower antimicrobial activity indicates that the thiosulfinate group plays an important role in allicin's activity. In this study, the antibacterial activities of AGE, allcin, and *in situ* allcin against gram-positive *Staphylococcus aureus* subsp. aureus Rosenbach and gram-nagative Escherichia coli were measured using bacterial plate-diffusion growth inhibition assay (Fig. 3). Data show that the maximum zone of inhibition (39 mm) was observed in *Staphylococcus aureus* and the minimum (17 mm) was observed for Escherichia coli. PBS was used as a negative control. The lowest concentration at which there is no growth was reported as minimum inhibitory concentration (MIC). Results from the dose–response study of allicin (data not shown) show that MICs were in the ranges of 56–140 and 280–560 µM in *Staphylococcus aureus* and Escherichia coli, respectively. The *in-situ* allicin is as active as allicin purified from AGE or allicin synthesized chemically in a dose-dependent manner. It has been reported by Miron et al. [14] that allicin readily diffuses across the phospholipid membranes. The basis for antimicrobial activity of the majority of antibiotics, allicin and the non-plant-derived antibiotics penicillin, citrinin, gliotoxin, clavacin and pyocyanines is their ability to react with cysteine. When allicin penetrates the cell, it reacts with free thio groups indiscreetly [13-15]. In addition, it is known that the antimicrobial efficacy of AGE is time and temperature dependent [7-9,57]. This explains why allicin is effective *in situ* because alliin and alliinase are more stable and less reactive toward free thio groups [58].

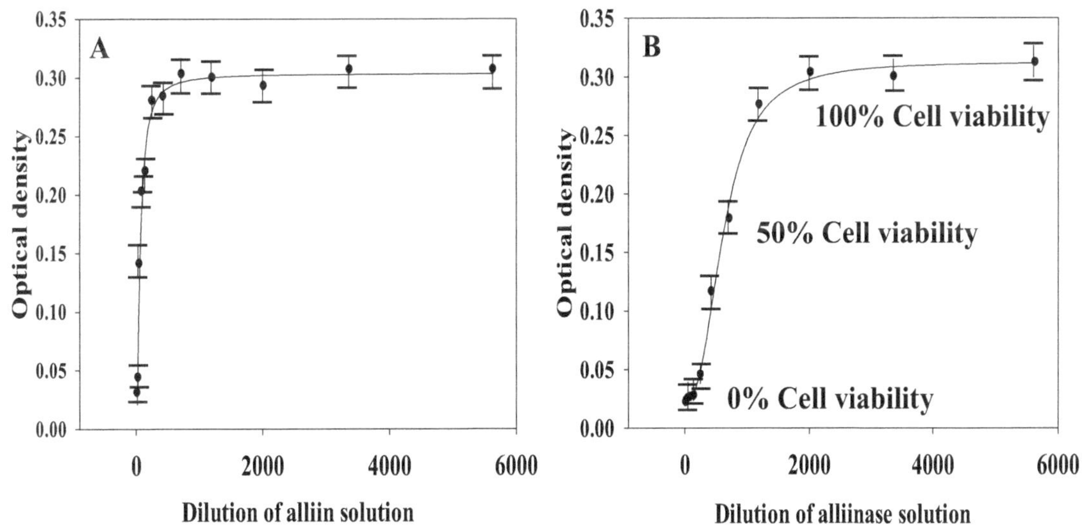

A, *in situ* allicin solution generated from different alliin concentrations with 0.58 µM alliinase;
B, *in situ* allicin solution generated from different alliinase concentrations with 0.56 mM alliin

Fig. 2. Dose–response curve of viability of *Mus musculuscolon* carcinoma cells CT26.WT assessed by MTT assay

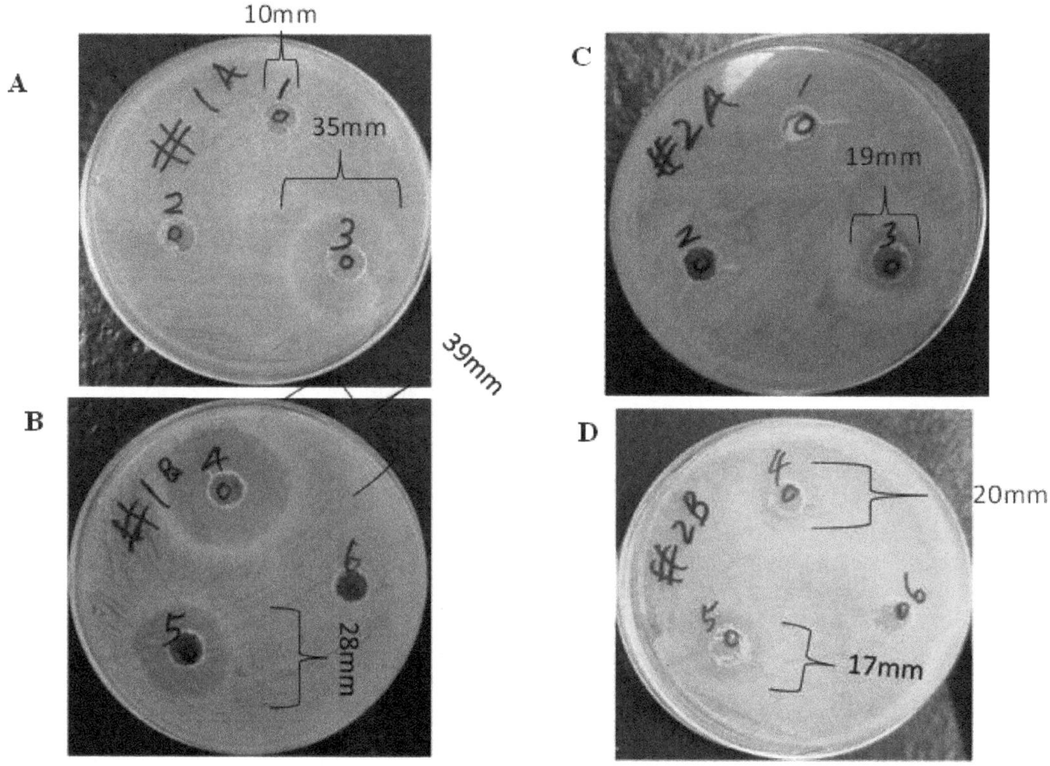

A, Staphylococcus aureus was grown in the presence of (1): alliin, (2): alliinase, or (3): mixture of alliin and alliinase;
B, Staphylococcus aureus was grown in the presence of (4): AGE, (5): allicin, or (6): PBS;
C, Escherichia coli was grown in the presence of (1): alliin, (2): alliinase, or (3): mixture of alliin and alliinase;
D, Escherichia coli was grown in the presence of (4): AGE, (5): allicin, or (6): PBS

Fig. 3. Effect of alliin, alliinase, allicin, AGE, mixture of alliin and alliinase, or PBS on the growth of gram-positive *Staphylococcus aureus* subsp. aureus Rosenbach (ATCC® 6538™) or gram-nagative *Escherichia coli* (Migula) Castellani and Chalmers (ATCC® 8739™) by a bacterial plate-diffusion growth inhibition assay

Flow cytometric univariate analysis of cellular DNA content allows discrimination among sub-G_1, G_1, S, and G_2/M phase cells and is the simplest and most commonly used method to reveal the frequency of cells in these cell cycle phases [40]. The cells are fixed or permeabilized with a detergent and then stained with a nucleic acid-specific fluorophore such as propidium iodide (PI). Results of fluorescence measurements are displayed as a cellular DNA content frequency histogram, which shows the proportions of cells in particular phases of the cycle based on the fluorescence intensity. The content of cellular DNA in G_2 and M phases have twice as much DNA as G_1 phase cells and S phase cells have intermediate DNA content. Flow cytometry can measure DNA content of 5000 cells per second and trace the dynamic changes in cell cycle. During apoptosis, a portion of nuclear DNA may be lost from the cell in shedding apoptotic bodies and some fragmented DNA may be extracted during fixation and staining. Consequently, apoptotic cells are often distinguished on frequency histograms by a deficit in DNA content (sub-G_1 cells). The actual percentages of G_1, S, G_2/M, and sub-G_1 cells are obtained by deconvolution of the histograms using a PeakFit program. Fig. 4 shows DNA content frequency histograms of the effect of PBS, *in situ* allicin, or AGE on the cell cycle of *Mus musculuscolon* carcinoma cells CT26.WT. The data show a notable increase in sub-G_1 populations of the cell cycle on the cells treated with allicin, indicating the induction of apoptosis. This agrees with previous finding that allicin induces apoptosis in cancer cells [59]. Again the results show that *in-situ* allicin is as active as allicin purified from AGE or synthesized chemically.

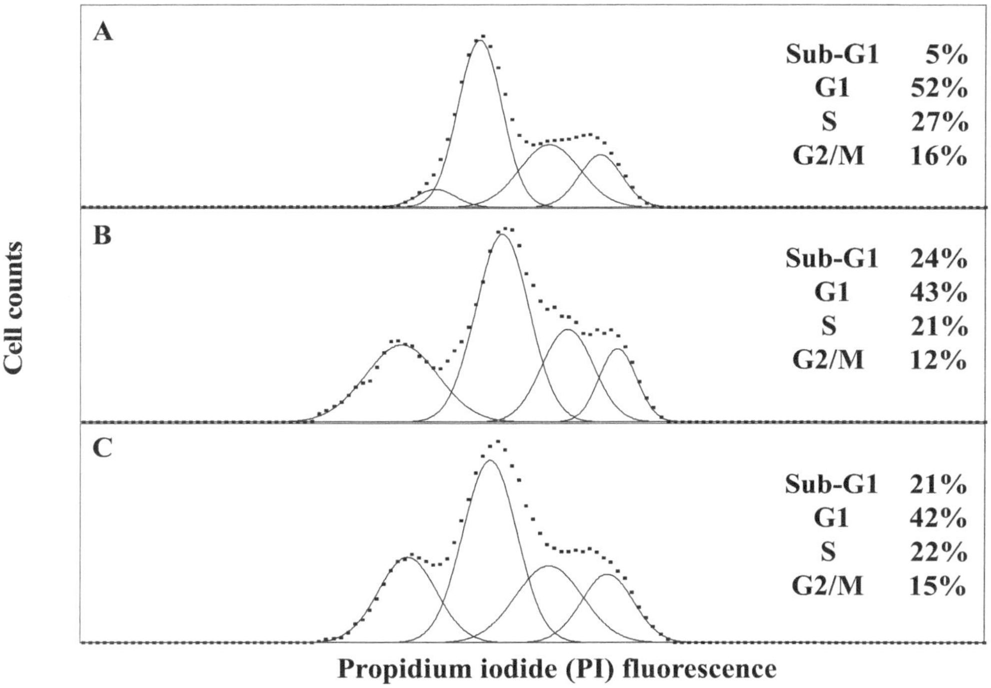

A, 8 μL of PBS was added into 200 μL cell culture;
B, 1.5 μL of alliin (56.18 mM) and 8 μL of alliinase (11.54 μM) were added into 200 μL cell culture;
C, 8 μL of AGE was added into 200 μL cell culture

Fig. 4. Flow cytometric analyses of the effect of PBS, *in situ* allicin, or aqueous garlic extract (AGE) on the cell cycle of *Mus musculus colon* carcinoma cells CT26.WT

4. CONCLUSION

We describe here a facile pathway to synthesize diastereomerically pure alliins and their conversion to allicin by enzymatic reaction of alliinase. Cancer cell viability assay MTT assay, bacterial plate-diffusion growth inhibition assay, and flow cytometry cell cycle analysis have been used to demonstrate the anti-cancerous/anti-pathogenic activities of the *in-situ* allicin. Protocols developed are simple and effective. Results demonstrate that the *in-situ* allicin is as active as allicin purified from AGE or allicin synthesized chemically in a dose-dependent manner. The data obtained here provide useful information on the design of the *in-situ* allicin strategy. Further research in modifying alliin structure (other substrates) and alliinase conjugated to a tumor targeting antibody against mouse cancer cells are also in progress.

CONSENT

The author hereby declares all experiments have been examined and approved by the appropriate ethics committee.

ETHICAL APPROVAL

It is not applicable.

ACKNOWLEDGEMENTS

Financial support from the Research Resources Center at UIC is gratefully acknowledged.

COMPETING INTERESTS

Authors have declared that no competing interests exist.

REFERENCES

1. Cavallito C, Bailey JH. Allicin, The antibacterial principle of *Allium sativum.* Isolation, physical properties and antibacterial action. J Am Chem Soc. 1944;66:1944-52.
2. Koch HP, Lawson LD. editors. Garlic: The science and therapeutic applications of *Allium sativum L.* and related species. Baltimore, USA: Williams & Wilkins; 1996.

3. Block E. Garlic and other alliums: The lore and the science. Cambridge, UK: The Royal Society of Chemistry; 2010.
4. Borlinghaus J, Albrecht F, Gruhlke MCH, Nwachukwu ID, Slusarenko AJ. Allicin: chemistry and biological properties. Molecules. 2014;19:12591-618.
5. Blumenthal M. Herb sales down 7.4 percent in mainstream market. HerbalGram, USA: American Botanical Council; 2005.
6. Tapiero H, Townsend DM, Tew KD. Organosulfur compounds from alliaceae in the prevention of human pathologies. Biomed Pharmacother. 2004;58:183-93.
7. Wang H, Li X, Liu X, Shen D, Qiu Y, Zhang X, Song J. Influence of pH, concentration, and light on stability allicin in garlic (*Allium sativum L.*) aqueous extract as measured by UPLC. J Sci Food Agric. 2014;95:1838-44.
8. Lee J, Gupta S, Huang J, Jayathilaka LP, Lee B. HPLC-MTT assay: Anti-cancer activity of aqueous garlic extract is from allicin. Anal Biochem. 2013;436:187-9.
9. Lee J, Gupta S, Huang J, Jayathilaka LP, Lee B. Preparation and purification of garlic-derived organosulfur compound allicin by green methodologies. In: Mohammad A, Inamuddin PM, editors. Green chromatographic techniques: separation and purification of organic and inorganic analytes. London, UK: Springer; 2013.
10. Lanzotti V. The analysis of onion and garlic. J Chromatogr A. 2006;1112:3-22.
11. Block E. The chemistry of garlic and onions. Sci Am. 1985;252:114–9.
12. Ichikawa M, Ide N, Yoshida J, Yamaguchi H, Ono K. Determination of seven organosulfur compounds in garlic by high-performance liquid chromatography. J Agric Food Chem. 2006;54:1535-40.
13. Rabinkov A, Miron T, Mirelman D, Wilchek M, Glozman S, Yavin E, Weiner L. S-allylmercaptoglutathione: The reaction product of allicin with glutathione possesses SH-modifying and antioxidant properties. Biochim Biophys Acta. 2000; 1499:144–53.
14. Miron T, Rabinkov A, Mirelman D, Wilchek M, Weiner L. The mode of action of allicin: its ready permeability through phospholipid membranes may contribute to its biological activity. Biochim Biophys Acta. 2000; 1463:20–30.
15. Freeman F, Kodera Y. Garlic chemistry—stability of S-(2-propenyl) 2-propene-1-sulfinothioate (allicin) in blood, solvents, and simulated physiological fluids. J Agric Food Chem. 1995;43:2332–38.
16. Cutler RR, Wilson P. Antibacterial activity of a new, stable, aqueous extract of allicin against methicillin-resistant *Staphylococcus aureus*. Br J Biomed Sci. 2004;61:71–4.
17. Miron T, Mironchik M, Mirelman D, Wilchek M, Rabinkov A. Inhibition of tumor growth by a novel approach: *In situ* allicin generation using targeted alliinase delivery. Mol Cancer Ther. 2003;2: 1295-301.
18. Arditti FD, Rabinkov A, Miron T, Reisner Y, Berrebi A, Wilchek M, Mirelman D. Apoptotic killing of B-chronic lymphocytic leukemia tumor cells by allicin generated *in situ* using a rituximab-alliinase conjugate. Mol Cancer Ther. 2005;4:325–32.
19. Fry FH, Okarter N, Baynton-Smith C, Kershaw MJ, Talbot NJ, Jacob C. Use of a substrate/alliinase combination to generate antifungal activity *in situ*. J Agric Food Chem. 2005;53:574–80.
20. Stoll A, Seebeck E. Chemical investigations on alliin, the specific principle of garlic. Adv Enzymol. 1951;11: 377-400.
21. Miron T, Shin I, Feigenblat G, Weiner L, Mirelman D, Wilchek M, Rabinkov A. A spectrophotometric assay for allicin, alliin, and alliinase (alliin lyase) with a chromogenic thiol: reaction of 4-mercaptopyridine with thiosulfinates. Anal Biochem. 2002;307:76–83.
22. Jayathilaka LP, Gupta S, Huang J, Lee J, Lee B. Preparation of (+)-trans-isoalliin and its isomers by chemical synthesis and RP-HPLC resolution. J Biomol Tech. 2014;25: 67-76.
23. Iberl B, Winkler G, Muller B, Knobloch K. Quantitative determination of allicin and alliin from garlic by HPLC. Planta Med. 1990;56:320-6.
24. Lancaster JE, Kelly KE. Quantitative analysis of the S-alk(en)yl-L-cysteine sulphoxides in onion (*Allium cepa L.*). J Sci Food Agric. 1983;34:1229-35.
25. Corbi PP, Massabni AC, Moreira AG, Medrano FJ, Jasiulionis MG, Cost-Neto CM. Synthesis, characterization, and biological activity of a new palladium(II) complex with deoxyalliin. Can J Chem. 2005;83:104-9.

26. Nikolić VD, Ilić DP, Nikolić LB, Stanković MZ, Stanojević LP, Savić IM. The synthesis and structure characterization of deoxyalliin and alliin. Advan Technol. 2012;1:38-46.
27. Yu TH, Wu CM, Rosen RT, Hartman TG, Ho CT. Volatile compounds generated from thermal degradation of alliin and deoxyalliin in an aqueous solution. J Agric Food Chem. 1994;42:146-53.
28. Stoll A, E. Seebeck E. Die synthese des naturlichen Alliins und seiner drei optisch aktiven isomeren. Helv Chim Acta. 1951; 34:481-7.
29. Koch I, Keusgen M. Diastereoselective synthesis of alliin by an asymmetric sulfur oxidation. Pharmazie. 1998;53:668-71.
30. Lajoie G, Crivici A, Adamson JG. A simple convenient synthesis of ω-tert-butyl ester of Fmoc-aspartic and Fmoc-glutamic acids. Synthesis. 1990;571-2.
31. Gawande MB, Branco PS. An efficient and expeditious Fmoc protection of amines and amino acids in aqueous media. Green Chem. 2011;13:3355-9.
32. Wehner JW, Lindhorst TK. S-Fluorenylmethyl protection of the cysteine side chain upon Nα-Fmoc deprotection. Beilstein J Org Chem. 2012;8:2149–55.
33. Mönch S, Rychlik M. Improved folate extraction and tracing deconjugation efficiency by dual label isotope dilution assays in foods. J Agric Food Chem. 2012; 60:1363−72.
34. Kowalak JA, Walsh KA. P-Methylthio-aspartic acid: Identification of a novel posttranslational modification in ribosomal protein S12 from *Escherichia coli*. Prot Sci. 1996;5:1625-32.
35. Kim TY, Brun YV, Reilly JP. Effects of tryptic peptide esterification in MALDI mass spectrometry. Anal Chem. 2005;77: 4185-93.
36. Ilić DP, Nikolić VD, Nikolić LB, Stanković MZ, Stanojević LP. Thermal degradation, antioxidant and antimicrobial activity of the synthesized allicin and allicin incorporated in gel. Hem Ind. 2010;64:85–91.
37. Fujisawa H, Watanabe K, Suma K, Origuchi K, Matsufuji H, Seki T, Ariga T. Antibacterial potential of garlic-derived allicin and its cancellation by sulfhydryl compounds. Biosci Biotechnol Biochem. 2009;73:1948–55.
38. Cutler RR, Wilson P. Antibacterial activity of a new, stable, aqueous extract of allicin against methicillin-resistant *Staphylococcus aureus*. Br J Biomed Sci. 2004;61:71–4.
39. Zaidan MR, Noor Rain A, Badrul AR, Adlin A, Norazah A, Zakiah I. In vitro screening of five local medicinal plants for antibacterial activity using disc diffusion method. Tropical Biomedicine. 2005;22: 165–70.
40. Darzynkiewicz Z, Bedner E, Smolewski P. Flow cytometry in analysis of cell cycle and apoptosis. Semin Hematol. 2001;38: 179-93.
41. Kuettner EB, Hilgenfeld R, Weiss MS. Purification, characterization, and crystallization of alliinase from garlic. Arch Biochem Biophys. 2002;402:192–200.
42. He Q. The alliinase and lachrymatory factor synthase systems in *Petiveria alliacea* (Doctoral Dissertation, University of Albany, NewYork), Ann Harbor, USA; UMI Dissertation Publishing; 2010.
43. Krest I, Keusgen M. Stabilization and pharmaceutical use of alliinase. Pharmazie. 1999;54:289-93.
44. Rabinkov A, Wilchek M, Mirelman D. Alliinase (alliin lyase) from garlic (allium sativum) is glycosylated at Asn146 and forms a complex with a garlic mannose-specific lectin. Glycocon J. 1995;12:690–8.
45. Musah RA, He Q, Kubec R, Jadhav A. Studies of a novel cysteine sulfoxide lyase from *Petiveria alliacea*: The first heteromeric alliinase. Plant Physiol. 2009; 151:1304–16.
46. Mallika T, Omer E, Lianfu Z. Separation and purification of alliinase and alliin from garlic (*Allium sativum*). J Acad Indus Res. 2014;2:599-605.
47. Manabe T, Hasumi A, Sugiyama M, Yamazaki M, Saito K. Alliinase [S-alk(en)yl-L-cysteine sulfoxide lyase] from *Allium tuberosum* (Chinese chive) Purification, localization, cDNA cloning and heterologous functional expression. Eur J Biochem. 1998;257:21-30.
48. Weik R, Franky A, Striedner G, Raspor P, Bayer K, Mattanovich D. Recombinant expression of alliinase lyase from garlic (*Allium sativum*) in bacteria and yeasts. Planta Med. 1998;64:387-8.
49. Wu X, Zhang T, Bian J, Xu J. Cloning of alliinase gene from garlic bulb and its expression in *Pichia pastoris*. Chin Trad Herb Drugs. 2012;43:143-7.
50. Crews P, Rodriguez J, Jaspars M. *Organic Structure Analysis*. 2nd ed. New York: Oxford University Press; 2009.

51. Li J, Sha Y. A convenient synthesis of amino acid methyl esters. Molecules. 2008;13:1111-9.
52. Smerdka J, Rademann J, Jung G. Polymer-bound alkyltriazenes for mild racemization-free esterification of amino acid and peptide derivatives. J Peptide Sci. 2004;10:603-11.
53. Ravilious GE, Jez JM. Structural biology of plant sulfur metabolism: From assimilation to biosynthesis. Nat Prod Rep. 2012:29: 1138-52.
54. Nock LP, Mazelis M. The C-S lyases of higher plants: direct comparison of the physical properties of homogeneous alliin lyase of garlic (*Allium sativum*) and onion (*Allium cepa*). Plant Physiol. 1987;85: 1079-83.
55. Cañizares P, Gracia I, Gómez LA, García A, De Argila CM, Boixeda D, De Rafael L. Thermal degradation of allicin in garlic extracts and its implication on the inhibition of the in vitro growth of *Helicobacter pylori*. Biotechnol Prog. 2004;20:32–7.
56. Hughes B, Lawson L. Antimicrobial effects of *Allium sativum L.* (garlic), *Allium ampeloprasum L.* (elephant garlic), and *Allium cepa L.* (onion), garlic compounds and commercial garlic supplement products. Phytother Res. 1991;5:154–8.
57. Durairaj S, Srinivasan S, Lakshmanaperumalsamy P. *In vitro* antibacterial activity and stability of garlic extract at different pH and temperature. Elec J Biol. 2009;5:5-10.
58. Appel E, Vallon-Eberhard A, Rabinkov A, Brenner O, Shin I, Sasson K, Shadkchan Y, Osherov N, Jung S, Mirelman D. Therapy of murine pulmonary Aspergillosis with antibody-alliinase conjugates and alliin. Antimicrob Agents Chemother. 2010; 54:898–906.
59. Oommen S, Anto RJ, Srinivas G, Karunagaran D. Allicin (from garlic) induce caspase-mediated apoptosis in cancer cells. Eur J Pharmacol. 2004;485:97–103.

7

Dysregulation of Signaling Pathways Plays a Role in the Development and Pathogenesis of Hepatocellular Carcinoma

Ayman Zaky Elsamanoudy[1,2*], Hatem A. El-Alfy[3], Moustafa Ahmed Mohamed Neamat-Allah[1], Dina Abd el Daim Ahmed Ibrahim[1] and Raymonde Hanna Assaf[1]

[1]Medical Biochemistry and Molecular Biology, Faculty of Medicine, Mansoura University, Egypt.
[2]Clinical Biochemistry, Faculty of Medicine, King Abdulaziz University (KAU), KSA.
[3]Tropical Medicine, Faculty of Medicine, Mansoura University, Egypt.

Authors' contributions

This work was carried out in collaboration between all authors. Authors RHA and AZE designed the study, wrote the protocol, managed the literature searches, edited the manuscript and supervised the work. Author DADAI collected data and wrote the first draft of the manuscript. Authors HAEA and MAMNA revised the manuscript. All authors read and approved the final manuscript.

Editor(s):
(1) Edward Eteshola, Department of Biomedical Engineering, The Ohio State University, Columbus, USA and Davis Heart and Lung Research Institute, The Ohio State University, Columbus, USA.
Reviewers:
(1) Mamdouh Moawad Ali, National Research Centre, Egypt.
(2) D. S. Sheriff, Melmaruvathur Adhiparasakthi Institute of Medical Sciences and Research, Tamil Nadu, India.
(3) Anonymous, University Sains Malaysia, Malaysia.

ABSTRACT

Background: Hepatocellular carcinoma (HCC) is the fifth most common solid malignancy worldwide and causes more than 600,000 deaths annually. Many risk factors predispose to HCC, these risk factors may present individually or collectively depending on the environmental situations. The risk factors for HCC include hepatitis B virus (HBV), hepatitis C virus (HCV), alcoholism, aflatoxin (AF), schistosomiasis and some hereditary diseases as haemochromatosis and haemophilia. Certain key regulatory signaling pathways integrated in the pathogenesis of HCC such as insulin-like growth factor (IGF) signaling pathway, transforming growth factor-β (TGF-β)

**Corresponding author: E-mail: ayman.elsamanoudy@gmail.com* signaling pathway, Toll-like receptor (TLR) signaling pathway and Notch signaling pathway. Dysregulation of different signaling pathways at any level including the receptors and downstream signaling pathway components represent a central protumorigenic principle in human hepatocarcinogenesis and this dysregulation can be targeted for new therapeutic modalities for HCC.

Aim of the Work: The aim of the present work was to study the recently discussed signaling pathways integrated in development of hepatocellular carcinoma in the high risk groups, and how genetic alterations in theses signaling pathways is a potentiating factor for HCC from the year 2000 until now. New diagnostic and therapeutic modalities for that disease could be provided by targeting these dysregulated signaling pathways.

Conclusion: Finally, it could be concluded that: 1) There are multiple mechanisms underlying the hepatocarcinogenesis and by these mechanisms, early diagnosis for HCC could be possible. 2) Understanding the signaling pathways dysregulated in high risk groups for HCC may provide new diagnostic and therapeutic modalities.

Keywords: Hepatocellular carcinoma; dysregulation; signaling pathways.

1. SIGNALING PATHWAYS INTEGRATED IN THE PATHOGENESIS OF HCC

Hepatocarcinogenesis is associated with a large accumulation of chromosomal, genetic and epigenetic alterations. Some of these alterations occur in different stages of hepatocarcinogenesis and the cause of these alterations could be dysregulation of important molecular cellular signaling pathways. Therefore, a malignant hepatocyte may be produced by alteration in gene(s) caused by dysregulation of different regulatory pathways [1].

Normal embryogenesis, organ development as well as tissue proliferation, regeneration and repair require balance of various molecular signaling pathways. Dysregulation of these signaling pathways and their components is the central principle in human tumorigenesis [2].

1.1 The Signaling Pathways Integrated in HCC Pathogenesis

There are multiple regulatory pathways with alterations that may have a role in HCC; these signaling pathways can be classified into:

1- Liver stem cell signaling pathways for example: TGF- β, Wnt/ β-catenin, IGF, TLR, Notch, Hepatocyte Growth Factor (HGF) and TGF-α/EGFR signaling pathways [2].
2- Signaling pathways that are associated with other types of cancers such as: nuclear factor-kappa B (NF- κB), janus kinase / signal transducer and activator of transcription (JAK/STAT) and mitogen activated protein kinase (MAPK) signaling pathways. All these are known as key contributory pathways to the cell transformation, proliferation, apoptosis and invasive behaviors in human HCC [3]

1.1.1 Liver stem cell signaling pathways

1.1.1.1 Transforming growth factor-beta signaling pathway

TGF-β is involved in cell differentiation, proliferation, migration and apoptosis. In the liver, TGF-β is a potent growth inhibitor of normal hepatocytes and induces apoptosis and cellular senescence in these cells [4].

The basic signaling cascade of TGF-β involves type I and type II transmembrane serine/threonine kinase receptors they are called TGF-β receptor (TBRI and TBRII). Activation of these receptors is caused by TGF-β factor binding to TBR leading to phosphorylation of the intracellular part of the receptor. The phosphorylated receptor recruits and activates intracellular signaling proteins by phosphorylation; these signaling proteins are called mothers against decapentaplegic protein (Smad) [5].

In vertebrates possess at least nine Smad proteins categorized into three functional classes: (1) Receptor activated Smads (R-Smads) as Smad1, Smad2, Smad3, Smad5 and Smad8. (2) Co-mediator Smads: Smad4 and Smad10 and (3) Inhibitory Smads: Smad6 and Smad7 [6].

Activation of TGF-β signaling pathway results in association of R-Smads as Smad 2 and Smad 3 (after their phophorylation) with Smad 4; this Smad complex then translocates to the nucleus where it interacts with various transcriptional factors such as cAMP response element binding protein (CREB) as shown in Fig. 1. These sequence of events leads to the enhancement of certain genes transcription such as p15 gene, p21 gene and E-cadherin gene [7].

There is an adaptor proteins named embryonic liver fodrin (ELF) these proteins play a critical role in modulating TGF-β signaling. ELF is a dynamic scaffolding protein that is required for Smad3 and Smad4 localization (from the cytoplasm to the nucleus). Disruption of ELF results in mislocalization of Smad3 and Smad4 leading to loss of the TGF-β dependent transcriptional response [9].

1.1.1.2 Wnt /β-catenin signaling pathway

Wnt signaling pathway plays a pivotal role in cell proliferation, cell/cell interactions, motility and tissue development [10]. Initiation of Wnt signaling involves the binding of Wnt proteins to a frizzled receptor (Fzd) and low density lipoprotein receptor related protein (LRP). The key intracellular component of this signaling pathway is cytoplasmic β-catenin protein [11].

In the absence of activated Wnt/β-catenin signaling, cytosolic β-catenin is rapidly phosphorylated by a complex of proteins collectively termed the "destruction complex" composed of the following proteins: Axin, adenomatous polyposis coli (APC), glycogen synthase kinase 3 (GSK3) and casein kinase 1 (CK1). The destruction complex phosphorylates the N-terminus of β-catenin, targeting the protein for proteasomal degradation and thereby maintaining low baseline cytosolic level of β-catenin [12].

The binding of Wnt ligand to Fzd receptor causes recruitment of the cytoplasmic dishevelled protein (DVL). The DVL protein causes disruption of the destruction complex leading to accumulation of cytoplasmic β-catenin. β-catenin, translocates to the nucleus where it serves as a coactivator to a number of genes by interacting with a family of transcription regulatory proteins called T-Cell Factor/ Lymphocyte Enhancer Factor (TCF/LEF) as shown in Fig. 2 [13].

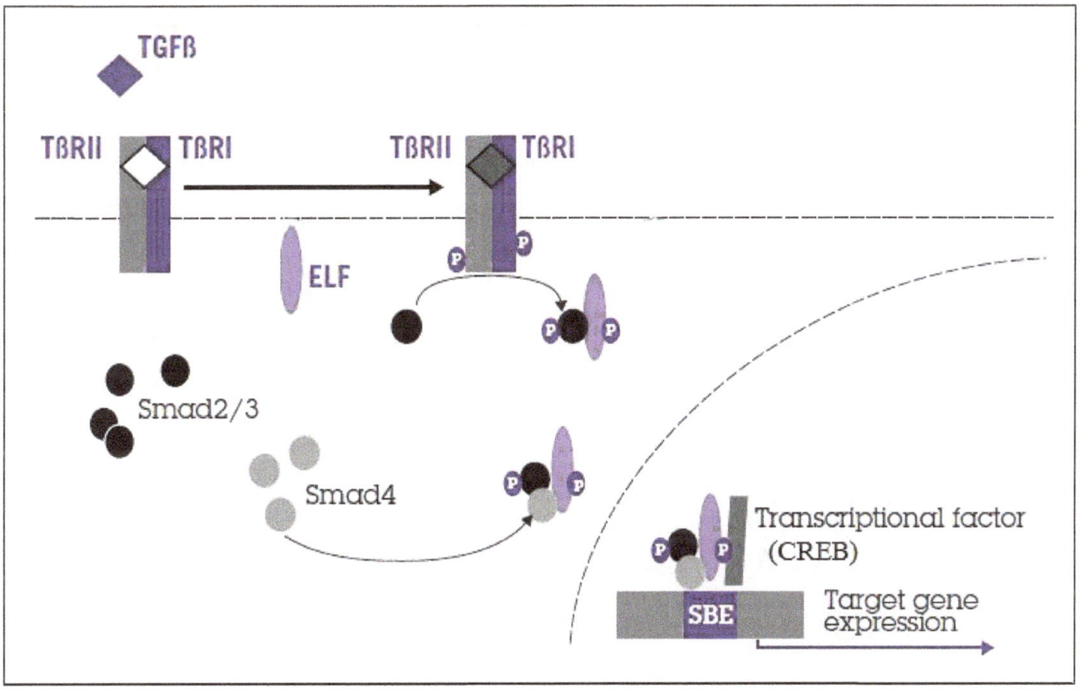

Fig. 1. Transforming growth factor-beta signaling pathway
(SBE: Smad binding element) (Quoted from Kitisin et al. [8])

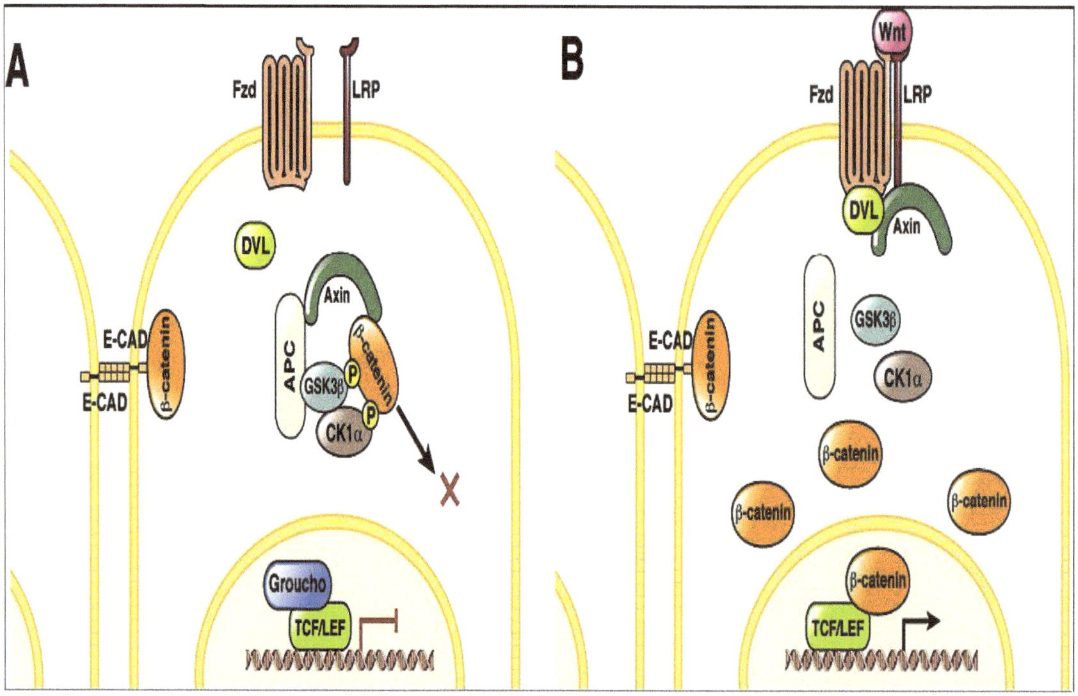

Fig. 2. Active and inactive Wnt signaling pathway. (A) In the absence of a Wnt ligand, β-catenin is bound to E-CAD at adherens junctions or is phosphorylated by the destruction complex. (B) Wnt ligand initiates signaling through FZD receptor and LRP coreceptor, activating and recruiting DVL and Axin to the membrane, thereby disrupting the destruction complex
(Quoted from White et al. [13])

Binding of β-catenin to TCF/LEF leads to stimulation of transcription of Wnt target gene to increase cell Proliferation. Grouch protein acts as corepressor to Wnt target genes by binding to and inhibiting TCF/LEF in the absence of β-catenin [14].

1.1.1.3 Insulin like growth factor signaling pathway

Insulin-like growth factor signaling pathway has highly conserved function in mammals and plays a critical role in metabolism and cell renewal [15]. IGF pathway is not only involved in cell growth but it also, promotes cell proliferation, migration and malignant transformation [16]. IGF pathway includes four essential components which are: ligands, receptors, substrates and ligand binding proteins [17].

1.1.1.4 The components of IGF signaling pathway

1.1.1.4.1 A- IGF ligands

IGF ligands include both IGF-1 and IGF-2. Their names are based on the observation that both IGF-1 and IGF-2 are peptides that have structure similarity to insulin and they share 40% homology with proinsulin [18]. However they are slightly different from insulin structurally by containing an additional domain which could account for their dramatically different role in neoplasms in comparison with insulin [19].

1.1.1.4.2 B- IGF receptor

IGF ligands bind to the second component of the IGF axis which is the receptors that includes IGF-1 receptor (IGF-1R), IGF-2R, insulin receptor and hybrid receptors. The hybrid receptors consist of IGF-1R hemireceptor and insulin hemireceptor. IGF-1 and IGF-2 ligands bind to IGF-1R with high affinity while IGF-2 is the only ligand for IGF-2R [15] as shown in Fig. 3.

1.1.1.4.3 C- substrates

The third component of the IGF axis refers to insulin receptor substrate (IRS) proteins mainly, which are the major signals downstream of IGF-1R activation. There are 4 types of IRS and the important ones among them include IRS-1 and IRS-2 [19].

1.1.1.4.4 C- ligand binding proteins

The last key component of the IGF axis consists of IGF binding proteins (IGFBPs). There are 6 members of IGFBPs with high affinities for IGF-1 and IGF-2. For example, IGFBPs1-4 bind both IGF-1 and IGF-2 with similar affinity but IGFBP-5 and 6 strongly prefer IGF-2 as their ligand [17].

IGF signaling is initiated by the binding of IGF ligand to the receptor (IGF-1R), which results in phosphorylation of intracellular portion of the receptors. Then, the phosphorylated receptor activates intracellular target proteins as IRS (by phosphorylation). IRS conveys the signal to specific downstream effectors such as phosphatidyl inositol 3-kinase (PI3K) and MAPK pathways. The result of activation of these pathways is transcriptional activation of various target genes such as p27, cyclin B and VEGF gene [20] (Fig. 4).

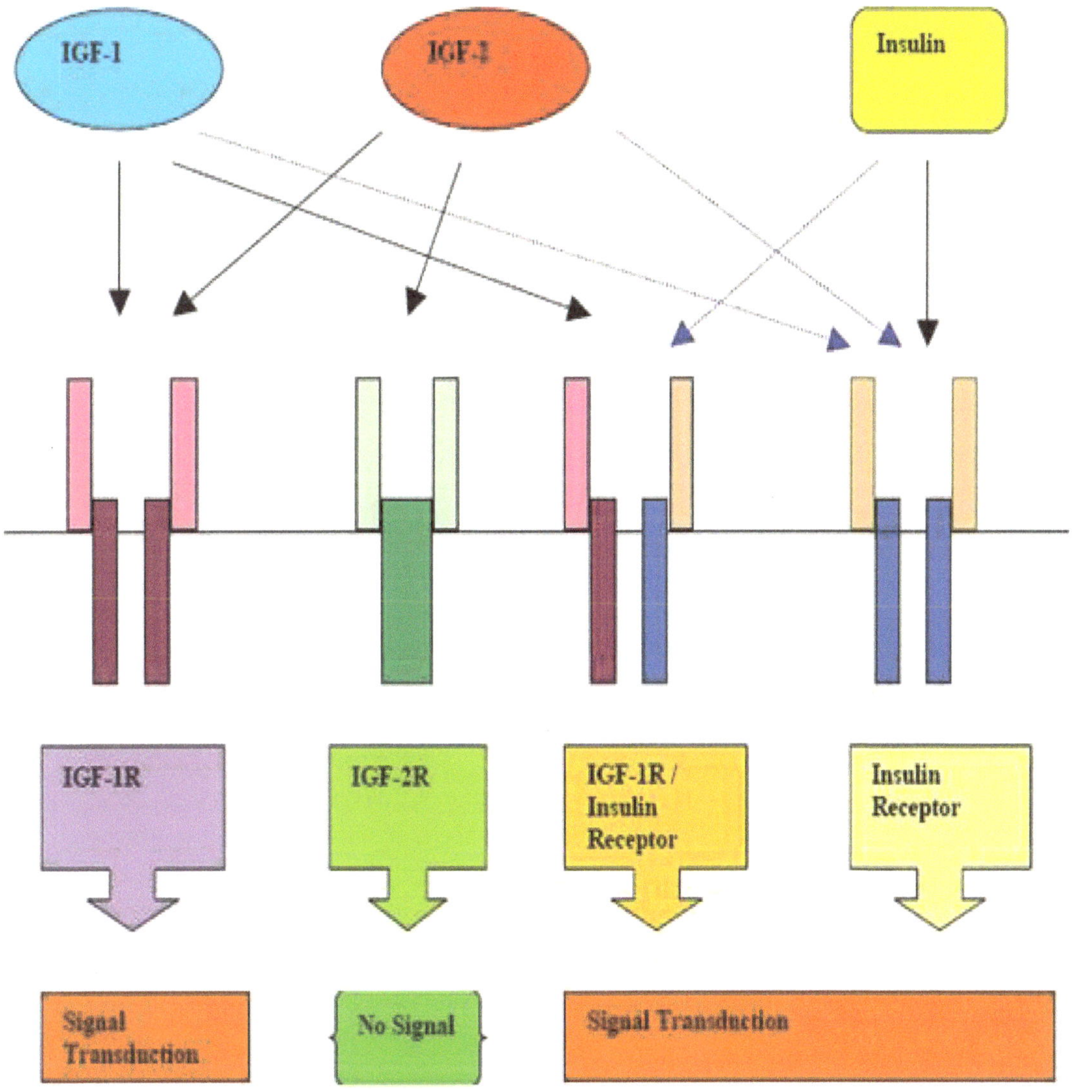

Fig. 3. Different types of IGF ligands and their receptors

Insulin receptor and IGF-1 receptor are both tyrosine kinases. IGF-2R functions as a clearance site for IGF-2. IGF-1 binds to IGF-1R and to IGF-1R/insulin receptor hemireceptor; it binds to insulin receptor only at very high concentrations. IGF-2 binds to IGF-1R, IGF-2R and binds to insulin receptor only during early fetal development. Insulin binds to insulin receptor and it binds to IGF-1R/Insulin hemireceptor at high concentration of the insulin. Signal transduction is activated after the activation of IGF-1R, IGF-1R/Insulin receptor hemireceptor and insulin receptor; however, IGF-2R activation results in no signal downstream. (Quoted from Wu and Zhu, [17])

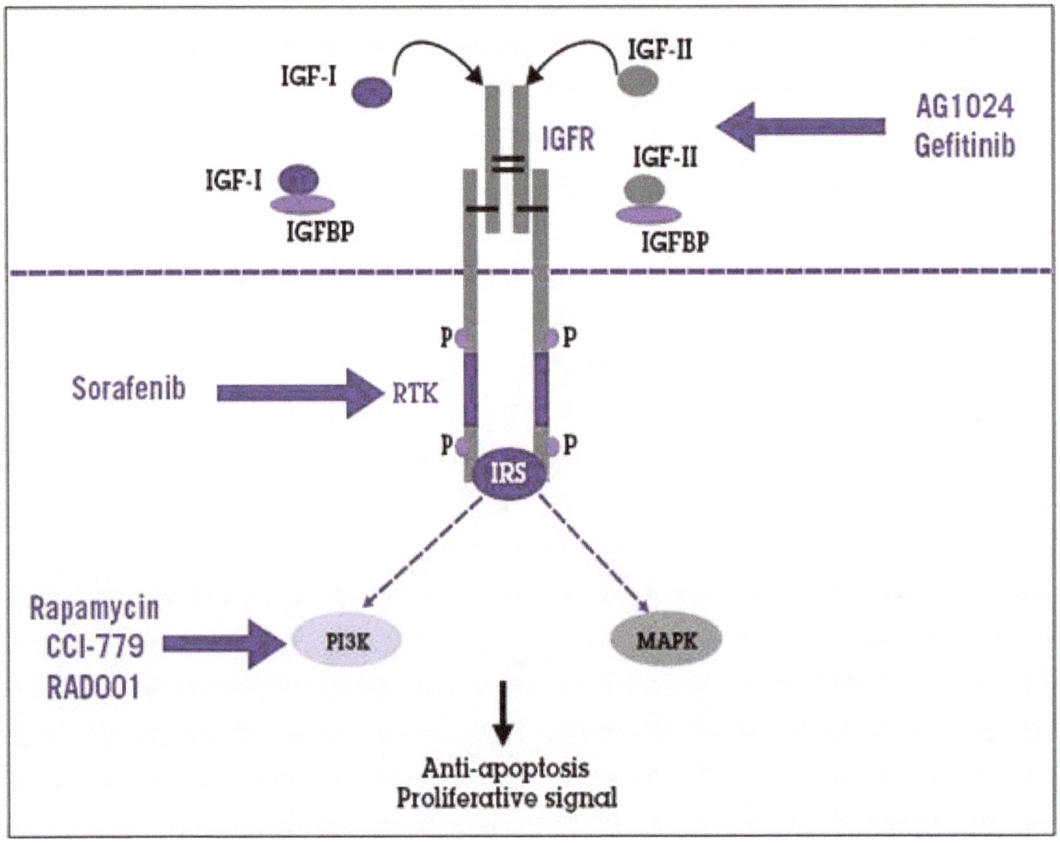

Fig. 4. Simplified schematic presentation of IGF signaling pathway

IGF-I and IGF-II bind to IGFR which is RTK with high affinity resulting in phosphorylation of intracellular proteins including IRS. The signal is then conveyed to specific downstream effectors such as PI3K and MAPK pathways. These pathways play crucial roles in antiapoptosis as well as cell proliferation. Therapeutic agents such as AG1024 and gefitinib aim to block this signaling pathway at the receptor level and downstream targeting agents such as rapamycin, CCI-779 and RAD001 are possible agents for HCC treatment. (RTK: receptor tyrosine kinase) (Quoted from Kitisin et al., [8])

1.1.1.5 Toll like receptor signaling pathway

TLR signaling is the first line of protection against microbial pathogens as TLR represents one of the important pathways that mediate the innate immunity. TLR can discriminate between a variety of pathogens through the function of pattern-recognition of the receptors and these receptors recognize certain microbial components known as pathogen associated molecular patterns [21].

Thirteen mammalian TLRs have been described and 10 are expressed in humans and each one of the receptors is responsible for recognizing distinct bacterial, viral or fungal structures as shown in Fig. 5 [22].

TLRs are characterized by two conserved regions: the extracellular leucine-rich region and the cytoplasmic Toll/IL-1 receptor (TIR) domain. TLRs share the initial common activation pathway mediated by ligand recognition by the extracellular domain of the receptor then activation of the TIR domain [23].

After TLRs activation, two signaling pathways mainly exist; the first is through the adaptor protein called myeloid differentiation factor 88 (MyD88) which is common to all TLRs and the other one is MyD88-independent pathway that is particular to the TLR3 and TLR4 signaling pathways [24]

1.1.1.6 MyD88-dependent TLR signaling pathway

Upon stimulation, specific ligand binding to the receptor, of the signaling pathway conformational changes occur to TIR domain of the receptor leading to recruitment of the MyD88 adaptor protein [25]. MyD88 recruits IL-1 receptor associated kinase 1 and 4 (IRAK1, 4). They associate with MyD88 leading to their activation then the activated IRAK associates with tumor necrosis factor receptor associated factor 6 (TRAF6) [24].

TRAF6 is recruited to the receptor complex and activated by IRAK-1 then TRAF6 dissociates from the receptor and associates with TGF-β activated kinase1 (TAK1) and TAK1-binding proteins (TAB) TAB1 and TAB2. The complex of TRAF6, TAK1, TAB1 and TAB2 moves into the cytoplasm [26].

The activated complex (TAK1, TRAF6, TAB1 and TAB2) activates the inhibitory kappa kinase complex (IKK), which consists of 2 catalytic subunit (IKKα, IKKβ) and one regulatory subunit (IKKγ), by phosphorylation at IKKβ subunit then the activated IKK phosphorylates the inhibitor of NF-κB (IκB) leading to degradation of IκB and thereby releases the inhibition on NF-κB [27] as shown in Fig. 6.

1.1.1.7 MyD88-independent TLR signaling pathway

TLR4 and TLR3 have been shown to induce activation of the signaling pathway in MyD88 knockout animals through activation of another pathway called MyD88 independent pathway [28].

This pathway is medited by another adaptor protein called TIR domain containing adaptor protein inducing interferon-beta (TRIF) that activates another downstream protein called interferon regulatory factors (IRFs) this protein passes to the nucleus and activates the gene coding for interferon-beta (IFN-β) leading to increased production of IFN-β that mediate the body antiviral response against viral pathogens [29] as shown in Fig. 6.

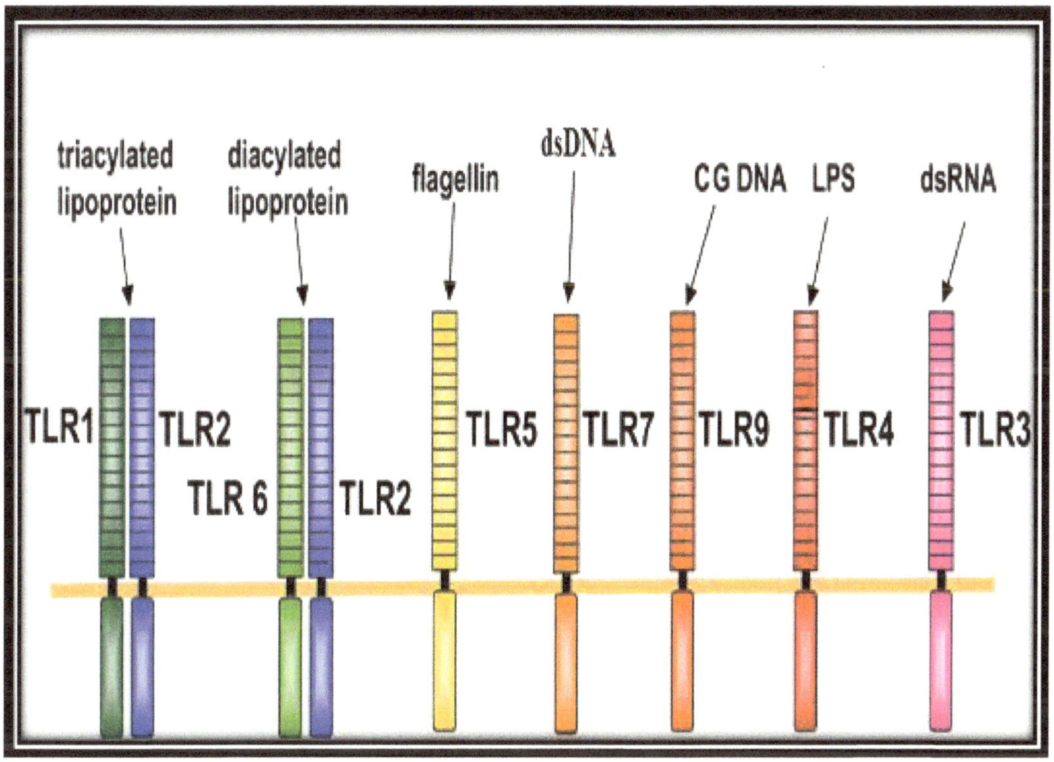

Fig. 5. Toll like receptors and the specific ligand for each receptor

TLR1–TLR9 have been characterized to recognize microbial components. TLR2 is present in the form of dimer with either TLR1 or TLR6 and discriminate the differences between triacyl and diacyl lipopeptides respectively. TLR4 recognizes bacterial LPS. TLR9 is the receptor for pathogen CG sequence in DNA, whereas TLR3 is implicated in the recognition of viral dsRNA. TLR5 is a receptor for bacterial flagellin. Thus, the TLR family discriminates between specific patterns of microbial components. (LPS: lipopolysaccharide, dsRNA= double stranded RNA) (Quoted from Takeda and Akira, [22]).

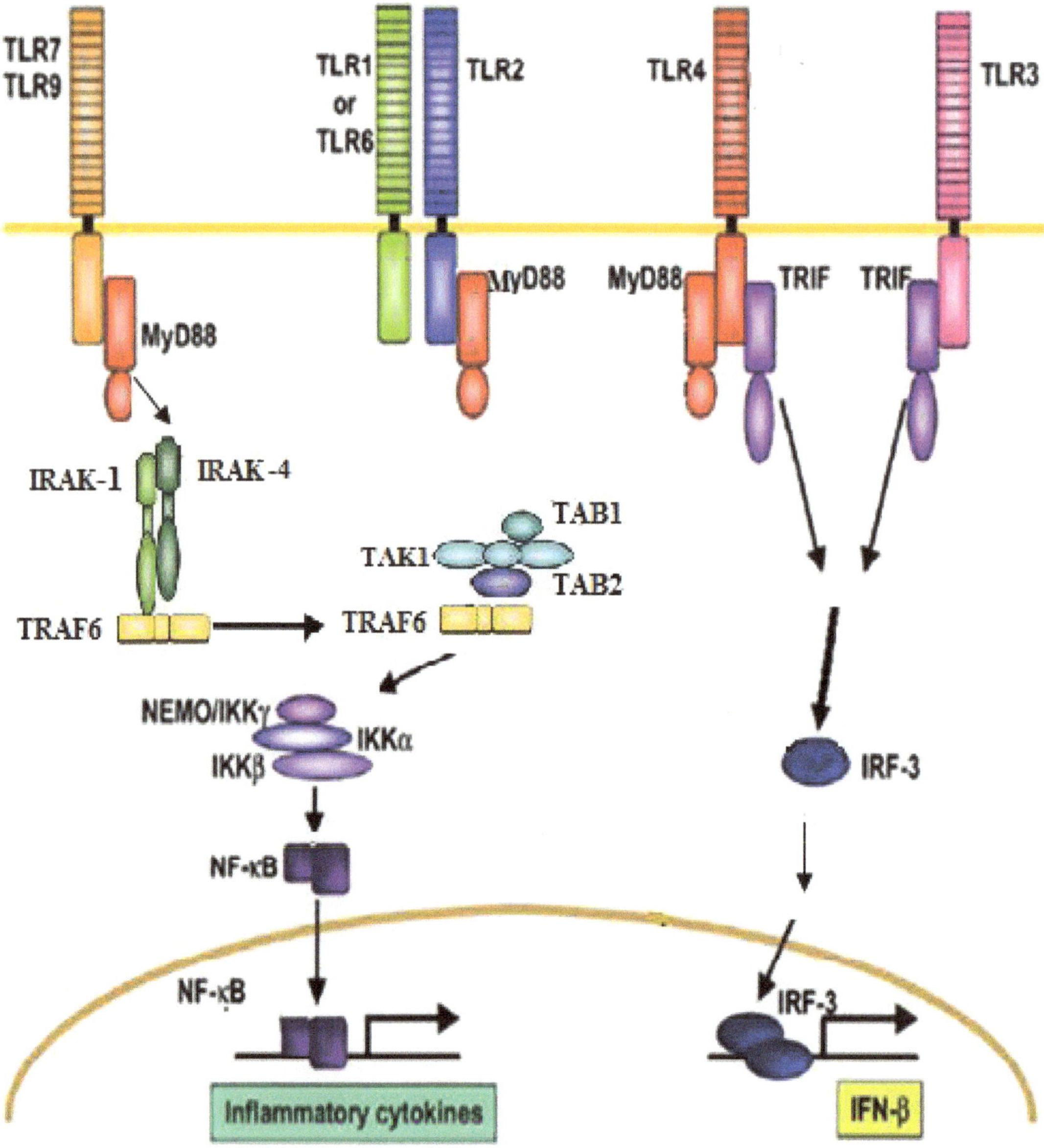

Fig. 6. MyD88 dependant and independent TLR signaling pathway
(Quoted from Takeda and Akira, [22])

1.1.1.8 Notch signaling pathway

Notch signaling pathway has a role in maintaining the balance between cell proliferation, differentiation and apoptosis [30]. Mammals possess four Notch receptors (N) from N1 to N4. These receptors are present on the surface of the receptor cells and can bind to five ligands present on the surface of the signal sending cells and these ligands are named as Delta-like1, 3, 4 (Dll1, 3 and 4), Jagged 1 and 2 (J1 and J2) [31] as shown in Fig. 7.

The extracellular part of the receptors contains EGF like repeats involved in ligand binding, followed by three cysteine-rich repeats that prevent ligand independent activation (LIN) and a hydrophobic stretch of amino acids mediating heterodimerization (HD) between the notch receptor extracellular subunit (N^{EC}) and the Notch receptor transmembrane subunit (N^{TM}). The cytoplasmic tail of the receptor harbors nuclear localization signals domain (NLS) [32].

Newly synthesized Notch receptors are proteolytically cleaved in the Golgi at site 1

(S1) by a furin-like protease during receptors transport to the cell surface. This cleavage generates a heterodimeric receptor consisting of an extracellular subunit that is noncovalently linked to a second subunit containing both the transmembrane domain and the cytoplasmic region of the Notch receptor [34].

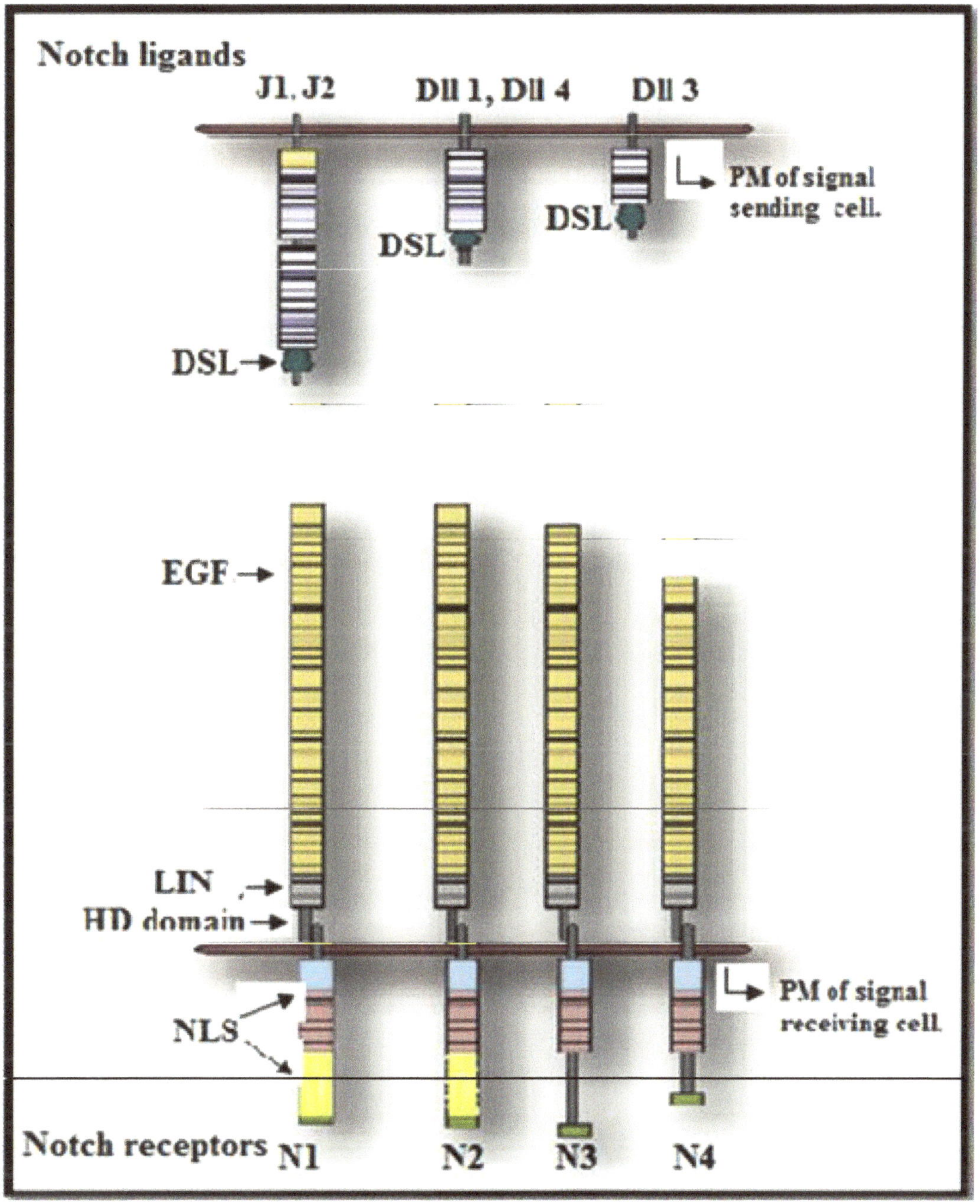

Fig. 7. The structure of Notch ligands and their receptors

Five conventional Notch ligands are known: J1, J2, Dll1, Dll3, and Dll4. A common structural feature of all ligands is an amino-terminal domain called DSL involved in receptor binding and there is structural similarity beween J1 and J2 and between Dll1and Dll4. There are four Notch receptors (N1–N4) and these receptors possess extracellular, transmembrane and intracellular domains. (DSL: Delta, Serrate and Lag-2) (Quoted from Radtke et al. [33])

Notch signaling is activated upon cell to cell contact as a result of interactions between Notch receptors on the signal receiving cell and their ligands on the signal sending cell [31]. Ligand receptor interaction between neighboring cells leads to two successive proteolytic cleavages of the receptor. The first cleavage is mediated by a disintegrin and metalloproteases (ADAM) which cleaves the receptors 12–13 amino acids external to the transmembrane domain at S2 site. The shedded extracellular domain is endocytosed by the ligand expressing cell [35].

Ligand binding to N^{EC} induces a conformational change within the notch receptors to expose the S2 cleavage site for proteolysis. After shedding of the extracellular domain, a second cleavage between intracellular and trans-membrane domain (at site S3) is mediated by the γ-secretase activity [36].

The third cleavage liberates the intracellular domain of Notch receptors which subsequently traffics to the nucleus and heterodimerizes with core binding factor-1 (CBF-1) which is DNA transcription factor in order to form a short-lived nuclear transcription complex; then Notch intracellular domain recruits other coactivators including mastermind like proteins (MAML1-3) in order to induce the transcription of downstream target genes (Fig. 8) [33].

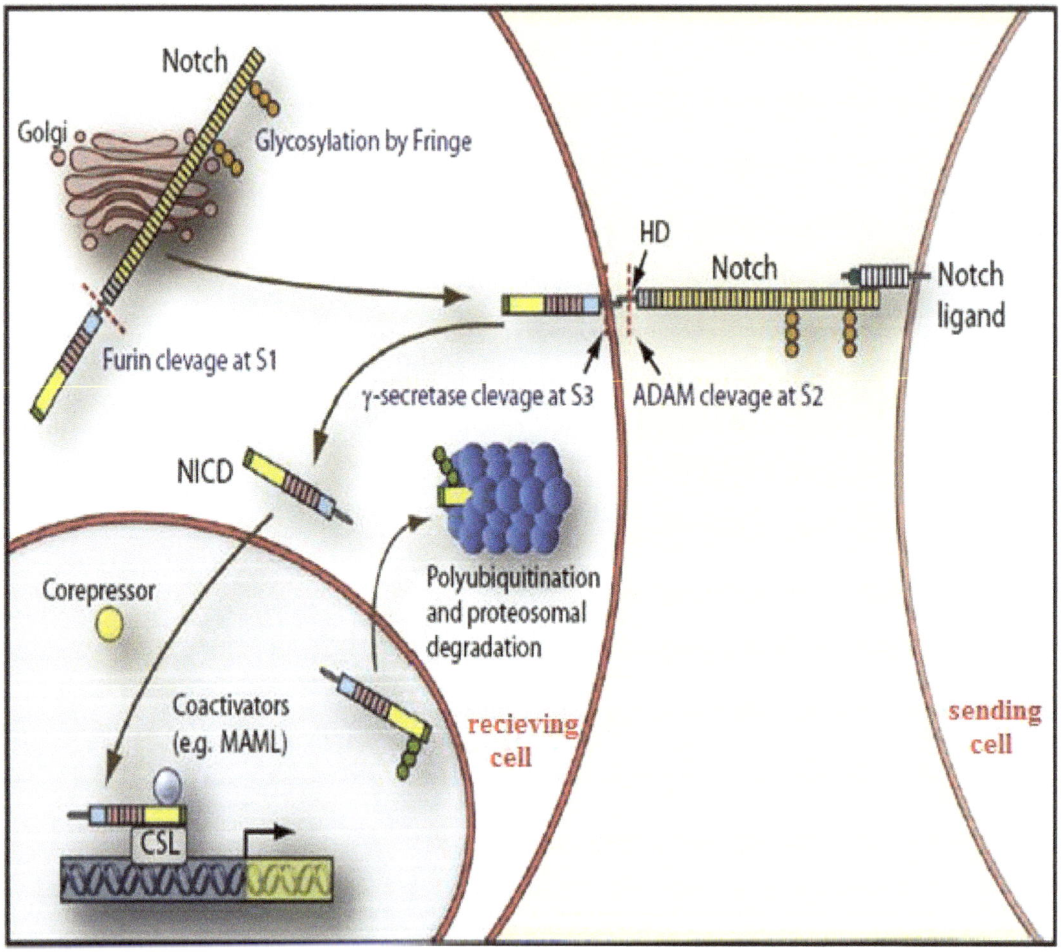

Fig. 8. Notch signaling pathway

Notch proteins are synthesized as single precursor proteins, which are cleaved in the Golgi at site S1 and EGF-like repeats are glycosylated by fringe proteins in the Golgi before receptors are transported to the cell surface. The second cleavage occurs at site S2 followed by a third cleavage at S3. Notch intracellular domain undergoes polyubiquitination and proteosomal degradation in the cytoplasm after finishing its action as a transcription co-activator. (NICD: Notch intracellular domain) (Quoted from Radtke et al., [33])

Notch target genes includes the hairy and enhancer of split 1 (Hes1), Hes5, Hes7 as well as Hes1-relate with YRPW motif (Hey1) and Hey2 [37]. Other Notch target genes are cyclin D1 and p21 gene. So notch signaling pathway is involved in the control of cell cycle, proliferation and apoptosis through its target genes [38].

There are emerging data suggesting that Notch can crosstalk or cooperate with other signaling pathways including NF-kB or TGF-β signaling pathway and thereby broaden the spectrum of target genes that are influenced by Notch signaling [39].

1.1.1.9 Hepatocyte growth factor signaling pathway

Hepatocyte growth factor is one of the most potent growth factors for hepatocytes and plays a role in tissue proliferation, migration, cell survival, angiogenesis and tissue regeneration [40].

Hepatocyte growth factor binds to the receptor tyrosine kinase called c-MET, this receptor is highly expressed in epithelial and endothelial cells. Binding of HGF to the c-MET receptor results in receptor autophosphorylation then phosphorylation of adaptor proteins such as growth factor receptor bound protein 2 (Grb2) and Grb2 associated binding protein 1 (Gab1). The HGF signaling is then conveyed to activation of various downstream effectors such as phospholipase C and PI3K as shown in Fig. 9 [2].

Activation of the effectors leads to subsequent activation of distinct transcription factors as activator protein 1 leading to enhancement of the expression of numerous target genes as matrix metalloprotease and urokinase plasminogen activator genes. The main function of both urokinase plasminogen activator and matrix metalloprotease is extracellular matrix degradation. So, HGF has a role in increasing tumor invasiveness [41].

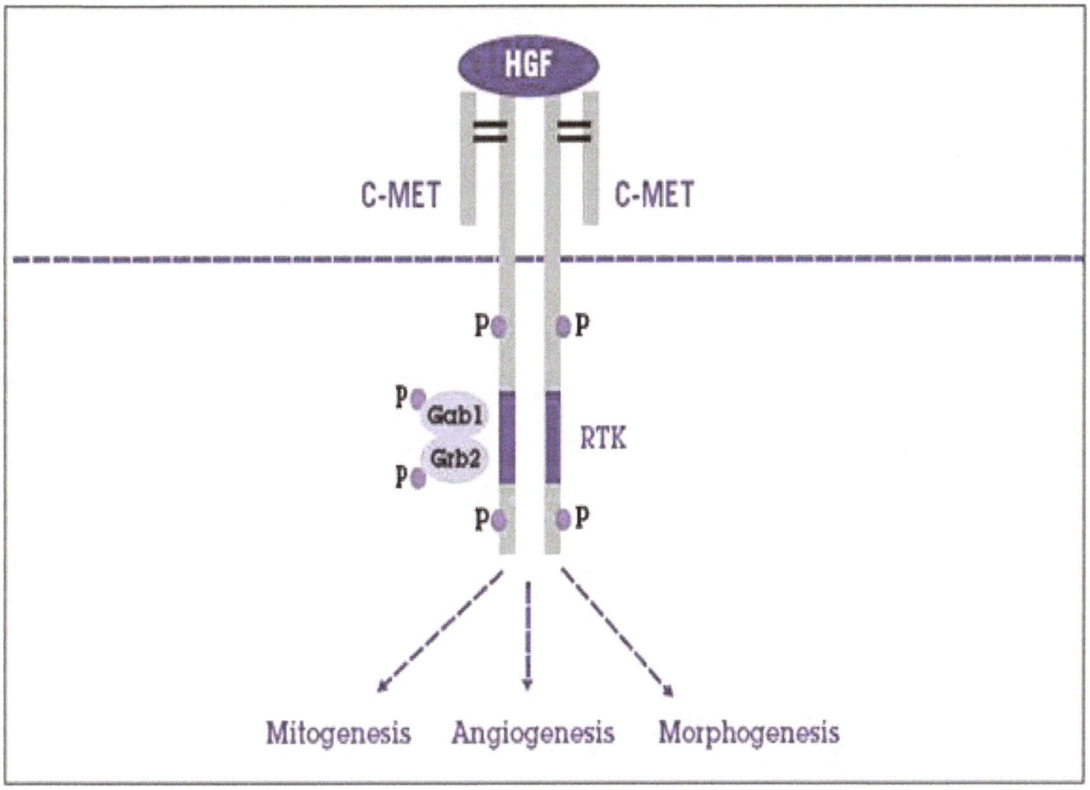

Fig. 9. Simplified schematic presentation of HGF signaling pathway

Binding of the HGF to the extracellular domain of the c-MET receptor leads to autophosphorylation of the intercellular part of the receptor then activation of adaptor proteins (Grb2 and Gab1). Activation of these adaptor proteins leads to activation of various downstream effectors resulting in transcription of various genes important in mitogenesis, angiogenesis and morphogenesis (Quoted from Kitisin et al. [2])

1.1.1.10 Transforming growth factor-α/ Epidermal growth factor signaling pathway

TGF-α/EGF signaling comprises at least eight ligands including TGF-α, EGF, heparin-binding EGF, amphiregulin, betacellulin, epiregulin, epigen and crypto [42]. TGF-α/EGF signaling conveys its signal through the receptor tyrosine kinase Family which has four different types and EGFR is the most important one [43].

Different ligand specificities and concentrations lead to differential phosphorylation of tyrosine residues at the cytoplasmic portion of the receptor [44]. Binding of the TGF-α to the receptor causes phosphorylation of the receptor's cytoplasmic portion that serves as a docking site for some proteins such as sarcoma homology 2 domains protein (Shc) and Grb2 [45].

The signal will then activate multiple downstream pathways which can directly or indirectly interact with each other such as MAPK and JAK / STAT pathway [46] (Fig. 10).

1.1.1.11 Hedgehog (Hh) signaling pathway

Hedgehog signaling pathway is a complex network of signaling molecules including both positive and negative regulatory proteins which plays a crucial role in cell differentiation and proliferation [47].

In the absence of Hh ligand, the membranous patched receptor (Ptch) antagonizes the pathway by preventing the activity of another transmembrane protein called smoothened (Smo) which is a 7-transmembranous protein receptor [48].

Binding of Hh ligands to Ptch relieves its inhibition on Smo. Smoothened protein activates some transcription factors called glioma-associated oncogenes (Gli-1, Gli-2, Gli-3). Different Gli proteins exhibit activating or repressing transcriptional activity. Gli-1 and Gli-2 mainly act as transcriptional activators while Gli-3 generates a repressor form in the absence or inhibition of Hh signaling as shown in Fig. 11 [49].

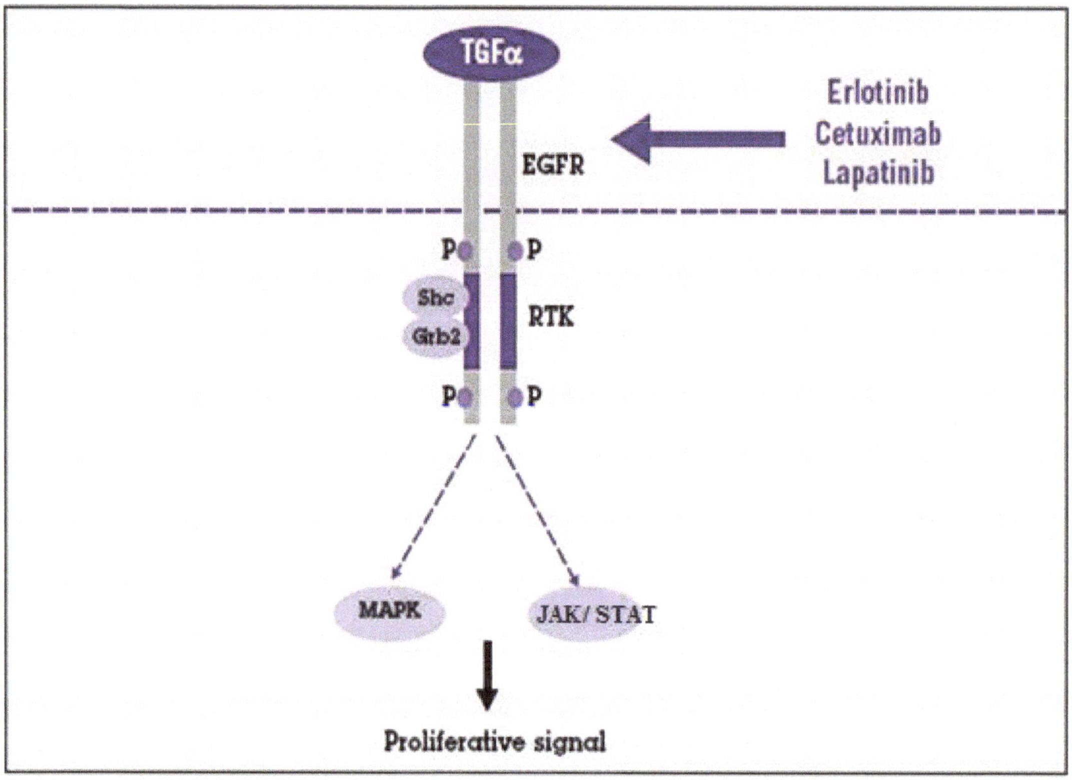

Fig. 10. Schematic presentation of TGF-α/EGF signaling pathway

New therapeutic agents for HCC such as erlotinib, cetuximab and lapatinib are aiming to inhibit HCC growth and metastasis formation by acting as tyrosine kinase receptor inhibitors (Quoted from Kitisin et al. [2])

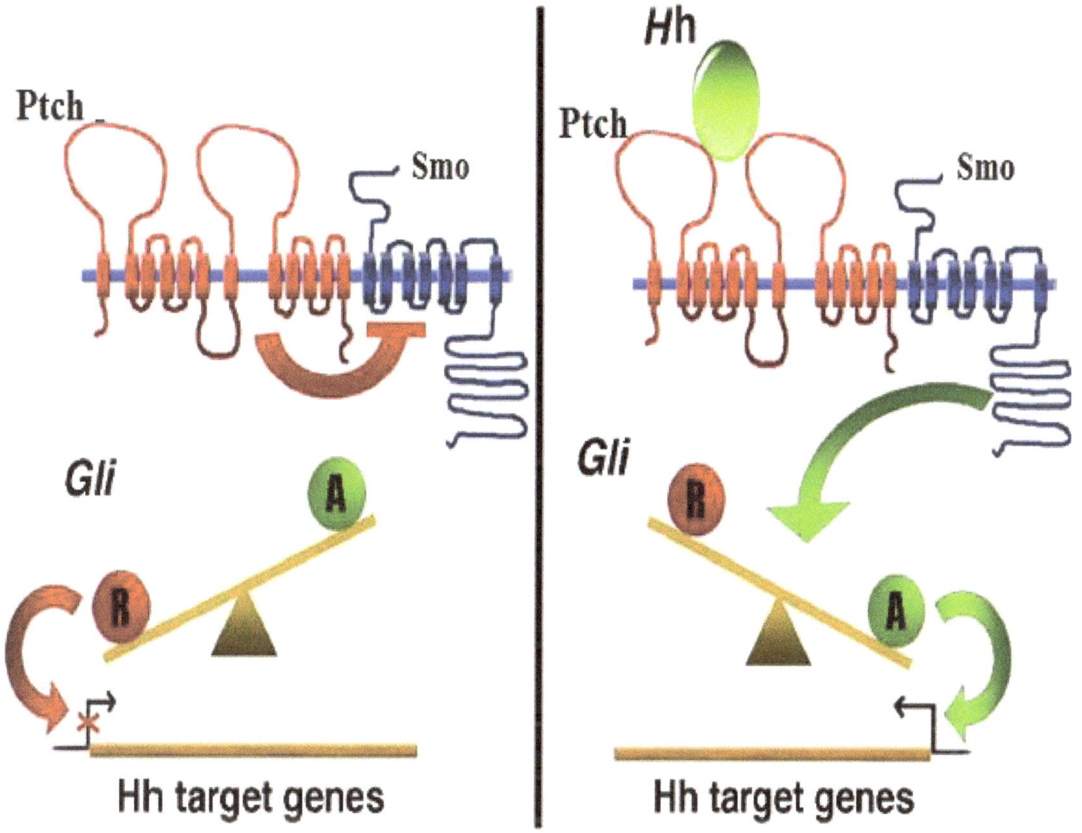

Fig. 11. Simplified presentation of hedgehog signaling pathway

In unstimulated cells (left), the activity of the transmembrane protein Smo is suppressed by the Ptch receptor and the majority of the Gli present is the transcription repressive (R). In stimulated cells (right) binding of Hh to Ptch activates Smo which in turn increase the level of the Gli transcription activator (A) resulting in the transcription of Hh target genes (Quoted from Parkin and Ingham, [49])

Like β-catenin, after ligand stimulation, Gli accumulates in the nucleus and induces the expression of numerous target genes that regulate proliferation, differentiation and extracellular matrix (ECM) interactions such as fibroblast growth factor, IGF2, EGF, Cyclin B, Cyclin D and Cyclin E genes [50].

1.1.2 Signaling pathways associated with other types of cancer

Each one of the incoming pathways is related to one or more of the previously mentioned liver stem cell signaling pathways.

1.1.2.1 NF-κB signaling pathway

NF-κB is a transcription factor that is considered to be a key regulator of the immune responses, inflammation and cell survival. Binding of the ligand to the TLRs is considered as a proinflammatory stimuli that activate NF-κB signaling mainly via IKK dependent phosphorylation then degradation of the phosphorylated IκB protein [51].

Once the inhibition on NF-κB is removed, it passes to the nucleus as shown in Fig. (12) and stimulates the transcription of genes encoding cytokines and antiapoptotic factors as genes coding for tumor necrosis factor-alpha, IL-1, IL-6, IL-10, IL-12 and many different chemokines [52].

1.1.2.2 MAPK pathway

MAPK pathway is a chain of proteins in the cell that communicates a signal from receptor tyrosine kinase (for example, EGFR) on the surface of the cell to the DNA in the nucleus of the cell. Thus, MAPK is a signaling component that is important in converting extracellular stimuli into a wide range of cellular responses [5].

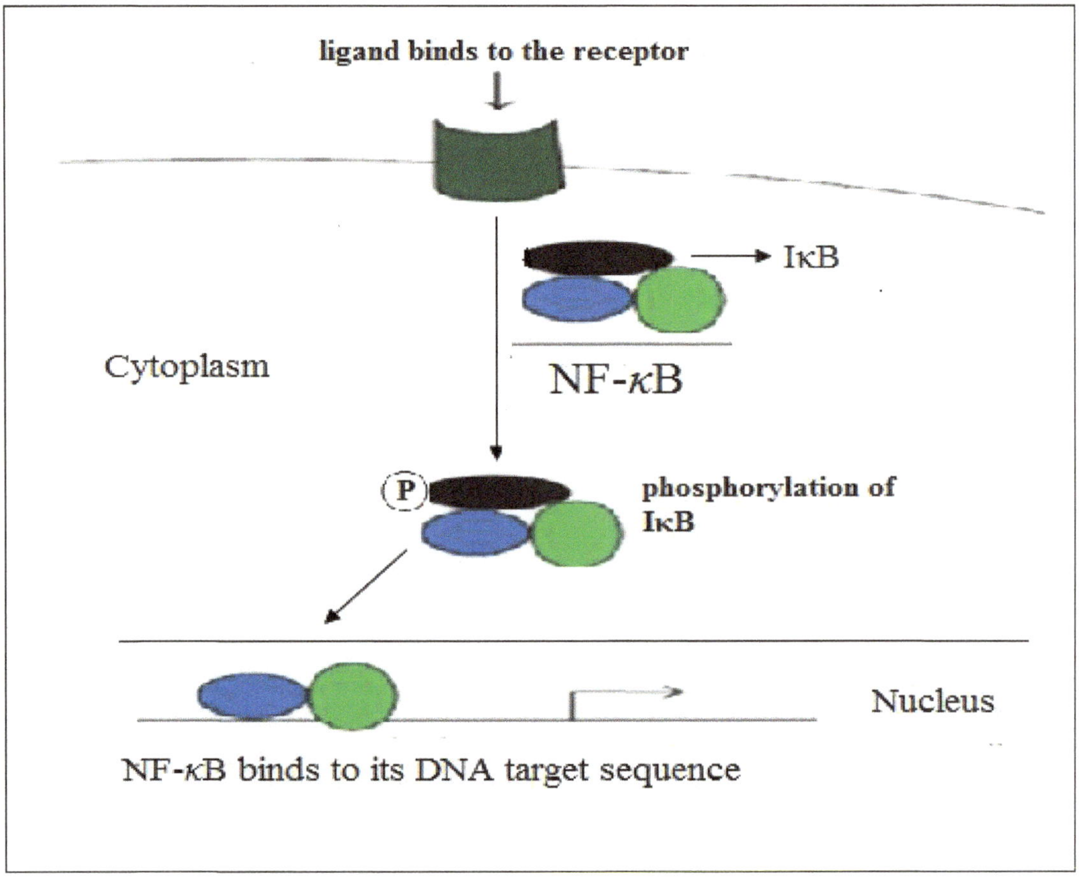

Fig. 12. NF- κB signaling pathway
(Quoted from Maeda, [52])

The signal starts when a signaling molecule binds to the receptor on the cell surface leading to MAPK activation by phosphorylation that is catalysed by MAPK kinases. After activation, MAPKs phosphorylate specific serine and threonine residues of target substrates, which include other protein kinases and many transcription factors affecting the expression of some proteins that controls cell cycle progression, apoptosis resistance and extracellular matrix remodeling [5].

1.1.2.3 JAK/STAT pathway

JAK/STAT pathway plays an important role in cellular processes like differentiation, proliferation, apoptosis and cell adhesion. STAT is located in the cytoplasm and become activated through tyrosine phosphorylation which typically occurs through JAK or growth factor receptor tyrosine kinases such as EGFR [5].

Activated STAT enter the nucleus and serve as transcription factors as it acts as a regulator of gene expression of various proteins including those for cell cycle progression, apoptosis resistance, extracellular matrix remodeling [5].

1.2 Cross-talk between Different Signaling Pathways Involved in HCC

Cytoplasmic downstream components of most growth factor signaling pathway complexes such as adaptor proteins, kinases and transcription factors are frequently utilized by more than one signaling pathway and numerous cross-talks between these pathways exist [43] for example:

1.2.1 Crosstalk between TGF-β and other signaling pathways

TGF-β signal is mediated through the Smad protein. The TGF-α/EGF as well as the HGF signaling causes phosphorylation of Smad2 potentially via some kinases leading to TGF-β signaling activation [5].

1.2.2 Crosstalk between Wnt/ β-catenin and other signaling pathways

Dysregulation of Wnt signaling is traditionally attributed to mutations in Axin, APC and β-catenin genes that lead to constitutive hyperactivation of the pathway. However, Wnt/ β-catenin signaling is also modulated through various other mechanisms in cancer, including cross talk with other signaling pathways [53-58] for example:

1.2.2.1 Cross talk between Wnt and TGF-β signaling pathway

This important association involves Wnt and TGF-β signaling pathways. Well-characterized association between β-catenin and E-cadherin is noted at the hepatocyte membrane and this connection has significant implications in cell-cell adhesion [59] as shown in Fig. 13.

In HCC cells, TGF-β signaling has been reported to correlate with the nuclear accumulation of β-catenin and may cause epithelial to mesenchymal transition (EMT). TGF-β induces the expression of snail and slug, two transcription factors, that induces EMT via downregulation of E-cadherin gene which in turn causes the disruption of the cell-cell adhesion [60].

This event has been shown to inhibit the β-catenin and E-cadherin association at the hepatocyte membrane causing nuclear accumulation of β-catenin and stimulation of β-catenin target genes so this is considered as a connection between Wnt and TGF-β signaling pathways [61].

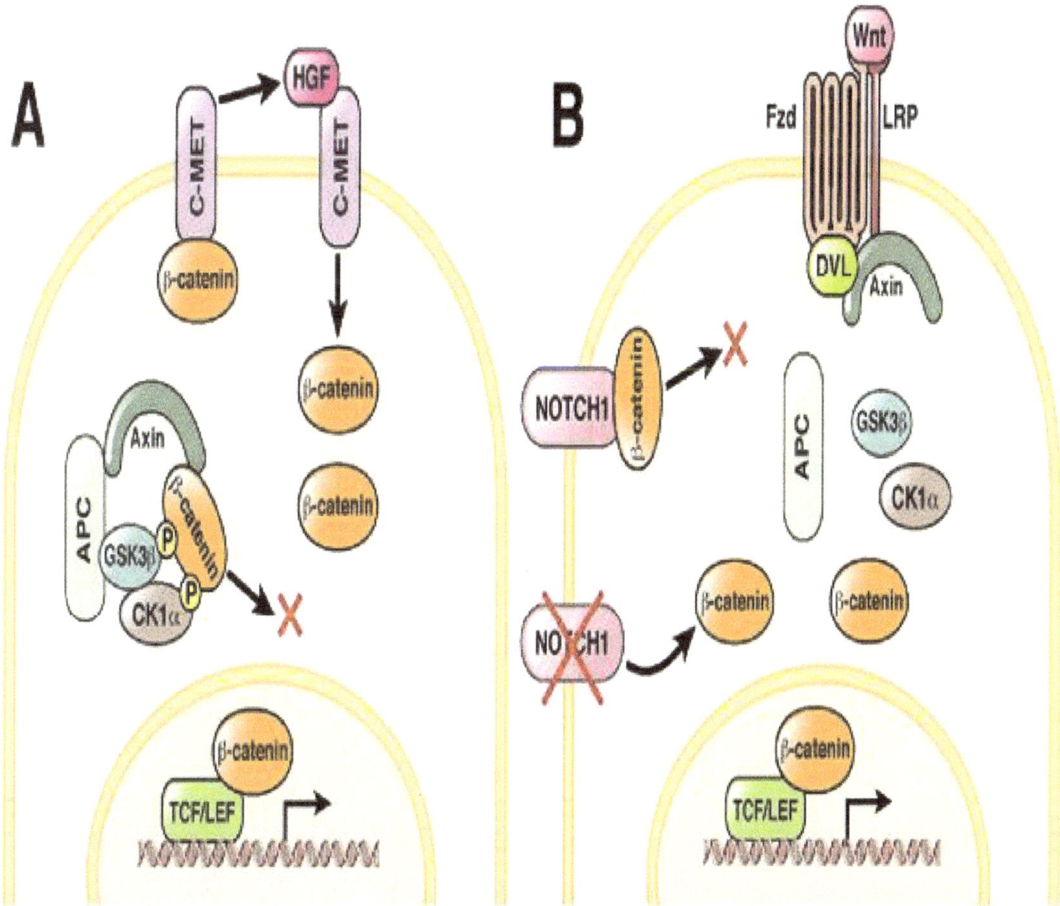

Fig. 13. Mechanisms by which Wnt/β-catenin signaling can be modulated in cancer. (A) In hepatocytes, β-catenin crosstalk with HGF pathway (B) Membrane-bound NOTCH1 can bind activated β-catenin and cause its lysosomal destruction

(Quoted from White et al. [13])

1.2.2.2 Cross talk between Wnt and HGF signaling pathway

In hepatocytes, β-catenin can be released from the membrane bound pool associated with the receptor c-MET. When c-MET engaged by HGF ligand, c-MET releases β-catenin into the cytoplasm with eventual β-catenin translocation into the nucleus and activation of its target genes as shown in Fig. 13 [13].

1.2.2.3 Cross talk between Wnt and Notch signaling pathway

The membrane bound Notch1 receptor can bind to the active free cytoplasmic β-catenin and negatively regulate it. Decrease in Notch1 expression may therefore potentiate Wnt/β-catenin signaling in certain types of cancer [62] as shown in Fig. 13.

2. DYSREGULATION OF SIGNALING PATHWAYS IN HCC

Dysregulation of different signaling pathways at any level including the receptors and downstream signaling pathway components represent a central protumorigenic principle in human hepatocarcinogenesis especially in liver stem cell signaling pathways [43].

2.1 Dysregulation of Signaling Pathways Associated with High Risk Groups for HCC

Exposure to risk factors for HCC highly predisposes to dysregulation in multiple signaling pathways including:

2.1.1 Transforming growth factor-beta signaling pathway dysregulation

There is marked attenuation of the TGF-β mediated antiproliferative response which has been shown in human HCC as there is marked reduction of TGF-β receptors on the surface of the hepatocytes in up to 70% of HCC patients [43].

It has been reported that Smad proteins play a role in HCC as Smad4 is mutated in 10% of HCC patients; also Smad2 mutations are identified in about 5% of HCC patients. Finally, inhibitory Smad7 is upregulated in 60% of advanced HCC [63].

TGF-β has antiproliferative role on the normal hepatocytes but in malignant hepatocytes, its expression increase and enhance malignant hepatocytes proliferation. This dual role of TGF-β signaling in HCC may be explained by its effect on the microenvironment of the tumor as tumor derived TGF-β could contribute to tumor growth indirectly by suppressing immune surveillance or stimulating production of angiogenic factors [64].

TGF-β signaling has also, been shown to induce an EMT in cells. This EMT process is characterized by decreased cell-cell adhesion through the decrease in E-cadherin protein production leading to enhanced migration and invasiveness [65].

Under normal conditions, TGF-β1 stimulates the proliferation of fibroblasts and induces the production of ECM protein as collagen in response to liver injury to repair the injured tissue. When liver injury is repeated, as occurs during persistent hepatitis virus infection (for example, HCV and HBV), the production of ECM continues and fibrous materials accumulate in the liver thus leading to cirrhosis that predisposes the liver to HCC as shown in Fig. 14 [66].

2.1.2 Wnt /β-catenin signaling pathway dysregulation

Alterations in Wnt signaling components have been described in HCC as up to 40% of HCCs exhibit accumulation of nuclear β-catenin [67]. Axin has been found to be mutationally inactivated in 3% to 14% of HCC cases [68] and the Fzd receptor is frequently overexpressed in HCC patients. These results suggest that more than one portion of the Wnt signaling pathway is dysregulated in HCC [69].

Studies implicate direct roles for hepatitis B virus and hepatitis C virus in modulating Wnt/ β-catenin signaling. Hepatitis C virus core protein correlates with increased Wnt ligand expression in HCC and genes inhibitory to Wnt/β-catenin signaling are preferentially methylated in hepatitis C virus related HCC [70]. Hepatitis B virus X protein is able to bind APC and displaces β-catenin from the destruction complex, resulting in activation of Wnt/ β-catenin signaling pathway [71].

2.1.3 Insulin like growth factor signaling pathway dysregulation

Dysregulation of IGF signaling in HCC occurs predominantly at the level of IGF-2 ligand bioavailability. IGF-2 ligand bioavailability is inversely dependent on IGF-2R expression level [43]. A reduced IGF-2R level is associated with less ligand-receptor binding and thus, a relative increase in local IGF-2 bioavailability. IGF-2R expression level is reduced in 63% of HCC patients and mutation at IGF-2R gene has been implicated as a pre-neoplastic lesion for HCC [72].

In chronic hepatitis B infection, HBx protein contributes to overexpression of IGF-2 gene [73] and in chronic hepatitis C infection HCV core protein also, induces overexpression of IGF-2 leading to induction of HCC development [74].

2.1.4 TLR signaling pathway dysregulation

There is evidence that the synergism between alcohol abuse with hepatitis C and hepatitis B infection in inducing hepatocellular carcinoma is due to activation of a common pathway which is Toll like receptor signaling causing induction of proinflammatory cytokine production by NF-κB activation [75].

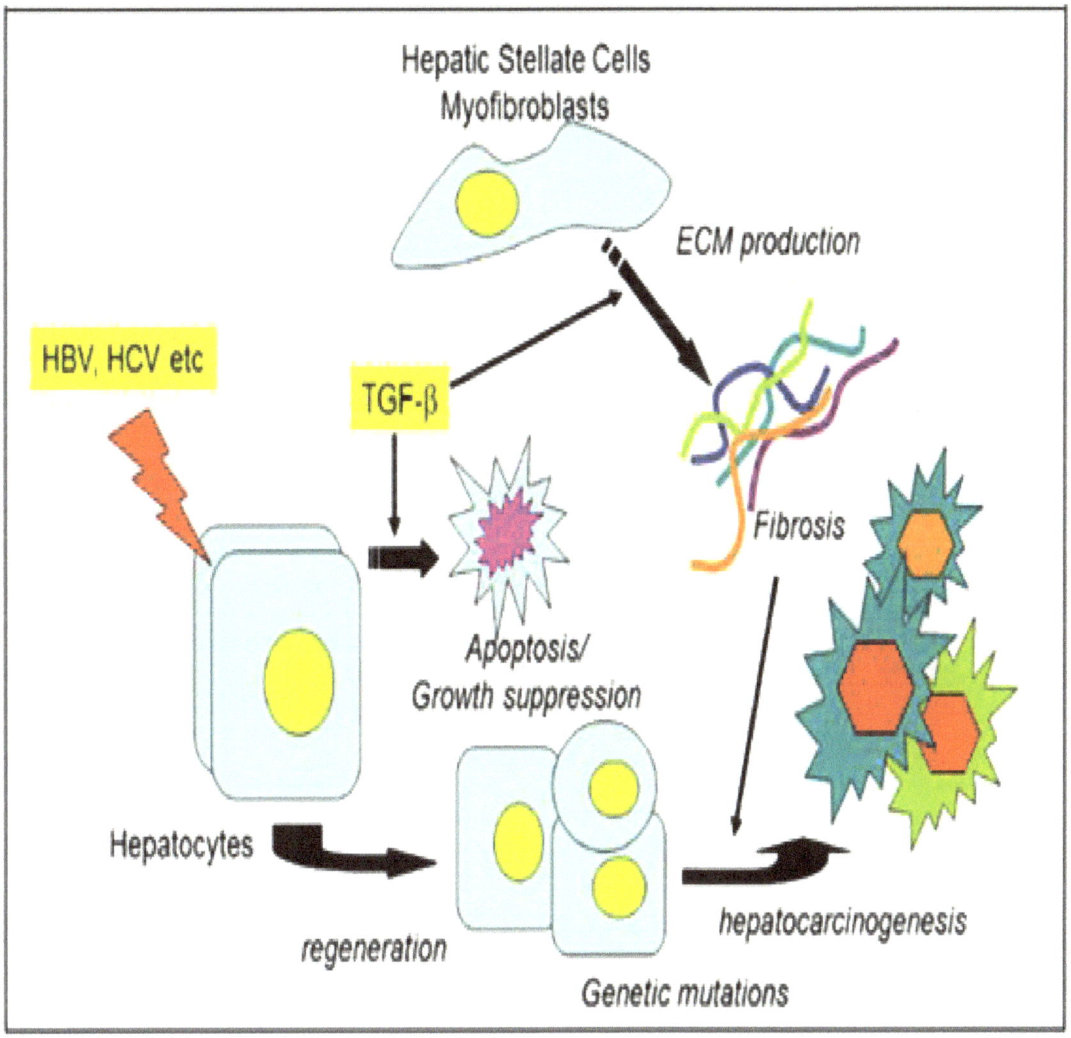

Fig. 14. Transforming growth factor-beta, fibrosis and HCC

In the earlier stages of the liver injury, TGF-β factor stimulates the proliferation of hepatocytes to restore the liver function or induces the apoptosis of injured hepatocytes to maintain tissue homeostasis. If the injurious agent persists TGF-β stimulates ECM protein production from the activated hepatic stallete cells and myofibroblasts. The persistent inflammation, the accumulated ECM proteins along with the genetic mutations lead to the development of HCC (Quoted from Ozaki et al. [66])

The HCV core and NS3 protein activate TLR2/TLR1 and TLR2/TLR6 leading to the production of inflammatory cytokines [28]. There is also marked upregulation of TLR2 and TLR4 detected in the hepatocytes and Kupffer cells in patients with chronic HCV. So hepatitis C virus infection contributes to the proinflammatory cytokine production by activation of TLR signaling pathway [76].

Alcohol ingestion decreases the intestinal mucosal barrier to LPS leading to the passage of the bacteria that has LP in its cell wall to the liver via the portal circulation. LPS leads to the activation of TLR4 on Kupffer cells causing pro-inflammatory cytokine production. Chronic alcohol consumption activates other TLRs, such as TLR1, 2 and 6-9 leading to further increase in the proinflammatory cytokines production [77].

2.1.5 Notch signaling pathway dysregulation

Notch has been shown to act as an oncogene in human cancer [78]. When acting as an oncogene Notch1 receptor and the signaling pathway are significantly upregulated which results in increased cellular proliferation, prevention of differentiation, decrease immune attack against the malignant cells and inhibition of apoptosis [36].

HCV infection causes induction of Notch signaling pathway by causing damage to the liver cell and the damaged liver cells activates Notch signaling pathway to induce liver cell proliferation [79].

2.1.6 Hepatocyte growth factor signaling pathway dysregulation

Patients in late stages of HCC had higher levels of serum HGF than healthy controls. So higher serum levels of HGF negatively correlate with patient survival time and positively correlate with tumor size [80].

c-MET receptor is overexpressed in HCC as compared with normal liver as c-MET is overexpressed in 20% to 48% of HCC patients [81]. This overexpression of c-MET could be partly due to growth factor dependent transcriptional activation of c-MET encoding gene [82].

Although, HGF has been shown to be overexpressed in the liver of HCC patients as compared to the normal liver but it is not expressed by tumor cells themselves [83]. Instead tumor cell products induce stellate cells and myofibroblasts to secrete HGF then HGF in turn stimulates tumor cell invasiveness [84].

2.1.7 Transforming growth factor-α signaling pathway dysregulation

TGF-α has been shown to be overexpressed in human HCC and TGF-α appears to act as potent mitogen for hepatocytes during the early stages of hepatocarcinogenesis and its level is correlated with tumor differentiation and proliferation [85]. It has been reported that elevated urinary level of TGF-α in patients suffering from HCC [86].

Interestingly, there are multiple links between chronic HBV infection and TGF-α signaling pathway. As the expression of TGF-α ligand in HCC patients correlate with the presence of viral polypeptides, HBs and HBc antigen, in the malignant liver cells and also, HBV-DNA induces TGF-α ligand expression [43].

2.1.8 Hedgehog signaling pathway dysregulation

The role of Hh pathway in the growth of tumors can be classified according to how the pathway is activated and these mechanisms include: Firstly, loss of function mutations in the pathway inhibitory proteins such as Ptch. Secondly, gain of function mutations in the pathway positive regulators such as Smo. Lastly: overexpression of the Hh ligands leading to autocrine or paracrine activation of the pathway [87].

It has been reported that there is increased expression of the transcription factor Gli1, Hh receptor and Smo in HCC tumor samples compared to normal liver tissue. It has been demonstrated that blocking the Hh pathway using cyclopomine, a steroid alkaloid that inhibits Hh signaling by binding to and inhibiting Smo action causes inhibition of cell proliferation, increases apoptosis and represses Hh induced genes such as cyclin D1 expression so Hh pathway blocker can be used for HCC treatment [88].

2.2 The Aim of Studying Dysregulated Signaling Pathways in HCC

Most HCC patients are diagnosed at a late stage so surgical treatment offers potential cure only for a minority of patients and therapeutic success of pharmacological approaches as chemotherapy is limited [89]. So numerous experimental strategies are aimed to trigger signaling pathway dysregulation at different levels for a wide range of curative non-surgical treatment for HCC [2].

Interestingly, functional studies revealed that dysregulation of some signaling pathways as Wnt, IGF, TGF-β and HGF signaling pathways may lead to HCC with specific clinical and biological features. Suggesting that receptor mediated signaling cascades may represent promising therapeutic targets [90].

Unfortunately, genetic heterogeneity of HCC severely complicates the development of effective and specific drugs which is reflected by the current lack of therapeutic options. For a small number of patients, partial hepatic resection or liver transplantation offers potential cure and only the multi kinase inhibitor such as sorafenib has been established as the first effective and approved systemic treatment for advanced HCC [91].

However, a number of clinical trials have indicated that the administration of antibodies and kinase inhibitors targeting different tumor relevant receptors are potential therapeutic agents but they did not further improve the situation. For this reason, the identification of different levels of regulation of different signaling pathways in HCC would help to direct drug development and are of high relevance for improvement of the current dismal and unsatisfactory situation [92] as shown in Fig. 15.

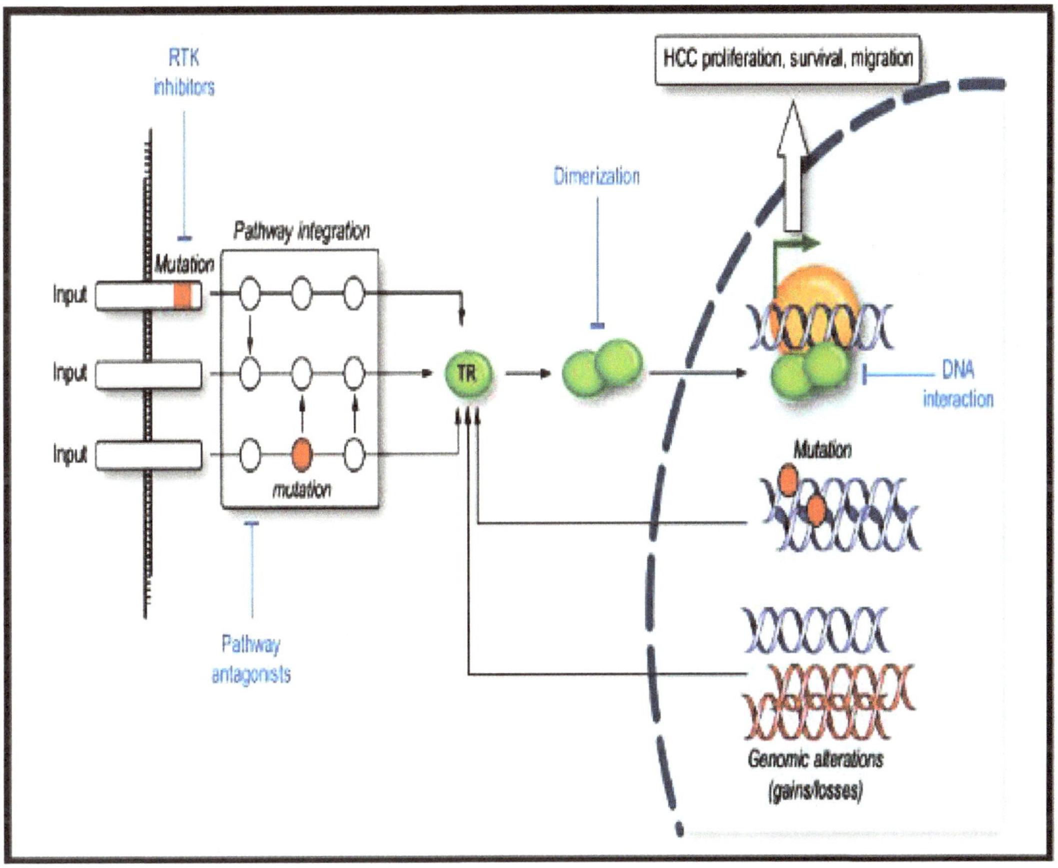

Fig. 15. Levels of dysregulation at signaling pathways and possibilities of interference in human HCC cells

The Levels of dysregulation at signaling pathways (in red) include: a- Mutations of upstream signaling pathway components (e.g., receptors and pathway constituents). b- Mutations and genomic alterations in genes coding for transcriptional regulators (TR). Possibilities of pathway interference (in blue) include the inhibition of receptor tyrosine kinase inhibitors, expression of pathway antagonists, inhibitors of TR-dimerization and inhibitors of TR/DNA interaction (Quoted from Malz et al. [89])

Some methodological approaches utilize small organic molecules that specifically inhibit the interaction between transcription factors and their respective DNA binding domains such as β-catenin/TCF interaction [93]. Many of these approaches have already been tested in HCC models showing promising preclinical anti tumorigenic effects [94].

In addition, several transcription factors form dimmers or multimers in order to recognize and bind their respective DNA binding sites as p53 protein. Disturbing these protein/protein interactions by specific substances may destabilize transcriptional complex assembly and thereby inhibit dysfunctional transcriptional activity [89].

Another possible approach is reconstitution of some tumor suppressor genes product such as p53 activity: A- If the cause of the defect is mutation at TP53 gene, small molecules that bind to the mutant p53 protein such as PRIMA-1 that restore its tumor suppressive function can be used [95]. B- When the cause of the defect is not TP53 mutation but is MDM2 overexpression, p53 reconstituted by substances that block MDM2 dependent degradation of p53 protein can be explored [96].

3. CONCLUSION

Hepatocellular carcinoma is the most common primary cancer of the liver and the incidence is increasing as HCC has risen to become the fifth commonest malignancy worldwide and the third leading cause of cancer related death. In addition, HCC carries a very bad prognosis as the 5-year survival rate is less than 5% in all HCC cases.

Hepatocarcinogenesis is a multistep process and this process may be induced by any of several risk factors that predisposes to HCC. Hepatocarcinogenesis may be mediated by multiple mechanisms but eventually all the mechanisms of HCC pathogenesis can create conditions that increases the chance of generating hepatocytes population containing critical combinations of structurally and functionally aberrant genes.

There are multiple risk factors for HCC and these risk factors can be classified into 1- Environmental risk factors such as chronic hepatitis B and hepatitis C virus 2- Metabolic risk factors such as diabetes mellitus and non-alcoholic fatty liver disease 3- Hereditary risk factors such as haemochromatosis and others.

Some of the HCC risk factors could lead to dysregulation of important molecular cellular signaling pathways. This dysregulation may lead to malignant transformation by inducing alteration in many important genes – for example, genes coding for proteins that control the cell cycle, cell growth, inflammatory response and others.

There are multiple regulatory pathways dysregulated in HCC, for example: TGF-β signaling pathway, Wnt/β-catenin signaling pathway, Toll like receptor signaling pathway, Notch signaling pathway, IGF signaling pathway, HGF signaling pathway and others.

The curative options for early stages of HCC are surgical resection followed by liver transplantation. However only 12% of patients are eligible for surgical intervention because most HCC patients are diagnosed at a late stage. So numerous experimental strategies are aiming to trigger the dysregulated signaling pathways in HCC for early diagnosis and a wide range of curative non surgical treatments.

Finally, it could be stated in concluding remarks that:

1) There are multiple mechanisms underlying the hepatocarcinogenesis and by understanding these mechanisms, early diagnosis for HCC could be possible.
2) Understanding the signaling pathways dysregulated in high risk groups for HCC may provide new diagnostic and therapeutic modalities.

COMPETING INTERESTS

Authors have declared that no competing interests exist.

REFERENCES

1. Ierardi E, Rosania R, Zotti M, Giorgio F, Prencipe S, Nicola Della Valle N, De Francesco V, Panella C. From chronic liver disorders to hepatocellular carcinoma: Molecular and genetic pathways. World J Gastrointest Oncol. 2010;2(6):259-264.
2. Kitisin K, Pishvaian M, Johnson L, Mishra L. Liver stem cells and molecular signaling pathways in hepatocellular carcinoma. Gastrointest Cancer Res. 2007;1(4):13-21.

3. Hopfner M, Schuppan D, Scherubl H. Growth factor receptors and related signalling pathways as targets for novel treatment strategies of hepatocellular cancer. World J Gastroenterol. 2008;14(1): 1-14.
4. Senturk S, Mumcuoglu M, Gursoy-Yuzugullu O, Cingoz B, Akcali K, Ozturk M. Transforming growth factor-beta induces senescence in hepatocellular carcinoma cells and inhibits tumor growth. Hepatology. 2010;52:966-974.
5. Giannelli G, Mazzocca A, Fransvea E, Lahn M, Antonaci S. Inhibiting TGF-β signaling in hepatocellular carcinoma. Biochem Bio phys Acta. 2011;1815:214-223.
6. Derynck R, Akhurst R, Balmain A. TGF-beta signaling in tumor suppression and cancer progression. Nat Genet. 2001;29: 117-129.
7. Liu X, Sun Y, Weinberg R, Lodish H. Ski/Sno and TGF-beta signaling. Cytokine Growth Factor Rev. 2001;12:1-8.
8. Kitisin K, Pishvaian M, Johnson L, Mishra L. Liver stem cells and molecular signaling pathways in hepatocellular carcinoma. Gastrointest Cancer Res. 2007;1(4):13-21.
9. Mishra B, Tang Y, Katuri V, Fleury T, Said AH, Rashid A, Jogunoori W, Mishra L. Loss of cooperative function of transforming growth factor beta signaling proteins, smad3 with embryonic liver fodrin, a beta-spectrin, in primary biliary cirrhosis. Liver Int. 2004;24:637-645.
10. Chien A, Conrad W, Moon R. A Wnt survival guide: From flies to human disease. J Invest Dermatol. 2009;129: 1614-1627.
11. Moon R, Shah K. Developmental biology: signaling polarity. Nature. 2002;417:239-240.
12. Kimelman D, Xu W. Beta-catenin destruction complex: Insights and questions from a structural perspective. Oncogene. 2006;25:7482-7491.
13. White B, Chien A, Dawson D. Dysregulation of Wnt/β-Catenin Signaling in Gastrointestinal Cancers. Gastroenterology. 2012;142:219-232.
14. Giles R, Van J, Clevers H. Caught up in a Wnt storm: Wnt signaling in cancer. Biochim Biophys Acta. 2003;1653:1-24.
15. Pollak M. Insulin and insulin-like growth factor signalling in neoplasia. Nature Rev Cancer. 2008;8:915-928.
16. Pollak M, Schernhammer E, Hankinson S. Insulin-like growth factors and neoplasia. Nature Rev Cancer. 2004;4:505-518.
17. Wu J, Zhu A. Targeting insulin-like growth factor axis in hepatocellular carcinoma. Journal of Hematology and Oncology. 2011;4(30):1-11.
18. Furstenberger G, Senn H. Insulin-like growth factors and cancer. Lancet Oncol. 2002;3:298-302.
19. LeRoith D, Roberts C. The insulin-like growth factor system and cancer. Cancer Lett. 2003;195(2):127-137.
20. Foulstone E, Prince S, Zaccheo O, Burns J, Harper J, Jacobs C, Church D, Hassan A. Insulin-like growth factor ligands, receptors and binding proteins in cancer. J Pathol. 2005;205:145-153.
21. Takeuchi O, Akira S. Toll-like receptors; their physiological role and signal transduction system. International Immunopharmacology. 2001;1(4):625-635.
22. Takeda K, Akira S. TLR signaling pathways. Seminars in Immunology. 2004; 16:3-9.
23. Dejardin E, Droin N, Delhase M, Haas E, Cao Y, Makris C, Li Z, Karin M, Ware C, Green D. The lymphotoxin-β receptor induces different patterns of gene expression via two NF-κB pathways. Immunity. 2002;17(4):525-535.
24. Akira S, Uematsu S, Takeuchi O. Pathogen recognition and innate immunity. Cell. 2006;124(4):783-801.
25. Meylan E, Tschopp J, Karin M. Intracellular pattern recognition receptors in the host response. Nature. 2006;442:39-44.
26. Deng L, Wang C, Spencer E, Yang L, Braun A, You J, Slaughter C, Pickart C, Chen Z. Activation of the I kappa B kinase complex by TRAF6 requires a dimeric ubiquitin-conjugating enzyme complex and a unique polyubiquitin chain. Cell. 2000; 103:351-361.
27. Wang C, Deng L, Hong M, Akkaraju G, Inoue J, Chen Z. TAK1 is a ubiquitin-dependent kinase of MKK and IKK. Nature. 2001;412:346-351.
28. Testro A, Visvanathan K. Toll-like receptors and their role in gastrointestinal disease. J Gastroenterol Hepatol. 2009;24: 943-954.
29. Liew F, Xu D, Brint E, O'Neill L. Negative regulation of toll-like receptor-mediated immune responses. Nature reviews. 2005;5:446-458.

30. Miele L. Notch signaling. Clin Cancer Res. 2005;12:1074-1079.
31. Bray S. Notch signaling: A simple pathway becomes complex. Nat. Rev. Mol. Cell. Biol. 2006;7:678-689.
32. Kopan R, Ilagan M. The canonical Notch signaling pathway: Unfolding the activation mechanism. Cell. 2009;137:216-233.
33. Radtke F, Fasnacht N, Robson H. Notch signaling in the immune system. Immunity. 2010;32:14-27.
34. Gordon W, Arnett K. Blacklow S. The molecular logic of Notch signaling: A structural and biochemical perspective. J. Cell. Sci. 2008;121:3109-3119.
35. Palomero T, Lim W, Odom D, Sulis M, Real P, Margolin et al. NOTCH1 directly regulates c-MYC and activates a feed-forward-loop transcriptional network promoting leukemic cell growth. Proc. Natl. Acad. Sci. USA. 2006;103(48):18261-18266.
36. Weng A, Millholland J, Yashiro-Ohtani Y, Arcangeli M, Lau A, Wai C. et al. c-Myc is an important direct target of Notch1 in T-cell acute lymphoblastic leukemia/ lymphoma. Genes Dev. 2006;20(15):2096-2109.
37. Kawamata S, Du C, Li K, Lavau C. Overexpression of the Notch target genes Hes in vivo induces lymphoid and myeloid alterations. Oncogene. 2002;21:3855-3863.
38. Borggrefe T, Oswald F. The Notch signaling pathway: Transcriptional regulation at Notch target genes. Cell Mol Life Sci. 2009;66:1631-1646.
39. Poellinger L, Lendahl U. Modulating Notch signaling by pathway intrinsic and pathway extrinsic mechanisms. Curr. Opin. Genet. Dev. 2008;18:449-454.
40. Efimova E, Glanemann M, Liu L, Schumacher G, Settmacher U, Jonas S. et al. Effects of human hepatocyte growth factor on the proliferation of human hepatocytes and hepatocellular carcinoma cell lines. Eur Surg Res. 2004;36(5):300-307.
41. Paumelle R, Tulasne D, Kherrouche Z, Plaza S, Leroy C, Reveneau S, et al. Hepatocyte growth factor /scatter factor activates the ETS1 transcription factor by a RAS-RAF-MEK-ERK signaling pathway. Oncogene. 2002;21(15):2309-2319.
42. Sunnarborg S, Hinkle C, Stevenson M, Russell W, Raska C, Peschon J, et al. Tumor necrosis factor-alpha converting enzyme (TACE) regulates epidermal growth factor receptor ligand availability. J. Biol. Chem. 2002;277(15):12838-12845.
43. Breuhahn K, Longerich T, Schirmacher P. Dysregulation of growth factor signaling in human hepatocellular carcinoma. Oncogene. 2006;25:3787-3800.
44. Guo L, Kozlosky C, Ericsson L, Daniel T, Cerretti D, Johnson R. Studies of ligand-induced site-specific phosphorylation of epidermal growth factor receptor. J Am Soc Mass Spectrom. 2003;14:1022-1031.
45. Zhang T, Ma J, Cao X. Grb2 regulates Stat3 activation negatively in epidermal growth factor signalling. Biochem J. 2003; 376(2):457-464.
46. Chang L, Karin M. Mammalian MAP kinase signalling cascades. Nature. 2001;410:37-40.
47. Rubin L, de Sauvage F. Targeting the Hedgehog pathway in cancer. Nat Rev Drug Discov. 2006;5:1026-1033.
48. Lum L, Beachy P. The Hedgehog response network: Sensors, switches and routers. Science. 2004;304:1755-1759.
49. Parkin C, Ingham P. The Adventures of Sonic Hedgehog in Development and Repair. I. Hedgehog signaling in gastrointestinal development and disease. AJP Gastrointest Liver Physiol. 2008;294: 363-367.
50. Lees C, Howie S, Sartor R, Satsangi J. The hedgehog signaling pathway in the gastrointestinal tract: Implications for development, homoestasis and disease. Gastroenterology. 2005;129:1696-1710.
51. Ghosh S, Karin M. Missing pieces in the NF-κB puzzle. Cell. 2002;109(2):81-96.
52. Maeda S. NF-κB, JNK, and TLR signaling pathways in hepatocarcinogenesis. Gastroenterol Res Pract. 2010;2010:3-9.
53. Wagner E, Nebreda A. Signal integration by JNK and p38 MAPK pathways in cancer development. Cancer. 2009;9:537-549.
54. Avruch J. MAP kinase pathways: The first twenty years. Biochim Biophys Acta. 2007; 1773:1150-1160.
55. Bromberg J, Darnell J. The role of STATs in transcriptional control and their impact on cellular function. Oncogene. 2000;19: 2468-2473.
56. Bowman T, Garcia R, Turkson J, Jove R. STATs in oncogenesis. Oncogene. 2000; 19:2474-2488.
57. Derynck R, Zhang Y. Smad-dependent and Smad-independent pathways in TGF-

beta family signaling. Nature. 2003;425: 577-584.
58. Shitashige M, Hirohashi S, Yamada T. Wnt signaling inside the nucleus. Cancer Sci. 2008;99:631-637.
59. Thompson M, Monga S. WNT/β-Catenin signaling in liver health and disease. Hepatology. 2007;45(5):1298-1305.
60. Miyoshi A, Kitajima Y, Sumi K, Sato K, Hagiwara A, Koga Y, Miyazaki K. Snail and SIP1 increase cancer invasion by upregulating MMP family in hepatocellular carcinoma cells. Br J Cancer. 2004;90: 1265-1273.
61. Xu Z, Shen M, Ma D, Wang L, Zha X. TGF-beta promoted epithelial-to-mesenchymal transformation and cell adhesion contribute to TGF-beta1-enhanced cell migration in SMMC-7721 cells. Cell Res. 2003;13(5):343-350.
62. Kwon C, Cheng P, King I, Andersen P, Shenje L, Nigam V, Srivastava D. Notch post-translationally regulates β-catenin protein in stem and progenitor cells. Nat. Cell. Biol. 2011;13:1-9.
63. Park Y, Chae K, Oh B, Choi J, Choi K, Park C. Expression of Smad7 in hepatocellular carcinoma and dysplastic nodules: Resistance mechanism to transforming growth factor-beta. Hepato-gastroenterology. 2004;51:396-400.
64. Kim H, Chung Y, Song B, Kim J, Yang S, Lee Y, Suh D. Expression of transforming growth factor beta-1 in chronic hepatitis and hepatocellular carcinoma associated with hepatitis C virus infection. Korean J Intern Med. 2000;15:165-170.
65. Giannelli G, Mazzocca A, Fransvea E, Lahn M, Antonaci S. Inhibiting TGF-β signaling in hepatocellular carcinoma. Biochem Bio phys Acta. 2011;1815:214-223.
66. Ozaki I, Hamajima H, Matsuhashi S, Mizuta T. Regulation of TGF-β1 induced proapoptotic signaling by growth factor receptors and extracellular matrix receptor integrins in the liver. Frontier in physiology. 2011;2:1-8.
67. Yau T, Chan C, Chan K, Lee M, Wong C, Fan S, et al. HDPR1, a novel inhibitor of the WNT/beta-catenin signaling, is frequently downregulated in hepatocellular carcinoma: involvement of methylation-mediated gene silencing. Oncogene. 2005; 24:1607-1614.
68. Taniguchi K, Roberts L, Aderca I, Dong X, Qian C, Murphy L, et al. Mutational spectrum of beta-catenin, AXIN1, and AXIN2 in hepatocellular carcinomas and hepatoblastomas. Oncogene. 2002;21: 4863-4871.
69. Merle P, de la Monte S, Kim M, Herrmann M, Tanaka S, Von Dem Bussche A, et al. Functional consequences of frizzled-7 receptor overexpression in human hepatocellular carcinoma. Gastroenterology. 2004;127:1110-1122.
70. Deng Y, Nagae G, Midorikawa Y, Yagi K, Tsutsumi S, Yamamoto S, et al. Identification of genes preferentially methylated in hepatitis C virus related hepatocellular carcinoma. Cancer Sci. 2010;101:1501-1510.
71. Hsieh A, Kim H, Lim S, Yu D, Jung G. Hepatitis B viral X protein interacts with tumor suppressor adenomatous polyposis coli to activate Wnt/β-catenin signaling. Cancer Lett. 2011;300:162-172.
72. Oka Y, Waterland R, Killian J, Nolan C, Jang H, Tohara K, et al. M6P/IGF2R tumor suppressor gene mutated in hepatocellular carcinomas in Japan. Hepatology. 2002; 35:1153-1163.
73. Kang-Park S, Lee J, Shin J, Lee Y. Activation of the IGF-II gene by HBV-X protein requires PKC and p44/p42 map kinase signalings. Biochem Biophys Res Commun. 2001;283:303-307.
74. Lee S, Park U, Lee Y. Hepatitis C virus core protein transactivates insulin-like growth factor II gene transcription through acting concurrently on Egr1 and Sp1 sites. Virology. 2001;283:167-177.
75. Machida K, Tsukamoto H, Mkrtchyan H, Duan L, Dynnyk A, Liu H, et al. Toll-like receptor 4 mediates synergism between alcohol and HCV in hepatic oncogenesis involving stem cell marker Nanog. Proc Natl Acad Sci. 2009;106:1548-1553.
76. Szabo G, Dolganiuc A, Mandrekar P. Pattern recognition receptors: A contemporary view on liver diseases. Hepatology. 2006;44:287-298.
77. Mandrekar P, Bala S, Catalano D, Kodys K, Szabo G. The opposite effects of acute and chronic alcohol on lipopolysaccharide induced inflammation are linked to IRAK-M in human monocytes. J Immunol. 2009; 183:1320-1327.
78. Maier M, Gessler M. Comparative analysis of the human and mouse Hey1 promoter: Hey genes are new Notch target genes. Biochem. Biophys. Res. Commun. 2000; 275:652-660.

79. Dotto G. Notch tumor suppressor function. Oncogene. 2008;27:5115-5123.
80. Vejchapipat P, Tangkijvanich P, Theamboonlers A, Chongsrisawat V, Chittmittrapap S, Poovorawan Y. Association between serum hepatocyte growth factor and survival in untreated hepatocellular carcinoma. J Gastroenterol. 2004;39(12):1182-1188.
81. Moinzadeh P, Breuhahn K, Stutzer H, Schirmacher P. Chromosome alterations in human hepatocellular carcinomas correlate with aetiology and histological-grade-results of an explorative CGH meta-analysis. Br J Cancer. 2005;92(5):935-941.
82. Pennacchietti S, Michieli P, Galluzzo M, Mazzone M, Giordano S, Comoglio P. Hypoxia promotes invasive growth by transcriptional activation of the met protooncogene. Cancer Cell. 2003;3(4): 347-361.
83. Stuart K, Riordan S, Lidder S, Crostella L, Williams R, Skouteris G. Hepatocyte growth factor/scatter factor-induced intracellular signalling. Int J Exp Pathol. 2000;81(1):17-30.
84. Guirouilh J, Le Bail B, Boussarie L, Balabaud C, Bioulac-Sage P, Desmoulière A, et al. Expression of hepatocyte growth factor in human hepatocellular carcinoma. J Hepatol. 2001;34(1):78-83.
85. Zhang J, Wang W, Li Q, Qiao Q. Expression of transforming growth factor-alpha and hepatitis B surface antigen in human hepatocellular carcinoma tissues and its significance. World J Gastroenterol. 2004;10(6):830-833.
86. Bruix J, Hessheimer A, Forner A, Boix L, Vilana R, Llovet J. New aspects of diagnosis and therapy of hepatocellular carcinoma. Oncogene. 2006;25:3848-3856.
87. Barakat M, Humke E, Scott M. Learning from Jekyll to control Hyde: Hedgehog signaling in development and cancer. Trends in Molecular Medicine. 2010; 30(10):1-12.
88. Patil M, Zhang J, Ho C, Cheung S, Fan S, Chen X. Hedgehog signaling in human hepatocellular carcinoma. Cancer Biology and Therapy. 2006;5(1):111-117.
89. Malz M, Pinna F, Schirmacher P, Breuhahn K. Transcriptional regulators in hepatocarcinogenesis - Key integrators of malignant transformation. J Hepatol. 2012; 57:186-195.
90. Hoshida Y, Nijman S, Kobayashi M, Chan J, Brunet J, Chiang D, et al. Integrative transcriptome analysis reveals common molecular subclasses of human hepatocellular carcinoma. Cancer Res. 2009;69:7385-7392.
91. Llovet J, Ricci S, Mazzaferro V, Hilgard P, Gane E, Blanc J, et al. Sorafenib in advanced hepatocellular carcinoma. N Engl J Med. 2008;359:378-390.
92. Villanueva A, Minguez B, Forner A, Reig M, Llovet J. Hepatocellular carcinoma: Novel molecular approaches for diagnosis, prognosis and therapy. Annu Rev Med. 2010;61:317-328.
93. Lepourcelet M, Chen Y, France D, Wang H, Crews P, Petersen F, et al. Small-molecule antagonists of the oncogenic Tcf/beta-catenin protein complex. Cancer Cell. 2004;5:91-102.
94. Shi H, Lambert J, Hautefeuille A, Bykov V, Wiman KG, Hainaut P, Caron de Fromentel C. *In vitro* and *in vivo* cytotoxic effects of PRIMA-1 on hepatocellular carcinoma cells expressing mutant p53ser249. Carcinogenesis. 2008;29: 1428-1434.
95. Essmann F, Schulze-Osthoff K. Translational approaches targeting the p53 pathway for anticancer therapy. Br J Pharmacol. 2012;165:328-344.
96. Shangary S, Qin D, McEachern D, Liu M, Miller RS, Qiu S, et al. Temporal activation of p53 by a specific MDM2 inhibitor is selectively toxic to tumors and leads to complete tumor growth inhibition. Proc Natl Acad Sci USA. 2008;105:3933-3938.

8

Biochemical Evaluation of the Antioxidant Nutrients on Blood Rheology among Prediabetics

Abousree Taha Ellethy[1*]

[1]Department of Oral and Medical Basic Sciences, Biochemistry Division, College of Dentistry, Qassim University, AL-Qassim, Saudi Arabia.

Author's contribution

The sole author designed, analyzed and interpreted and prepared the manuscript.

Editor(s):
(1) Carmen Lucia de Oliveira Petkowicz, Federal University of Parana, Curitiba, Parana, Brazil.
Reviewers:
(1) Boguslaw Lipinski, Joslin Diabets Center, Boston, USA.
(2) Jevas Chibuike Ozougwu, Rhema University, Nigeria.

ABSTRACT

Background: Oxidative stress is involved in the developing of several diseases including diabetes followed by the majority of severe acute illness and/or contra complications. Current studies indicated increasing plasma viscosity with progressing microangiopathy among diabetic cases.
Objectives: To evaluate whether antioxidant nutrients biomarkers of selenium, vitamin A and vitamin E are associated with alterations of blood viscosity in pre-diabetes (PD).
Materials and Methods: Whole blood samples were collected from (20) PD non obese subjects with blood HbA c1 (5.7-6.4%) and (20) control healthy subjects. Different biomolecules were assessed including selenium concentration in both plasma and red blood cells (RBCs), glutathione peroxidase activity in RBCs, plasma lipids, vitamin A and vitamin E plasma concentration, and the blood viscosity.
Results: Comparing to control healthy subjects, blood viscosity was significantly increased, selenium was not altered in the plasma of PD, subjects but it was markedly decreased in RBCs and negative correlation with the elastic component of whole blood viscosity. Mean RBCs, glutathione activity was reduced. Vitamin A and vitamin E plasma levels were correlated with blood viscosity. Plasma viscosity correlated strongly with increased cholesterol level (r=0.67, P 0.002) and triglycerides concentration (r=0.66 , P 0.005).

**Corresponding author: E-mail: abousree.ellethy@yahoo.com*

Conclusion: Blood viscosity increased in PD and this may be contributed to reduced nutritional antioxidants.

Keywords: Prediabetes; selenium; glutathione peroxidase; vitamin E; vitamin A; blood viscosity.

ABBREVIATIONS

(PD) pre-diabetes; (RBCs) red blood cells; (OS) oxidative stress; (GPx) glutathione peroxidases; (HbA1c) glycosylated hemoglobin; (WBV) whole blood viscosity .

1. INTRODUCTION

By the end of 2030, diabetes mellitus will be the seventh leading cause of death [1]. The PD is a triggering stage between normal health and diabetes in which fasting plasma glucose is increased with impaired glucose tolerance. PD populations are willing to get diabetes mellitus within 10 years or less [2].

Current researchers investigated increasing oxidative stress (OS) among hypoglycemic populations. In diabetes mellitus, OS plays a part acute endothelial dysfunction and activation of inflammation in the blood vessels of diabetic patients. It takes part in the development of different microvascular complications [3] but there are no data for early stages of the disease. In turn, it causes overproduction superoxides in mitochondria within endothelial cells of different blood vessels. It increases the biosynthesis of glycated end products, expression of their receptors, activation of protein kinase C and the overabundance of hexosamine pathways [4]. The OS stimulates intravascular blood coagulation in diabetes-associated atherosclerotic complications by increasing fibrin deposits that in turn proliferates plasma viscosity [5,6]. It passes reactive oxygen derivatives in small amounts that possess free radical properties mediated lipid peroxidation with increasing antioxidant enzyme levels [7]. It takes part in decreasing plasma levels of vitamins A, C and E that in turn points decreasing antioxidative barrier in patients with different metabolic syndromes [8].

Increasing reactive oxygen products affect plasma antioxidant activity and disturbed oxidation-reduction balance and lead to severe alterations of cellular function with subsequent cell damage [9]. Mainly, glutathione peroxidases (GPx) play an antioxidant to protect of erythrocyte against peroxides from oxidative damages by increasing the level of reduced glutathione. GPx is involved in the detoxification of hydrogen and lipid peroxides. Low levels of GPx have been correlated with free radical related disorders that in turn damage DNA, proteins, and lipids of the cell. Accordingly, it leads to cell dysfunction in different tissues, causing pathogenesis of vascular complications [10,11]. As the deficiency of antioxidants may increase the risk of late complications in diabetics [8]. Nutrients with the antioxidant action, such as selenium, vitamin A and vitamin E have been suggested to exist protective effects in diabetes mellitus.

Diabetes, OS obesity, inflammation, aging, smoking and cancer are important factors for getting deterioration of blood rheology [12]. Mainly, impairment of blood rheological parameters affects both viscosity and RBCs elasticity which subsequently diminishes both blood flow rate and tissue oxygenation. In addition, RBCs elasticity reflects their blood aggregability and deformability [13,14]. Little information about blood rheology in the early stages before getting diabetes and/or PD cases are known. Consequently, monitoring of blood rheology is urgently useful for those PD populations.

Therefore, the main aim of the present study was to examine the state of antioxidant nutrients like selenium together with selenium-dependent GSH, vitamin A, and vitamin E in PD cases and to establish whether they are associated with alterations of blood viscosity.

2. MATERIALS AND METHODS

Whole venous blood samples were collected from 20 PD. All were none obese (BMI < 30) and none smokers. In addition, 20 healthy persons matched for sex, age, and body mass index was considered as (control group). All participants were recruited from the outpatient clinic of the Suez Canal University Hospital for the study. PD was considered with HbA1c (5.7-6.5)%. Both PD and control groups were subjected to the same laboratory measurements. Applicants were asked to fast for at least 8hs earlier morning blood collection. Aliquots of all samples were centrifuged at 3000 rpm then collecting plasma to keep at -70C till biochemical analysis. The study protocol was confirmed to the ethical guidelines of the declaration of Helsinki (Sixth revision, 2008). The laboratory investigations were done for the followings:

2.1 Glycosylated Hemoglobin (HbA1c)

Blood glucose and plasma lipids were measured via enzyme immunoassay technique (ELISA). Fasting serum glucose was measured by the glucose oxidase method (Roche Diagnostics GmbH, Germany) [15]. Glucose was measured by the glucose oxidase method using the YSI analyzer (Yellow Springs, OH, USA). HbA1c was measured by HPLC (Biorad, Hercules, CA, USA). Numerous of biochemical measurements were performed on fasting plasma collected. Plasma fasting glucose, HDL, cholesterol and triglycerides levels were measured by marketable test kit methods using the automatic clinical chemistry analyzer (Architect CI8200, Abbott Diagnostics.

2.2 Selenium

Selenium concentrations in plasma and RBCs were measured with hydride generation atomic absorption spectrophotometry (Perkin-Elmer 3030, Perkin-Elmer Corp., Norwalk, CT) [16]. Mainly, the whole blood –anticoagulant with EDTA-was centrifuged, then plasma and buffy coat were removed. The RBCs washed three times with isotonic saline solution; 1 ml of packed RBCs. Plasma and/or RBCs samples were diluted 1:4 with a triton-X (Sigma Chemical, St. Louis, MO) and nitric acid solution (Fisher Scientific, Pittsburgh, PA) and incubated at room temperature for 10 min, then adding 70% perchloric acid with boiling 3 mins, then adding v/v 20 mmol/l HC1 using HPLC deionized water then treated with 30% sodium borohydride, and 15% sodium hydroxide reagent. Treated plasma and RBCs Se levels absorbance were measured at 196 nm.

2.3 Glutathione Peroxidase Activity

Glutathione peroxidase (GPx) activities in RBCs were measured enzymatically based on Pagalia and Valantine method [17] using GPx assay (RANSEL) assay kit, Randox Laboratories Ltd, Crumlin, BT29 4QY, UK. Packed washed cells were lysed with 1 ml of demineralized water, diluted 1:1 with double strength Drapkin reagent and used for both the determination of GPx activity and hemoglobin. Samples were frozen at – 40 C until determination of GPx activity. The reaction solution was: 0.2 mM NADPH, 4 mM GSH, 4 mM EDTA, 4 mM NaN3, 1 I.U glutathione reductase, H202, and the sample in 100 mM phosphate buffer PH 7.0. The oxidation of NADPH was followed at 37 Co using a 340 nm filter with a spectrophotometer (Hitachi U –2000). Enzymatic activity is expressed as ug Hb/L. Hb in the RBCs lysate was measured by the cyanmethemoglobin method [18]. The coefficient of variation (between-run) for GPx activity ranged from 2.2% to 7.5%. The reproducibility was 6.5%.

2.4 Antioxidative Vitamins

Plasma concentrations of vitamin A (all-trans-retinol) and vitamin E (a –tocopherol) were determined by using high-performance liquid chromatography [19]. The coefficient of variation of the method was 2.5% and 2.2% for retinol and a-tocopherol, respectively.

2.5 Whole Blood Viscosity

Whole blood viscosity (WBC) was determined in EDTA anticoagulant blood with a capillary viscometer (OCR-O Anton Paar Cokg, Graz, Austria) at a constant oscillation frequency of 2 Hz for shear rates between 1 S to 100 S. Blood samples were kept in the rotation until estimation to avoid sedimentation. Experiments were done within 1 – 2 h after blood collection at a constant temperature of 37 + 0.1 Co with blood samples adjusted to the hematocrit of 45 + 1% [20].

2.6 Statistical Analysis

Data are presented in terms of mean, standard deviation (SD) of the mean, and percentages. Statistical analysis was carried out by a computer program (SPSS Ver. 20). Student t-test, Chi-square, the Mann-Whitney U test, and correlation tests were used to evaluate the results. Data are presented as mean + SD. Statistical significance in figures is denoted with P value less 0.05.

2.7 Ethical Consideration

Before the initiation of the study, informed consent was obtained from all individuals selected for the study. The aim and the value of the work were explained to them in a simplified manner. There was no harm being inflicted on them. On the contrary, all would have the benefits of follow-up and the results of the study.

3. RESULTS

Twenty patients with PD, 60% were males with average age 46.7±5.2 years old, were studied. Glycated hemoglobin (Hb A1c) was 6.1+1.1%. No significant difference between patients and control group regarding the age and sex distribution.

Table 1 shows that both the patient and the control groups were well matched regarding age, sex, body mass index and blood pressure. There was a significant difference as regards the Hb A1c level between the both groups. Regarding the lipid profile, there was a significant difference between the patient and control groups as regard total cholesterol level and triglycerides level. Regarding the blood viscosity, there was a significantly increased in the patient group (4.23+0.68 vs 3.51+0.6, P< 0.001).

Table 2 shows that the mean plasma selenium concentration in the patient group was not different from that in the control groups. In contrast, the concentration of selenium in RBCs of the patient group was significantly reduced compared with control groups. Selenium-dependent glutathione peroxidase activity (GPx) was slightly but significantly reduced in the patient group. Vitamin A and E concentrations differed slightly, but significantly between the groups.

In the Table 3, significant correlations were found between whole blood viscosity and RBCs selenium, vitamin A, vitamin E and GPx values and in addition to both hematocrit and HbA1c. On the other hand, no correlations were recorded with lipids plasma selenium.

4. DISCUSSION

Intensive organization for PD are required based on relevant evidence to help physicians in selecting the best possible management strategies for PD, especially those suffering from specific conditions. The PD is monitored by the level of hyperglycemia and associated with impaired fasting glucose and/or impaired glucose tolerance. Increases in HbA1c levels are related to an increasing evidence point to a progression of PD to frank diabetes [21,22].

Recent research studies in this area of PD support a central role for reactive oxygen species in both the formation of AGEs and in AGE-induced pathologic alterations in gene expression. AGEs alter matrix-matrix, matrix cell, and cell-cell interaction with subsequent changes in plasma viscosity and severity of PD [23].

Table 1. Clinical and biochemical PD of both patient and control groups

Pattern	PD (n=20)	Control (n=20)	*P*
Age (years)	46.7±5.2	41.5±7.4	n.s
Sex F/M	8/12	9/11	n.s
Body mass index (BMI) %	24.2±0.6	23.5±0.8	n.s
Systolic B.P mm Hg	122.4±13.4	121.9±6.7	n.s
Diastolic B.P mm Hg	81.4±4.2	79.9±4.7	n.s
Hb A_{1c} %	6.1±1.1	5.1±0.1	<0.001
Plasma protein gm %	7.1±0.4	7.5±0.4	n.s
Haematocrit %	44.8±3.7	45.3±2.4	n.s
Total cholesterol mg %	216.7±12.5	198.4±16.5	<0.01
Triglyceride mg %	185.4±18.9	145.8±17.9	<0.01
Whole blood viscosity cP	4.23±0.68	3.51±0.6	<0.001

n.s: non significant

Table 2. Antioxidant nutrients in both PD and control groups

Antioxidant nutrients	PD (n=20)	Control (n=20)	*P*
Plasma selenium umolL^{-1}	1.29±0.3	1.31±0.7	n.s
RBCs selenium umolL^{-1}	0.79±0.3	1.35±0.3	<0.01
GPx ugHb^{-1}	6.77±0.4	8.2±0.9	<0.05
Vitamin A umolL^{-1}	1.88±0.6	1.99±0.7	<0.05
Vitamin E umolL^{-1}	29.1±6.9	31.9±1.2	<0.05

n.s: non significant

Table 3. Correlations between the severity of antioxidant nutrient, blood chemistry and blood viscosity in PD group

Prameters	Whole blood viscosity	
	r	*P*
Plasma selenium	0.31	n.s
RBCs selenium	-0.59	<0.05
GPx	-0.63	<0.05
Vitamin A	-0.71	<0.01
Vitamin E	-0.68	<0.05
Haematocrit	+0.72	<0.05
HbA_{1c}	0.41	<0.05
Cholesterol	0.39	n.s
triglyceride	0.41	n.s
Plasma protein	0.46	n.s

n.s: non significant

Many studies using a fluorescence spectrophotometer techniques informed reduced membrane fluidity for RBCs in type 2 diabetes mellitus. These data, documents the represented data in which oxidative deregulation is represented with decreased membrane fluidity of submitochondrial particles and increased mitochondrial dysfunction with diabetes mellitus type 2 [24]. Therefore, the problem, whether nutritional antioxidants have a similar possible role in RBCs membrane fluidity and blood viscosity in PD and the correlation with plasma lipids. The present data considering measurements done on whole blood, approve concomitant increased viscous components in the PD group. Possible mechanisms for the reduction of membrane fluidity include increased cholesterol/phospholipids ratio, or enzymatic changes such as reduction of Na+, K+ ATPase activity [25-27].

However, the current results suggest that the selenium concentration in RBCs may also contribute to the fluidity of the blood. The cause of significant reduction of selenium concentration in RBCs of PD group in our study is questionable. Dietary deficiency of selenium can be excluded as there was no detectable difference in plasma selenium, a marker for nutritional intake relative to the control.

Significant reduction in GPx activity was recorded in the current study. Only a borderline significant correlation between GPx activity and selenium in RBCs in PD group was seen, probably due to the reduced selenium concentration in RBCs. Plasma selenium levels drop independently of selenium level during the acute phase response and GPx. The resulted data conclude that changed concentrations of selenium in PD with critical illness may due to the effects of the OS. The place of reduced GPx activity in the development of PD changes in blood rheology must be defined by further studies.

There was a significantly increased blood viscosity in the PD group in comparison with the control group. From this evidence, it can also hypothesize that PD patients with higher blood viscosity may prolong insulin resistance with advances to frank diabetes. It is in agreement with Zhao et al. [28] study. However, elevated whole blood viscosity is attended with a prevalence of insulin resistance in middle-aged and elderly Chinese population [29]. Therefore, further studies are recommended for studying the relationship between insulin sensitivity and blood viscosity among PD cases.

Borderline differences were found in vitamin A and vitamin E between PD patient and control group. Regarding vitamin A, a different result was recorded by Godala et al. [30]. They recorded decreased level of vitamins A and E points to the weakening of the antioxidative barrier in patients with the metabolic syndrome. In addition, García et al. [31] reported the association of low concentrations of vitamins A and E in children with lipids, inflammation, and insulin resistance and attributed these observations to a reduced mobilization of vitamin A from the liver by an unknown mechanism.

A significant negative correlation was found between vitamin A, vitamin E and whole blood viscosity. This observation may be attributed to the hypothesis that vitamin A and vitamin E are highly effective as an antioxidant for suppressing the peroxidative lysis of liposomes of PD erythrocyte lipids as in selenium and GPx activity.

5. CONCLUSION

The contribution of increasing blood viscosity based different biochemical parameters in developing PD cases to be a question of debate. The present data show that whole blood viscosity was increased in PD and affected by the RBCs concentrations of nutrient antioxidants, selenium, vitamin A, and vitamin E. Increased levels of the OS may motivate some of the increased risks of getting the development of increased whole blood viscosity in PD. Interventions to decrease levels of the OS by methods such as antioxidants therapy, glycemic and lipid control are indicated. Whether or not nutritional antioxidant replacement improves the concentration of selenium, vitamin A, and vitamin E whole blood viscosity with a subsequent improvement in insulin resistance in the next future among prediabetics. Such studies are required to prevent progression of PD to frank diabetes mellitus.

ACKNOWLEDGEMENTS

The author acknowledges the imperative contributions of many individuals to this study, Medical and nursing staff members in the diabetes clinic, clinical pathology and internal medicine departments for their scientific technical support. The author is indebted to Professor Adel Morshedy, the chairman of the Clinical Epidemiology Unit, Suez Canal University, Egypt, for his valuable guide and a great help in revising the manuscript.

COMPETING INTERESTS

Author has declared that no competing interests exist.

REFERENCES

1. World Health Organization, Global Status Report on Noncommunicable Diseases 2010, WHO, Geneva, Switzerland; 2011.
2. Buysschaert M, Medina JL, Buysschaert B, Bergman M. Definitions (and Current Controversies) of diabetes and prediabetes. Curr Diabetes Rev; 2015.
3. Wang JS, Lin SD, Lee WJ, Su SL, Lee IT, Tu ST, et al. Effects of acarbose versus glibenclamide on glycemic excursion and oxidative stress in type 2 diabetic patients inadequately controlled by metformin: A 24-week, randomized, open-label, parallel-group comparison. Clin Ther. 2011;33(12): 1932-42
4. Farabi SS, Carley DW, Smith D, Quinn L. Impact of exercise on diurnal and nocturnal markers of glycaemic variability and oxidative stress in obese individuals with type 2 diabetes or impaired glucose tolerance. Diab Vasc Dis Res; 2015.
5. Lipinski B. Pathophysiology of oxidative stress in diabetes mellitus. J Diabetes Complications. 2001;15(4):203-10.
6. Otto C, Richter WO, Schwandt P. Contribution of fibrinogen and lipoproteins to plasma viscosity in hyper-cholesterolemia and hypertriglyceridemia: Evaluation by selective depletion of low-density lipoproteins or fibrinogen. Metabolism. 2000;49(6):810-3.
7. Likidlilid A, Patchanans N, Peerapatdit T, Sriratanasathavorn C. Lipid peroxidation and antioxidant enzyme activities in erythrocytes of type 2 diabetic patients. J Med Assoc Thai. 2010;93(6):682-93.
8. Godala M, Materek-Kuśmierkiewicz I, Moczulski D, Rutkowski M, Szatko F, Gaszyńska E, et al. Estimation of plasma vitamin A, C and E levels in patients with metabolic syndrome. Pol Merkur Lekarski. 2014;36(215):320-3.
9. Eshaq RS, Wright WS, Harris NR. Oxygen delivery, consumption, and conversion to reactive oxygen species in experimental models of diabetic retinopathy. Redox Biol. 2014;2:661-6
10. Yu KJ, Zhang MJ, Li Y, Wang RT. Increased whole blood viscosity associated with arterial stiffness in patients with non-alcoholic fatty liver disease. J Gastroenterol Hepatol. 2014;29(3):540-4.
11. Zhao HY, Li J, Xu M, Wang TG, Sun WW, Chen Y, et al. Elevated whole blood viscosity is associated with insulin resistance and non-alcoholic fatty liver. Clin Endocrinol (Oxf); 2015.
12. Hitsumoto T. Factors affecting impairment of blood rheology in obese subjects. J Cardiol. 2012;60(5):401-6.
13. Li G, He H, Yan H, Zhao Q, Yin D. Does carbonyl stress cause increased blood viscosity during storage? Clin Hemorheol Microcirc. 2010;44(2):145-54.
14. Simmonds MJ, Meiselman HJ, Marshall-Gradisnik SM, Pyne M, Kakanis M, Keane J, et al. Assessment of oxidant susceptibility of red blood cells in various species based on cell deformability. Biorheology. 2011;48:293–304
15. Gabbay KH, Hasty K, Breslow JL, Ellison RC, Bunn HF, Gallop PM. Glycosylated hemoglobins and long-term blood glucose control in diabetes mellitus. J Clin Endocrinol Metab. 1977;44(5):859-64.
16. Aansson L, Pettersson J, Eriksson L, Olin A. Atomic absorption spectrophotometric determination of selenium in human blood components. Clin Chem. 1989;35:537–542. Rosella LC, Lebenbaum M, Fitzpatrick T, Zuk A, Booth GL. The prevalence of undiagnosed and prediabetes diabetes in Canada (2007-

2011) According to Fasting Plasma Glucose and HbA1c Screening Criteria. Diabetes Care; 2015.

17. Paglia DE, Valentine WN. Studies characterization of erythrocyte glutathione peroxidase. J lab Clin Med 1967;70:158–169.
18. Von Kampen EJ, Zijlstra WG. Standardaization of hemoglobinometry. The hemoglobincyanide method. Clin Chem Acta. 1961;6:538–544. Tsai CJ, Hsieh CJ, Tung SC, et al. Acute blood glucose fluctuations can decrease blood glutathione and adiponectin levels in patients with type 2 diabetes. Diabetes Res Clin Pract. 2012;98:257–263.
19. Bieri JG, Tolliver TJ, Citignami GL. Simultaneous determination of a – Tocopherol and retinol in plasma or red cells by high performance liquid chromatography. Am J Clin Nutr. 1979;32: 2143–2149.
20. Schneditz D, Ribitsch V, Kenner T. Rheological discrimination between nature, rigid and aggregated red blood cells in oscillatory flow. Biorheol. 1985;22:209–219.
21. Gradinaru D, Borsa C, Ionescu C, Margina D. Advanced oxidative and glycoxidative protein damage markers in the elderly with type 2 diabetes. J Proteomics. 2013;92: 313-22
22. Huang M, Liang Q, Li P, Xia J, Wang Y, Hu P, et al. Biomarkers for early diagnosis of type 2 diabetic nephropathy: A study based on an integrated biomarker system. Mol Biosyst. 2013;9(8):2134-41.
23. Kellow NJ, Coughlan MT, Savige GS, Reid CM. Effect of dietary prebiotic supplementation on advanced glycation, insulin resistance and inflammatory biomarkers in adults with pre-diabetes: A study protocol for a double-blind placebo-controlled randomised crossover clinical trial. BMC Endocr Disord. 2014;14:55
24. Rodríguez-Carrizalez AD, Castellanos-González JA, Martínez-Romero EC, Miller-Arrevillaga G, Villa-Hernández D, Hernández-Godínez PP, et al. Oxidants, antioxidants and mitochondrial function in non-proliferative diabetic retinopathy. J Diabetes. 2014;6(2):167-75.
25. Likidlilid A, Patchanans N, Peerapatdit T, Sriratanasathavorn C. Lipid peroxidation and antioxidant enzyme activities in erythrocytes of type 2 diabetic patients. J Med Assoc Thai. 2010;93(6):682-93.
26. Jacobs S, Schiller K, Jansen E, Fritsche A, Weikert C, di Giuseppe R, et al. Association between erythrocyte membrane fatty acids and biomarkers of dyslipidemia in the EPIC-Potsdam study. Eur J Clin Nutr. 2014;68(4):517-25.
27. Kröger J, Jacobs S, Jansen EH, Fritsche A, Boeing H, Schulze MB. Erythrocyte membrane fatty acid fluidity and risk of type 2 diabetes in the EPIC-Potsdam study. Diabetologia. 2015;58(2):282-9.
28. Zhao HY, Li J, Xu M, Wang TG, Sun WW, Chen Y, Bi YF, Wang WQ, Ning G. Elevated whole blood viscosity is associated with insulin resistance and non-alcoholic fatty liver. Clin Endocrinol (Oxf); 2015.
29. Vayá A, Hernández-Mijares A, Suescun M, Solá E, Cámara R, Romagnoli M, et al. Metabolic alterations in morbid obesity. Influence on the haemorheological profile. Clin Hemorheol Microcirc. 2011;48(4):247-55.
30. Godala M, Materek-Kuśmierkiewicz I, Moczulski D, Rutkowski M, Szatko F, Gaszyńska E, et al. Estimation of plasma vitamin A, C and E levels in patients with metabolic syndrom. Pol Merkur Lekarski. 2014;36(215):320-3.
31. García OP, Ronquillo D, del Carmen Caamaño M, Martínez G, Camacho M, López V, et al. Zinc, iron and vitamins A, C and E are associated with obesity, inflammation, lipid profile and insulin resistance in Mexican school-aged children. Nutrients. 2013;5(12):5012-30.

9

Carotenoid Levels, Total Phenolic Content and Antioxidant Activity Variations in Varieties of *Citrullus lanatus* under Storage at Room Temperature

J. Y. Jr. Hena[1,2], D. M. Andala[3], H. N. Nyambaka[1] and M. P. Nawiri[1*]

[1]Kenyatta University, Nairobi, Kenya.
[2]University of Liberia, Monrovia, Liberia.
[3]Multimedia University of Kenya, Nairobi, Kenya.

Authors' contributions

This work was carried out in collaboration between all authors. Author HNN designed the study, wrote the protocol and supervised the work. Author JYJH carried out all laboratories work and performed the statistical analysis. Author DMA managed the analyses of the study. Author MPN wrote the first draft of the manuscript. Author JYJH managed the literature searches and edited the manuscript. All authors read and approved the final manuscript.

Editor(s):
(1) Carmen Lúcia de Oliveira Petkowicz, Federal University of Parana, Curitiba, Parana, Brazil.
Reviewers:
(1) Anonymous, University of Missouri-St. Louis, USA.
(2) Yongchun Zhu, Shenyang Normal University, China.

ABSTRACT

Introduction: Non-communicable diseases (NCDs) are economically burdening and are globally projected to increase deaths upto 75% by 2030. With the cause partly attributed to unhealthy diets, the major contributing factor is production of reactive oxygen and nitrogen species in biochemical reactions in human cells. Through a dietary approach, this can be countered by carotenoid and phenolic antioxidants in *Citrullus lanatus* (watermelon). The challenge presented is on the effect of room temperature (RT) storage on the levels of β- carotene, lutein and lycopene carotenoids and phenolics in watermelon. Varieties, of watermelon grown in Mwea, Kenya are reported for their carotenoid levels; antioxidant activity (AA) and total phenolic content (TPC) within 14 days storage period at RT.

**Corresponding author: E-mail: nawiri.mildred@ku.ac.ke*

Methods: The methods employed were HPLC, FCR and DPPH assay.
Results: The ranges of carotenoid levels were; β-carotene; 0.13±0.03- 3.15±0.02 µg/100 g, lutein; 10357.58±62.98-30573.99±434.40 µg/100 g and lycopene 113.42±0.39-522.76±1.36 µg/100 g. The AA ranged between 3.10±0.04-41.35±0.29 and 59.12±0.15- 79.93±0.18% in fresh and dried watermelons, while the TPC ranged between 266.00±5.00-896.67±7.37 mg/100 g GAE. The effect of storage on the nutritional values was shown by a general decrease on levels of β-carotene, lycopene, AA and TPC while lutein levels increased. There were significant differences ($p<0.05$) in measurements of carotenoids, AA and TPC between all sample varieties.
Conclusion: Storing watermelons for upto 14 days has both negative and positive implications on the carotenoid and total phenolic levels but clearly reduces the antioxidant activity and hence the effectiveness on the fight against NCDs.

Keywords: Citrullus lanatus; carotenoids; total phenolics; antioxidant activity.

1. INTRODUCTION

About 60% of global mortality, morbidity and disability cases are caused by non communicable diseases (NCDs) such as cardiovascular diseases (CVDs) and cancer among others [1]. In Kenya, over 1.6 million and nearly 7% of those affected are estimated to die annually from CVDs and cancer [2,3]. The statistics on NCDs are mainly pegged to unhealthy diets, tobacco use, exposure to environmental carcinogens and physical inactivity [1]. Fundamentally, these diseases are due to the presence of reactive oxygen and nitrogen species; superoxide, hydroxyl radical and nitric oxide radical, that are a consequence of aerobic respiration [4]. Medication for NCDs is generally unaffordable especially developing nations and therefore a dietary approach, given its efficacy and affordability would be a sustainable approach. As such, specific attention has been drawn to antioxidants (free radical scavengers) including carotenoid and phenolic nutraceuticals that are found in fruits such as *Citrullus lanatus* [5,6].

Watermelon is a member of the cucurbit family and one of the most widely cultivated crops in the world. Its global consumption is greater than that of any other cucurbit. It accounts for 6.8% of the world area devoted to vegetable production and this is expected to increase due to the increasing demand in the European market due to unfavorable weather experienced in those countries [7]. The recommended varieties are Charleston Gray, Congo, Crimson Sweet, Moon and Stars, Orangeglo, Sugar Baby, Sunday Special. In Kenya, the fruit is a good cash crop with very good market opportunities, particularly in urban areas [8]. The levels of nutraceuticals as well as antioxidant activity in fruits vary depending on factors including climatic conditions, part of the plant used (example, the peel, seeds and red/pink fleshy portion),variety or species, storage period and condition [9-11]. The effect on β- carotene, lutein and lycopene carotenoids as well as the total phenolic content and the antioxidant activity with storage of varieties of *Citrullus lanatus* under normal room temperature were investigated.

2. MATERIALS AND METHODS

2.1 Chemicals and Reagents

Acetonitrile, Dichloromethane, Methanol (all HPLC grade), Hexane, Acetone, Ethanol (all analytical grade), All-trans-lycopene and Lutein standards, Gallic acid, Ascorbic acid (AscA), Butylated hydroxytoluene (BHT), 2,2-Diphenyl-1-picrylhydrazyl (DPPH), Folin-Ciocalteu reagent, Anhydrous Sodium Carbonate (Na_2CO_3), Anhydrous Sodium Sulphate (Na_2SO_4) and Magnesium Carbonate ($MgCO_3$) were sourced from Sigma-Aldrich Chemical Co., USA. All-trans-beta-carotene standard was obtained from Fluka Chemicals Corp., Switzerland. All chemicals were supplied by Kobian Kenya Limited.

2.2 Experimental Procedures

Varieties of *Citrullus lanatus* were obtained from one geographical region, Mwea in Kenya and identified by a taxonomist. Purposive sampling was employed to obtain mature-ripe samples which were then transported to the laboratory for analysis. Samples were shelved open and stored at normal laboratory room temperature and sun light for 14 days as carotenoid levels, total phenolic content and antioxidant activity were measured. During dried samples analysis of antioxidant activity, samples were diced, air dried on sieve and stored in the laboratory under similar conditions as the fresh samples for two months prior to DPPH assay. Storage in refrigerator at 4°C was not considered as this may be unaffordable by many consumers especially in developing countries.

The procedures were carried out in subdued light within the shortest possible time to prevent isomerization and photo-degradation of the carotenoids and other phytochemicals [10]. Extraction procedures were done according to [10,12] with minimal modifications.

2.3 Extraction and Analysis of Carotenoids

Two clean samples of pata negra, sugar baby and Charleston grey varieties of *Citrullus lanatus* variety were diced quarterly, opposite quarters homogenized by blending using an electric blender. A 10.0 g (±0.05) of homogenized sample with 0.6 g of $MgCO_3$ (acid neutralizing agent) and 0.8 g of BHT (antioxidant) was sonicated for two minutes and mixed for five minutes with 30 mL of extraction solvents mixture (hexane/acetone/ethanol, 2:1:1 v/v) using an orbital shaker on ice. The mixture was vacuum filtered through filter paper using a 250 mL Buchner flask and funnel and the filtrates transferred into a 250 mL separatory funnel. This process was repeated several times using 10 mL of the extraction solvents until no appreciable colour was observed in both filter paper and the solid residue. A 50 mL aliquot of 15% saturated NaCl solution was added to the filtrate and mixed well for two minutes to enhance phase partitioning. Distilled water was added to themixture to completely wash the organic layer from any impurities. The aqueous phase was decanted and the organic phase was recovered and filtered over 8.0 g of anhydrous $NaSO_4$ which was rinsed three times with 3.5 mL of extraction solvent. The filtrate was evaporated to near dryness under nitrogen using a rotatory evaporator attached to a vacuum pump at 40°C and reconstituted in 5 mL methanol. Extracts were filtered under nitrogen gas into dark reagent vials covered with aluminium foil using a 0.45 µm single membrane filter and stored in the fridge prior to HPLC injection. All extractions and analysis were done in triplicates. Fresh mobile phase consisting of ACN:MetOH:DCM (70:10:20 v/v) was prepared each time by measuring 700 mL (ACN), 100 mL (MetOH) and 200 mL (DCM), filtered and degassed (to remove any air bubble) for 30 minutes using an ultrasonic agitator before use.

Samples and standards (20 µL aliquot) were introduced into the reserved phase HPLC column (Gemini-NX 5uc 18110A, 250 × 4.6 mm id × 5 µm particle size) through an autosampler. The HPLC also consisted of CTO-10ASVP oven and a LC-20AD pump. At a mobile phase flow rate of 0.8 mL/min on a C_{18} column for 18 minutes, the analytes were isocratically eluted. Using a deuterium lamp UV-visible detector (SPD-20A) with sensitivity set at 0.05-absorbance units full scale (aufs), beta-carotene (BC) was detected at 450 nm, lutein at 452 nm while lycopene at 471 nm. The carotenoids peaks were identified by comparing the retention time to that of the standards and by spiking the samples with known concentration of the standards. Quantification was done based on linear calibration curves using peak areas. Concentrations in mg/100g of fresh samples were calculated as shown in Equation (1).

$$C_X(mg/100g) = \left(\frac{P_X - B}{S}\right) \times D_f \times 100g \quad (1)$$

Where, C_X= concentration of analyte in mg/100 g, P_X = Peak area of analyte, B = intercept, S = slope and D_f= dilution factor

From BC concentration, the amount of retinol equivalents was calculated based on 12:1 conversion for BC/VA (FNB, 2001).

2.4 Extraction of Total Phenolics and Antioxidant Activity

Similar extraction procedure was carried out fort TPC and AA as was done for carotenoids. However, a 3.0 g of homogenized samples with 0.3 g $MgCO_3$ and 0.5 g BHT was sonicated and mixed well with 20 mL of extraction solvents. The mixture was vacuum filtered repeatedely and dried over 5.0 g $NaSO_4$. The filtrate was evaporated to near dryness and redissolved in 5 mL methanol. It was stored in the dark in air tight vials covered with aluminium foil until DPPH and FCR assays.

2.5 Determination of Total Phenolic Content

Total phenolic content of *Citrullus lanatus* extracts was determined by employing the Folin-Ciocalteu reagent (FCR) method with some modifications [13]. A 0.5 mL of extract was mixed with 5 mL of Folin-Ciocalteu regent (diluted 10 times with distilled water) and allowed to stand for 10 minutes at room temperature. A 4 mL of Na_2CO_3 (200 g/L) was added to the mixture, mixed thoroughly and incubated for 60 minutes in the dark. Triplicate measurements were made and absorbance measured at 765 nm against a blank (containing all the reagents except the sample) and compared to a gallic acid calibration curve. Total phenolics were estimated as gallic acid equivalent (GAE) in milligram per 100 g of extractive (mg of GAE/ 100 g of extractive) using Equation 2.

$$TPC(mg/100gGAE) = \left(\frac{A_E - C}{M}\right) \times D_F \times 100g \quad (2)$$

Where,

TPC_E= total phenolic content of sample extract in mg/100g of Gallic acid equivalent, A_E = absorbance of sample extract, C = intercept, M = slope and D_F = dilution factor.

2.6 Determination of Antioxidant Activity

The free radical scavenging effects (antioxidant activity) was determined at initial storage period (day 2) at extract concentration of 100 µg/mL. This was determined based on the extracts scavenging activity on the stable 2, 2-diphenyl-1-picrylhydrazyl (DPPH) free radical with slight modifications as described [14,15]. Extract concentration of 100 µg/mL for both fresh and dried samples (100 µg extract in 1 mL of 95 % methanol) was added to 2 mL of 20 ppm DPPH methanolic solution. The reaction mixture was shaken vigorously and incubated for 30 minutes at room temperature in the dark. The absorbance was read at 516 nm against DPPH solution (2 mL DPPH in 2 mL of 95% methanol) as control. The ability of the extracts to scavenge the DPPH radical (antioxidant activity or radical scavenging activity) was calculated as a decrease in the absorbance of DPPH radical using Equation 3.

$$\%RSA = \left(1 - \frac{A_{TS}}{A_C}\right) \times 100 \quad (3)$$

Where, %RSA= percentage radical scavenging activity as a measure of antioxidant activity, A_{TS} = absorbance of test sample, A_C = absorbance of control.

The effective concentration (EC_{50}) of extracts required to decrease the initial DPPH radical concentration by 50% was determined by plotting scavenging activity against concentration of extracts. In similar procedures, the antioxidant activity of two synthetic antioxidants (AscA and BHT) was determined for purpose of comparison.

2.7 Data Analysis

Values obtained for the carotenoids, TPC and antioxidant activity at room temperature storage in the varieties of *Citrullus lanatus* were expressed as means of three replicate measurements ± standard deviation and compared by One-Way ANOVA at 95% using SPSS version 20 for windows. Values of $p<0.05$ were considered significantly different and SNK-test at $\alpha=0.05$ indicated differences among means. Student's T-test was applied for mean comparison of antioxidant activity of fresh and dried samples.

3. RESULTS AND DISCUSSION

3.1 Levels and Trends of Carotenoid and Total Phenolics

The levels of lutein, β-carotene, lycopene, total phenolic content (TPC) and retinol equivalent (RE) determined at day one in the three varieties of *Citrullus lanatus* are presented in Table 1. The trends in level of the carotenoids and total phenolic content during the 14 day storage are presented in Figs. 1-4.

The levels ranged between 10357.58-16287.07 µg/100 g (lutein), 0.49-3.15 µg/100 g (β-carotene), 113.42-522.76 µg/100 g (lycopene) and 0.038-0.242 µg/100 g (RE). The levels of the carotenoids (lutein, β-carotene, lycopene), TPC and RE differred significantly ($p<0.05$) between varieties, which is attributed to among other factors that of variation in the varieties. The highest level of lutein was found in Charleston grey while that of β-carotene, lycopene and RE were found in sugar baby variety. As per Britton and Khachik (2009), the levels of lutein were very high (>2 mg/100 g), whereas β-carotene and lycopene were low (0-0.1 mg/100 g) and moderate (0.1-0.5 mg/100 g) respectively. Calculations show that the daily recommended intake of VA (Retinol equivalent of 1-3 mg/day) can be realized by consuming 10 kg/day of pata negra and sugar baby and 30 kg/day of Charleston grey as this will provide 1.1, 2.42 and 1.14 mg/day RE respectively [16]. The levels of TPC ranged from 266.00 to 896.67 mg/100 g GAE. Charleston grey had the highest TPC level among the three varieties. Although a number of factors not within the scope of this study such as extent of maturity and ripening stage are key, the differences is attributed to be that of differences in the varieties [9,17,18].

The carotenoids and total phenolic content had varying trends during the storage period in the different varieties of *Citrullus lanatus*. Lutein (Fig. 1) showed a regular and significant increase in levels ($p<0.05$), an increase that may be attributed to the conversion of other carotenoids to lutein or to the presence of ε-cyclase and ε-hydroxylase enzymes in *Citrullus lanatus* [6]. Beta-carotene and lycopene (Figs. 2 & 3) did not show a general trend, however a significant change ($p<0.05$) in content was observed with storage period possibly due to the presence and activation of β-cyclase and hydroxylase enzymes, while its reduction in pata negra and in sugar baby may be due to said enzymes inactivation [6]. An identical trend of total phenolic content was shown in the three varieties over the 14 days of storage period, with the levels comparatively high at day 4 (Fig. 4). An extreme decrease in levels was observed between days 4 and 6 while day 6-14 remained relatively low and statistically unchanged ($p>0.05$) in each variety. The degradation of TPC is attributed to the presence of phenolic acids that decrease with storage and ripening of fruits [17]. However, its increase at day 4 could be due to the synthesis of more anthocyanins and other phenolic compounds at said day, which later declined perhaps owing to physiological changes [17].

Table 1. Levels of carotenoids, total phenolic content and retinol equivalent in varieties of *Citrullus lanatus*

	Concentration (Mean±SD; n=3)				
Variety	**Lutein (µg/100 g)**	**β-carotene (µg/100 g)**	**Lycopene (µg/100 g)**	**TPC (mg/100 g GAE)**	**RE (µg/100 g)**
Pata negra	10357.58±62.98[a]	1.43±0.02[b]	113.42±0.39[a]	266.00±5.00[a]	0.110±0.01[b]
Sugar baby	15972.27±155.14[b]	3.15±0.02[c]	522.76±1.36[c]	693.00±6.56[b]	0.242±0.03[c]
Charleston grey	16287.07±157.81[c]	0.49±0.01[a]	248.70±0.70[b]	896.67±7.37[c]	0.038±0.01[a]
p-value	<0.001	<0.001	<0.001	<0.001	<0.001

Mean values with different small letters within the same column are significantly different ($p<0.05$, SNK-test)

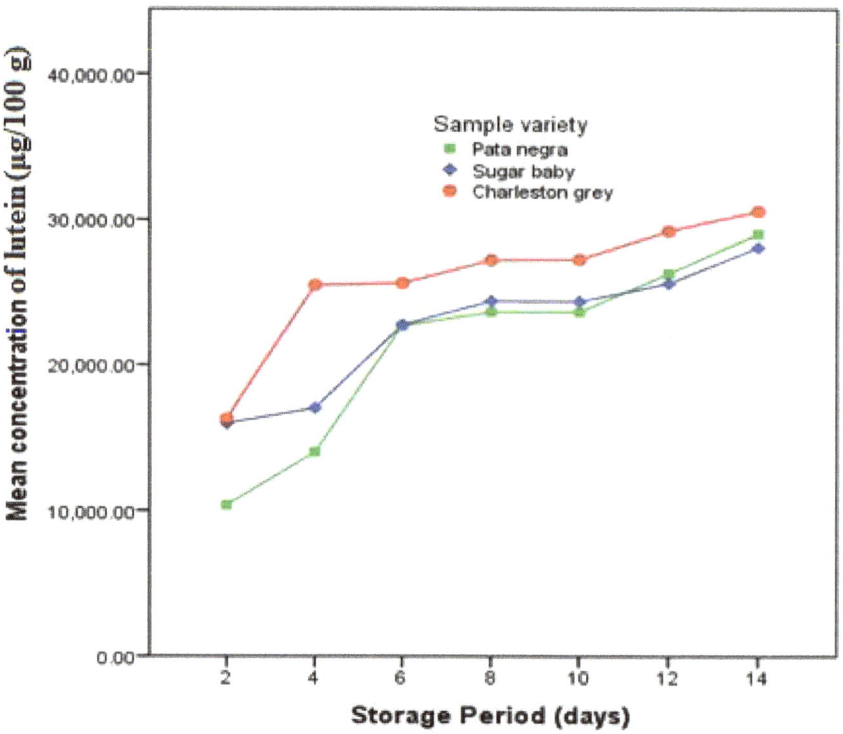

Fig. 1. Trends of levels of lutein in varieties of *Citrullus lanatus* stored at room temperature for 14 days

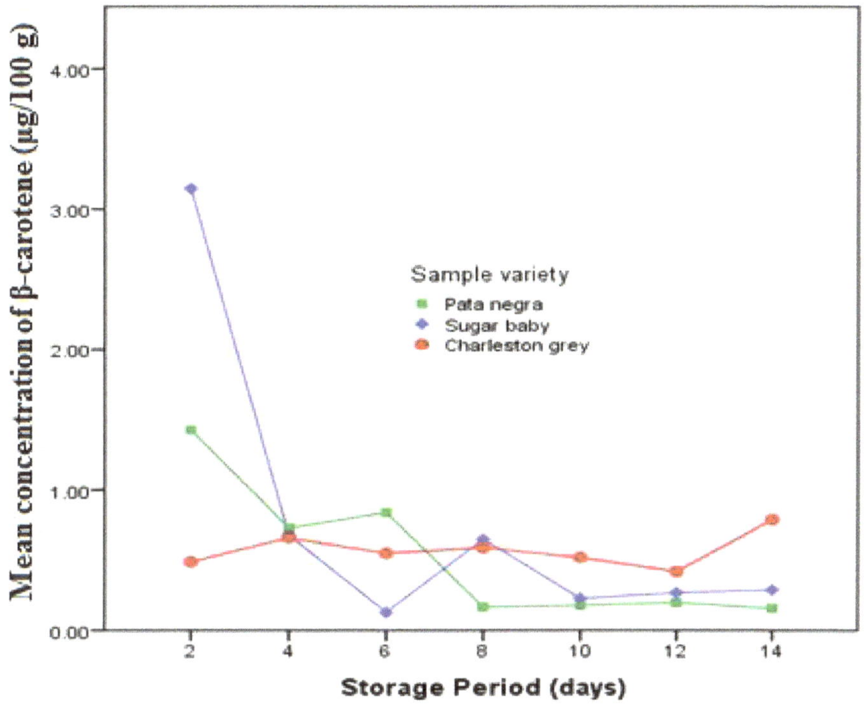

Fig. 2. Trends of levels of β-carotene in varieties of *Citrullus lanatus* stored at room temperature for 14 days

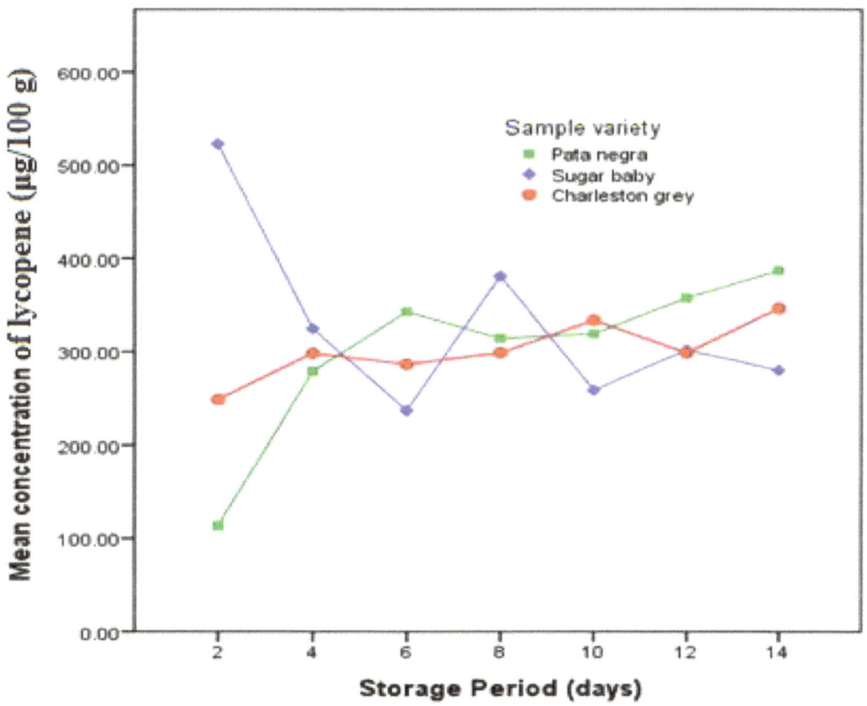

Fig. 3. Trends of levels of lycopene in varieties of *Citrullus lanatus* stored at room temperature for 14 days

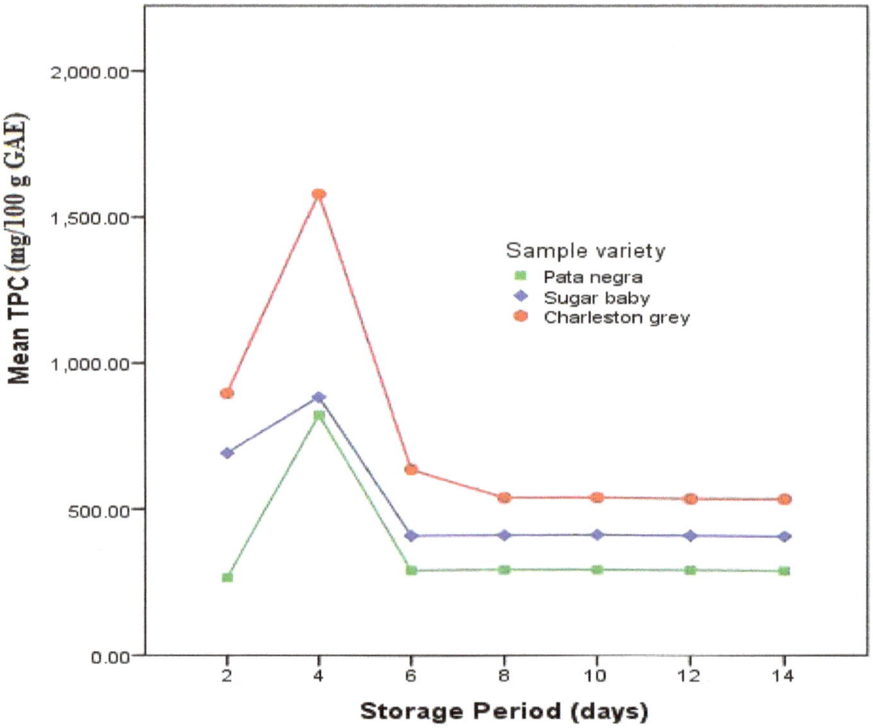

Fig. 4. Trends of levels of total phenolic content in varieties of *Citrullus lanatus* stored at room temperature for 14 days

3.2 Levels and Trends of Antioxidant Activity

The levels, expressed as percentage radical scavenging activity (%RSA) and the effective concentration (IC_{50}) of fresh and dried samples of *Citrullus lanatus* varieties and synthetic antioxidants; ascorbic acid (AscA) and BHT are presented in Table 2. The trend in levels of antioxidant activity during stored at room temperature for 14 days is shown in Fig. 5.

The % RSA of fresh and dried samples of *Citrullus lanatus* ranged from 17.10±0.16-28.78±0.17% and 59.12±0.15-79.93±0.18%, whereas IC_{50} ranged between 14.96-25.70 µg/mL (low % RSA) and 2.38-4.73 µg/mL (high % RSA), respectively. There were significant variations ($p<0.05$) among varieties due to the different levels of phytochemicals among other factors mentioned for differences in carotenoids and TPC [6,9,17,18]. Charleston grey and sugar baby varieties had the lowest and highest % RSA respectively. This was in both fresh and dried samples, where dried samples were comparatively high, perhaps due to pre-concentration/drying of samples. The radical scavenging efficiencies (IC_{50}) of the dried samples were similar to the synthetic antioxidants. The differences between these results and from other authors can be explained due to the use of different extraction solvents and cultivars among other factors [9,18].

There were significant variations ($p<0.05$) in the antioxidant activity with storage period for each sample variety, with the degradation trends being identical in pata negra and sugar baby varieties ($p<0.05$).The antioxidant activity levels were high at day 4 for pata negra and sugar baby and at day 2 for Charleston grey, which possibly could be due to favorable physiological changes and increase in the synthesis of some phytochemicals at said days, but thereafter declined with progression of storage period [17,19]. The extent of degradation in antioxidant activity was reasonably high in Charleston grey, resulting to a loss of 55.0%, followed by 47.6% in sugar baby, while 19.8% in pata negra had the least reduction. Geometric isomerization of carotenoids, different content of antioxidants and the physiological changes that accompany the phytochemicals during fruit storage may be responsible for the degradation in the antioxidant activity [6,19].

Table 2. Levels of % RSA and IC_{50} of *Citrullus lanatus* varieties and synthetic antioxidants

Variety	% RSA (Mean±SD), (n = 3)		IC50 (in mg/mL)		Synthetic antioxidant
	Fresh samples	Dried samples	Fresh samples	Dried samples	
Patanegra	20.67±0.16[b]	61.90±0.07[b]	23.34	3.96	AscA: %RSA = 70.00±0.17 IC_{50}= 5.22
Sugar baby	28.78±0.15[c]	79.93±0.18[c]	14.96	2.38	BHT: %RSA =
Charleston grey	17.10±0.16[a]	59.12±0.15[a]	25.70	4.73	77.47±0.10
P-value	<0.001	<0.001			IC_{50} = 2.60

Mean values with different small letters within the same column are significantly different (p<0.05, SNK-test)

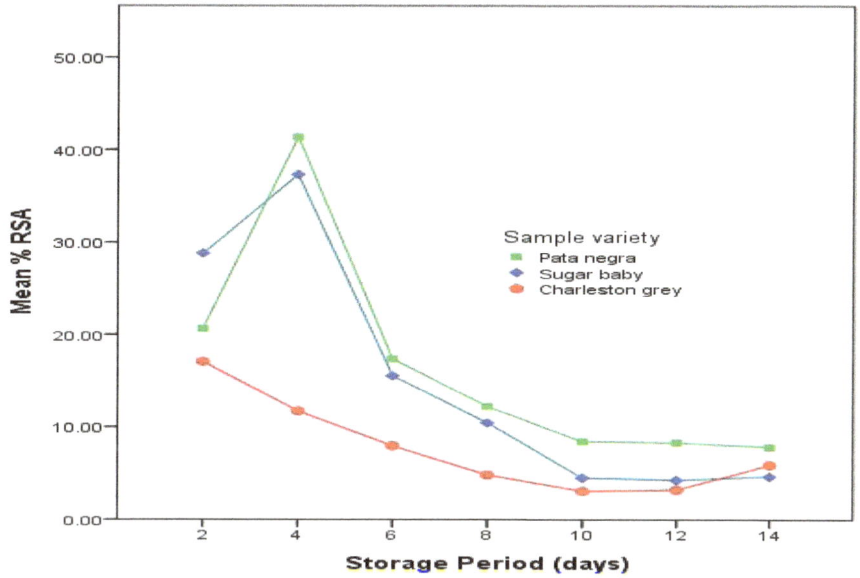

Fig. 5. Trends of levels of antioxidant activity in varieties of *Citrullus lanatus* stored at room temperature for 14 days

4. CONCLUSION

Storing watermelons for upto 14 days has both negative and positive implications on the carotenoid and total phenolic content levels but clearly reduces the antioxidant activity and hence the effectiveness on the fight against NCDs.

COMPETING INTERESTS

Authors have declared that no competing interests exist.

REFERENCES

1. WHO Global strategy on diet, physical activity and health; 2004. (WHA 57.17).
2. Alwan A, Maclean DR, Riley LM, d'Espaignet ET, Matthers CD, Stevens GA, Bettcher D. Monitoring and surveillance of chronic non-communicable diseases: Progress and capacity in high-burden countries; 2010.
3. MPHS (Ministry of Public Health and Sanitation (2011-2016). Kenya National Cancer Control Strategy.
4. Shon MY, Kim TH, Sung NJ. Antioxidants and free radical scavenging activity of *Phellinus baumii* (Phellinus of Hymenochaetaceae) extracts. Food Chemistry. 2003;82:593-597.
5. Miniati E. Assessment of phenolic compounds in Biological samples. Ann 1st Super Sanita. 2007;43:362-368.
6. Britton G, Khachik F. Carotenoids in food; nutrition and health. Birkhauser Verlag, 5. Basel-Boston-Berlin; 2009. Available:www.birkhauser.ch
7. NAFIS (National Farmers Information Services). Watermelons Fruit (*Citrullus lanatus*).

Available:http://www.nafis.go.ke/fruits/watermelons/ (Accessed, 09/15/2015)
8. AIRC. Fruits and Vegetables Technical Handbook Second Edition (Revised). Ministry of Agriculture and Livestock Production, Nairobi, Kenya; 2003. ISBN: 6633-764-01-1
9. Martınez-Valverde I, Periago MJ, Provan G, Chesson A. Phenolic compounds, lycopene and antioxidant activity in commercial varieties of tomato (*Lycopersicon esculentum*). Journal of the Science of Food and Agriculture. 2002; 82:323-330.
10. Kimura M, Rodriguez-Amaya DB. Harvest plus handbook for carotenoids analysis. Harvest plus-IFPRI, Washington DC; 2004.
11. Ahmad I, Shahzad T, Choudhry S, Saeed MK, Khan MN. DPPH free radical scavenging activity of tomato, cherry tomato and watermelon: Lycopene extraction, purification and quantification. International Journal of Pharmacy and Pharmaceutical Sciences. 2014;6:223-228.
12. Wystmanstraat J, Cucu T, Loco JV, Anne van Den Bergh J, Vandevijvere S. Assessment of dietary intake of lycopene by the Belgian adult population. Emerging chemical Risk-operational Direction, Food, Medicine and consumer safety. Epidemiology-operational Direction public Health surveillance.1050 Brussels, Belgium; 2011. Available:www.wiv.isp.be
13. Majhenic L, Skerget M. and Knez Z. Antioxidant and antimicrobial activity of guarana see extracts. Food Chemistry. 2007;104:1258-1268.
14. Lee WY, Hainida EKI, Maleyki AMJ. and Amin I. Antioxidant capacity and phenolic content of selected commercially available cruciferous vegetables. Malaysian Journal of Nutrition. 2007;13:71-80.
15. Sonboli A, Mojarrad M, Ebrahimi SN. and Enayat S. Free radical scavenging activity and total phenolic content of methanolic extracts from male inflorescence of *Salix aegyptiaca* grown in Iran. Iranian Journal of Pharmaceutical Research. 2010;9:293-296.
16. Murkovic M, Muelleder U, Neunteufl H. Carotenoid content in different varieties of pumpkins. Journal of Food Composition Analysis. 2002;15:633-638.
17. Manach C, Scalbert A, Morand C, Rémésy C, Jimenez L. Polyphenols: food sources and bioavailability. American Journal of Clinical Nutrition. 2004;79:727–747.
18. Rahman H, Manjula K, Anoosha T, Nagaveni K, Eswaraiah CM, Bardalai D. *In-vitro* anti-oxidant activity of *Citrullus lanatus* seed extracts. Asian Journal of Pharmaceutical and Clinical Research. 2013;6:152-157.
19. Adedayo BC, Oboh G. and Akindahunsi AA. Changes in the total phenol content and antioxidant properties of pepper fruit (Dennettiatripetala) with ripening. African Journal of Food Science. 2010;4:403–409.

10

Synthesis and Evaluation of *in vitro* Antibacterial Properties of Secoisolariciresinol Diglucoside

Sadiq Mareai[1,2*], Shaukath Khanum[3] and Javaraiah Rajesh[1]

[1]*Department of Biochemistry, Yuvaraja's College, University of Mysore, Mysore, Karnataka, India.*
[2]*Department of Biochemistry, Applied Science College, University of Thamar, Thamar, Yemen.*
[3]*Department of Chemistry, Yuvaraja's College, University of Mysore, Mysore, Karnataka, India.*

Authors' contributions

This work was carried out in collaboration between all authors. Authors JR, SK and SM designed the study, wrote the protocol and supervised the work. Author SM carried out all laboratories work, performed the statistical analysis, managed the analyses of the study and wrote the manuscript. All authors read and approved the final manuscript.

Editor(s):
(1) Mohamed Fawzy Ramadan Hassanien, Biochemistry Department, Zagazig University, Egypt.
Reviewers:
(1) Ala Uddin, Bacha Khan University, Pakistan.
(2) Augusto Lopes Souto, Federal University of Paraiba, Brazil.
(3) Pongsak Rattanachaikunsopon, Ubon Ratchathani University, Thailand.

ABSTRACT

In the present study, Secoisolariciresinol diglucoside (SDG) was synthesized from commercially available 3,4-dimethoxytoluene, butanediol and penta acetyl glucose. This includes bromination, alkylation, glycosylation, deacetylation and demethylation reactions to yield the interested title compound (SDG), further the assignment of the chemical structure of the synthetic SDG was performed with IR, MS-TOF, ^{1}H and ^{13}C NMR. The antibacterial property of the synthetic SDG was investigated against seven different Gram positive and Gram negative bacteria. The results of the present study proved that SDG can be obtained with good yield (56.2%) under this scheme. Moreover, it exerts an exceptional antibacterial activity where it showed maximum zone of inhibition with 30 and 27.3 mm against *E. coli* and minimum 7 and 6.1 mm against *S. pyogens* and *S. typhimurium* in agar well and disk diffusion respectively. It also showed maximum activity (18 mm at MIC 200 ppm) against *S. typhimurium*, while minimum inhibitory activity was 4 mm with

**Corresponding author: E-mail: sadiq98ye@gmail.com*

MIC at 300 ppm against *S. pyogens*. The scheme of synthesis was found to be an efficient method and the synthetic SDG was also proved to possess a broad spectrum of activity against a panel of bacteria responsible for causing most common diseases.

Keywords: Synthetic secoisolariciresinol diglucoside; IR; MS-TOF; ^{1}H and ^{13}C NMR and anti-bacterial activity.

1. INTRODUCTION

With the development and clinical use of antibiotics in the nineteenth century, there has been a drastic reduction in the incidences of bacterial infections diseases, which were a leading cause of human mortality world-wide. But, there has been an alarming increase in bacterial resistance to the present chemotherapeutic agents largely due to their injudicious use and this has turned out to be an issue of global concern [1,2]. Emergence of multidrug-resistant pathogens threatened the clinical efficacy of many existing antibiotics [3], leading the scientists to increasingly turn their attention to finding out new antimicrobial compounds with diverse chemical structures and novel mechanisms of action for the new infectious diseases [4]. Phytoestrogens have been found *via* several *in vivo* and *in vitro* studies to exert a potent biological activity besides their major estrogenic and anti-estrogenic activities due to their structural similarity with estrogens [5,6]. Lignans are second major class of phytoestrogens that have been proved to exert antioxidant, anti-inflammatory, anticancer, anti-diabetic and many other biological properties. Isoflavones and coumestans are the other classes of phytoestrogens [7]. Lignans are natural poly phenolic compounds which have long been recognized as challenging targets for organic synthesis due to their complex and diverse chemical structure as well as their important pharmacological properties. Epidemiological studies have reported the chemo preventive effects of lignans on tumors of colon, skin and mammary glands [8]. Lignans also exert antibacterial and cytotoxic activities [9]. Lignans also showed antimicrobial activity against several important pathogenic bacteria with different minimum inhibitory concentrations for each microorganism [10]. Additionally, there are not any reports on side effects and toxicity on lignan usage [11]. Flaxseed, the richest source of SDG, has gained importance in food industry as a component in designer food, functional food and in value added products because of its high content of lignans, which exert nutraceutical and therapeutic properties [12]. SDG from different cultivars of Indian flax seed have been found to exert potential anti-bacterial activity against microbial pathogens [13]. It is also found to possess potential properties such as both *in vivo* and *in vitro* antioxidant activities [14], anti-diabetic type I and II [15], anti-cancer effect on the breast [16], prostate [17], and anti-inflammatory agents [18].

It is believed that isolation of SDG from natural sources is a very laborious, sensitive and time consuming process. Additionally, isolation methods involve use of chemicals and reagents which may cause deleterious effects on animals including humans [19]. Apart from this, the yield of the product is less on isolation, purity will not be 100% and bioavailability of this SDG is low in animals [20]. Moreover, isolated and commercially available SDG is quite expensive. Lignan synthesis relies on limited number of integral reactions that are required for the construction of the basic 18-carbon skeleton whose modification (e.g. cyclization, reduction, oxidation and hydration) can generate the entire set of lignan compounds. Despite their attractive and proven biological properties, only few studies have approached the chemical synthesis of lignans, for instance alphahydroxylated lactone lignans [21], benzopyran [22], erythronordihydroguaiaretic acid [23] and olive type lignans [24]. Therefore, the objective of the present investigation was to design, synthesize and corroborate the antibacterial activity of SDG, a principal lignan of flaxseed against a panel of Gram positive and Gram negative bacteria responsible for causing the most prevalent infectious diseases.

2. MATERIALS AND METHODS

The chemicals used were purchased from Sigma Aldrich, USA and the general solvents used in the experiments were of analytical grade. Melting points were determined by using open capillary method with KOFLER Microscope and are uncorrected, IR spectra were obtained in (cm^{-1}) JASCO FT-IR spectrometer using KBr pellet technique. ^{1}H and ^{13}CNMR spectra were recorded (in ppm) on BRUKER AV 400MHz spectrometer using deuterated methanol as solvent and TMS as internal standard. Mass Spectroscopy analysis was carried out by ESI-TOF MS. Elemental analysis was performed on a Perkin Elmer 2400 Elemental Analyser to confirm the structure of the synthetic compounds. Flash column chromatography was generally performed on silica gel (60 - 200 mesh) for purification of the compounds with the solvent mixture of chloroform/hexane. Precoated alumina with silica gel 60 F_{254} (E. Merck) plates were used for monitoring of reactions and purity determination. Spots were detected by UV (254 nm).

2.1 Chemistry

Synthesis of SDG was reported in this researcher's earlier study wherein the yield was less from the adopted scheme [25]. Therefore, herein a new scheme was designed for the synthesis of SDG.The chemical synthesis of SDG was accomplished through a novel five-step synthesis sequence (Fig. 1) starting from the bromination of the commercially available compound 3,4-dimethoxytoluene (**1**) with N-bromosuccinimide (NBS) in the presence of carbon tetrachloride to achieve 1,2-dimethoxy-4-bromomethylbenzene (**2**). 2,3-bis(3,4-dimethoxybenzyl) butane-1,4-diol (**3**) was afforded by stirring compound **2** with 1,4-butanediol in the presence of n-butyl lithium in 15 mL of DMF. Condensation of compound **3** with 2,3,4,6-tetra-*O*-acetyl-alpha-D-glucopyranosyl bromide in presence of catalytic amount of 2N hydrochloric acid and ethanol as solvent gave 2,3- bis (3,4-dimethoxybenzyl)butane-1,4-*O*-tetra acetyl glucopyranoside (**4**). Compound, 2,3-bis(3,4-dimethoxybenzyl)butane-1,4-*O*-glucose(**5**) was obtained in good yield by deacetylation of compound **4** with sodium hydroxide in presence of ethanol. Finally, the synthesis of the title compound (**6**), i.e. secoisolariciresinol diglucoside (SDG) was attained *via*regioselective partial demethylation of compound **5** using stannous chloride in presence of dichloromethane and 2N hydrochloric acid.

2.1.1 Synthesis of 1,2-dimethoxy-4-bromomethylbenzene (2)

The 3,4-dimethoxytoluene (**1**) was selectively brominated according to the regioselective method of Hong et al. [26]. Using NBS as brominating agent. NBS (5.83 g, 32.80 mmol) was added to a stirred solution of 3,4-dimethoxytoluene (5.00 g, 32.80 mmol) in 70 mL of CCl_4 and the reaction mixture was refluxed for 11 h. The reaction mixture was cooled; succinimide was filtered off and washed with CCl_4. The solvent was evaporated under reduced pressure to yield the crude product (**2**). Recrystallization from the mixture of hexane and diethyl ether (1:1) afforded white needles of compound **2**.

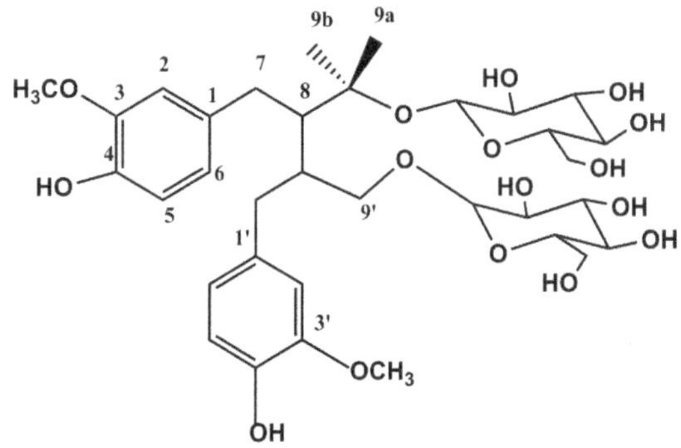

Fig. 1. Chemical structure of SDG

Chemical structure of SDG skeleton represents the position of the carbon and hydrogen atoms

Yield: 5.054 g (66.6%), mp 50.0 - 52.1℃. ^{1}H-NMR (CD_3OD) δ 6.90 (H, s, C_2-H $_{arom}$), 6.76 (H, d, J= 8.1 Hz, C_5-H $_{arom}$), 6.68 (H, d, J= 7.8 Hz, C_6-H $_{arom}$), 4.54 (2H, s, CH_2), 3.38 (6H, s, 2-OCH_3). ^{13}C-NMR (CD_3OD) 150.1 (C4), 150 (C3), 131(C1), 122.1 (C6), 114 (C2), 112.5 (C5), 33.6 (C7). MS: m/z 231.0874(M + 1). $C_9H_{11}BrO_2$ (231.09) Anal. Cal. C, 46.78; H, 4.80. Found: C, 46.72; H, 4.78.

2.1.2 Synthesis of 2,3-bis (3,4-dimethoxybenzyl)butane-1,4-diol (3)

The alkylation of compound **2** was carried out based on the modified procedure of Garonski and Garonska [25]. One equivalent of compound **2**(0.97 g, 10.67 mmol) was added to butanediol (5.00 g, 21.60 mmol) and n-butyl lithium (2.20 g, 16.00 mmol) in 15 mL of DMF. The mixture was stirred at (- 60℃ to 05℃) for 3 h and then the reaction mixture was stirred at 20℃ for 10 h. The reaction mixture was cooled to room temperature, extracted with hexane (3 x 30 mL), and the solvent was evaporated to yield compound **3** (Oil).

Yield: 3.50 g (83.8%), IR (nujol, cm^{-1}): 3430, 3320 (OH). ^{1}H NMR (CD_3OD) δ 6.83 (2H, d, J= 8.2 Hz, $C_{5,5'}$- H $_{arom}$), 6.78 (2H, s, $C_{2,2'}$ -H $_{arom}$), 6.70 (2H, dd, J= 1.9, 8.0 Hz, $C_{6,6'}$-H $_{arom}$), 3.85 (12H, s, 4-OCH_3), 3.64 (2H, dd, J= 7.6, 7.8 Hz, 2

-OH), 3.62 (2H, dd, J= 5.6, 9.5 Hz, $C_{9a,9a'}$-H $_{aliph}$), 3.37 (2H, dd, J= 6.6, 9.8 Hz, $C_{9b,9b'}$-H $_{aliph}$), 2.62 (2H, dd, J= 8.1, 13.6 Hz, $C_{7a,7a'}$-H $_{aliph}$), 2.40 (2H, dd, J= 6.9, 13.7 Hz, $C_{7b,7b'}$-H $_{aliph}$), 1.92 (2H, m, $C_{8,8'}$-H $_{aliph}$). ^{13}C NMR (CD_3OD) 149.2 (C3,3'), 147 (C4,4'), 132 (C1,1'), 122.4 (C6,6'), 113.4 (C2,2'), 112.6 (C5,5'), 65.6 (C9,9'), 56.2 (C-OCH_3),42.8 (C8,8'), 35.3 (C7,7'). MS: m/z 390.4721(M+ 1). $C_{22}H_{30}O_6$ (390.47) Anal. Cal. C, 67.67; H, 7.74. Found: C, 67.51; H, 7.66.

2.1.3 Synthesis of 2,3-bis (3,4 - dimethoxybenzyl)butane-1,4-*O*-tetra acetyl glucose (4)

Compound **3** (3.50 g, 9.00 mmol) was added to a solution of 2,3,4,6 tetra-*O*-acetyl α-D glucopyranosyl bromide (7.12 g, 18.00 mmol) in 10 mL ethanol. The reaction mixture was stirred and refluxed for 19 h. After the completion of the reaction monitored by TLC, the reaction mixture was kept overnight to achieve crystalline product. Finally, the crystalline product was extracted with ether (3 × 30 mL), and the ether layer was passed through anhydrous sodium sulphate. The ether was evaporated and the product was recrystallized from the mixture of hexane and diethyl ether (1:1) to give compound **4** as white crystalline solid.

Yield: 5.07 g (52.9%), mp 65 - 66.2℃, IR (nujol, cm^{-1}): 1710 (C=O for COO). ^{1}H NMR (CD_3OD) δ 6.83 (2H, d, J= 8.0 Hz, $C_{5,5'}$-H $_{arom}$), 6.78 (2H, s, $C_{2,2'}$-H $_{arom}$), 6.74 (2H, dd, J= 1.8,8.1Hz, $C_{6,6'}$-H $_{arom}$), 3.80 (12H, s, 4-OCH_3), 3.65 (2H, d, J= 7.1 Hz, $C_{9,9'}$-β-*O*-Gluanomeric), 3.62 (2H, dd, J= 5.4, 10 Hz, $C_{9a,9a'}$- H $_{aliph}$), 3.38 (2H, dd, J= 6.5,10 Hz, $C_{9b,9b'}$-H $_{aliph}$), 2.20 - 4.52 (38H, $C_{9,9'}$-*O*-Glu), 2.60 (2H, dd, J= 8.6,13.8 Hz, $C_{7a,7a'}$-H $_{aliph}$), 2.41 (2H, dd, J= 8.2,13.4 Hz, $C_{7b,7b'}$-H $_{aliph}$), 2.18 (2H, m, $C_{8,8'}$-H $_{aliph}$).^{13}C NMR (CD_3OD) 149.2 (C3,3'), 147 (C4,4'), 132 (C1,1'), 122.6 (C6,6'), 113.2 (C2,2'), 112.6 (C5,5'), 72.7 (C9,9'), 59.8 (C9,9'-O-Glu anomeric), 56.1 (C-OCH3), 39.7 (C7,7'), 37.8 (C8,8'), 20.8 - 170.0 (C9,9'-O-Glu).MS: m/z 1051.0498(M + 1).$C_{50}H_{66}O_{24}$ (1051.04) Anal. Cal. C, 57.14; H, 6.33. Found: C, 56.42; H, 6.65.

2.1.4 Synthesis of 2,3-bis (3,4- dimethoxybenzyl) butane-1,4 -*O*- glucose (5)

Deacetylation of compound **4** (5.00 g, 4.76 mmol) was achieved by adding 0.75 g NaOH in 10 mL water and 15 mL ethanol. The reaction mixture was stirred and refluxed on water bath for 30 min. The reaction mixture was cooled to room temperature and the crude product was extracted with diethyl ether (3 × 30 mL), and the ether layer was passed through anhydrous sodium sulphate. The solvent was evaporated to yield compound **5** as a pasty mass.

Yield: 2.01 g (59.23%), IR (nujol, cm^{-1}): 3200, 3400 (OH). ^{1}H NMR (CD_3OD) δ 6.80 (2H, d, J= 8.2 Hz, $C_{5,5'}$-H $_{arom}$), 6.78 (2H, s, $C_{2,2'}$-H $_{arom}$), 6.76 (2H, dd, J= 2,8.0 Hz, $C_{6,6'}$-H $_{arom}$), 4.2 (2H, d, J= 7.1, $C_{9,9'}$-β-*O*Gluanomeric), 3.83 (12H, S, 4-OCH_3), 3-3.8 (22H, m, $C_{9,9'}$-*O*-Glu), 3.44 (2H, dd, J= 5.6, Hz, $C_{9a,9a'}$-H $_{aliph}$), 3.20 (2H, dd, J= 6.5, 10 Hz, $C_{9b,9b'}$-H $_{aliph}$), 2.64 (2H, dd, J= 8.4, 13.8 Hz, $C_{7a,7a'}$-H $_{aliph}$), 2.38 (2H, dd, J= 8.4, 13.6 Hz, $C_{7b,7b'}$-H $_{aliph}$), 2.18 (2H, m, $C_{8,8'}$-H $_{aliph}$).^{13}C NMR (CD_3OD) 149.1 (C3,3'), 147 (C4,4'), 132.4 (C1,1'), 122.6 (C6,6'), 113.3 (C2,2'), 112.4 (C5,5'), 104 (C9,9'-O-Glu anomeric), 71.8 (C9,9'), 64 - 79.5 (C9,9'-O-Glu), 56.2 (C-OCH3), 40.6 (C8,8'), 39.8 (C7,7'),. MS: m/z 714.7548(M+ 1). $C_{34}H_{50}O_{16}$ (714.75) Anal. Cal. C, 57.13; H, 7.05. Found: C, 58.50; H, 7.86.

2.1.5 Synthesis of secoisolariciresinol diglucoside (6)

Partial demethylation of compound **5** was carried out according to the regioselective method of Arvind et al. [27]. With minor modification in the reagent quantities. To a solution of compound **5** (1.90 g, 2.66 mmol) in 20 mL of dichloromethane, stannous chloride (1.00 g, 5.27 mmol) was added. The reaction mixture was stirred for 3 h at 40℃. Then 10 mL of 2 N HCl was added to the reaction mixture and stirring was continued for 10 min to obtain the crude demethylated compound **6**. The reaction mixture was washed with water (3 × 25 mL), the organic layer was dried over anhydrous sodium sulphate, and the solid product was chromatographed using silica gel column and hexane/chloroform (1:1) as eluent to obtain yellow crystal of title compound **6** with chemical structure represented in Scheme 1.

Yield: 1.30 g (56.2%) mp 119 - 121℃ (literature m.p118 - 120℃). IR (nujol, cm^{-1}): 3150, 3450 (OH). ^{1}H-NMR (CD_3OD) δ 6.6 (2H, s, $C_{2,2'}$-H $_{arom}$), 6.64 (2H, d, J= 8.1 Hz, $C_{5,5'}$-H $_{arom}$), 6.54 (2H, dd, J= 2, 8.1 Hz, $C_{6,6'}$-H $_{arom}$), 4.22 (2H, d, J= 7.8 Hz, $C_{9,9'}$ –β-*O*Gluanomeric), 4.1 (2H, dd, J= 5.6, 10 Hz, $C_{9a,9a'}$-H $_{aliph}$), 3.73 (6H, s, 2-OCH_3), 3.48 (2H, dd, J= 6.5, 9.9 Hz, $C_{9b,9b'}$-H $_{aliph}$), 3.12 - 3.94 (22H, m, $C_{9,9'}$-*O*-Glu) 2.7 (2H, dd, J= 7, 13.8 Hz, $C_{7b,7b'}$-H $_{aliph}$), 2.62 (2H, dd, J= 8.2, 13.8 Hz, $C_{7a,7a'}$-H $_{aliph}$), 2.10 (2H, m, $C_{8,8'}$-H $_{aliph}$).^{13}C-NMR (CD_3OD) δ 148.8 (C3/3'), 145.4 (C4/4'), 134 (C1/1'), 123 (C6/6'), 115.5 (C5/5'), 113.7 (C2/2'), 104.8 (C-*O*-Glu'.), 71.4 (C9/9'), 62.7 - 78.3

(C9/9'-*O*- Glu.), 56.4 (C-OCH3), 41.2 (C8/8'), 35.8 (C7/7'). MS: m/z: 686.7014(M + 1). $C_{32}H_{46}O_{16}$ (686.71) Anal. Cal. C, 55.97; H, 6.75. Found: C, 57.88; H, 7.80.

2.2 Determination of Antibacterial Effects

The antibacterial activity of the synthetic SDG was tested against *Salmonella typhimurium* (MTCC 98), *Escherichia coli* (MTCC 40), *Shigellaflexineri* (H57), *Staphyllococcusaureus* (MTCC 96), *Xanthomonas* (XL), *Bacillus subtilis* (MTCC 619), and *Streptococcus pyogens* (1925) which were obtained from the stock culture maintained at Department of Studies in Microbiology, University of Mysore, Mysore, India.

All the MTCC cultures employed in this current task were maintained on nutrient agar slants (Hi Media chemicals, India) and preserved at 4℃. Broth cultures of *S. typhimurium, E. coli, S. flexineri, S. aureus, Xanthomonas, B. subtilis,* and *S. pyogens* were prepared by inoculating the bacterial pathogens in nutrient broth (Hi-Media chemicals, India) and incubated at 28℃ for 24 h. Spread plate method was followed by inoculating 100 μL of broth cultures on a nutrient agar plate.

The agar medium was purchased from Hi-Media laboratories Ltd., Mumbai, India. Nutrient broth, subculture, base layer medium, agar medium and peptone water were prepared as per the standard procedure.

2.2.1 Agar–well diffusion assay

The antimicrobial activity was measured by agar well diffusion assay following the method of Bauer et al [28]. Sample dissolved in ethanol (2.5 mg/mL) was used for the assay. The wells were prepared in the plates with the help of a cork-borer (0.85 cm). About 50 μL of the test compound was introduced into the wells and allowed to diffuse for 2 h at 4℃. Plates were incubated at 37℃ for 48 h and the microbial growth was determined by measuring the diameter of inhibition zones. Methanol alone was used as a negative control and streptomycin as a positive control. The assay was carried out in triplicate and the mean values are presented.

2.2.2 Agar–disc diffusion method

Synthetic SDG was screened for antimicrobial activity by disc diffusion based on the method of Parekh et al [29]. Using Mueller-Hinton agar medium to study the preliminary antibacterial activity against *S. typhimurium, E. coli, S. flexineri, S. aureus, Xanthomonas, B. subtilis,* and *S. pyogens*. Synthetic SDG was dissolved in 5 mL of dimethyl sulfoxide (DMSO) (1000 ug/mL). 0.1 mL of this compound was introduced on the disc (0.7 cm) (Hi-Media) and allowed to dry. Then the disc was impregnated on the seeded agar plate. DMSO was used as a negative control, whereas, Streptomycin was used as a positive control. The plates were done in triplicates and were incubated. The antimicrobial activity was taken on the basis of the diameter of zone of inhibition.

2.2.3 Minimum Inhibitory Concentration (MIC)

The MIC was determined by the modified method of Dufouret al., and Gary et al. [30,31] Different concentrations (100 ppm to 300 ppm) of synthetic SDG and Chloramphenicol standard (positive control) and 100 μL of the bacterial suspension (10^5 CFU/mL) were placed aseptically in 10 mL of nutrient broth and incubated for 24 h at 37℃. The growth was observed both visually and by measuring O.D. at 600 nm at regular intervals followed by plating with nutrient agar. The lowest concentration of test sample showing no visible growth was recorded as the minimum inhibitory concentration. Sample tubes were maintained for each concentration of test sample and the readings were plotted against O.D at 600 nm as growth curves.

3. RESULTS AND DISCUSSION

3.1 Chemistry

The reaction sequence for title compound 6 was outlined in **Scheme 1**.This synthesis begins with the regeoselective bromination of the commercially available compound **1** (3,4-dimethoxy toluene) obtained from Sigma Aldrich Chemicals Co. (St. Louis, MO, USA). Compound **3** was achieved by the alkylation of compound **2** with butanediol. The methoxy groups in compound **3** played an important role in hindering the condensation reaction with pentaacetyl glucose which occurs at phenyl rings. Finally, compound **6** was achieved by the deacetylation and partial demethylation of compound **4** and **5** respectively. All the synthetic compounds were obtained in moderate yield and their structures were confirmed by IR, NMR and MS-TOF. The current synthesis scheme of SDG was found to be an efficient approach. The great advantage of this approach is that the compounds used are readily synthesized in quick time with less quantity of chemicals. Moreover, a good yield of the titled compound was achieved in this approach for the first time in contrast to previous procedures. Furthermore, purification process needed less time, as all the starting materials can be removed easily by flash chromatography, and purification of the products can be carried out by column chromatography.

3.2 Antibacterial Activities

3.2.1 Agar-well diffusion assay

The antibacterial activity of synthetic SDG was evaluated *in vitro* by agar well diffusion method against seven bacterial species. It exhibited significant antibacterial activity against all tested bacterial strains with various degrees of inhibition. The maximum activity (30 mm) was against E. coli, whereas, it was minimum (7 mm) against *S. pyogens*. On the other hand the activity was moderate for the rest of the tested bacterial strains as shown in Table 1.

3.2.2 Agar-disc diffusion method

The evaluation of the antibacterial activity of synthetic SDG was also carried out using another *in vitro* method which is agar disc diffusion.

The synthetic SDG was found to exert high and most pronounced antibacterial activity with inhibition zone of 27.3 mm against *E. coli* and lesser activity with minimum inhibition zone of 6.1 mm against *S. typhimurium. S. flexineri, S. aureus, Xanthomonas, B. subtilis,* and *S. pyogens* were also found to be moderately inhibited by the synthetic SDG as displayed in the Table 2.

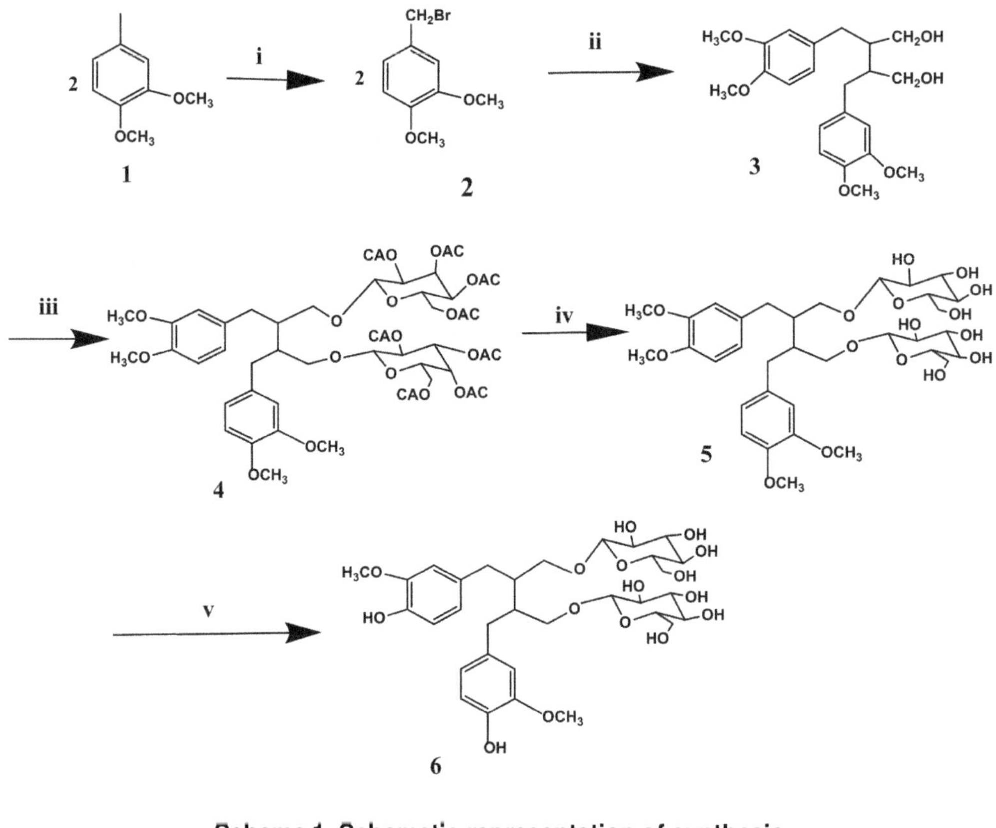

Scheme 1. Schematic representation of synthesis

Schematic representation for the Reagents and conditions used in the synthesis of SDG and various intermediates: (i) NBS, CCl_4, reflux, 11 h; (ii) 1,4- butanediol, n-But Li, DMF, mix together - 60°C – 05°C, 13 h; (iii) 2,3,4,6 tetra-o-acetyl α-D glucopyranosyl bromide,HCl, ethanol, reflux, 19 h; (iv) NaOH, ethanol, diethylether; (v) dichloromethane, stannous chloride, stirred, 3 h, 40°C, 2 N HCl

Table 1. Antibacterial activity of synthetic SDG against some bacteria using agar-well diffusion method

Compound	Conc.	Microorganisms Diameter of zone of inhibition (mm)						
		S. typhimurium	*E. coli*	*S. flexineri.*	*S. aureus.*	*Xanthomonas*	*B. subtilis*	*S. pyogens*
Streptomycin	30 μg/mL	25±0.12	34±0.09	34±0.16	26±0.20	32±0.15	23±0.23	30±0.05
Syn. SDG	2.5 mg/mL	18±0.34	30±0.21	17±0.31	23±0.14	12±0.17	21.4±0.53	7±0.11
Methanol	50 μL	-	-	-	-	-	-	-

Each value represents mean of three different observations ± S.D.(-) - no zone of inhibition observed, Streptomycin - as positive control, Methanol - as negative control

Table 2. Antibacterial activity of synthetic SDG against some bacteria using agar-discdiffusion method

Compound	Conc.	Microorganisms Diameter of zone of inhibition (mm)						
		S. typhimurium.	*E. coli*	*S. flexineri.*	*S. aureus.*	*Xanthomonas.*	*B. subtilis*	*S. pyogens*
Streptomycin	30 μg/mL	27.2±0.10	29.1±0.20	25.8±0.09	26.3±0.30	27.1±0.17	20.6±0.24	32.2±0.43
Syn. SDG	2.5 mg/mL	6.1±0.32	27.3±0.10	23±0.27	15.7±0.54	21.2±0.22	19.3±0.19	10.2±0.51
DMSO	50 μL	-	-	-	-	-	-	-

Each value represents mean of three different observations ± S.D. (-) - no zone of inhibition observed, Streptomycin - as positive control, DMSO- as negative control

Table 3. MIC values for Chloramphenicol as well as synthetic SDG against several bacteria

Compound	Microorganisms MIC (ppm)						
	S. typhimurium	*E. coli*	*S. flexineri.*	*S. aureus.*	*Xanthomonas.*	*B. subtilis*	*S. pyogens.*
Chloramphenicol	100	150	200	200	250	150	100
Syn. SDG	200	100	300	150	100	200	200

Chloramphenicol was used as positive control

3.2.3 Minimum Inhibitory Concentration (MIC)

The agar dilution method was used to determine the MIC of the synthetic SDG. MIC was defined as the lowest concentration of the test agent that inhibited bacterial growth, as indicated by the absence of turbidity. MIC valuesrange from 100 ppm to 300 ppm as given in Table 3. The test compound was effective against *E. coli* and *Xanthomonas* with MIC of 100 ppm and also it inhibited the growth of *S. aureus* at 150 ppm. *S. typhimurium, B. subtillis*and *S. flexineri* were completely inhibited by synthetic SDG at 200, 200, and 300 ppm respectively.The minimum concentration of the synthetic SDG that inhibits the growth of the tested organism in comparison to the antibacterial standard chloramphenicol was determined and listed in the Table 3. All the concentrations of the synthetic SDG showed strong activity against all the test organisms on concentration dependent manner.

The assessment of antibacterial properties associated with synthetic SDG was accomplished against important pathogenic bacteria in comparison to some antibacterial standards. The synthetic SDG was found to be a worthy antibacterial agent as it showed higher activity against *E. coli* when compared with other microorganisms either in agar well or disk diffusion assay. It also showed high inhibitory activities at minimum concentration against E. coli as well as against *Xanthomonas*. The activities of synthetic SDG against bacteria may be indicative to the broad spectrum activities. Similarly, closely related structure matairesinol, the second major lignans with traces of pinoresinol, lariciresinol and isolariciresinol showed antibacterial activity against several important pathogenic bacteria [9,10]. The phenolic chemical structure of SDG helps justify its antibacterial activity against various bacterial strains. In other words, presence of electron donating groups, methoxy group at Meta position and hydroxyl group at Para position in our target compound (SDG) have revealed good antimicrobial activity against these sorts of bacteria.As reported earlier, phenolic compounds at low concentration affect enzyme activity, especially of those enzymes associated with energy production, while at greater concentrations they cause protein denaturation. Furthermore, Mundt et al. [32] reported that, the effect of phenolic compounds on microbial growth could be the result of the ability of these compounds to alter microbial cell permeability, permitting the loss of macro-molecules from the interior and these compounds could also interact with membrane proteins causing a deformation in their structure and functionality as well as affecting cellular activity.

4. CONCLUSION

From the present study it can be concluded that, an efficient route for the synthesis of SDG was achieved and also the characterization of these compounds was discussed. The preliminary antimicrobial activities of the synthetic SDG showed a broad spectrum of activity against clinically important pathogenic bacteria responsible for causing most common bacterial diseases. This antibacterial effect of the synthetic SDG against these panels of vital pathogenic bacteria can be a preferred supplement due to its known health benefits as an antibacterial agent. It would also assist for development of a new alternative medicine system and opens the possibility of finding new clinically effective antimicrobial compounds and production of new antibiotics. It is essential that the research should be continued to use this promising compound in experimental animals and carefully investigate and elucidate its mechanism of action.

COMPETING INTERESTS

Authors have declared that no competing interests exist.

REFERENCES

1. Service F. Antibiotics that resist resistance. Scien. 1995;270:724-727.
2. Westh H, Zinn S, Rosdahl T. An international multicenter study of antimicrobial consumption and resistance in *Staphylococcus aureus* isolates from 15 hospitals in 14 countries. Microb. Drug Resist. 2004;10:169-176.
3. Bandow E, Brotz H, Leichert I. Proteomic approach to understanding antibiotic action. Antimicrob. Agents Chemother. 2003;47:948-955.
4. Rojas R, Bustamante B, Bauer J. Antimicrobial activity of selected Peruvian medicinal plants. J. Ethnopharmacol. 2003;88:199- 204.
5. Sharon R, Thompson U. Phytoestrogens and Lignans: Effects on reproduction and chronic disease. Antinutrients and Phytochemicals in Food. Am. Chem. Soc. 1997;16:273-293.

6. Marina T, Anne M, Agnès F, Maryvonne N, Marie-Christine R, Françoise CJ. Dietary lignan intake and postmenopausal breast cancer risk by estrogen and progesterone receptor status. Natl. Cancer Inst. 2007; 99:475-86.
7. Saleem M, Kim H, Ali M, Lee Y. An update on bioactive plant lignans. Nat. Prod. Rep. 2005;22:696-716.
8. Thompson L. Experimental studies on lignans and cancer. BaillieresClin. Endocrin. Metabo. 1998;12:691-705.
9. Moujir L, Seca A, Silva A, López M, Padilla N, Cavaleiro J, Neto C. Cytotoxic activity of lignans from *Hibiscus cannabinus*. Fitoterapia. 2007;78:385-387.
10. Barbary M, El-sohaimy A, El-saadani A, Zeitoun M. Antioxidant, antimicrobial and anti-HCV activities of lignan extracted from flaxseed. J. Agri. Biol. Sci. 2010;6:247-256.
11. Obermeyer WR, Musser SM, Betz JM, Casey RE, Pohland AE, Page SW. Chemicalstudies of phytoestrogens and related compounds in dietary supplements. Flax and Chaparral. Proc. Soc. Exp. Biol. Med. 1995;208:6-12.
12. Rickard SE, Thompson LU. Health effects of flaxseed mucilage. L. Inform. 1997;8:860-865.
13. Rajesha J, Ranga A, Madhusudhan B, Karuna M. Antibacterial properties of secoisolariciresinol diglucoside isolated from Indian flaxseed cultivars. Curr. Trends Biotechnol. Pharm. 2010;4:551-560.
14. Rajesha J, Kotamballi N, Chidambara M, Karun M, Basavaraj M, Gokare R. Antioxidant potentials of flaxseed by *in vivo* model. J. Agric. Food Chem. 2006;54: 3794-3799.
15. Prasad K. Oxidative stress as a mechanism of diabetes in diabetic BB prone rats: Effect of secoisolariciresinol diglucoside (SDG). Mol. Cell. Biochem. 2000;209:89-96.
16. Serraino M, Thompson L. Flaxseed supplementation and early markers of colon carcinogenesis. Cancer Lett. 1992; 63:159-65.
17. Serraino M, Thompson L. The effect of flaxseed supplementation on early risk markers for mammary carcinogenesis. Cancer Lett. 1991;60:135-142.
18. Malcolm O, Neda, C, Hope W, Harold A. Effects of flaxseed derivatives in experimental polycystic kidney disease vary with animal gender. Lipids. 2006; 41:1141-1149.
19. Stasevich O, Mikhalenok S, Kurchenk V. Isolation of secoisolariciresinol diglucoside from lignan-containing extracts of *Linum usitatissimum* seeds. Khim. Prir. Soedin. 2009;1:21-23.
20. Kuijsten A, Arts IC, Vree TB, Hollman PC. Pharmacokinetics of enterolignans in healthy men and women consuming a single dose of secoisolariciresinol diglucoside. J Nutr. 2005;135(4):795-801.
21. Khamlach K, Dhal R, Brown E. Heterocycles. 1990;31:2195-2199.
22. Singh. P, Singh D. An elegant synthesis of 6H-benzofuro [3,2-c][1]benzopyran-6-ones. Heterocycles. 1985;23:903-90.
23. Yamu X, Yuanyuan Z, Wei W, Yining D, Rui H Synthesis and bioactivity of erythro-nordihydroguaiaretic acid, threo-(-)-saururenin and their analogues. J. Serb. Chem. Soc. 2010;75:1325-1335.
24. Satoshi Y, Yoshimasa H, Takuya K, Toshiya M. Synthesis and antioxidant activity of olivil-type lignans. Biosci. Biotechnol. Biochem. 2005;69:113-122.
25. Sadiq M, Shaukath K, Rajesha. Secoisolariciresinol Diglucoside - A Phytoestrogen Nutraceutical of Flaxseed: Synthesis and Evaluation of Antioxidant Potency. J. Free Rad. Antiox. 2011;1:31-38.
26. Hong A, Zhan U, Shi C, Chin. Chem. Lett. 2003;14:371-374.
27. Arvind N, Sunil C, Sachin S, Asish B. Prosess for regioselective demethylation of p-methoxy group in phenolic ester and diaryl ketone moieties. US. 2006;7:218B2.
28. Bauer W, Kirby M, Sherris C. Antibiotic susceptibility testing by a standardized single disk method. Am. J. Clin. Pathol. 1966;45:493-496.
29. Parekh J, Chanda S. *In-vitro* antimicrobial activities of extracts of Launaea procumbens Roxb. (Labiateae), *Vitis vinifera* L. (Vitaceae) and *Cyperus rotundus* L. (Cyperaceae) Afric. J. Biomed. Res. 2006;9:89-93.
30. Dufour M, Simmonds S Bremer. Development of a method to quantify in vitro the synergistic activity of natural

antimicrobials. J. Int. J. Food Microbiol. 2003;85:249-258.

31. Gary D, Ryszard A, Ronald P. Food Microbiol. 2003;20:211-216.

32. Mundt S, Kreitlow S, Jansen R. Fatty acids with antibacterial activity from the cyanobacterium Oscillatoriaredekei HUB 051. J. Appl. Phycol. 2003;15:263-267.

11

Leaf Ethanolic Extract of *Bauhinia monandra* Increases Insulin Secretion in Rats Subjected to Intrauterine Malnutrition

A. C. C. Argolo[1], V. C. R. Dantas[2], A. C. R. D. Saturnino[2], A. F. S. Santos[3], J. Brandão-Neto[2], A. D. O. Paixão[4] and L. C. B. B Coelho[1*]

[1]*Departamento de Bioquímica, Centro de Ciências Biológicas, Universidade Federal de Pernambuco, UFPE, Recife, PE, Brasil.*
[2]*Programa de Pós-Graduação em Ciências da Saúde, RN, Brasil.*
[3]*CEB-Centre of Biological Engineering, University of Minho, Braga, Portugal.*
[4]*Departamento de Fisiologia e Farmacologia, Centro de Ciências Biológicas, Universidade Federal de Pernambuco, UFPE, Recife, PE, Brasil.*

Authors' contributions

This work was carried out in collaboration between all authors. Authors ADOP and LCBBC designed the study, wrote the protocol and supervised the work. Authors ACCA and ACRDS carried out all laboratory work and performed the statistical analysis. Author VCRD managed the analyses of the study. Authors JBN and LCBBC wrote the first draft of the manuscript. Author AFSS managed the literature searches and edited the manuscript. All authors read and approved the final manuscript.

Editor(s):
(1) Yi-Ren Hong, College of Medicine, Kaohsiung Medical University, Taiwan.
(2) Mohamed Fawzy Ramadan Hassanien, Biochemistry Department, Zagazig University, Egypt.
Reviewers:
(1) Anonymous, Gunma University, Japan.
(2) Mathew Folaranmi Olaniyan, Achievers University, Owo, Nigeria.
(3) Upendarrao Golla, Indian Institute of Science Education and Research (IISER) Bhopal, India.

ABSTRACT

Aims: This work aimed to investigate the effects of *Bauhinia monandra* ethanolic extract (E) on blood glucose levels (G), insulin and lipids of control (C) and prenatal malnourished (M) rats.
Samples: Animals (from 90 days) were on standard or isocaloric high glucose diet (67%, w/w) during 30 days.

Corresponding author: E-mail: lcbbcoelho@gmail.com

Study Design: Animals received 0.1% DMSO (2.5 ml/kg, po.) or E (500 mg/kg, po.). Glucose tolerance was evaluated in blood following glucose overload (1.5 g/kg, po.).
Place and Duration of Study: Departamento de Bioquímica and Departamento de Fisiologia e Farmacologia, Universidade Federal de Pernambuco, between January 2006 and February 2007.
Methodology: In acute tests, glucose tolerance and serum insulin were evaluated in C and M receiving DMSO or E. Lipids and hepatic enzymes were analyzed before DMSO or E application. Glucose was orally administered by gavage, for glucose tolerance; thirty minutes later, E or DMSO was given. Blood samples were collected for glucose and insulin measurements. In the chronic assay, glucose tolerance, lipids and hepatic enzymes were studied in groups C or M receiving DMSO or E. The group M receiving E had a diet rich in glucose (30 days). Glucose tolerance, lipids, and hepatic enzymes were evaluated before and after diet rich in glucose treatment with E. Serum insulin was measured by chemiluminescence; also, aspartate and alanine aminotransferases were analyzed after high dietary glucose. Serum total cholesterol and triglycerides were estimated with an auto analyzer.
Results: In control rats, acute E administration reduced G, but did not change insulin secretion; in M rats, it induced insulin peak. In chronic study, animals on high glucose diet with E showed lower G; prenatal malnourished rats revealed higher insulin levels. Cholesterol, triglycerides and alanine aminotransferase did not change by intrauterine malnutrition or E.
Conclusion: Results indicated that E may act on beta-cells and stimulate insulin secretion.

Keywords: Bauhinia monandra; leaf ethanolic extract; hypoglycemic activity; blood glucose; serum insulin.

ABBREVIATIONS

ALT: Alanine Aminotransferases; AST: Aspartate Aminotransferases; BmoLL: Bauhinia Monandra Leaf Lectin; BmoRoL: B. Monandra Root Lectin; C: Control; E: B. Monandra Ethanolic Extract; G: Blood Glucose Levels; GCE: Group Control Receiving a Diet Rich in Glucose and Treated with Extract; Group CE: Control Receiving B. Monandra Ethanolic Extract; Group CV: Control Receiving Vehicle; GCV: Group Control Receiving a Diet Rich in Glucose and Treated with Vehicle; GME: Group of Prenatal Malnourished Wistar Rats Receiving a Diet Rich in Glucose and Treated with Extract; Group ME: Prenatal Malnourished Wistar Rats Receiving B. Monandra Ethanolic Extract; Group MV: Prenatal Malnourished Wistar Rats Receiving Vehicle; GMV: Group of Prenatal Malnourished Wistar Rats Receiving a Diet Rich in Glucose and Treated with Vehicle ; GTT: Glucose Tolerance; HDL: High Density Lipoprotein; M: Prenatal Malnourished Wistar Rats; NIDDM: Non-insulin Dependent Diabetes Mellitus; V: Vehicle.

1. INTRODUCTION

Plants from the genus *Bauhinia* are widely distributed in most countries and have been frequently used in folk medicine to treat several ailments, including diabetes [1,2,3]. In recent years, with the higher incidence of diabetes in the world, the interest in these plants has increased considerably [4,5]. Medicinal plants have been an alternative for treatment of diabetes in developing countries. There are more than 350 traditional plants already used with this aim [6] and the antidiabetic activity is mainly due to presence of the secondary metabolite in plants [7]. Therefore, active compounds of these plants may be used to develop new drugs for treatment of diabetes. The therapeutic properties of *Bauhinia* have been supported by studies using plant extracts [8,2]. Aqueous, ethanolic and hexanic extracts [2,4] from *B. forficata* leaves have induced hypoglycemia in rats. The *Bauhinia* genus presents terpenes, steroids, lactones and flavonoids [8]. *B. monandra* Kurz ("pata-de-vaca", "orquidea del pobre", pulse or "Napoleon's plume") is an ornamental Fabaceae species with an expressive content of galactose specific proteins, namely *B. monandra* leaf lectin, BmoLL [1] and *B. monandra* root lectin, BmoRoL [9]. BmoLL showed hypoglycemiant activity without cytotoxic or genotoxic effects [10]. In Northeastern Brazil, the leaves of *B. monandra* are largely used in folk medicine to reduce blood glucose levels.

In this way, type 2 diabetes mellitus previously called non-insulin dependent diabetes mellitus (NIDDM) or maturity-onset diabetes is a combination of tissue resistance to insulin action with an abnormal insulin secretion in response to glucose [6] and beta cell dysfunction, strongly associated with obesity and a sedentary lifestyle [11]. Since 2010, it was estimated that 280 million of people had diabetes, with NIDDM making up about 90% of the cases globally. The incidence of this disease is increasing rapidly and at the end of 2030, the number of cases will be double, because of increasing longevity and obesity [6]. Diabetes is an important problem to health policy in all world and its prevalence is on a steady increase worldwide. This disease is now identified as one of the main threats to human health in the 21st century [11] considering its high incidence and chronic complications, such as coronary heart disease [12], nephropathy, retinopathy [13], chronic kidney disease [14], renal failure, blindness, amputations and hospitalizations [15].

Also, glucose intolerance and hyperinsulin can be induced by prenatal malnutrition [16]. Furthermore, when prenatal malnutrition is experimentally induced in rats, oligonephronia [17,18] and its consequences such as hypertension, renal hemodynamic and glomerular morphological changes [17] have been observed. The factors above are compatible with the development of chronic renal failure, commonly accompanying diabetes. In accordance with these findings, prenatal malnutrition and low birth weight in humans have been correlated with renal disease during adult life [19].

Currently available oral agents for the treatment of type 2 diabetes mellitus include a variety of compounds from five different pharmacologic classes with distinct mechanisms of action: the sulfonylureas, metiglinides, biguanides, α-glucosidase inhibitors, and thiazolidinediones [12]. In general, α-glucosidase inhibitors delay carbohydrate absorption, metiglinides and sulfonylureas increase insulin supply; biguanides and thiazolidinediones enhance insulin action [12]. However, these drugs have limited use because of pathogenically undesirable conditions and high rates of secondary failure [6].

The aim of this work was to investigate the acute and chronic effects of the ethanolic extract from *B. monandra* leaves to treatment of rats subjected to intrauterine malnutrition. The use of this plant tissue in the popular medicine to diabetes treatment was the main motivation of the present study. The extract effects were evaluated on blood glucose and insulin in rats; lipids and hepatic enzymes were also investigated.

2. MATERIALS AND METHODS

2.1 Preparation of the Extracts

All chemicals were of analytical grade. Solvents for the extraction were purchased from Merck (Darmstadt, Germany). Leaves of *B. monandra* Kurz were harvested from ornamental trees in Recife (State of Pernambuco, Northeast of Brazil). A sample of the collected material is archived as voucher specimen number 57462, at the herbarium Dárdano de Andrade Lima, Empresa Pernambucana de Pesquisa Agropecuária, Recife City, Brazil. Dried and powdered leaves of *B. monandra* (200 g) were extracted with 95% ethanol (1 L) until exhaustion. After filtration, the combined extracts were concentrated at 40°C to dryness, the ethanolic extract (E).

2.2 Diets

Intrauterine malnutrition was induced using a multideficient diet, as previously described [20,17]. The content of its main nutrients was (g/g%): proteins 9, carbohydrates 78, lipids 1.1, fiber 7, minerals 4, sodium chloride 0.17 and Kcalories 356. No vitamin supplement was added. However, part of the diet was supplemented with 0.2% (g/g) sodium chloride. The same nutrient in the standard diet (as indicated by the manufacturer, Purina, Agribands do Brasil, Paulínia, SP, Brazil) contains the following contents (g/g%): protein 23, carbohydrates 41, lipids 2.5, fibers 9, minerals 8, sodium chloride 0.37 and Kcalories 278.

The content of the diet high in glucose was (g/g%): glucose 66.73, standard diet 33.27 (Purina, Agribands do Brasil, Paulínia, SP, Brazil) to obtention of an isocaloric diet.

2.3 Animals

All experiments were performed on control (C) and intrauterine malnourished (M) Wistar rats of both genders (250-300 g). Animals were kept under a constant 12 h light and dark cycle and an environmental temperature of 23°C. Rats, C and M were obtained of dams on a standard or on a multi deficient diet, respectively, during mating and pregnancy. After parturition, all rats had free access to food standard diet and water. At adult age, part of the rats was maintained from 13-16 weeks on a standard diet, the groups C and M and another part was maintained for the same period on a diet high in glucose (G), the groups GC and GM. All authors hereby declare that "Principles of laboratory animal care" (NIH publication No. 85-23, revised 1985) were followed, as well as specific national laws where applicable. All experiments have been examined and approved by the the Ethic Committee from the Centro de Ciências Biológicas, Universidade Federal de Pernambuco (document number 35/05).

2.4 Experimental Protocol

The biological assays were performed into two series, the acute test and the chronic test. In the first series, glucose tolerance test (GTT), and serum insulin were evaluated in C and M receiving vehicle (V, 0.1% DMSO), the groups CV (n=8) and MV (n=8); and in C and M receiving E. (500 mg/kg, po.), the groups CE (n=8) and ME (n=7). Additionally, lipids and hepatic enzymes were studied, before V or E was administered in all animals. Animals were fasted for 12 h to collect blood samples for biochemical measurements. Blood withdrawal was performed from retro-orbital plexus in the animals under ether anesthesia. Glucose (1.5 g/kg) was orally administered by gavage, for GTT study, after a basal blood sample collection. Thirty minutes later, E or V (2.5 mL/kg, po.) were administrated, also by gavage. Subsequently, blood samples were collected after 1, 2 and 4 h, for glucose and insulin measurements.

In the second series, GTT, lipids and hepatic enzymes were studied in GCV (n=9), GMV (n=13), GCE (n=9) and GME (n=13) receiving a diet rich in glucose and simultaneously treated with E (500 mg/kg, po) or V (2.5 mL/kg, po), during 30 days. GTT, lipids, and hepatic enzymes were evaluated before and after diet or E treatment. Body weight, food intake and blood glucose levels were monitored weekly during 30 days of treatment.

2.5 Analytical Methods

Blood glucose levels were determined by the glucose oxidase method [21] immediately after collection. Assay of serum insulin in blood was measured by chemiluminescence (Beckman Coulter™, Access®, USA). Serum activities of aspartate and alanine aminotransferases, AST and ALT, respectively, were measured by the kinetic method [22] after high dietary glucose. Serum total cholesterol and triglycerides were estimated by enzymatic method with an autoanalyzer using commercial reagents.

2.6 Statistical Analysis

Data were expressed as mean ± standard error of the mean (SEM). Statistical analysis was evaluated by using one way ANOVA, followed by Student-Neuman-Keuls test to determine the significant difference between the groups. *P* values of = .05 were considered to be statistically significant.

3. RESULTS AND DISCUSSION

3.1 Body Weight

Prenatal malnutrition was based on the low birth weight from offspring of dams submitted to multideficient diet [23]. Higher basal serum glucose and insulin levels as well as alterations in glucose homeostasis after an acute glucose overload indicated that this was a valuable model to study hypoglycemic activity in leaf extracts from *B. monandra*. The body weight at birth was 19% lower in M than in C rats (5.2±0.2 vs. 6.4±0.08 g, *p* = .05, respectively), while at adult age the values of body weight were the same for both groups (364±24 vs. 323±19 g, respectively). In the groups chronically treated with E, basal values of body weight were similar for both GC and GM groups; at the end of E administration there was a not significant tendency of body weight loss in both GC and GM groups (Fig. 1).

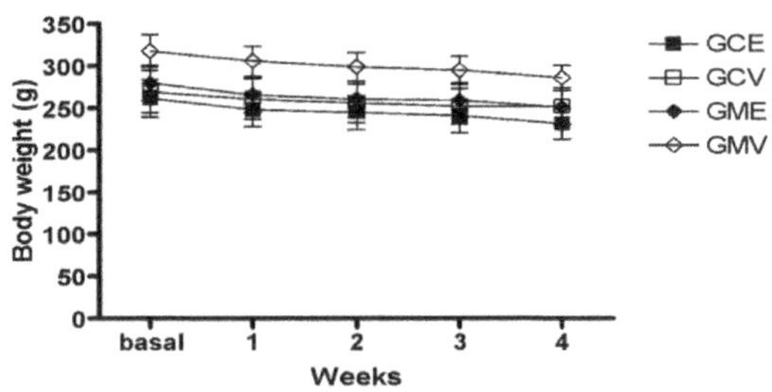

Fig. 1. Effect of leaf ethanolic extract from *B. monandra* on body weight of control, GCE (▪), GCV (▫), and prenatal malnourished rats, GME (◆) and GMV (◇) for 4 weeks.

Values are given as mean ±S.E. n=7/8

The present findings corroborate with previous studies showing glucose intolerance [19]. Furthermore, the prenatal malnutrition is also a condition in which growth factors occur in serum [24] that may contribute with atherosclerotic disease, a common chronic complication present in non-insulin dependent diabetes mellitus [25].

3.2 Acute and Chronic Effect of E on Serum Glucose and Insulin Levels

In the acute study, the effect of E on serum glucose levels is shown in Figs. 2 (A and B).

Basal fasting serum glucose was higher (18%, p = .05) in M than in C rats (122±4 vs. 103±6 mg/dL, respectively). Both CV and MV presented a glucose peak in the first hour after glucose overload (188±6 and 171±8 mg/dL, respectively). During the peak, the glucose level was lower in CE and ME as compared with their respective control groups (161±7 and 156±5 mg/dL, respectively). In the fourth hour after V or E administration, the serum glucose levels returned to their basal values in all groups. Basal serum insulin in CV was lower than in MV rats (0.32±0.11 vs. 2.17±0.82 ∫IU/mL, p = .05, respectively, Fig. 2 C and D). Different from CV, the CE group did not show a peak of insulin secretion in the first hour after glucose overload (1.36±0.45 vs. 0.52±0.26 ∫IU/mL, respectively). However, ME presented a peak of insulin secretion in the first hour (1.93±1.03 ∫IU/mL), while CE did in the second hour after glucose overload (1.40±0.57 ∫IU/mL). Acute E administration reduced the glucose level in C rats, however the peak of insulin secretion in CE rats was retarded to 2 h (Fig. 2 C).

Serum glucose evaluated weekly during chronic treatment was found to be reduced in GC and GM rats (p = .05, Fig. 3).

At the end of treatment, during GTT evaluation, basal fasting serum glucose was higher in GMV and GCV groups than that seen in GME and GCE groups, respectively (137±4 vs. 100±5 mg/dL and 119±5 vs. 94±5 mg/dL, respectively; p = .05, Figs. 4A and B). Also, serum glucose was high in GMV than in GCV group in the first and fourth hour, respectively (205±6 vs. 184±4 mg/dL and 144±4 vs. 125±4 mg/dL, respectively; p = .05, Figs. 4A and 4B). In the two groups chronically treated with E, GME and GCE, the serum glucose peak after acute glucose overload was lower than that in their respective control groups, GMV and GCV (164±8 vs. 205±6 and 153±6 vs. 184±4 mg/dL, respectively; p = .05, Fig. 4 A and B).

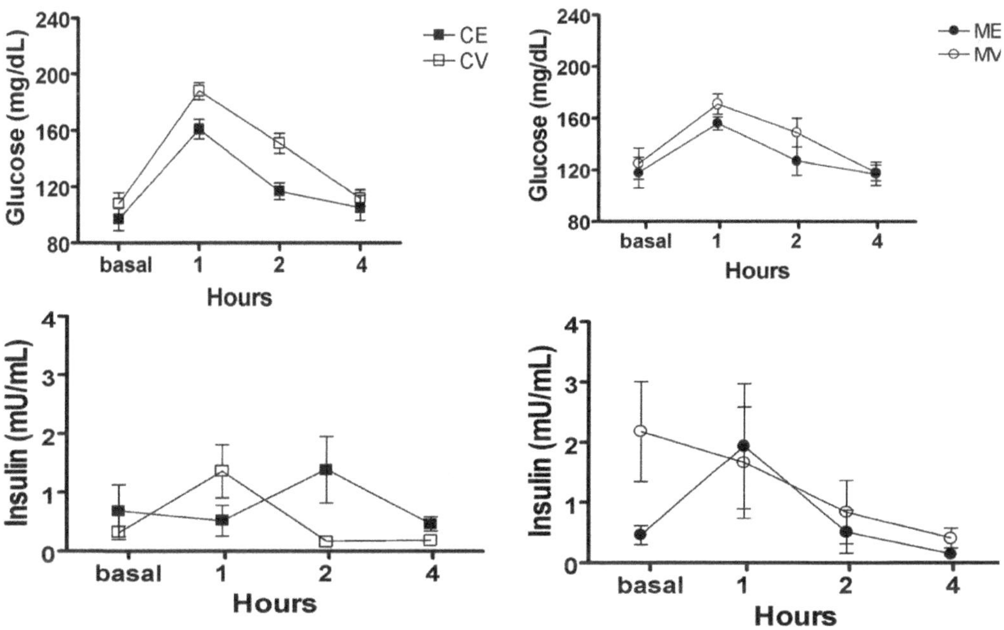

Fig. 2. Acute effect of leaf ethanolic extract from *B. monandra* on serum glucose (A and B) and insulin (C and D) levels of control (CE and CV, n=8) and prenatal malnourished (ME, n=7 and MV, n=8) rats. CE (•), CV (•), ME (•) and MV (○)

*Values are given as mean ±S.E. * P = .05 (CE vs. CV), # P =.05 (ME vs. MV).*

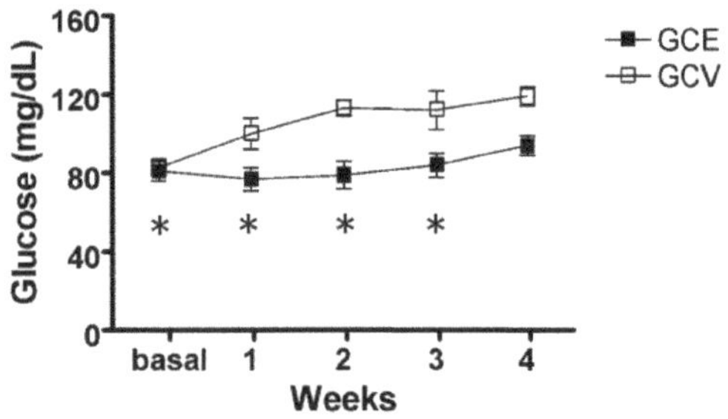

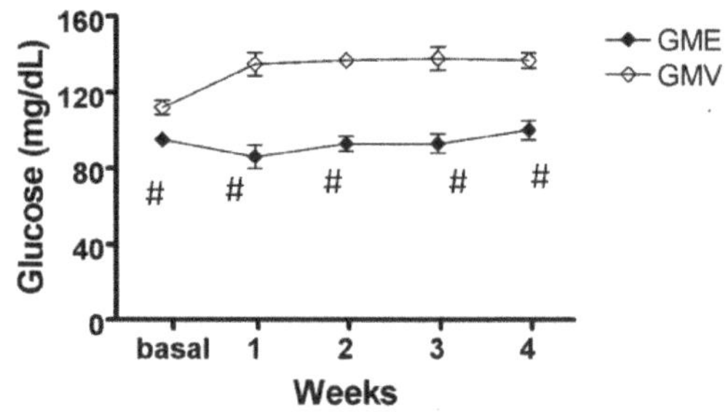

Fig. 3. Effect of leaf ethanolic extract from *B. monandra* on weekly glucose level of control (A; GCE and GCV, n=9) and prenatal malnourished (B; GME and GMV, n=13) rats. GCE (▪), GCV (▪), GME (◆) and GMV (◇)

*Values are given as mean ±S.E. * P = .05 (GCE vs. GCV), # P = .05 (GME vs. GMV)*

At the end of treatment, during GTT evaluation, basal level of serum insulin was similar for both groups, GME and GCE rats (0.89±0.31 μIU/mL, and 0.51±0.08 μIU/mL, Figs. 4C and 4D). However, at the first hour, the levels of serum insulin were higher in GME than in GMV rats (0.44±0.11 μIU/mL vs. 0.167±0.032 μIU/mL, respectively; p = .05). Different from GCE, GME group did not show a peak of insulin secretion in the first hour after glucose overload (1.64±0.43 vs. 0.44±0.11 μIU/mL, respectively; p = .05). The values of serum insulin peak were similar for both groups GCE and GCV. Although, neither GMV nor GME had presented an insulin peak, GME presented higher insulin secretion during all points of glucose tolerance curve. Thus, E seemed to restore the capacity of M rats to secrete insulin.

It was reported by Lino et al. [2] that aqueous, ethanolic and hexanic extracts of *Bauhinia forficata* leaves showed antidiabetic activity; so, this plant extracts can be used in the treatment of diabetes mellitus type II. Antihyperglycemic activity was also revealed when it was used aqueous extract of fresh leaves of *B. forficata* (1 g/kg, p.o.) [4]. In our study, based on the hypoglycemic effect of E on control and prenatal malnourished rats, and on the levels of plasma insulin, it may be hypothesized that the mechanisms involved include an augmentation on insulin secretion, like the sulfonylureas, an oral agent for the treatment of type 2 diabetes mellitus compound, lower blood glucose in normal and type 2 diabetic animals by stimulating insulin release from pancreatic cells [6].

These findings are in agreement with hypoglycemic activity induced by extracts from other *Bauhinia* species, such as *B. candicans, B. forficata, B. variegata* and *B. tomentosa* [2,3,5].

3.3 Profile of Cholesterol, Triglycerides, AST and ALT in Control and Prenatal Malnourished Rats

Plasma levels of AST and ALT, withdrawn before V or E administration in the acute study, were similar for both groups, M (150±10.5 and 71±2 IU/L, respectively) and C (164±8 and 59±5 IU/L, respectively), while the levels of total cholesterol and triglycerides, also withdrawn before V or E administration in the acute study, were higher in C (100.4±5.3 and 254±37 mg/dL, respectively; p<0.05) than M (77.8±7 an *B. variegata* leaf ethanolic extractd 140±15 mg/dL, respectively; p<0.05). Table 1 shows the effect of chronic E treatment on the plasma levels of AST, ALT, total cholesterol and triglycerides. It was remarkable that AST was lower in GMV than the one observed in GCV rats (139±8 vs. 187±9 IU/L, respectively; p< 0.05). Regarding E effects, the group GME presented AST 55% higher (p < 0.05) than that one seen in GMV.

Thiruvenkatasubramaniam and Jayakar [3] and Aljobouri et al. [5] studied the hypoglycemic effect of *Bauhinia variegata* leaf ethanolic and aqueous extract and there was significant reduction in total cholesterol, triglycerides and elevation of high density lipoprotein (HDL) in diabetic mice. Also, Lino et al. [2] study showed reductions in plasma glucose, triglycerides, total cholesterol and HDL-cholesterol in a model of alloxan-induced diabetes in rats treated by extracts of *B. forficata.*

Table 1. Aspartate aminotransferase (AST), alanine aminotransferase (ALT), triglycerides and cholesterol levels in the serum of control (C) and malnourished (M) rats treated with vehicle (V) and extract (E)

	GCE (n=9)	GCV (n=9)	GME (n=13)	GMV (n=13)
AST (IU/L)	205±15	187±9.4 #	216.7±13.4#	139.1±7.6
ALT (IU/L)	74.7±6	70.1±3.8	73.5±4.3	70.1±1.9
Triglycerides (mg/dL)	186±34.1	188.4±34.5	122.3±16.4	129.6±13.6
Cholesterol (mg/dL)	82.3±6.8	90±4.7	80±4.7	84.2±4.8

Results are expressed as means ± SEM. # P = .05 (vs. GMV). n=7/8. GCE: Group control receiving a diet rich in glucose and treated with extract; GCV: Group control receiving a diet rich in glucose and treated with vehicle; GME: Group of prenatal malnourished Wistar rats receiving a diet rich in glucose and treated with extract; GMV: Group of prenatal malnourished Wistar rats receiving a diet rich in glucose and treated with vehicle

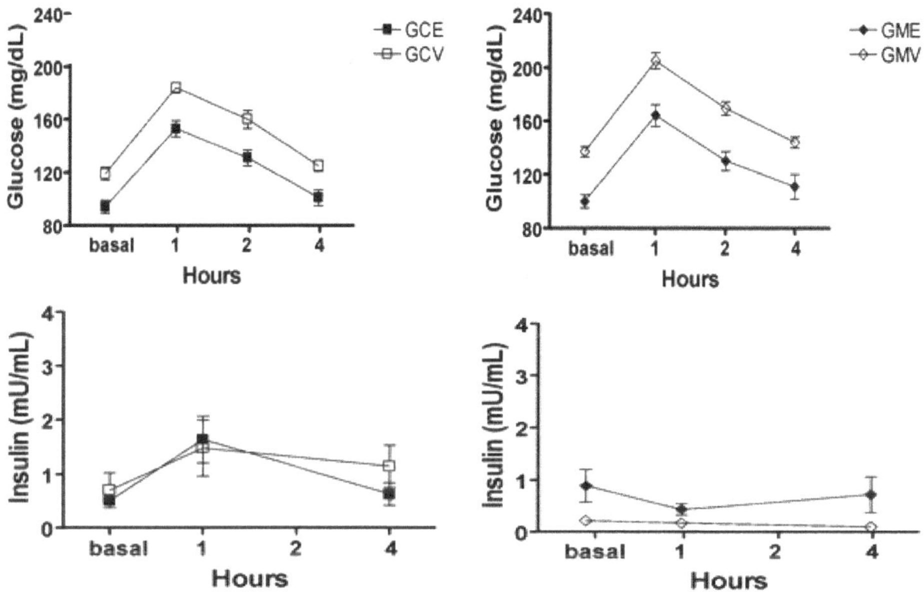

Fig. 4. Chronic effect of leaf ethanolic extract from *B. monandra* on serum glucose (A and B) and insulin (C and D) level of control (GCE and GCV, n=9) and prenatal malnourished (GME and GMV, n=13) rats. GCE (▪), GCV (▫), GME (◆) and GMV (◇)

*Values are given as mean ±S.E. * P = .05 (GCE vs. GCV), # P = .05 (GME vs. GMV)*

4. CONCLUSIONS

The results obtained in this work provide evidence for a direct pancreatic action of *B. monandra* leaf ethanolic extract, probably through an activity on endocrine pancreatic β cells. This is an innovative study that reports for the first time that a hypoglycemic effect of *B. monandra* leaf ethanolic extract may act on beta-cells and stimulate insulin secretion. The property should be explored with good prospects for pharmacological application.

COMPETING INTERESTS

Authors have declared that no competing interests exist.

REFERENCES

1. Coelho LCBB, Silva MBR. Simple method to purify milligram quantities of the galactose specific lectin from the leaves of *Bauhinia monandra*. Phytochem Analysis. 2000;11:295-300.
2. Lino CS, Diógenes JPL, Pereira BA, Faria RAPG, Andrade Neto M, Alves RS, Queiroz MGR, Sousa FCF, Viana GSB. Antidiabetic activity of *Bauhinia forficata* extracts in alloxan-diabetic rats. Biol Pharm Bull. 2004;27(1):125-27.
3. Thiruvenkatasubramaniam R, Jayakar B. Anti-hyperglycemic and anti-hyperlipidaemic activities of *Bauhinia variegata* l

on streptozotocin induced diabetic rats. Scholars Research Library. Der Pharmacia Lettre. 2010;2(5):330-34.

4. Vasconcelos F, Sampaio SV, Garofalo MA, Guimaraes LF, Giglio JR, Arantes EC. Insulin-like effects of *Bauhinia forficata* aqueous extract upon *Tityus serrulatus* scorpion envenoming. J Ethnopharmacol. 2004;95:385-92.
5. Aljobouri AM, Rashid KI, Ibrahim SA, Zyer AAJ, Abbas ZN. Study the effect of *Bauhinia variegata* Linn. ethanolic extract on reducing glucose and lipid levels of white albino mice. Int J Curr Microbiol App Sci. 2015;4(3):652-58.
6. Surya S, Salam AD, Tomy DV, Carla B, Kumar RA, Sunil C. Diabetes mellitus and medicinal plants-a review. Asian Pac J Trop Dis. 2014;4(5):337-47.
7. Patel DK, Kumar R, Laloo D, Hemalatha S. 2012. Diabetes mellitus: An overview on its pharmacological aspects and reported medicinal plants having antidiabetic activity. Asian Pac J Trop Biomed. 2012;2(5):411-20.
8. Argolo ACC, Sant'Ana AEG, Pletsch M, Coelho LCBB. Antioxidant activity of leaf extracts from *Bauhinia monandra*. Bioresource Technol. 2004;95:229-33.
9. Souza JD, Silva MBR, Argolo ACC, Napoleão TH, Sá RA, Correia MTS, Paiva PMG, Silva MDC, Coelho LCBB. A new *Bauhinia monandra* galactose-specific lectin purified in milligram quantities from secondary roots with antifungal and termiticidal activities. Int Biodeterior Biodegradation. 2011;65:696-702.
10. Sisenando HAACN, Macedo MFS, Saturnino ACRD, Coelho LCBB, Medeiros SRB. Evaluation of the genotoxic potential of *Bauhinia monandra* leaf lectin (BmoLL). Food Chem Toxicol. 2009;47: 303-08.
11. Ezuruike UF, Prieto JM. The use of plants in the traditional management of diabetes in Nigeria: Pharmacological and toxicological considerations. J Ethnopharmacol. 2014;155(2):857-924.
12. Anselmino M. Cardiovascular prevention in type 2 diabetes mellitus patients: The role of oral glucose-lowering agents. J Diabetes Complicat. 2009;23:427-33.
13. Takao T, Matsuyama Y, Suka M, Yanagisawa H, Kikuchi M, Kawazu S. Time-to-effect relationships between systolic blood pressure and the risks of nephropathy and retinopathy in patients with type 2 diabetes. J Diabetes Complications. 2014;28(5):674-78.
14. Pugliese G, Solini A, Bonora E, Fondelli C, Orsi E, Nicolucci A, Penno E. Chronic kidney disease in type 2 diabetes: Lessons from the renal insufficiency and cardiovascular events (RIACE) italian multicentre study. Nutr Metab Cardiovasc Dis. 2014;24:815-22.
15. Defossa E, Wagner M. Recent developments in the discovery of FFA1 receptor agonists as novel oral treatment for type 2 diabetes mellitus. Bioorg Med Chem Lett. 2014;24:2991-3000.
16. Zou M, Arentson EJ, Teegarden D, Koser SL, Onyskow L, Donkin SS. Fructose consumption during pregnancy and lactation induces fatty liver and glucose intolerance in rats. Nutr Res. 2012; 32(8):588-98.
17. Paixão ADO, Maciel CR, Teles MBB, Figueiredo-Silva J. Regional Brazilian diet-induced low birth weight is correlated with changes in renal hemodynamics and glomerular morphometry in adult age. Biol Neonate. 2001;80:239-46.
18. Almeida JR, Mandarim-de-Lacerda CA. Maternal gestational protein-calorie restriction decreases the number of glomeruli and causes glomerular hypertrophy in adult hypertensive rats. Am J Obstet Gynecol. 2005;192(3):945-51.
19. Painter RC, Roseboom TJ, van Montfrans GA, Bossuyt PM, Krediet RT, Osmond C, Barker DJ, Bleker OP. Microalbuminuria in adults after prenatal exposure to the Dutch famine. J Am Soc Nephrol. 2005;16(1):189-94.
20. Teodósio NR, Lago ES, Romani SA, Guedes RC. A regional basic diet from northeast Brazil as a dietary model of experimental malnutrition. Arch Latinoam Nutr. 1990;40:533-47.
21. Trinder P. Determination of glucose in blood using glucose oxidase with an alternative oxygen acceptor. Ann Clin Biochem. 1969;6:24.
22. Karmen A, Wróblewski F, LaDue JS. Transaminase activity in human blood. J Clin Invest. 1955;35:126-33.
23. Magalhães JC, da Silveira AB, Mota DL, Paixão ADO. Renal function in juvenile rats subjected to prenatal malnutrition and chronic salt overload. Exp Physiol. 2006; 91(3):611-19.
24. Paixão ADO, Alessio ML, Martins JP, Leger CL, Monnier L, Pares-Herbute N. Regional

Brazilian diet-induced pre-natal malnutrition in rats is correlated with the proliferation of cultured vascular smooth muscle cells. Nutr Metab Cardiovasc Dis. 2005;15(4):302-09.

25. Kooistra M, Geerlings MI, Mali WTPM, Vincken KL, van der Graaf Y, Biessels GJ. Diabetes mellitus and progression of vascular brain lesions and brain atrophy in patients with symptomatic atherosclerotic disease. The SMART-MR Study. J Neurol Sci. 2013;332:69-74.

12

Purification and Characterization of a Small Chito-specific Lectin from *Datura innoxia* Seeds Possessing Anti-microbial Properties

Ruby Singh[1] and C. G. Suresh[1*]

[1]Division of Biochemical Sciences, CSIR-National Chemical Laboratory, Pune, 411008, India.

Authors' contributions

This work was carried out in collaboration between both authors. Authors RS and CGS have envisaged the study, designed the protocols and performed the data analysis. Author RS carried out the experiments and wrote the first draft of the manuscript. Author CGS checked the analyses and edited the manuscript. Both authors read and approved the final form of manuscript.

Editor(s):
(1) Carmen Lúcia de Oliveira Petkowicz, Federal University of Parana, Curitiba, Parana, Brazil.
Reviewers:
(1) Luana Cassandra Breitenbach Barroso Coelho, Federal University of Pernambuco, Brazil.
(2) Volodymyr Chernyshenko, Palladin Institute of Biochemistry NAS of Ukraine, Ukraine.

ABSTRACT

Aims: To purify and characterize the smallest lectin from *Datura innoxia* (*Di*L9) and to evaluate its antimicrobial application potential.
Study Design: Protein purification and characterization using biochemical and biophysical techniques. MALDI-MS/MS for peptide mass finger printing. *In vitro* assay of antimicrobial activity.
Place and Duration of Study: Biochemical Sciences Division, CSIR-National Chemical Laboratory, Pune, India, From Jan 2014 to July 2015.
Methodology: Dried *Datura* seeds were taken as source of lectin and purified using ion-exchange and gel filtration chromatographic techniques. The rabbit erythrocytes were used for the study of hemagglutinating activity. Structural properties were studied using Fluorescence and Circular dichroism spectroscopic methods. The studied lectin was also examined for its antimicrobial activity using agar disc diffusion method and minimum inhibitory concentration (MIC) was calculated using microdilution method.

**Corresponding author: E-mail: cg.suresh@ncl.res.in*

Results: The lectin was purified from dried seeds of *Datura innoxia*. It was characterized as a chito-specific glycoprotein with 4% neutral carbohydrate content and molecular weight of 9 kDa. Circular dichroism and fluorescence studies showed it to be structurally different from other known lectins isolated from *Datura*. Protein was found to be highly thermostable with sugar binding capacity spread over a broad pH range (1-12), and showed resistance to effects of organic solvents, chemical denaturation and proteolytic digestion. *Di*L9 also demonstrated antibacterial activity against both Gram-positive and Gram-negative bacteria; however, more effective against *Enterococcus faecalis* and *Bacillus cereus* with MIC values at 0.325 and 0.15 mg ml^{-1}, respectively.
Conclusion: The present study revealed that the lectin purified from *D. innoxia* is a glycoprotein and agglutinates pronase treated rabbit erythrocytes. The lectin is highly thermostable with respect to its structure and activity at all denaturing conditions, also showed significant antibacterial activity. This data also supports the fact of its traditional use in folk or alternative medicines for treatment of bacterial infections.

Keywords: Lectin; Datura innoxia; chito-specific; thermostable; circular dichroism; fluorescence; antibacterial activity.

ABBREVIATIONS

DiL9: *Datura innoxia lectin 9 kDa;* ***GDn-HCl:*** *Guanidine hydrochloride;* ***GDn-SCN:*** *Guanidine-thiocyanate;* ***GlcNAc:*** *N-acetyl glucosamine;* ***DMSO:*** *Dimethyl sulfoxide;* ***ACN:*** *Acetonitrile.*

1. INTRODUCTION

Plants have been used as a remedy for human diseases owing to the presence of various therapeutic components in them capable of acting as antimicrobial agents. *Datura innoxia* (Family: *Solanaceae*) or toloache, is one such plant which has been used for various purposes such as pain killer, fever reducing agent, cure for diarrhea and skin diseases [1]. All parts of the plant possess hallucinogenic, hypnotic and narcotic effects and are considered traditional medicines around the world. An Indian traditional medical practice, despite toxicity, recognized the medicinal effects of *Datura* as depended on the dosage and method of medicine preparation. The plant is originally a native of Central and South America; it is later introduced in Asia. It is cultivated in southeastern Europe mainly for the production of scopolamine and other alkaloids [2].

Among the many therapeutic components present in plants, lectin is one of the most potent antimicrobial agents known. Lectin is a hemagglutinating protein which can bind sugars present on cell surfaces and can precipitate as glycoconjugates [3]. Lectins are mostly extracted from the plant storage organs like seeds, and may constitute upto 10% of its total protein content [4]. Other parts of the plant also contain lower amounts of lectins, sometimes with different carbohydrate specificities or having enzymatic function. In the field of biomedicine, lectins have been used as antimicrobial and diagnostic tools. Lectin through its interaction with bacterial cell wall components like teichoic acids, peptidoglycans and lipopolysaccharides, inhibits the growth of Gram-positive and Gram-negative bacteria by agglutinating and immobilizing the microorganisms [5,6]. Lectins are also known to inhibit the growth of fungi by attaching to hyphas, and thus interfering with its nutrient absorption and spore germination process [7], disturbing the synthesis or deposition of chitin monomers in the cell wall. Many pathogens have become resistant to multiple antibiotics and threaten to develop immunity against the currently used antimicrobial drugs and therefore infection becomes untreatable. Thus, the search for new antimicrobial agent is of utmost importance. In the present work we have purified and characterized *Datura innoxia* lectin and explored its antimicrobial activity.

There are only few reports available on the characterization of lectins from *Solanaceae* family and their medical applications. For instance, a lectin purified from *Solanum tuberosum* showed antimicrobial and antibiofilm activities [8] whereas one from *Datura stramonium* was shown mitogenic [9]. Previously, few lectins from *Datura innoxia* have been reported which are of high molecular weight, around 150 and 300 kDa [10]. Other high molecular weight *Datura innoxia* lectins (82, 54, 49 and 41 kDa) are also reported having N-acetyl glucosamine oligomers specificity [11]. This is the first report of a small chito-specific agglutinin purified from the seeds of *Datura innoxia* (*Di*L9) and characterized using biochemical and biophysical techniques, which is part of an effort to expand the knowledge of lectins for novel applications.

2. MATERIALS AND METHODS

2.1 Biological Materials

The seeds of *Datura innoxia* were collected locally from the premises of CSIR-National Chemical Laboratory (NCL), Pune, India. The rabbit blood was procured from National Toxicology Centre (NTC), Pune, India. The bacterial and fungal strains were obtained from National Collection of Industrial Microorganism (NCIM), CSIR-NCL, Pune, India.

2.2 Chemicals and Reagents

Q-sepharose, Sephacyrl S-200, bovine serum albumin (BSA), urea, guanidine hydrochloride, guanidine thiocyanate, chemical modifiers like diethylpyrocarbonate (DEPC), 2,4,6-trinitro-benzenesulphonic acid (TNBS), Woodward's reagent K (WRK); phenylglyoxal (PGO), N-acetylimidazole (NAI), phenylmethylsulfonyl fluoride (PMSF), 5,5'-dithiobis-2-nitrobenzoic acid (DTNB); gel filtration markers, sugars and sugar derivatives were purchased from Sigma Chemicals Co. (USA). Markers used in SDS-PAGE were procured from Bio-Rad. All other reagents and buffers used in the study were of analytical grade.

2.3 Isolation and Purification of Lectin from *Datura innoxia* Seeds

Twenty grams of dried seeds of *Datura innoxia* were soaked and ground in 20 mM sodium phosphate buffer, pH 7.4. The homogenate was filtered through double layered muslin cloth and centrifuged at 10000x rpm for 20 min at 4°C. The supernatant was harvested and precipitated with 60% ammonium sulphate at 4°C, dialyzed against 20 mM sodium phosphate buffer, pH 7.4 and then subjected to Q-sepharose column chromatography, column equilibrated with the same buffer. The lectin was eluted in the unbound fractions. The eluted fractions showing hemagglutinating activity were pooled, dialyzed against TBS (50 mM Tris-Cl, pH 8.0 containing 150 mM NaCl) and loaded on gel filtration column (Sephacyrl S-200) equilibrated previously with TBS. On the basis of absorbance at 280 nm (A_{280}) and detectable hemagglutinating activity the fractions were pooled, dialyzed against deionized water and stored at -20°C for further use.

2.4 Determination of Protein and Carbohydrate Content

The amount of protein was monitored by measuring the absorbance at 280 nm of the column eluate. Protein concentration was estimated using Lowry method with BSA as standard [12]. Phenol-Sulphuric method was used to determine the neutral carbohydrate content with slight modification using mannose as standard, as reported [13].

2.5 Molecular Weight Determination

The molecular weight was estimated by running 12% SDS-PAGE and native subunit structure was determined using Gel filtration chromatography (Sephacryl S-200) and MALDI-TOF/TOF.

2.6 Hemagglutination and Sugar Specificity Test

A 2-fold serial dilution of purified *Datura* lectin in 10 mM phosphate buffered saline (PBS, pH 7.4, 50 µl) were mixed with 50 µl of 3% rabbit erythrocyte suspension in 96-well U-shaped microtiter plates at room temperature and the agglutination was observed after 1 h of incubation. Reciprocal of the highest dilution exhibiting detectable agglutination gives hemagglutination titre (HU) and the specific activity was calculated as HU mg^{-1} protein. Sugar specificity of *Di*L9 was examined by estimating the ability of sugars to inhibit agglutination. Various glycoproteins, carbohydrates and their derivatives were incubated with *Di*L9 and the assay was carried out.

2.7 Chemical Modification Studies

Chemical modification is one of the most widely used techniques to study ligand/sugar binding sites and identification of the residues involved in substrate binding. To determine the role played by tryptophan residues in lectin activity, NBS modification method was attempted [14]. Modification of histidine residues with DEPC prepared in absolute ethanol was carried out

[15]. PMSF was used for serine modification in 50 mM Tris-HCl buffer [16]. Lysine residues were modified with TNBS [17]. Modification of the carboxylate groups with WRK was carried out in 50 mM citrate-phosphate buffer pH 6.0, with different concentration range (5-20 mM) of WRK for 30 min [18]. Arginine residues were modified with phenylglyoxal [19]. Free cysteine residues were modified using DTNB [20]. NAI was used as chemical modifier for tyrosine residues [21]. The native lectin without any reagent was used as a positive control for hemagglutination activity in each case of chemical modification. Each assay was performed in triplicates.

2.8 Effect of pH on Lectin Activity and Stability

Various buffers were used to determine the lectin stability in range of pH 1-12 and activity at room temperature for 12 h. Buffers used were: 20 mM glycine-HCl (pH 1-3), 20 mM citrate-phosphate (pH 4-6), 20 mM sodium- phosphate (pH 7), 20 mM tris-HCl (pH 8-9) and 20 mM glycine-NaOH (pH 10-12). The lectin sample at room temperature served as positive control and buffer was taken as negative control. Each assay was performed in triplicates.

2.9 Thermal Inactivation Assay

The effect of temperature on lectin activity was determined by incubating the sample at different temperatures (25-100ºC) for 30 min. The kinetics of inactivation to deduce thermostability of the lectin was carried out by first incubating the sample at higher temperature in PBS for fixed intervals of time (20-120 min) followed by cooling to room temperature and assaying the activity, as mentioned earlier. For lectin heat denaturation process, the free energy change (ΔG) of activation was also determined using Arrhenius equation [22], a simple way to understand temperature dependence of reaction rates.

$$\Delta G = RT \ln(kT/k_1h) \quad (1)$$

where $\boldsymbol{R}$ is the gas constant (1.987 cal mol^{-1} K^{-1}), $\boldsymbol{T}$ is the absolute temperature (K), $\boldsymbol{k}$ is the Boltzmann constant (1.37 x 10^{-6} erg K^{-1}), $\boldsymbol{k_1}$ is the velocity constant and $\boldsymbol{h}$ is Planck's constant (6.25 x 10^{-27} erg s^{-1}).

The velocity constant ($\boldsymbol{k_1}$) can be calculated from the slope of the curve obtained from the following equation,

$$k_1t = - \ln (A/A_o) \quad (2)$$

where $\boldsymbol{t}$ is the time of heat treatment (s), $\boldsymbol{A}$ is the residual hemagglutinating activity and $\boldsymbol{A_o}$ is the initial hemagglutinating activity before heat treatment.

2.10 Effect of EDTA, Metal Ions and Chemical Denaturants

*Di*L9 was incubated with EDTA, various divalent metal ions, chemical denaturants like urea, guanidine hydrochloride (GDn-HCl) and guanidine thiocyanate (GDn-SCN) and reducing agents like β- mercaptoethanol (βME) and dithiothreitol (DTT) for 24 h at different concentrations. After incubation, 50 µl aliquot was withdrawn and residual activity was scored after 1 h in triplicates as described above.

2.11 Stability of *Di*L9 in Organic Solvents

To study the stability of *Di*L9 in various organic solvents, agglutination assay was carried out. The lectin was incubated in ethanol, methanol, acetonitrile (ACN), dimethylsulphoxide (DMSO) and propanol (25% and 50% v/v) at pH 7.4 for 24 h. All these solvents were completely miscible in deionized water. Presence of organic solvents can modify/alter the sugar binding sites of the lectin. The incubation was performed at room temperature in tightly closed vials to prevent evaporation. 50 µl aliquots were withdrawn at regular intervals of time for the hemagglutination assay.

2.12 Resistance to Proteolysis

Proteases such as pepsin, trypsin, chymotrypsin and pronase were used for proteolytic digestion of *Di*L9 at 37°C in 20 mM sodium phosphate buffer at pH 7.4. Lectin and each proteolytic enzyme were incubated at 10:1 molar ratio for 24 h. 50 µl aliquots were removed at regular intervals and assayed for hemagglutination activity.

2.13 Spectroscopic Measurements

2.13.1 Fluorescence studies

The intrinsic fluorescence of *Di*L9 was recorded using a Perkin Elmer Life Sciences LS50 spectrofluorimeter. The protein solution (0.05 mg ml^{-1}) was excited at 295 nm and emission was measured in the range of wavelengths 310-400 nm with slit width of 7 nm for both monochromators. The spectrum was recorded at

28°C and the contribution of the solvent was subtracted from the protein spectra for further analysis. The barycentric mean (BCM) was calculated according to the equation:

$$BCM_\lambda = \sum(F_\lambda) x \lambda / \sum F_\lambda \quad (3)$$

where F_λ is the fluorescence emission at wavelength λ, and the summation was carried out over the range of appreciable values of F.

2.13.2 Circular dichroism (CD) studies

CD measurements of *Di*L9 were performed on a JASCO J-815-150S (Jasco, Tokyo, Japan) spectropolarimeter. Far-UV CD spectrum of the lectin (0.2 mg ml^{-1}) was recorded in a rectangular quartz cell of 1 mm path-length in the wavelength range 190-250 nm. Each recorded spectrum was an average of 3 accumulations. The baselines (buffer spectra) were subtracted from the protein spectra for further analysis. Results were expressed in terms of mean residual ellipticity (MRE) in deg cm^2 $dmol^{-1}$ defined as:

$$MRE = M\, \theta_\lambda / 10\, d\, c\, r \quad (4)$$

where $\boldsymbol{M}$ is the molecular weight of the protein, $\boldsymbol{\theta_\lambda}$ is the observed ellipticity in millidegree, $\boldsymbol{d}$ is the cell pathlength in cm, $\boldsymbol{c}$ is the protein concentration in mg ml^{-1} and $\boldsymbol{r}$ is the average number of amino acid residues in the protein. CDPro software was used to determine the secondary structure elements of the lectin (http://lamar.colostate.edu/~sreeram/CDPro/main.html). Low NRMSD values were observed for analysis with CONTINLL. The scans were also taken for lectin incubated in the presence of chemical modifiers and organic solvents as described in section 2.7 & 2.11.

2.14 Mass Spectrometry

To identify the protein, the in gel digestion protocol was carried out using mass spectrometry [23]. The trypsin digested peptides were mixed with a solution of α-cyano-4-hydroxycinnamic acid (10 mg ml^{-1}) prepared in 50% acetonitrile and 0.3% trifluoroacetic acid. The samples were analyzed in a mass spectrometer MALDI-TOF/TOF in reflection mode. On an average 100 MS/MS spectra were measured for the digested protein and were analyzed using Protein pilot software. The plant lectin database was created using Uniprot online software.

2.15 Evaluation of Antimicrobial Activity

2.15.1 Antibacterial assay

Antimicrobial activity of the lectin was examined using disc diffusion method [24] on sterile nutrient agar (NA) plates. Total of 6 species (listed in Table 2) of pathogenic bacteria were selected and were seeded separately onto the surface of NA plates. The lectin sample was aseptically added on the sterile disc placed on the nutrient agar plates and the organisms were allowed to grow at 37°C for 12 h. The transparent ring around the disc shows zone of inhibition measured in millimeters (mm) indicating the antimicrobial activity of the lectin. Anti-mycobacterial activity of lectin was also evaluated for its *in vitro* effects against dormant and active phases of *M. tuberculosis* H37Ra using XTT Reduction Menadione Assay (XRMA) protocol [25]. All experiments were performed in triplicates.

2.15.2 Antifungal Assay

Antifungal bioassays were performed against *Candida albicans, Trichoderma viride, Gibberella saubinetii, Fusarium oxysporum, Cephalosporium* sp, *Saccharomyces cerevisiae, Fusarium moniliforme* and *Aspergillus falvus* using agar well diffusion method [24], in order to determine the susceptibility of these pathogenic fungi towards *Di*L9. Amphotericin B (as positive control) and lectin were dissolved in deionized water and added to the well (8 mm in diameter) of potato dextrose agar plates, incubated at 28°C. The diameter of the clear zone around the well was measured and considered as the inhibitory power of the lectin against the particular test pathogen.

3. RESULTS AND DISCUSSION

Since lectins are known for their potential in applications such as anti-microbials, anticancer; characterization of lectins and understanding the structure-function relationship are highly essential to determine their applications in clinical microbiology and pharmaceutical industry. A combination of various biochemical and biophysical techniques were used to characterize the smallest lectin purified from *Datura innoxia* to understand the protein structure-function relationship. Antimicrobial assays were also carried out to explore its antibacterial and antifungal activity.

3.1 Purification and Structural Characterization of *Di*L9

The *Di*L9 agglutinin was purified using Q-sepharose ion exchange and gel filtration chromatography. On gel filtration column, the protein was resolved into two peaks, P1 and P2 (Fig. 1) and lectin activity was observed for both the peaks. The lectin from P1 peak corresponded to a 88 kDa lectin which was already reported from *Datura* seeds [11], while lectin (*Di*L9) obtained from P2 peak was of the size 9 kDa. We understand that this is the smallest agglutinin, which is characterized in this study. Table 1 summarizes the purification step; specific activity and enrichment fold of *Di*L9 lectin. It is important to note that *Datura* seeds are rich in different types of lectins, and hence we observed a decrease in activity per unit mass of protein at each step of chromatographic techniques presumably by the removal of other lectins from the preparation.

3.1.1 Molecular weight determination and estimation of carbohydrate content

The purity of *D. innoxia* lectin was evaluated at each step of purification using SDS-PAGE under reducing conditions. The ammonium sulphate precipitated protein showed many bands ranging from 6-97 kDa. After Q-sepharose chromatography, the unbound fractions gave only two bands of 9 and 88 kDa possessing lectin activity. As already discussed, the 88 kDa lectin was previously reported, so we focused mainly on the small agglutinin which was finally purified using sephacryl S-200 (Fig. 2A). The native polyacrylamide gel electrophoresis showed that *Di*L9 moved as a single band. MALDI-TOF/TOF provided the exact molecular weight measured to be 9656.6 Da, a monomeric lectin (Fig. 2B). Hence, *Di*L9 is representation for *Datura innoxia* lectin of molecular weight 9 kDa. It is a glycoprotein containing 4% neutral carbohydrate estimated using phenol-sulphuric acid assay. This lectin has lower content of sugar as compared to the other lectins characterized from the same source. For instance, the sugar content reported for *Datura innoxia* seed agglutinin (DIA) of molecular weight 88 kDa was 25 % [11]. The other Solanaceae lectins like potato lectin [26] and Jimson weed lectin [9] had 50% and 40% carbohydrate content, respectively.

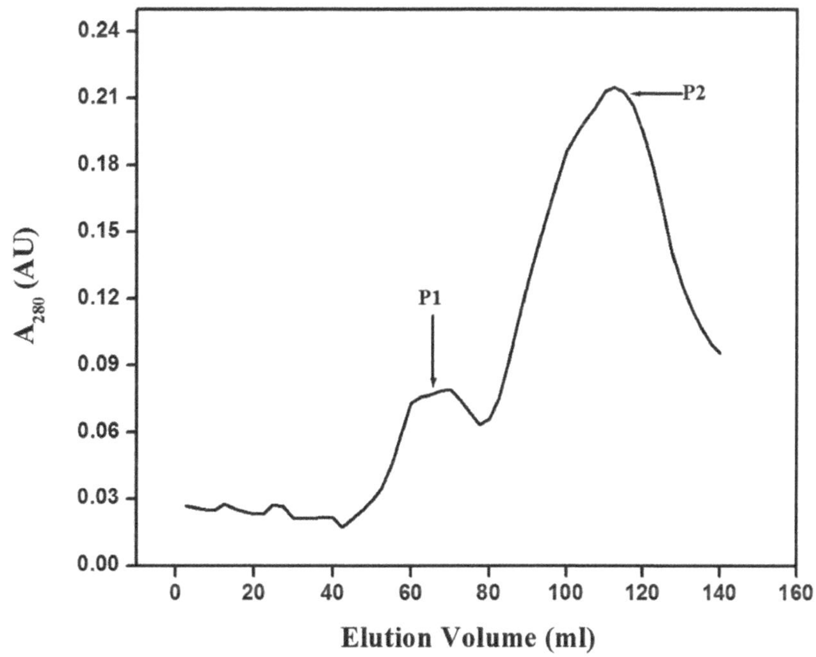

Fig. 1. Purification profile of *Di*L9 using Sephacryl S-200 gel filtration chromatography, equilibrated with TBS at flow rate of 1 ml/min. The column resolved into two different peaks (P1 & P2). P2 corresponds to *Di*L9. Data shown are representative of three such repeats

Table 1. Summary of *Di*L9 purification from 20 g seeds of *D. innoxia*

Fraction	Volume (ml)	Total protein(mg)[1]	Total activity	Specific activity[2]	Fold purification[3]	% yield
Crude extract	230	529	36800	69.56	1.0	100
Q- Sepharose	70	62	11200	180.6	8.5	30.4
Sephacryl S-200	22	26.4	880	34	20	2.4

[1] *Crude protein extract from 20 g seeds of D. innoxia*

[2] *Specific activity is defined as the hemagglutination unit (HU) divided by the protein concentration (mg ml^{-1}) of the assay solution*

[3] *Purification fold was calculated as the ratio of total protein in the extract to the purified protein at each purification step*

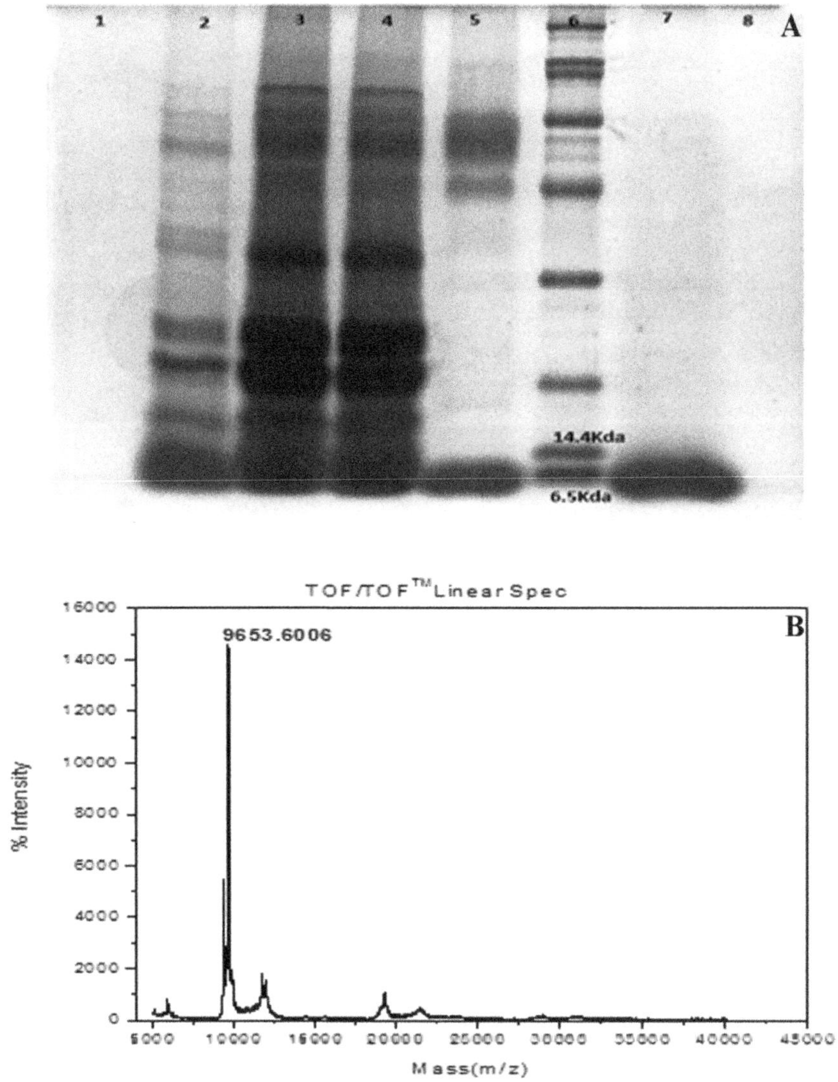

Fig. 2. Estimation of molecular weight of *Di*L9. (A) Purity check on 12% SDS-PAGE. Lane 2: crude extract; lane 3-4: 60% ammonium sulphate precipitate; lane 5: loaded on Sephacryl S-200 column; 6-Protein molecular weight marker; 7- Pure *Di*L9. (B) MALDI-TOF/TOF spectra

3.1.2 Mass spectrometry analysis

The MALDI TOF-MS/MS data analysis gave us the average masses of several peptides which were obtained after trypsin digestion. However, few sequence similarity was observed with Con A lectin and lectin fragment from *Caragana frutex* and *Penaeus japonicas,* which are distinctly related (Table 2). Due to lack of sufficient homology data for protein sequences in the databases, it was not possible to elucidate primary structure of the lectin by peptide mass fingerprinting. Hence, different techniques need to be applied to determine the primary sequence.

3.1.3 Characterization using fluorescence and CD

The intrinsic fluorescence of native *Di*L9 gave λ_{max} (emission wavelength) of 354 nm indicating that the tryptophan residues were exposed on the protein surface. A total of 4 Trp residues have been estimated through NBS titration, of which three are present on the surface and one is buried. In the presence of 6M GDn-HCl, a shift of 3 nm was observed in the barycentric mean (Fig. 3A). This shows that the lectin has been unfolded and the buried Trp residues have become exposed to the polar solvent resulting in the shifting of λ_{max} towards longer wavelengths.

The far-UV CD spectrum of *Di*L9 is shown in Fig. 3B. The calculation of secondary structure elements by CONTINLL program of CDPro analysis showed α-helix: 31.5%, β-sheet: 17.9%, turns: 20.7% and random coil: 29.8%. Previously, two high molecular weight lectins (150 kDa and 300 kDa) from *Datura* have been characterized using CD spectral studies and were shown to have conformations of polyproline II type [27], which was found to be absent in *Di*L9. Thus, these lectins are different with respect to their structural features.

3.2 Biochemical Characterization

3.2.1 Carbohydrate specificity of *Di*L9

*Di*L9 exhibited low agglutination activity when assayed with rabbit erythrocytes even at higher concentration of the protein whereas a minimal concentration of the protein was required for agglutination with pronase treated erythrocytes. Pronase is a nonspecific protease that cleaves membrane proteins present on erythrocytes and exposes other receptor glycoproteins. This difference in the agglutination activity could be due to unavailability of or weak binding to the receptor sites on erythrocytes by *Di*L9. No change in lectin agglutination activity was observed when incubated with EDTA and divalent metal ions, thus suggesting that it could be a metal ion independent protein for hemagglutination. The sugar specificity of *Di*L9 was tested against several sugars and glycoproteins by hemagglutination inhibition assay. Of all the carbohydrates tested, only N-acetyl glucosamine (GlcNAc) oligomers, i.e., *N,N',N''-triacetylchitotriose* (2.5 mM) showed inhibitory action. The monomer glucosamine, GlcNAc and N,N'-diacetylchitobiose did not show inhibition even at higher concentrations (Table 3). This confirms that the sugar binding site requires at least three β (1→ 4)-linked N-acetylglucosamine units. Similarly, other lectins from *Datura stramonium* [28], *Solanum tuberosum* [29] and *Urtica dioica* [30] *were* also known to show GlcNAc oligomer sugar specificity. The glycoproteins tested such as fetuin, ovalbumin, mucin, casein, etc., also showed hemagglutination inhibition at low concentrations.

3.2.2 Chemical modification studies

As a consequence of protein microenvironment within the folded structure, protein side chains frequently exhibit reactivity different from those expected from studies with model compounds. Here, purified lectin, *Di*L9, was subjected to various chemical modifications in order to determine the amino acid residues that are responsible for the hemagglutinating activity. It was observed that in the presence of 2 mM NBS, there was a complete loss of activity, implying direct involvement of tryptophan in sugar binding. This observation could be correlated with the intrinsic fluorescence study which showed presence of exposed tryptophan residues which might be directly involved in sugar binding. About 50% and 25% residual hemagglutination activity was observed on NAI and WRK treatment indicating possible role of tyrosine and aspartate/glutamate residues near the sugar binding region of the protein (Table 4). The far-UV CD spectra also showed loss of secondary structure in the presence of these chemical modifiers (2 mM, 4 h incubation), thus resulting in the loss of lectin activity. Incubating the lectin with NBS resulted in complete loss of structure, and hence no activity was observed as mentioned before (Fig. 4). Other chemical modifiers did not show any significant change in the hemagglutination activity of *Di*L9. From these findings it can be concluded that Trp, Tyr and Asp/Glu are involved in carbohydrate binding site of *Di*L9 lectin.

3.2.3 Effect of pH on lectin activity

The hemagglutination activity of *Di*L9 was found to be stable in the pH range 1-12, retaining full activity even after incubating for 24 h. Similar behavior was observed in other lectins as well, like lectin from *Allium sativum* [31] *and Hericium erinaceum* [32]

3.2.4 Effect of temperature on lectin activity and stability

*Di*L9 was found to be thermostable till 90°C. The lectin retained 100% of its hemagglutination activity at 60°C for 2 h of incubation. With increase in temperature, the lectin showed sigmoidal decay curve with 50%, 25% and 12.5% activity at 70, 80 and 90°C, respectively (Fig. 5A). The thermostability of *Di*L9 in terms of hemagglutinating activity varies with both temperatures as well with incubation time i.e., lectin retained 50% of its maximum activity at 70°C even after 2 h of incubation, but rapidly inactivated at 100°C. This implies that the lectin undergoes conformational changes under extreme temperatures resulting in the loss of activity (Fig. 5B). Examples of other lectins with similar thermostability with respect to hemagglutinating activity includes lectin from edible mushroom *Agaricus arvensis* [33], *Pinellia ternate* [34] and *Trichosanthes dioica* [35].

Table 2. Analysis of MALDI-MS/MS data using protein pilot software

Sl.no	Species	Sequence	Accessions	Homology (%)
1.	Lectin fragment from Caragana frutex	DWDPTGDR	tr|Q9ZRN7|Q9ZRN7_9FABA	9
2.	Calcineurin B-like protein 2 OS=Oryza sativa subsp. japonica	GLEDPQVLAR	sp|Q3HRP5|CNBL2_ORYSJ	4.4
3.	C-type lectin from Penaeus japonicus	GGVWFDTTMR	tr|I2C0B3|I2C0B3_PENJP	3
4.	REVERSED Concanavalin A-like lectins	EILDYTMGR	RRRRRtr|S3DKM5|S3DKM5_GLAL2	1.2
5.	Coatomer subunit beta OS=Setaria italica GN=Si000204m.g PE=4 SV=1	LLCNTGDDVR	tr|K3XE87|K3XE87_SETIT	1

Table 3. Hemagglutination inhibition activity of *Di*L9 by saccharides and glycoproteins

S.no	Sugar / Glycoprotein	Minimum concentrations required to inhibit *Di*L9 activity (μg ml^{-1})
1.	Glucosamine	NI*
2.	N-acetyl glucosamine	NI
3	N,N'-Diacetylchitobiose	NI
4.	N, N',N"- Triacetylchitotriose	750
5.	Fetuin	62.5
6.	Ovalbumin	78
7.	Bovine submaxillary mucin	156
8.	Thyroglobulin	9

**NI: No inhibition*

Table 4. Effect of chemical modification on *Di*L9 activity

Sl.no	Chemical modifier	Concentration (mM)	Residual hemagglutinating activity (%) #
1.	Native	0	100
2.	DTNB	10	100
3.	TNBS	10	100
4.	WRK	5	25
5.	NAI	10	50
6.	NBS	2	0
7.	PMSF	20	100
8.	PGO	10	100
9.	DEPC	10	75

#Full activity (100%) of DiL9 corresponds to titre of 2^6

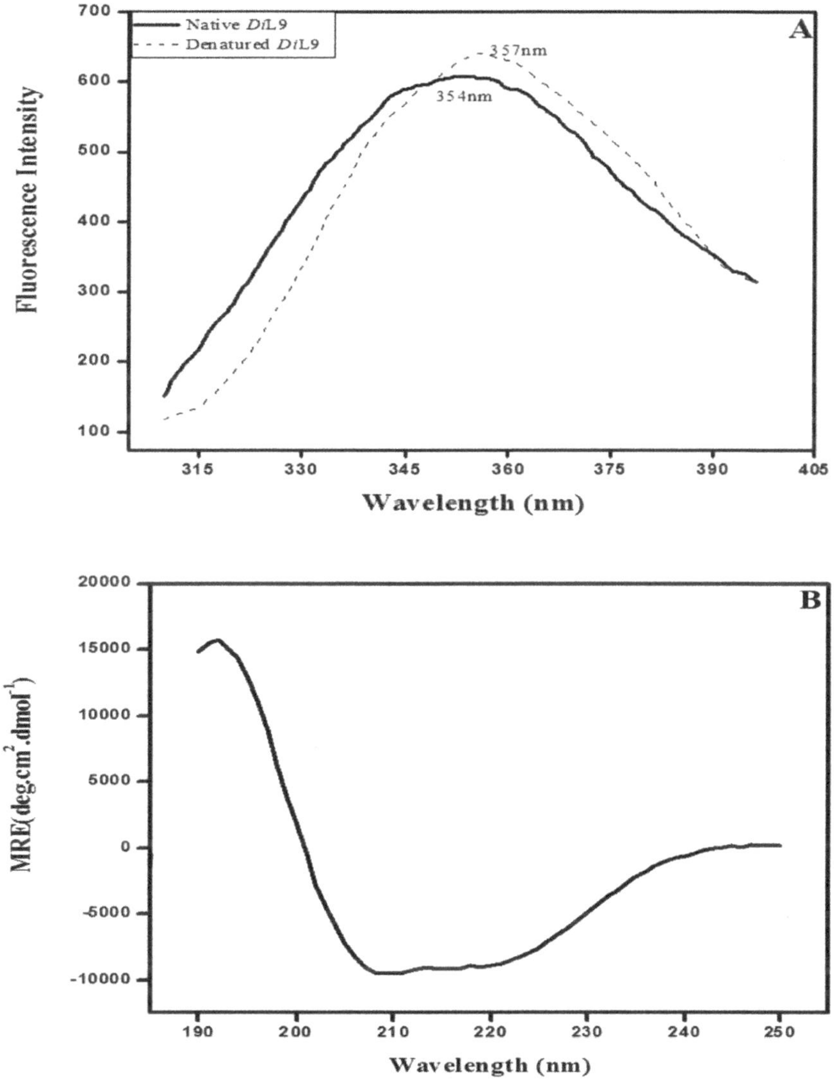

Fig. 3. Biophysical characterization of *Di*L9. (A) Intrinsic fluorescence spectra of native (λ_{max} at 354 nm) and denatured (λ_{max} at 357 nm) *Di*L9 (0.05 mg ml^{-1} at pH 7.4, 28°C). (B) Far-UV CD spectra of *Di*L9 (0.2 mg ml^{-1})

The ΔG value or the Gibb's free energy of thermostability at different incubation temperatures and time intervals were calculated. For example, the ΔG value at 60°C was found to be zero, due to no change in hemagglutination activity implying no thermal enthalpy (ΔH) change. But, with increase in temperature the ΔG values were found to be positive indicating that energy is absorbed in a non-spontaneous process. Considering this phenomenon, the maximum residual hemagglutinating activity observed at 90°C for 20 min corresponded to ΔG of 43 kcal mol^{-1}; the lectin is considered stable at this temperature where the entropies (S) of the native and denatured states are equal. Since many lectins are known to be toxic and show anti nutritional effects on mammals, including humans, the free energy of activation of lectin denaturation process remains an important physicochemical parameter when considering lectin's applications in various fields like medicine, agriculture and other related fields of human and animal nutrition [36].

3.2.5 Effect of chemical denaturants

Chaotropic agents such as urea, GDn-HCl and GDn-SCN at high concentrations disrupt the hydrophobic interactions in the interior of lectin that are essential to maintain the native conformation. Urea had no effect on *Di*L9 activity. However, complete loss of hemagglutination activity of *Di*L9 was observed in the presence of GDn-SCN (2M) but retained 12.5 % of residual activity in the presence of GDn-HCl (6M) for 24 h (Fig. 6). It could be concluded that GDn-SCN is the most potent denaturing agent for this lectin. In the presence of βME and DTT (2-5 mM), complete loss of lectin activity was observed implying presence of disulphide bridges that play important role in maintaining the possible tertiary structure of the protein.

3.2.6 Stability of *Di*L9 in organic solvents

Physical properties like miscibility, hydrophobicity and hydrogen bonding capacity of organic solvents can have drastic influence on the structural and functional aspects of proteins. Here, *Di*L9 showed good tolerance towards high concentrations of organic solvents. The hemagglutinating activity of the lectin was retained completely in the presence of methanol and ACN (all 50% v/v), and remained constant up to 24 h of incubation. The far-UV CD spectra of *Di*L9 in presence of organic solvents also showed increase in negative ellipticity implying enhancement in the compactness of the lectin structure.

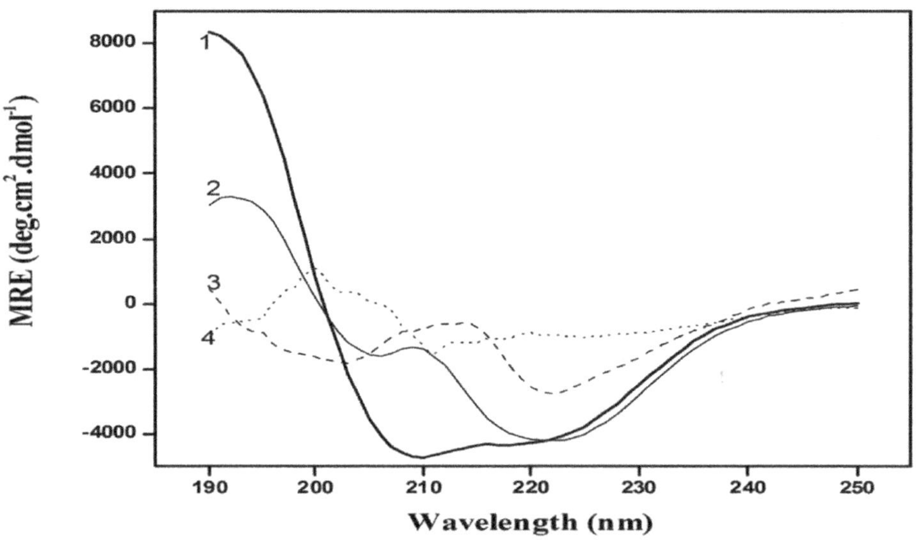

Fig. 4. Far-UV CD spectra of chemical modification effects on *Di*L9; 1- Native *Di*L9, 2-NAI, 3-WRK and 4- NBS treated *Di*L9. The lectin (0.2 mg ml^{-1}) was incubated with 2 mM of each chemical modifier for 4 h

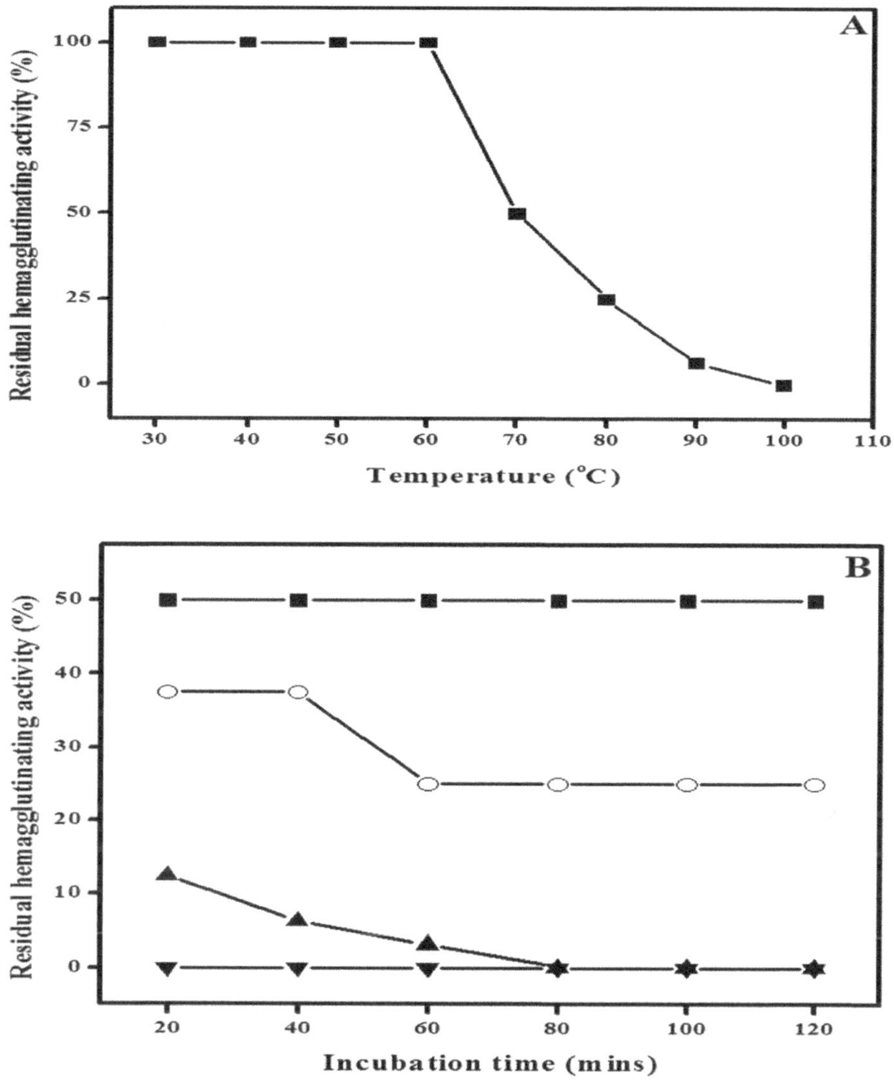

Fig. 5. Thermal denaturation of *Di*L9. (A) Effect of temperature on agglutinating activity of purified *Di*L9 towards a rabbit erythrocyte suspension in PBS. (B) Thermostability of *Di*L9 at: (-■-) 70°C; (-○-) 80°C; (-▲-) 90°C; and (-▼-) 100°C. Each point on the line is the mean (+1 SD) derived from three replicates. Full activity (100%) corresponds to a titre of 2^6

The hemagglutinating activity of *Di*L9 in organic solvents was observed to decrease with an increase in the carbon chain length of the solvent like ethanol as well as in DMSO. The possible reason could be that the solvents compete for hydrogen bonding interaction with protein atoms, thus disrupting the structure [37]. These observations were also reflected in the far-UV CD spectra of *Di*L9 incubated with 50% organic solvents for 4 h, where *Di*L9 in the presence of methanol and ACN retained its structure, on the other hand lost complete structure on incubating with ethanol and DMSO (Fig. 7).

3.2.7 Proteolytic effects on *Di*L9 activity

As part of interrelationship between the structure and activity of the lectins, it is important to understand the conditions under which the molecule resists the conformational changes in order to retain its native conformation and biological activity. Capacity of the *D. innoxia* lectin for being digested by various proteolytic enzymes with different specificity was also investigated. No change in the hemagglutination activity of the lectin was observed when incubated with trypsin (EC 3.4.21.4), chymotrypsin (EC 3.4.21.1), pepsin (EC 3.4.23.1) and papain (EC 3.4.21.62) at 37°C for 24 h (data not shown). However, in the presence of protease (Nagarse; 3.4.21.62), 50% loss in lectin activity was observed. This feature was found to be in accordance with reported susceptibility of other *Datura* lectins to proteolysis [38]. The resistance of *Di*L9 and other lectins to proteolysis was attributed to the fact that it's a glycoprotein and is well stabilized by numerous disulphide bonds, because of which the native structure is retained under denaturing conditions.

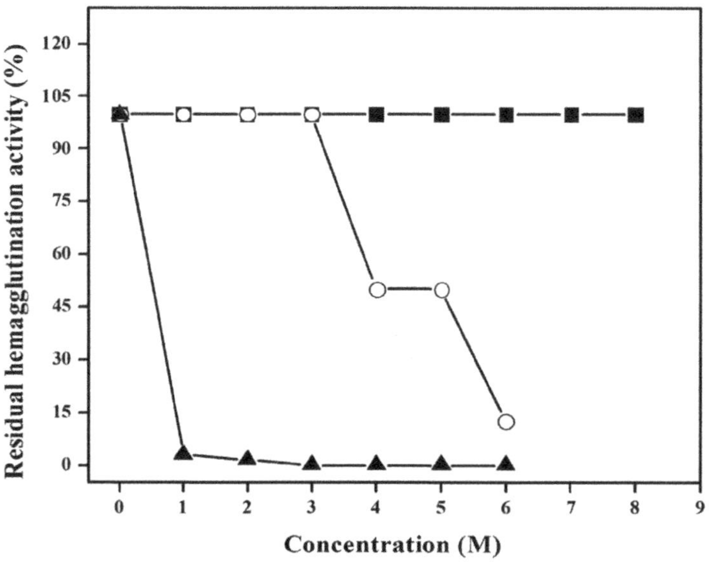

Fig. 6. Effect of chemical denaturants on *Di*L9. Lectin was incubated with required denaturant concentration, 20mM phosphate buffer, pH 7.4 for 16 h at room temperature. (-■-) Urea; (-O-) GDn-HCl and (-▲-) GDn-SCn

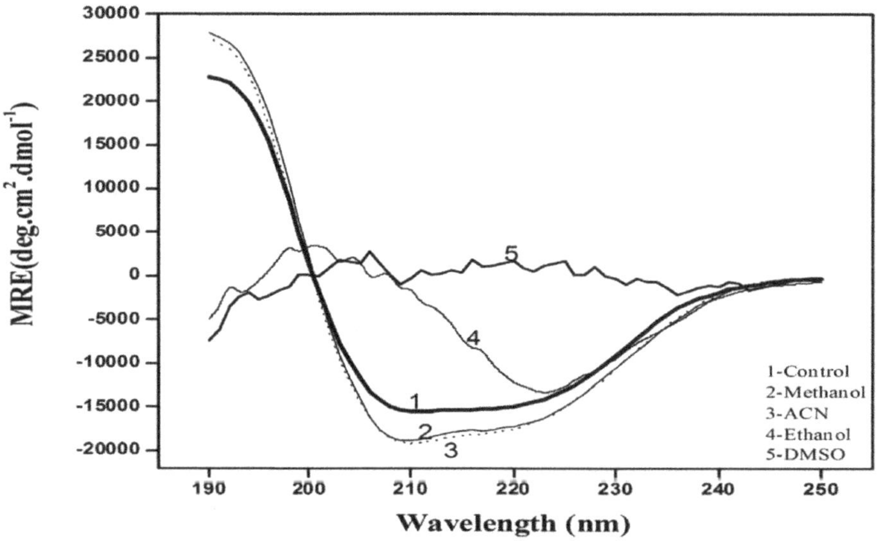

Fig. 7. Effect of organic solvents on far-UV CD spectra of *Di*L9. The lectin (0.2 mg ml^{-1}) was incubated with 50% of organic solvent for 4 h at 28°C

3.2.8 Antimicrobial activity

Lectins are known to play defensive role against bacteria, fungi and insects. Their ability to inhibit microbial growth varies from species to species. In the present study, *Di*L9 was subjected to screening for *in vitro* antibacterial inhibition growth against 6 pathogenic bacteria and the inhibitory action was compared to that of ampicillin, an antibacterial antibiotic. The lectin agglutinated both Gram-positive and Gram-negative bacteria (data not shown). *Di*L9 exhibited significant antibacterial activity against *E. faecalis* and *B. cereus* with 13 and 11 mm of diameter (Fig. 8) of inhibition zone, and MIC of 150 and 250 µg ml^{-1}, respectively (Table 5). No activity was found against mycobacterium, indicating *Di*L9 has specificity towards Gram positive/negative bacteria only. This antibacterial function of *Di*L9 could be related to its observed ability to bind to GlcNAc oligomers as already discussed, being responsible for bacterial recognition, where GlcNAc is a part of peptidoglycan and murien constituting the cell wall components. *In vitro* antifungal susceptibility by *Di*L9 was also determined against eight pathogenic fungi by considering amphotericin B, an anti-fungal agent used as positive control. But *Di*L9 showed some, though not significant, inhibition of growth only against *F. oxysporum.* From these findings, we would suggest that *Di*L9 has potential antibacterial activity though less of antifungal activity. The resistance to currently used antibiotics by microorganisms posed threat towards prevention of infectious diseases, and thus stimulating researchers to discover new antimicrobial agents. Owing to the bacteriostatic effect, *Di*L9 characterized here could be considered for use as an alternative antimicrobial agent. The results from this study imply the application potential of this lectin in clinical microbiology and its plausible therapeutic potential. The present study also gave useful insights into structure-function relationship of this small lectin and its high stability under denaturing conditions. Further studies might be required to explore more of its antibacterial action and design lectin preparation for potential application to control pathogens. With the advancement in protein engineering and nanotechnology, the protein bioavailability can be increased to optimize for target delivery strategies with more efficacies.

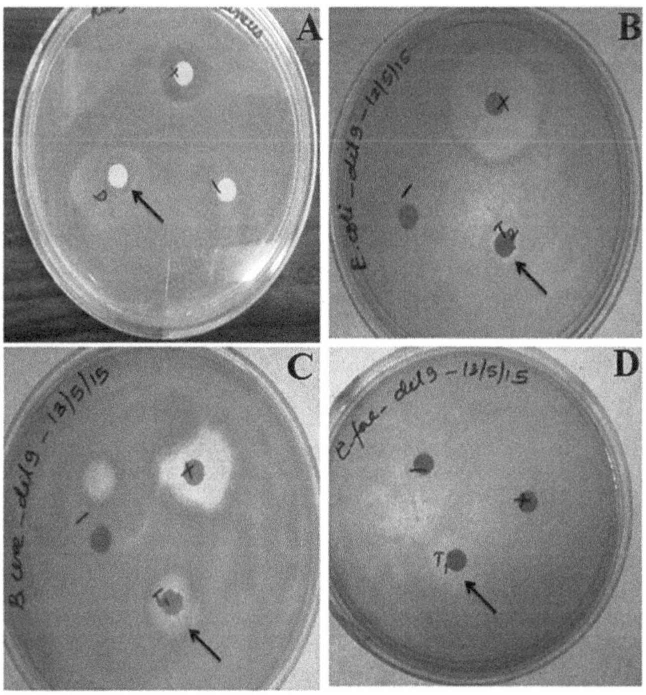

Fig. 8. Inhibitory effect of purified *D.innoxia* seed lectin on the *in vitro* growth on nutrient agar plates of (A) *S. aureus*, (B) *E. coli,* (C) *B. cereus* and (D) *E. faecalis*. For each plate 5 mm diameter discs were seeded with 10 µl of PBS (-) alone as the negative control, (+) 5 µg of ampicillin and (T_1) purified *Di*L9. Data shown are representative of three repeat assays

Table. 5 *In vitro* antibacterial activity of *Datura innoxia* Lectin

SL.no	Pathogen name	Gram staining feature	Zone of inhibition (mm)*	MIC (mg ml^{-1})
1	*Staphylococcus aureus*	Gram positive	10	0.325
2	*Bacillus cereus*	Gram positive	11	0.25
3	*Enterococcus faecalis*	Gram positive	13	0.15
4	*Escherichia coli*	Gram negative	7	0.5
5	*Salmonella typhimurium*	Gram negative	-	-
6	*Pseudomonas aeruginosa*	Gram negative	-	-

**Zone of inhibition (in mm diameter): includes 5 mm diameter of disc placed in agar disc diffusion assay*

4. CONCLUSION

In summary, a chito-specific agglutinin was purified from the seeds of *D. innoxia* using ion exchange and gel filtration chromatography. MALDI-TOF/TOF analysis gave 9653.6 Da monomeric protein with 4% neutral carbohydrate content. The far-UV CD spectra showed structural features different from other *Datura* lectins, thus, showing presence of diverse lectins in the same plant species. Only chitotriose sugar and glycoproteins showed hemagglutination inhibition implying presence of extended sugar binding site in the lectin. *Di*L9 was found to be highly stable in a wide range of pH, at high temperature, in the presence of chemical denaturants, organic solvents and resistant to proteolytic digestion, implying its structural and functional rigidity. *Di*L9 also showed significant antibacterial activity against certain pathogenic bacteria.

ACKNOWLEDGEMENTS

RS is Senior Research Fellow of the CSIR, New Delhi, India. RS is a registered Ph.D. student and CGS is faculty of AcSIR, New Delhi, India.

COMPETING INTERESTS

The authors certify that they have no conflict of interests with or involvement in any organization or entity with any financial interest in the subject matter or materials discussed in this manuscript.

REFERENCES

1. Yadava RN, Jain S. A new bioactive flavone glycoside from the seeds of *Melilotus indica* All. J Asian Nat Prod Res. 2005;7:595-9.
2. Cucu V, Paun E. On the changes of alkaloid content in herbs of *Datura innoxia* Mill. Planta Med. 1968;16:338-42.
3. Goldstein IJ, Hughes RC, Monsigny M, Osawa T, Sharon N. What should be called a lectin? Nature. 1980;285:66-66.
4. Arockia Jeyaprakash A, Jayashree G, Mahanta SK, Swaminathan CP, Sekar K, Surolia A, Vijayan M. Structural basis for the energetics of jacalin-sugar interactions: Promiscuity versus specificity. J Mol Biol. 2005;347:181-8.
5. Gomes FS, Procopio TF, Napoleao TH, Coelho LC, Paiva PM. Antimicrobial lectin from *Schinus terebinthifolius* leaf. J Appl Microbiol. 2013;114:672-9.
6. Paiva PMG, Gomes FS, NapoleÃ£o TH, SÃ¡ RA, Correia MTS, Coelho L. Antimicrobial activity of secondary metabolites and lectins from plants. Current Research, Technology and Education Topics in Applied Microbiology and Microbial Biotechnology. 2010;1:396-406.
7. Bourne Y, Ayouba A, Rouge P, Cambillau C. Interaction of a legume lectin with two components of the bacterial cell wall. A crystallographic study. J Biol Chem. 1994; 269:9429-35.
8. Hasan I, Ozeki Y, Kabir SR. Purification of a novel chitin-binding lectin with antimicrobial and antibiofilm activities from a Bangladeshi cultivar of potato (*Solanum tuberosum*). Indian J Biochem Biophys. 2014;51:142-8.
9. Crowley JF, Goldstein IJ. *Datura stramonium* lectin. Methods Enzymol. 1982;83:368-73.
10. Levitskaya SV, Asatov SI, Yunusov TS. Isolation of two forms of a lectin from the seeds of *Datura innoxia*. Chemistry of Natural Compounds. 1985;21:240-242.
11. Petrescu SM, Petrescu AJ, Rudiger HEF. Purification and partial characterization of a lectin from *Datura innoxia* seeds. Phytochemistry. 1993;34:343-348.
12. Lowry OH, Rosebrough NJ, Farr AL, Randall RJ. Protein measurement with the

folin phenol reagent. Journal of Biological Chemistry. 1951;193:265-275.
13. DuBois M, Gilles KA, Hamilton JK, Rebers PA, Smith F. Colorimetric method for determination of sugars and related substances. Analytical Chemistry. 1956; 28:350-356.
14. Spande TF, Witkop B. [58] Determination of the tryptophan content of proteins with N-bromosuccinimide. In Methods Enzymol (Hirs, C.H.W., ed.^eds). Academic Press; 1967;498-506.
15. Ovaldi J, Libor S, Elodi P. Spectrophotometric determination of histidine in protein with diethylpyro-carbonate. Acta Biochem. Biophys. (Budapest). 1967;2:455-458.
16. Gold AM, Fahrney D. Sulfonyl fluorides as inhibitors of esterases. II. Formation and Reactions of Phenylmethanesulfonyl α-Chymotrypsin*. Biochemistry. 1964;3:783-791.
17. Habeeb AFSA. Determination of free amino groups in proteins by trinitrobenzenesulfonic acid. Analytical Biochemistry. 1966;14:328-336.
18. Sinha U, Brewer JM. A spectrophotometric method for quantitation of carboxyl group modification of proteins using Woodward's Reagent K. Anal Biochem. 1985;151:327-33.
19. Takahashi K. The reaction of phenylglyoxal with arginine residues in proteins. Journal of Biological Chemistry. 1968;243:6171-6179.
20. Habeeb AFSA. [37] Reaction of protein sulfhydryl groups with Ellman's reagent. Methods Enzymol. 1972;25:457-64.
21. Riordan JF, Wacker WEC, Vallee BL. N-Acetylimidazole: A reagent for determination of “Free” tyrosyl residues of proteins*. Biochemistry. 1965;4:1758-1765.
22. Leite YF, Silva LM, Amorim RC, Freire EA, de Melo Jorge DM, Grangeiro TB, Benevides NM. Purification of a lectin from the marine red alga Gracilaria ornata and its effect on the development of the cowpea weevil *Callosobruchus maculatus* (Coleoptera: Bruchidae). Biochim Biophys Acta. 2005;1724:137-45.
23. Shevchenko A, Tomas H, Havlis J, Olsen JV, Mann M. In-gel digestion for mass spectrometric characterization of proteins and proteomes. Nat. Protocols. 2007;1: 2856-2860.
24. Cole MD. Key antifungal, antibacterial and anti-insect assays— A critical review. Biochemical Systematics and Ecology. 1994;22:837-856.
25. Singh U, Akhtar S, Mishra A, Sarkar D. A novel screening method based on menadione mediated rapid reduction of tetrazolium salt for testing of anti-mycobacterial agents. J Microbiol Methods. 2011;84:202-7.
26. Allen AK, Neuberger A. The purification and properties of the lectin from potato tubers, a hydroxyproline-containing glycoprotein. Biochem J. 1973;135:307-14.
27. Levitskaya SV, Yunusov TS. A study of the lectins of *Datura innoxia* seeds II. deglycosylation with trifluoromethane sulfonic acid. Chemistry of Natural Compounds. 1995;31:129-133.
28. Desai NN, Allen AK, Neuberger A. Some properties of the lectin from Datura stramonium (thorn-apple) and the nature of its glycoprotein linkages. Biochemical Journal. 1981;197:345-353.
29. Kilpatrick DC. Isolation of a lectin from the pericarp of potato (*Solanum tuberosum*) fruits. Biochemical Journal. 1980;191:273-275.
30. Peumans WJ, De Ley M, Broekaert WF. An unusual lectin from stinging nettle (Urtica dioica) rhizomes. FEBS Letters. 1984;177:99-103.
31. Clement F, Venkatesh YP. Dietary garlic (Allium sativum) lectins, ASA I and ASA II, are highly stable and immunogenic; 2010.
32. Li Y, Zhang G, Ng TB, Wang H. A novel lectin with antiproliferative and HIV-1 reverse transcriptase inhibitory activities from dried fruiting bodies of the monkey head mushroom Hericium erinaceum. J Biomed Biotechnol. 2010;716515.
33. Zhao JK, Zhao YC, Li SH, Wang HX, Ng TB. Isolation and characterization of a novel thermostable lectin from the wild edible mushroom *Agaricus arvensis*. J Basic Microbiol. 2011;51:304-11.
34. Zuo Z, Fan H, Wang X, Zhou W, Li L. Purification and characterization of a novel plant lectin from Pinellia ternata with antineoplastic activity. Springerplus. 2012; 1:13.
35. Dharkar PD, Anuradha P, Gaikwad SM, Suresh CG. Crystallization and preliminary characterization of a highly thermostable lectin from Trichosanthes dioica and

comparison with other Trichosanthes lectins. Acta Crystallogr Sect F Struct Biol Cryst Commun. 2006;62:205-9.
36. Vasconcelos IM, Oliveira JT. Antinutritional properties of plant lectins. Toxicon. 2004; 44:385-403.
37. Knubovets T, Osterhout JJ, Klibanov AM. Structure of lysozyme dissolved in neat organic solvents as assessed by NMR and CD spectroscopies. Biotechnol Bioeng. 1999;63:242-8.
38. Levitskaya SV, Yunusov TS. Study of the structure of the lectins of*Datura innoxia* seeds. III. Resistance of the lectins to the action of various proteases. Chemistry of Natural Compounds. 1996;32:74-76.

13

Adipose Tissue: A Regulator for Obesity and Its Complications

Shaik Rahiman[1,2*], Tarek H El-Metwally[1,3] and Divya Shrivastava[2]

[1]Departments of Medical Biochemistry, Al Jouf University, Saudi Arabia.
[2]Department of Life and Basic Sciences, Jaipur National University, Rajasthan, India.
[3]Departments of Medical Biochemistry, Assiut University, Egypt.

Authors' contributions

This work was carried out in collaboration between all authors. Author SR collected references and drafted the manuscript. Authors THEM and DS edited the manuscript for final submission. All authors read and approved the final manuscript.

Editor(s):
(1) Luisa Di Paola, Chemical and Biochemical Engineering Fundamentals, Faculty of Engineering Università Campus Biomedico, Via Alvaro del Portillo, Roma, Italy.
Reviewers:
(1) P. G. Raman, Devi Ahilya University, Indore, India.
(2) Ds sheriff, Benghazi University, Libya.
(3) Anonymous, Punjab Institute of Medical Sciences, India.

ABSTRACT

Adipose tissue is a key player in whole body metabolism and excess adipose tissue poses a major risk factor for the development of metabolic disorders such as type 2 diabetes. In response to nutritional overload, de novo adipocyte differentiation can serve as an adaptive mechanism by increasing the storage capacity of adipose tissue and maintaining normal adipocyte function. This in turn prevents systemic lipid overload, which is a major cause for insulin resistance. Adipose tissue is of two types; the fat storing white adipose tissue and the thermogenic brown adipose tissue. While the former is implicated in obesity, insulin resistance and diabetes, the latter is a physiological anti-obesogenic and antidiabetic through adaptive thermogenesis by uncouplers proteins. Obesity results from the imbalanced energy intake for expenditure with excessive fat accumulation in adipose tissue. Obesity is associated with numerous metabolic disorders, including hyperlipidemia, diabetes mellitus, hypertension, stroke, osteoarthritis, infertility and certain types of cancer. Obesity is associated with chronic subclinical inflammation in which the metabolism of

**Corresponding author: E-mail: rahimhi@gmail.com*

adipose tissue plays an important role. The adipose tissue is an endocrine organ which has a fundamental role in metabolic, inflammatory, cardiovascular homeostatic regulation through lipogenesis, lipolysis, steroidogeneis, and secretion of several biologically active adipokines and pro-inflammatory cytokines with diverse protein structures and functions. This review article will mainly focus on the pathophysiological changes of adipose tissue fat during obesity in relation to energy expenditure towards prevention or development of obesity and its complications.

Keywords: Obesity; metabolic syndrome; insulin resistance; diabetes; adipose tissue; white adipose tissue; brown adipose tissue; adipokines; adaptive thermogenesis; uncoupler proteins; bile acids.

1. OBESITY AT GLANCE

Obesity is gaining attention as one of the major leading public health problems. The terms overweight and obesity are both labels for ranges of weight that are greater than generally considered healthy for a given height. Obesity is considered a situation of positive energy balance due to energy intake that exceeds energy consumption (surplus) leading to the storage of triacylglycerols in adipose tissue, ready to be mobilized when expenditure exceeds intake [1]. International Diabetes Federation defined obesity as a metabolic syndrome describing a cluster of factors associated with an increased risk for diabetes and atherosclerotic cardiovascular health problems. The causes of insulin resistance related to obesity are diverse. However the current modern sedentary life style, with lack of physical activity and an uncontrolled over-nutrition and genetic predisposing factors supports metabolic diseases [2]. The prevalence of overweight and obesity has also increased in children and adolescents in developing countries from 1980 to 2013 [3]. International Obesity Taskforce reported that 1.1 billion people are overweight and 312 million are classified as obese [4]. Among the global obesity prevalence, the US has attracted broad public attention as the highest officially reported prevalence of obesity [33.8%] [5] followed by European and less developed countries as a factor of over-nutrition [6]. According to World Health Organization (WHO), obesity is generally more common among women than men [7]. Although reports confirmed the higher rates of obesity in female than in males [8], studies on college students revealed higher rates of obesity in males than in females [9].

Obesity is considered a disorder of multifactorial aetiology since it is the sum of genetic, behavioral, life style and environmental factors, such as fetal programming, control of appetite, energy expenditure and availability, and nutritional content of food [10]. Obesity is commonly defined as body mass index (BMI = weight (in kg) divided by height (in m) squared) of 30 kg/m^2 or higher. This distinguishes it from being overweight as defined by a BMI of between 25-29.9 kg/m^2 (WHO, 2000). The values above this are categorized as clinically obese. Within the defined `obese' category there is further sub-division, as demonstrated in Table 1 [11].

Table 1. The body mass index (BMI) scale, highlighting the classified limits

WHO classification for Europids	BMI (kg/m^2)
Underweight	<18.5
Healthy/Lean	18.5-24.9
Overweight	25-29.9
Obese I	30-34.9
Obese II	35-39.9
Morbid Obesity	>40

2. ADIPOSE TISSUE AND OBESITY

There are different adipose tissue depots within the body. One special subclass is brown adipose tissue (BAT), which is especially developed in infants and small rodents, where it can be induced by cold exposure [12]. Brown adipocytes have many small lipid droplets, in contrast to white adipocytes, which have one large lipid core. The main function of brown adipocytes is the production of heat through expression of uncoupling-protein 1 (UCP1). It has long been believed that only children and rodents have brown adipose tissue, which is located in distinct pads adjacent to the shoulder blades. However, it has recently been shown that brown adipocytes exist in adult humans and are interspersed into the common white fat pads and might also be inducible by cold [13]. Inducing brown adipocytes by specific agents could therefore be an interesting novel therapeutic intervention to increase energy expenditure and treat obesity.

White adipose tissue (WAT) represents the majority of body fat. Based on the localization, adipose tissue depots can be classified as subcutaneous adipose tissue underneath the dermis or visceral adipose tissue within the cavities of the body. Most visceral adipose tissue lies within the abdominal cavity with the gonadal, the mesenteric and the retroperitoneal fat compartments being the most prominent [14]. This distinction is not only of anatomical interest but also has functional implications. It has been shown that visceral and subcutaneous adipose tissue show distinct gene expression patterns [15]. Furthermore, visceral adipose tissue poses a greater risk for metabolic complications than subcutaneous adipose tissue [16]. The reasons for this are currently not clear. Potential explanations include that factors secreted from the visceral adipose tissue drain directly into the portal vein or that adipocyte function itself is altered in different depots, which will be discussed further below. Due to hormonal differences, overweight women tend to have more subcutaneous adipose tissue, whereas overweight men show increased central adiposity and are thus more prone to metabolic complications [17]. Moreover, treatment with glitazones leads, amongst other things, to redistribution of fat to the subcutaneous compartment, thereby ameliorating the metabolic profile without causing weight loss [18].

In human body adipose tissue acts as an endocrine organ and plays an important role in the human metabolism [19]. It has essential metabolic functions in the body, although excess amount of adipose tissue predisposes to several metabolic complications. About 21% of adipose tissue accounts for an average man of 70 kg [20]. The excess of energy is stored in the form of triglycerides in the adipocytes in adipose tissue. Besides its obvious storage function, adipose tissue has mechanical functions such as insulation and protection against mechanical forces. Adipose tissue plays a potential role in obesity-related chronic mild inflammation by secretion of several biologically active adipocytokines (Figure 1) including adiponectin, cytokines (tumor necrosis factor-α [TNF-α], leptin, interleukin-6 [IL-6], C-C motif chemokine ligand-2 [CCL2], and chemo-attractant proteins [CCL receptor-2 [CCR2]) and other proteins with potential role in the regulation of angiogenesis or vascular homeostasis, and blood pressure (plasminogen activator inhibitor-1 [PAI-1] and angiotensinogen). Adipose tissue expresses essential substances that also regulate the carbohydrate, lipid metabolism [adiponectin and resistin], thermogenesis (uncoupling proteins [UCPs], coactivator-1α [PGC-1] and PPAR-γ) and adipogenesis (peroxisome proliferator-activated receptor-γ [PPAR-γ]) [21].

Obesity is associated with a state of chronic, low-grade inflammation characterized by abnormal cytokine production and the activation of inflammatory signaling pathways in WAT [22]. Increased inflammatory cytokines play a critical role in obesity-related inflammation and metabolic pathologies. For example, TNF-α is an important cytokine that induces the production of IL-6 [23] that is one of the major determinant of the acute phase response in the chronic inflammation and also has effects on lipid metabolism, insulin action and glucose transport [24]. Previous studies stated that reduction of adipose mass leads to decrease in the expression levels of TNF-α and IL-6 in obese individuals and animal models [24]. Additionally, TNF-α can also modulates leptin secretions whereas, reduction in body fat positively associated with decreased leptin circulating levels as wells as its gene expression [25].

In cases of lipodystrophy the majority of normal adipose tissue depots are ectopically stored in extra-adipose tissue, leading to risk for a number of diverse conditions such as hypertension atherosclerosis, cancer and insulin resistance [26]. Understanding the pathological and molecular changes in the adipose tissue in the context to obesity and related complications has become complicated. However, adipose tissue malfunction in obesity causes increased lipolysis with concomitant release of excess non-esterifed free fatty acids (FFAs) resulting in the inhibition of glucose uptake in muscle as well as suppression of hepatic glucose output, leading to insulin resistance as a secondary complications [26]. Elevated FFAs also affect the fatty acid-based signaling cascade [27], leads to the decreased vasodilatation and hypertension [28]. Furthermore, intracellular fatty acids lead to activation of inflammatory pathways [29]. Moreover, serum elevated FFAs increase risk for myocardial infarction, cardiomyopathy [30] and enhance atherosclerosis [31]. Whereas, β-cells in the pancreas suffer impaired glucose-stimulated insulin secretion for long term exposure to the elevated FFAs [32,33].

In obesity, the adipocytokines profile of adipose tissue also undergoes diverse changes apart from lipolysis. Hukshorn et al. [34] reported that

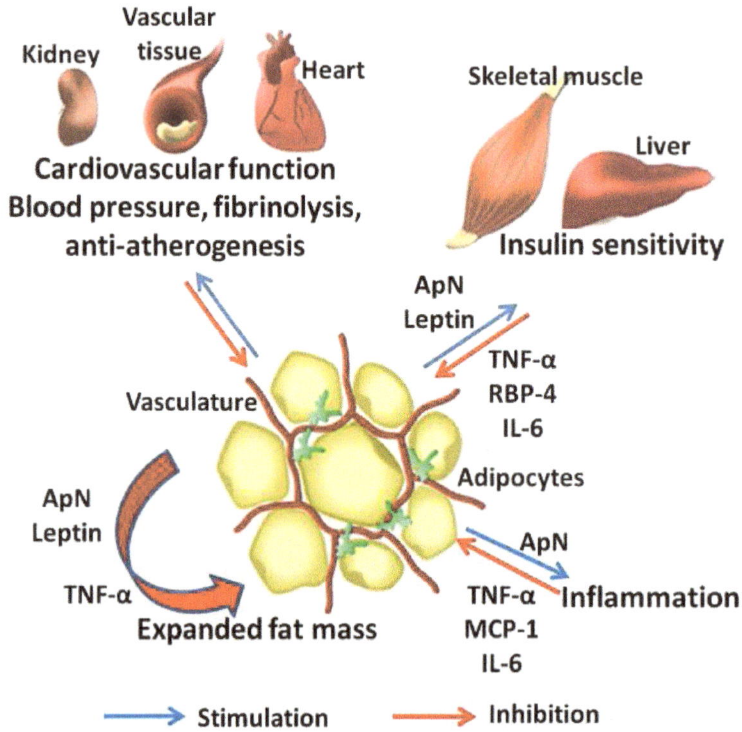

Fig. 1. Regulation of Adipokines in obesity and its complications
Adiponectin (ApN), PAI-1, plasminogen activator inhibitor type 1; AGT, angiotensinogen; TNF-α, tumor necrosis factor-α ; RBP-4, retinol-binding protein-4; MCP-1, monocyte chemoattractant protein-1; IL-6, interleukin-6 [56]

elevated levels of leptin in chronic inflammation state of obesity concomitantly impaired leptin signaling hence, unattractive target for the treatment of obesity. However, the levels of adiponectin are diminished in obesity as a exceptional adipokine [35,36]. In obese individuals the inflammatory cytokines are highly upregulated causing massive infiltration of macrophages due to secretions of high levels of monocyte chemo-attractant protein 1 [MCP-1] from adipocytes [37]. The other proinflammatory cytokines either released from adipocytes or macrophages directly are IL-1β, IL-6, TNF-α, and C-reactive protein [CRP]. It is stated that visceral fat has more severe metabolic effects than subcutaneous fat because most of the cytokines are secreted to a greater extent from visceral adipose tissue [38]. Cytokines released from adipose tissue induce a systemic low-grade inflammation promoting insulin resistance [39]. Moreover, in obesity, TNF-α inhibits triglyceride synthesis through downregulation of PPARγ and its target gene lipoprotein lipase (LPL) as well as glucose transporter GLUT4, and also inhibits insulin-mediated attenuation of lipolysis, down regulates the lipid droplet protein perilipin and up-regulates cAMP pool; all of which further enhances the release of FFAs [40].

3. ADIPOSE TISSUE AND ENERGY METABOLISM

Adipogenesis defined as a development of mature adipocyte from its multipotent mesenchymal stem cell. PPARγ is an adipose tissue secreted molecule that acts as a master regulator of adipogenesis [41,42]. The isoform of PPARγ such as PPAR-γ2 is expressed in adipose tissue and responsible for the differentiation and proliferation of adipocytes leads to an increase in adipose mass and its gender dependency [43-45]. While BAT plays a potential role though adaptive thermogenesis, WAT act mainly as fat storage. BAT contains high number of mitochondria and small, multilocular lipid droplets aided in the adaptive thermogeneisis [46].

Several studies have focused on UCP families [UCP1, UCP2, and UCP3] role in the adaptive thermogenesis as a physiological defense against diabetes and obesity [47]. Argyropoulos and Harper [48] reported a significant UCP-1 expression in BAT for whole body energy expenditure and its dysfunction leads to the development of obesity. UCP1 is a unique mitochondrial membranous protein devoted to adaptive thermogenesis, a specialized function performed by brown adipocytes. Among the family of mitochondrial metabolite carriers, UCP1 is the only member able to translocate protons through the inner membrane of brown adipocyte mitochondria (Figure 2). UCP1 provokes energy dissipation in the form of heat by uncoupling respiration from ATP synthesis. UCP1 in turn also stimulates high levels of fatty acid oxidation [49]. UCP2 has high degree of sequence similarity with UCP1 [55–60%] and have been identified in both rodent and human tissues [50]. UCP3 is mainly expressed in human skeletal muscle and BAT [51].

The imbalance between energy intake and expenditure represent the major predisposing factor to contributing to obesity and its complications. Nowadays, most of anti-obesity drugs mainly target the energy intake through gastrointestinal mechanisms without altering any central mechanisms such as nutrient digestion and absorption inhibitors. Previous studies showed that administration of bile acids (BAs) to mice increases energy expenditure in BAT, hence, BAs came to light. BAs have long been known to be essential in dietary lipid absorption and cholesterol catabolism. However, BAs are established now as signaling molecules that activate mitogen-activated protein kinase pathways [52], are ligands for the G-protein-coupled receptor (GPCR) TGR5 [53,54]. BAs promote energy expenditure in BAT by activation of TGR5 leads to prevention of obesity and insulin resistance [55]. TGR5 plays an important role in energy expenditure to regulate body metabolism. The cAMP-dependent thyroid hormone activating enzyme type 2 iodothyronine deiodinase [D2] expression is induced by BAs in human skeletal muscle and thermogenic tissues of mouse brown fat. In this tissue, binding of BAs to the plasma membrane G-protein–coupled bile acid receptor 1 [Gpbar1, also known as TGR5] triggers an increase in cAMP formation and, subsequently, D2 expression. D2 subsequently activates thyroxine [T4] into tri-iodothyronine [T3]. T3 is predicted to induce uncoupling protein [UCP] expression in BAT. UCP is known to dissipate the proton gradient in the electron transport chain. Therefore, the activation of BAs/TGR5/cyclic AMP/D2/T3/UCP pathway cascade causes a decrease in the synthesis of ATP while enhancing substrate oxidation and thus regulates energy homeostasis [55].

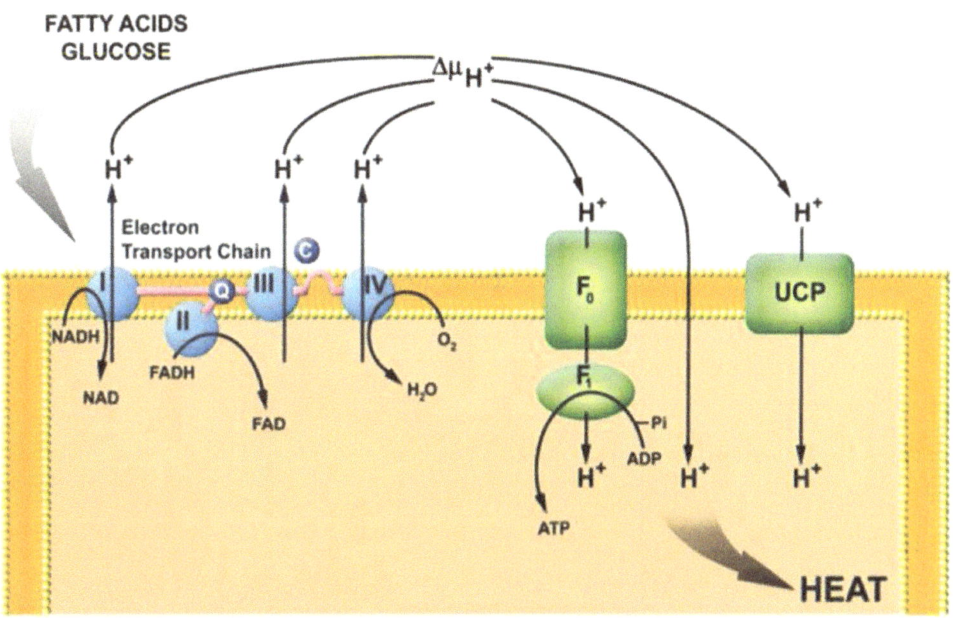

Fig. 2. Mitochondrial ATP metabolism through proton transport chain and adaptive thermogenesis through uncoupler proteins

Ubiquinone (Q), cytochrome c (C), complexes I, II III, and IV of the electron transport chain, uncoupler protein (UCP), Nicotinamide adenine dinucleotide (NAD), Flavin adenine dinucleotide (FAD) [57]

4. OBESITY AND COMPLICATIONS

Obesity as major health concern leads to a reduction in life expectancy of nearly 8 years, and to several co-morbidities, including the metabolic syndrome consisting of insulin resistance, type 2 diabetes, cardiovascular disease, fatty liver disease, cancer, hypertension, stroke, dementia, and obstructive sleep apnoea [10,58-60]. A positive correlation between pain and obesity has also been reported in relation to several pathological conditions [61], and a negative relationship between obesity and quality of life has been identified using health-related quality of life scores, which include measures of pain, fatigue and physical functioning [62]. A number of studies have shown that obese individuals also have increased susceptibility to systemic infections, are more likely to develop infectious complications after surgery, and are more susceptible to nosocomial diseases [63]. In addition, obesity has been found to have a serious negative impact on the individual's psychology and social life as a consequence of abnormal body image, prejudice, discrimination and limitations affecting daily life activities due to increased body weight (Obesity: preventing and managing the global epidemic. WHO consultation, 2000). Clearly the spread of obesity and its association with several complications, e.g., physical impairment, pain and other co-morbid disorders, including diabetes and cardiovascular diseases, strengthens the necessity to further understanding of the mechanisms underlying this pathological state and to achieve better quality of life for obese patients.

5. CONCLUSION

Adipose tissue as an endocrine organ acts as a primary storage of excess energy (in WAT) as well as potential in adaptive thermogenesis (in BAT). Adipose tissue secretes various adipocyte-specific factors known as adipokines such as leptin and adiponectin with beneficial effects on insulin action and lipid metabolism. The derived cytokines (tumor necrosis factor-α (TNF-α), interleukin-6 (IL-6) from adipose tissue in obese individuals make understand the molecular the link between inflammation, obesity and insulin resistance. The secretion of adipokines are directly proportional to the adipose tissue mass, promoting the idea that reducing total adipose mass may be a strategy for the treatment of obesity and its complications. Additionally, understanding the signaling pathways in target tissues such as the skeletal muscle, brain, and liver by which adipokines control metabolism, the signaling activation cascade of TGR5 by BAs and its role in energy expenditure in BAT, the molecular mechanisms responsible for macrophage polarization and targeting the mechanisms involved in macrophage recruitment from the circulation may also help in the development of novel therapies for the obesity and its related complications.

COMPETING INTERESTS

Authors have declared that no competing interests exist.

REFERENCES

1. Joosen AM, Westerterp KR. Energy expenditure during overfeeding. Nutr Metab. 2006;3:25.
2. O'Rahilly S. Human genetics illuminates the paths to metabolic disease. Nature. 2009;462:307-314.
3. Marie Ng, et al. A systematic analysis for the global burden of disease study 2013. The Lancet. 2014;384(9945):766-781.
4. Haslam DW, James WP. Obesity. Lancet. 2005;366:1197-1209.
5. Flegal KM, Carroll MD, Ogden CL, Curtin LR. Prevalence and trends in obesity among US adults, 1999-2008. JAMA. 2010;303:235-241.
6. Hossain P, Kawar B, El Nahas M. Obesity and diabetes in the developing world - A growing challenge. N Engl J Med. 2007; 356:213-215.
7. World Health Organization: Obesity: Preventing and managing the global epidemic. Report of a WHO consultation. World Health Organ Tech Rep Ser. 2010; 894:1-253.
8. Berghofer A, et al. Obesity prevalence from a European perspective: A systematic review. BMC Public Health. 2008;8:200.
9. Yahia N, Achkar A, Abdallah A, Rizk S. Eating habits and obesity among Lebanese university students. Nutr J. 2008;7:32.
10. Bluher M. Adipose tissue dysfunction in obesity. Exp Clin Endocrinol Diabetes. 2009;117(6):241-250.
11. Weisell RC. Body mass index as an indicator of obesity. Asia Pac J Clin Nutr. 2002;11(8):S681-4.

12. Janniger CK, Schwartz RA, Szepietowski JC, Reich A. Intertrigo and common secondary skin infections. Am Fam Physician. 2005;72(5):833-838.
13. Cannon B, Nedergaard J. Brown adipose tissue: function and physiological significance. Physiol Rev. 2004;84:277-359.
14. Cypess AM, et al. Identification and importance of brown adipose tissue in adult humans. N Engl J Med. 2009;360: 1509-1517.
15. Shen W, et al. Adipose tissue quantification by imaging methods: A proposed classification. Obesity Research. 2003;11:5-16.
16. Vohl MC, et al. A survey of genes differentially expressed in subcutaneous and visceral adipose tissue in men. Obesity Research. 2004;12:1217-1222.
17. Kissebah AH, Krakower GR. Regional adiposity and morbidity. Physiol Rev. 1994;74:761-811.
18. Krotkiewski M, Bjorntorp P, Sjostrom L, Smith U. Impact of obesity on metabolism in men and women. Importance of regional adipose tissue distribution. J Clin Invest. 1983;72:1150-1162.
19. Coelho M, Oliveira T, Fernandes R. Biochemistry of adipose tissue: an endocrine organ. Archives of Medical Science: AMS. 2013;9(2):191-200.
20. Shen W, et al. Adipose tissue quantification by imaging methods: A proposed classification. Obesity Research. 2003;11:5-16.
21. Ebtesam A AL-Suhaimi, Adeeb Shehzad. Leptin, resistin and visfatin: The missing link between endocrine metabolic disorders and immunity. Archives of Medical Science: AMS. 2013;9(2):191-200.
22. Jung UJ, Choi M-S. Obesity and its metabolic complications: The role of adipokines and the relationship between obesity, inflammation, insulin resistance, dyslipidemia and nonalcoholic fatty liver disease. International Journal of Molecular Sciences. 2014;15(4):6184-6223.
23. Hajri T, Tao H, Wattacheril J, Marks-Shulman P, Abumrad NN. Regulation of adiponectin production by insulin: Interactions with tumor necrosis factor-α and interleukin-6. Am. J. Physiology - Endocrinology and Metabolism. 2011; 300(2):E350-E360.
24. McArdle MA, Finucane OM, Connaughton RM, McMorrow AM, Roche HM. Mechanisms of obesity-induced inflammation and insulin resistance: Insights into the emerging role of nutritional strategies. Frontiers in Endocrinology. 2013;4:52.
25. Sophie Carter, Alexandre Caron, Denis Richard, Frédéric Picard. Role of leptin resistance in the development of obesity in older patients. Clin Interv Aging. 2013;8: 829–844.
26. Jianping Ye. Mechanisms of insulin resistance in obesity. Frontiers of Medicine. 2013;7(1):14-24.
27. Roberts CK, Hevener AL, Barnard RJ. Metabolic syndrome and insulin resistance: Underlying causes and modification by exercise training. Comprehensive Physiology. 2013;3(1):1-58.
28. Yong Wu, Hillary Zhou, Ke Wu, Sangkyu Lee, Ruijin Li, Xuan Liu. PTEN phosphorylation and nuclear export mediate free fatty acid-induced oxidative stress. Antioxidants & Redox Signaling. 2014;20(9):1382-1395.
29. Hellmann J, Zhang MJ, Tang Y, Rane M, Bhatnagar A, Spite M. Increased saturated fatty acids in obesity alter resolution of inflammation in part by stimulating prostaglandin production. Journal of Immunology. 2013;191(3):1383-1392.
30. Pilz S, März W, Wellnitz B, et al. Association of vitamin D deficiency with heart failure and sudden cardiac death in a large cross-sectional study of patients referred for coronary angiography. J Clin Endocrinol Metab. 2008;93:3927–3935.
31. Bamba V, Rader DJ. Obesity and atherogenic dyslipidemia. Gastroenterology. 2007;132:2181-2190.
32. Jun Ho Kim, Hyung Taek Cho, Young Jun Kim. The role of estrogen in adipose tissue metabolism: Insights into glucose homeostasis regulation. Endocrine Journal. 2014;61(11):1055-1067.
33. Carpentier A, et al. Acute enhancement of insulin secretion by FFA in humans is lost with prolonged FFA elevation. Am J Physiol. 1999;276:E1055-1066.
34. Hukshorn CJ, et al. The effect of pegylated recombinant human leptin [PEG-OB] on weight loss and inflammatory status in obese subjects. Int J Obes Relat Metab Disord. 2002;26:504-509.
35. Michael Rosenbaum, Rudolph L Leibel. 20 years of Leptin: What we know and what the future holds. J Endocrinol. 2014;223: T83–T96.

36. Yatagai T, et al. Hypoadiponectinemia is associated with visceral fat accumulation and insulin resistance in Japanese men with type 2 diabetes mellitus. Metabolism. 2003;52:1274-1278.
37. Panee J. Monocyte Chemoattractant Protein 1 (MCP-1) in obesity and diabetes. Cytokine. 2012;60(1):1-12.
38. Maury E, et al. Adipokines oversecreted by omental adipose tissue in human obesity. American Journal of Physiology 2007; 293:E656-665.
39. Makki K, Froguel P, Wolowczuk I. Adipose tissue in obesity-related inflammation and insulin resistance: Cells, cytokines, and chemokines. ISRN Inflammation. 2013; 139239.
40. Guilherme A, Virbasius JV, Puri V, Czech MP. Adipocyte dysfunctions linking obesity to insulin resistance and type 2 diabetes. Nat Rev Mol Cell Biol. 2008;9:367-377.
41. Mori S, Kiuchi S, Ouchi A, Hase T, Murase T. Characteristic expression of extracellular matrix in subcutaneous adipose tissue development and adipogenesis; Comparison with visceral adipose tissue. International Journal of Biological Sciences. 2014;10(8):825-833.
42. Rosen ED, Spiegelman BM. Molecular regulation of adipogenesis. Annu Rev Cell Dev Biol. 2000;16:145–171.
43. Garin-Shkolnik T, Rudich A, Hotamisligil GS, Rubinstein M. FABP4 attenuates PPARγ and adipogenesis and is inversely correlated with PPARγ in adipose tissues. Diabetes. 2014;63(3):900-911.
44. Ogawa R, Mizuno H, Watanabe A, Migita M, Hyakusoku H, Shimada T. Adipogenic differentiation by adipose-derived stem cells harvested from GFP transgenic mice-including relationship of sex differences. Biochem Biophys Res Commun. 2004; 319:511–517.
45. Tontonoz P, Hu E, Spiegelman BM. Stimulation of adipogenesis in fibroblasts by PPAR gamma 2, a lipid-activated transcription factor. Cell. 1994;79:1147–1156.
46. Kajimura S, Saito M. A new era in brown adipose tissue biology: Molecular control of brown fat development and energy homeostasis. Annual Review of Physiology. 2014;76:225-249.
47. Cypess, Aaron M, and Ronald Kahn C. Brown fat as a therapy for obesity and diabetes. Current Opinion in Endocrinology, Diabetes, and Obesity 2010;17(2):143–149.
48. Argyropoulos G, Harper ME. Uncoupling proteins and thermoregulation. J Appl Physiol. 2002;92:2187–2198.
49. Daniel Ricquier. Uncoupling protein1 of brown adipocytes, the only uncoupler: A historical perspective. Cellular Endocrinology. 2011;2.
50. Fleury C, Neverova M, Collins S, et al. Uncoupling protein-2: A novel gene linked to obesity and hyperinsulinemia. Nat Genet 1997;15:269–72.
51. Vidal-Puig A, Solanes G, Grujic D, et al. UCP3: An uncoupling protein homologue expressed preferentially and abundantly in skeletal muscle and brown adipose tissue. Biochem Biophys Res Commun. 1997; 235:79–82.
52. Gupta S, Stravitz RT, Dent P, Hylemon PB. Down-regulation of cholesterol 7a-hydroxylase (CYP7A1) gene expression by bile acids in primary rat hepatocytes is mediated by the c-Jun N-terminal kinase pathway. J. Biol. Chem. 2001;276:15816-15822.
53. Kawamata Y, et al. A G protein-coupled receptor responsive to bile acids. J. Biol. Chem. 2003;278:9435-9440.
54. Maruyama T, et al. Identification of membrane-type receptor for bile acids (M-BAR). Biochem. Biophys. Res. Commun. 2002;298:714-719.
55. Watanabe M, et al. Bile acids induce energy expenditure by promoting intracellular thyroid hormone activation. Nature. 2006;439:484–489.
56. Maury E, Brichard SM. Adipokine dysregulation, adipose tissue inflammation and metabolic syndrome. Molecular and Cellular Endocrinology. 2010;314:1–16.
57. Bruce M Spiegelman, Jeffrey S Flier. Obesity and the regulation of energy balance. Cell. 2001;104:531–543.
58. Trayhurn P, Wood IS. Signalling role of adipose tissue: Adipokines and inflammation in obesity. Biochem Soc Trans. 2005;33(Pt 5):1078-1081.
59. Galic S, Oakhill JS, Steinberg GR. Adipose tissue as an endocrine organ. Mol Cell Endocrinol, 2010;316(2):129-139.
60. Ragab SMM, Abd-Elghaffar SKH, Tarek H, El-Metwally, Badr G, Mahmoud MH, Omar HM. Effect of a high fat, high sucrose diet

on the promotion of non-alcoholic fatty liver disease in male rats: The ameliorative role of three natural compounds. Lipids in Health and Disease. 2015;14(83):1-11.
61. Grotle M, Hagen KB, Natvig B, Dahl FA, Kvien TK. Obesity and osteoarthritis in knee, hip and/or hand: an epidemiological study in the general population with 10 years follow-up. BMC Musculoskelet Disord. 2008;9:132.
62. Anandacoomarasamy A, Smith G, Leibman S, Caters on I, Giuffre B, Fransen M, Sambrook P, March L. Cartilage defects are associated with physical disability in obese adults. Rheumatology. 2009;48(10): 1290-1293.
63. Smith SR, et al. Effect of pioglitazone on body composition and energy expenditure: a randomized controlled trial. Metabolism. 2005;54:24-32.

14

Changes in Antioxidant Status between Different Cooking Methods in Eggs Yolk

Parisa Sadighara[1*] and Rooholla Ghanbari[1]

[1]Department of Environmental Health Engineer, Food Safety Division, Faculty of Public Health, Tehran University of Medical Sciences, Tehran, Iran.

Authors' contributions

This work was carried out in collaboration between both authors. Author PS designed the study, wrote the protocol and manuscript. Author RG carried out all laboratories work.

Editor(s):
(1) Mohamed Fawzy Ramadan Hassanien, Biochemistry Department, Zagazig University, Egypt.
Reviewers:
(1) Zhao Chen, Department of Biological Sciences, Clemson University, USA.
(2) S. Ramesh, Department of Chemistry, Saveetha University, India.
(3) Anonymous, University of Bologna, Italy.
(4) Armagan Hayirlim, Ataturk University, Turkey.

ABSTRACT

Background: The method of cooking causes important changes in quality and safety of food. The purpose of this study was to survey the relation between nutritional- qualitative factors of eggs in different of cooking methods.
Methods: The eggs were exposed in different cooking including water cooking, cooking in a microwave and frying. Lipid oxidation, Vitamin E and carotenoid of yolk were measured.
Results: The changes in lipid peroxidation were not significant. There was significant change in Vitamin E level in water cooking method. The levels of vitamin E in water cooking and frying method were 0.82±0.05 and 2.6±0.6 mg/g respectively. The carotenoids content also reduced in microwave method.
Conclusion: This study showed that significant differences may exist in some nutritional-qualitative factors such as vitamin E and carotenoids levels between common cooking methods of egg. Our findings offer frying as suitable method to maintain their active components.

Keywords: Vitamin E; carotenoids; lipid peroxidation; cooking methods.

**Corresponding author: E-mail: sadighara@farabi.tums.ac.ir*

1. INTRODUCTION

Cooking is a process for preparing food with the use of heat. Heating is applied to food to enhance flavor, taste, inactivate pathogenic microorganisms and increase shelf life [1]. The kind of cooking influences on food component. It can lead to undesirable modifications, such as the loss of the nutritional value and lipid oxidation [2]. There are evidences that different method of cooking display different levels of nutritional-qualitative factors in food. The nutritional-qualitative factors in food including antioxidants can reduce degenerative disease. The best known natural antioxidants that have proven important in the food and in human health are Vitamin E and carotenoid [3]. Furthermore, Vitamin E and carotenoid are considered to be critical in protection against oxidative damages. It is suggested that accumulation of natural antioxidant Vitamin E and carotenoid protect unsaturated fatty acid against peroxidation during cooking.

Among the dietary constituents of food, lipids have critical roles of food safety and quality. The unsaturated fatty acids are sensitive to oxidative damages. The polyunsaturates fatty acids are vulnerable to free radical damages. Their autooxidation is also catalysed by heat [1]. LDL cholestrol are also sensitive to oxidation. Cholesterol oxidation products seem to have important biological effects such as atherogenicity, cytotoxicity and modification of enzyme activity [3]. The yolk from an egg contains about 213 mg cholesterol [4].

Eggs have been recognized as valuable sources of high nutritional components. Eggs are a key ingredient in many countries food. There are different methods of cooking to prepare eggs such as water cooking, frying, and cooking in a microwave. The water cooking and frying are traditional method. The use of microwave has been increased because of its advantages which include speed and convenience [5]. The objective of this study was to determine the effects of different method of cooking on some qualification factors include vitamin E and carotenoids plus lipid peroxidation.

2. MATERIALS AND METHODS

2.1 Experimental Design

We used three method of cooking. To prepare water boiling, the egg is placed in water and boiled at 100°C for 20 min. The yolk was removed for assessing. Then, the microwave was adjusted to 180°C for 5 min used for cooking eggs. In frying method, the yolk separated and placed on a pan without vegetable oils. The temperature reached to 80°C for 2 min.

2.2 Measurement of Lipid Peroxidation

The formation of thiobarbituric acid in samples was assessed for the measurement of lipid peroxidation according to an original method [6]. Briefly, the supernatant of the sample homogenate was mixed with 20% trichloroacetic acid and the mixture was centrifuged. Then, thiobarbituric acid was added to the supernatant and heated. The absorbance of the supernatant was measured at 532 nm and TBARS concentration was calculated using MDA standard curve.

2.3 Measurement of Vitamin E Content

The yolk vitamin E content was measured according to a published method. Samples were exposed to Fe^{3} solution, TPTZ and acetate buffer (pH 4). Then, the standard curve was prepared with appropriate vitamin E concentrations. The absorbance of samples was read at 595 nm wavelength [7].

2.4 Measurement of Total Carotenoids

Total carotenoids of samples were measured using β-carotene standard curve and spectrophotometry at 470 nm wavelength. The total carotenoid content of the samples was calculated on the basis of the standard curve of β-carotene [8].

2.5 Statistical Analysis

The evaluation was made by comparing group. The difference more than 95 %($p \leq 0.05$) was considered significant. The data values were expressed as mean±SD.

3. RESULTS

Level of the nutritional-qualitative factors in different method of cooking was shown in Table 1. The data values were expressed as mean±SD. The evaluation was made by comparing group by analysis of T-Test in SPSS software.

Table 1. Level of nutritional-qualitative parameters of egg yolk

	Level of lipid peroxidation (mg MDA /g)	Vitamin E mg/g	Carotenoids level mg/g
Water cooking	3±0.47	0.82±0.05	1.8±0.1
Frying	3.2±0.16	2.6±0.6*	1.96±0.05
Microwave	3± 0.19	1.6±0.5	1.56±0.11

The changes in lipid peroxidation were not significant. The level of vitamin E significantly decreased in this water method cooking

The level of vitamin E was significantly different between water cooking compared to frying (*p* = 0.04). The P values are not more than 0.05 in lipid peroxidation. Table 1 shows no significant difference in lipid peroxidation level. The level of vitamin E was highest in frying ($p<0.05$).

The changes in level of carotenoids were observed between frying method and microwave cooking (*p*=0.05).

4. DISCUSSION

The egg is usually treated by various processes. These processes could change its composition. In the current study, the methods of cooking which can affect the nutritional- qualitative factors were surveyed.

The egg yolk is susceptible to oxidation because of high level of polyunsaturated fatty acid. Oxidative stress results from excessive levels of reactive oxygen species (ROS) and damages macromolecules such as DNA, lipids, proteins and carbohydrates. ROS attacks polyunsaturated fatty acids and causes lipid peroxidation. MDA is a product of lipid peroxidation that has been used as an indicator in oxidative damages. This component is highly reactive and can cause deterioration of biological molecules such as DNA. MDA is mutagenic and tumorigenic [9]. Major changes that occur during heating are due to oxidation. The polyunsaturated fatty acids are considered to be susceptible to oxidation during heating [1]. The yolk is extremely susceptible to lipid peroxidation because of the accumulation of polyunsaturated fatty acids. The level of lipid peroxidation didn't change in conventional methods of cooking (Table 1). MDA products did not increase in these methods of cooking. Therefore there is no hazard as results of lipid peroxidation in different cooking.

The oxidative stability of foods during cooking depends on the level of antioxidant. Presence of antioxidants has an important role on the prevention oxidative changes. This time require a considerable degree antioxidant against peroxidation [10].

The antioxidants in foods decrease the formation of heterocyclic amines (HCAs) that form during the frying of protein rich food. HCAs are recognized as animal carcinogens [11]. The vitamin E level didn't alter in frying and microwave heating. The results suggested that changes of vitamin E level depended to processing time. The exposure time in frying method was 2 min. In previous study, the amount of vitamin E in animal fats destroyed after 8 min of heating [12]. This vitamin at high temperature is susceptible to degradation. Therefore, the degradation of vitamin E is low in frying method.

Carotenoids have also been documented for health promoting function. These components are known to provide a range of biological effects (provitamin, antioxidant, coloring etc.). The role of carotenoids as a source of pigment and in immune defense system has been established. Because of their oxidation sensitivity, carotenoids must be protected from light, heat, and oxygen. In this study, carotenoids levels were significantly increased in frying cooking and microwave method ($P<0.05$) (Table 1). In previous study, the bioaccessibility of egg carotenoids can be affected by different food preparation methods [13].

5. CONCLUSIONS

The antioxidants content of fry cooked egg were considered more than other method. The level of vitamin E was highest in this method compared to water cooking and microwave method. The eggs that prepare with frying pan have the potential to be a good source of active components. These active components are considered to be powerful antioxidants in eggs. In this study, the frying of eggs is recommended among different of cooking method.

COMPETING INTERESTS

Authors have declared that no competing interests exist.

REFERENCES

1. Weber J, Bochi VC, Ribeiro CP, Victorio AM, Emanuelli T. Effect of different cooking methods on the oxidation, proximate and fatty acid composition of silver catfish fillets. Food Chemistry. 2008;106:140-146.
2. Rodriguez-Estrada MT, Penazzi G, Caboni MF, Bertacco G, Lercker G. Effect of different cooking methods on hamburgers. Meat Science. 1997;3:365-375.
3. Iqbal S, Bhanger MI, Anwar F. Antioxidant properties and components of some commercially available varieties of rice bran in Pakistan. Food Chemistry. 2005;93:265-272.
4. Awad AC, Bennink MR, Smith DM, Composition and functional properties of cholesterol reduced egg yolk. Poultry Science. 1997;76:649–653
5. Cossignani L, Simonetti MS, Neri A, Damiani P. Changes in olive composition due to microwave heating. JAOCS. 1998; 75:931-937.
6. Sicinska P, Bukowska B, Michalowicz J, Duda W. Damage of cell membrane and ant-oxidative system in human erythrocytes incubated with microcystin-LR in vitro. Toxicon. 2006;47:387-397.
7. Devi I, Memon SA, Khuhawar MY. Spectrophotometric analysis of vitamin E using Cu (I)-bathocuproine or/and Fe (II)-2, 4, 6-tris-(2-pyridyl)-s-triazine complexes. Journal of Chemical Society of Pakistan. 2004;26:239-242.
8. Thaipong K, Boonprakob U, Zevallos LC, Byrne DH. Hydrophilic and lipohilic antioxidant activities of guava fruits. The Southeast Asian Journal Tropical Medicine and Public Health. 2005;36:254-257.
9. Zhang Y, Chen SY, Hsu T. Immunohistochemical detection of malondialdehyde–DNA adducts in human oral mucosa cells. Carcinogenesis. 2001; 23:207–211.
10. Surai PF. Tissue specific changes in the activities of antioxidant enzymes during the development of the chicken embryo. British poultry science. 1999;40:397-405.
11. El-Abassy RM, Donfack P, Materny A. Assessment of conventional and microwave heating induced degradation of carotenoids in olive oil by VIS raman spectroscopy and classical methods. Food Research International. 2010;43:696-700.
12. Anjum F, Anwar F, Jamil A, Iqbal M. Microwave roasting effects on the physic-chemical composition and oxidation stability of sunflower seed oil. JAOCS. 2006;83:777-784.
13. Nimalaratne C, Savard P, Gauthier SF, Schieber A, Wu J. Bioaccessibility and digestive stability of carotenoids in cooked eggs studied using a dynamic in vitro gastrointestinal model. J Agric Food Chem. 2015;25;63:2956-62.

15

(1H-Pyrrolo [2,3-b]pyridine)7-Azaindole Derivatives and Their Antiurease, Phosphodiesterase and β-glucuronidase Activity

Zafar S. Saify[1], Nighat Sultana[2*], Ajmal Khan[1] and Shazia Haider[3]

[1]HEJ Research Institute of Chemistry, University of Karachi, Karachi-75270, Pakistan.
[2]Pharmaceutical Research Center, PCSIR Laboratories Complex, Karachi-75280, Pakistan.
[3]Department of Pharmaceutical Chemistry, Faculty of Pharmacy, University of Karachi, Karachi-75270, Pakistan.

Authors' contributions

This work was carried out in collaboration between all authors. Author ZSS designed the study, wrote the protocol, and author NS wrote the first draft of the manuscript, managed the literature searches. Author AK performed the statistical analysis and author SH managed the analyses of the study. All authors read and approved the final manuscript.

Editor(s):
(1) Rosario Gomez Garcia, Department of Biochemistry, Loyola University, USA.
Reviewers:
(1) Nura Suleiman Gwaram, Chemistry, University of Malaya, Malaysia.
(2) Anonymous, University of Jyväskylä, Finland.
(3) Anonymous, Poznań University of Life Sciences, Poland.
(4) Anonymous, Sun Yat-sen University, China.

ABSTRACT

A variety of 7-azaindole analogs **1-14** with variable substituents on phenyl ring of phenacyl moiety were synthesized and evaluate for their urease, phosphodiesterase and β-glucuronidase Inhibitory potential. Compound **9** (IC_{50} = 2.19±0.37μM) showed potent urease inhibitory potential than standard thiourea (IC_{50} = 21.00±0.01μM). However, while compounds 10 (IC_{50} = 255.11±6.62μM) and 8 (IC_{50} = 133.3±0.46μM), exhibited moderate urease potential. Moreover, compound **2** (IC_{50} = 20.83± 0.234μM) showed potent phosphodiesterase inhibitory potential than standard EDTA (IC_{50} = 274.00+0.007μM). Compound **8** (IC_{50} 192.6±3.53μM) was found to be moderate β-glucuronidase inhibitor, as compare to standard 1,4, lactone D saccharic acid (IC_{50} = 48.41±1.24μM).

**Corresponding author: Email: nighat2001us@hotmail.com, w_80ahmed@yahoo.com*

Nevertheless, compounds **13** (36.81% inhibition) and **14** (47.11% inhibition) showed less than 50% β-glucuronidase inhibition, therefore they were not further evaluated for their IC_{50} values. The size of the substituent, electron donating or withdrawing affect of substituents as well as the position of substituent on phenyl affects the activity.

Keywords: 7-Azaindole; urease; β-glucuronidase *inhibition; synthesis; phosphodiesterase.*

ABBREVIATIONS

Thin layer chromatography, TLC; VDCs, voltage dependent Ca21 channels, Phosphodiesterases-I, PDE;

1. INTRODUCTION

7-Azaindoles are the most widely studied nitrogen analogue of the indole ring system [1]. The 7-azaindole nucleus has proved to be an important model for working on its own synthesis and synthesizing its different analogues [2-5]. In the past, indole derivatives have attracted the interest of scientists for their potential biological activities like [6-10] analgesic, blood pressure lowering, anti-inflammatory activities and acting as effective coronary vasodilator, cardiovascular and potent hypotensive agent [11-14]. Saify et al. synthesized a series of 7-azaindole derivatives having important antimicrobial response against gram positive and gram negative bacteria [15-18]. It has been used as an indole bioisostere to improve physicochemical and pharmacokinetic properties of several drug candidates such as dopamine D4 ligands [19]. It is also an emerging pharmacophore in ATP competitive kinase inhibitors as it contains the typical motif to dock into the adenine binding pocket [20,21].

Activities of ureases have been shown to be an important virulence determinant in the pathogenesis of many clinical conditions. Urease is directly involved in the formation of infectious stones and contributes to the pathogenesis of urolithiasis, pyelonephritis, ammonia and hepatic encephalopathy, hepatic coma, urinary catheter encrustation [22]. It plays an important role in the pathogenesis of gastric and peptic ulcer including cancer [23]. In agriculture, high urease activity causes significant environmental problems by releasing large amount of ammonia into the atmosphere [24]. Therefore strategies based on urease inhibition are now considered as the first line of treatment for infections caused by urease producing bacteria [25].

β-Glucuronidase is an exoglycosidase enzyme that catalyzes the cleavage of glucuronosyl-*O*-bonds. Liver damage causes an increase in this enzyme level in blood. Liver cancer is also suspected to be related to the over expression of this enzyme. Many β-glucuronidase inhibitor such as 8-hydroxytriceine glucuronide, scoparic acid A and C have already been isolated from different plants and some are used clinically [26-28].

Phosphodiesterases-I (PDE) are a type hydrolases that act on di-esters of phosphoric acid and releases nucleoside-5-monophosphate from various nucleosides diphosphate and triphosphate bonds. Phosphodiesterases are widely distributed in mammalian intestinal mucosa, liver cells, serum, and snake venom, as well as in various plants [29]. PDE inhibitors have been identified as new potential therapeutic agent in pulmonary arterial hypertension, coronary heart disease, dementia, depression and schizophrenia.

Our research group is working on drug discovery, and affianced in the search of new enzyme inhibitors since a decade [13-17,30,31,32]. We have recently reported 7-aza-indoles, as an excellent anaphylactic inhibitors [33]. In view of the biological significance of 7-aza-indole and its structural resemblance with compounds which were found in enzyme inhibitors. It was of immense interest to synthesize and evaluate new derivatives of 7-aza-indole against urease, β-glucuronidase and phosphodiesterase enzyme [32]. Herein, we describe a synthesis of derivatives of 1H pyrrolo[2,3-b] pyridine (**1-14**). Compounds **1-14** were evaluated for their inhibition against urease, β-glucuronidase and phosphodiesterase enzyme as well as structure-activity relationship is also described.

2. MATERIALS AND METHODS

2.1 Experimental

2.1.1 General procedure

Reagents were procured from Aldrich Chemical Company. The entire solvents were reagents grade. Reactions were monitored by thin layer chromatography TLC using pre-coated silica gel, GF-254 and analytical TLC was executed on silica gel (Kieselgel 60, 254, E. Merck) pre-coated 0.25 mm plates. Visualization was accomplished with ultraviolet light at 254 and 365 nm UVP UVLS-26 Series (Cambridge). Iodine vapors were also employed for the recognition of spots. All melting points were recorded on Gallenkamp melting point apparatus and were corrected. Silica granules from E. Merck was used for drying reaction product after workup. Ultraviolet spectra were recorded in methanol on a Hitachi U-3200 spectrophotometer. Infra Red spectra were measured on a Shimadzu IR 460 spectrophotometer using KBR disc. Mass spectra were determined on Varian Massen spectrometer MAT 311A spectrometer. Nuclear magnetic resonance spectra were recorded in DMSO-D_6 and MeOD on AVANCE AV 300 spectrometer operating at 300, 400, and 500 MHz.

Phosphodiesterase enzyme activity was monitored spectrophotometrically at 37°C on a microtiter plate spectrophotometer (Spectra Max, Molecular Devices).

β-Glucuronidase (E.C. 3.2.1.31) from *E. coli*, (*p*-nitrophenyl-β-D-glucuronide) was obtained from Sigma Chemical Co. Other reagents were also purchased from commercial sources.

Phosphodiesterase from both ropsatrox crude dried venom (P4631), substrate bis (*p*-nitrophenyl) phosphate sodium salt (N-3002), and magnesium acetate were purchased from Sigma, while ammonium acetate was purchased from Merck. Tris buffer was a product from Scharlau (TR 0423). The standard inhibitor EDTA was purchased from Avon Chem limited.

2.1.2 General method for preparation of compounds

7-azaindole derivatives were synthesized by refluxing equimolar quantity of 7-azaindole in acetone (20-25 ml) and was added successively and 2-bromo-2-acetonapthone, 1-adementile-bromo ethyl-ketone, 2-bromo-3-nitro-acetophenone, 2-bromo-4-nitro-acetophenone were dissolved separately in 20-25 ml of acetone. Reaction mixture was stirred for 24 hours at 52-54°C. The process of reaction was monitored through thin layer chromatography. The crude solid product was filtered and washed with acetone. The product thus obtained was purified through re-crystallization by using ethanol and ether. The pure compound was dried in desicators over anhydrous calcium sulphate. Melting point and spectral data were obtained to confirm the structure of compound.

2.1.3 β-Glucuronidase enzyme inhibition

β-Glucuronidase activity was determined by measuring the absorbance at 405 nm of *p*-nitrophenol formed from the substrate by the spectrophotometric method. The total reaction volume is 250 μl. The reaction mixture contained 185 μl of 0.1 M acetate buffer, 5 μl of test compound solution, 10 μl of enzyme solution incubated at 37°C for 30 min. The plates were read on a multiplate reader (SpectraMax plus 384) at 405 nm after the addition of 50 μl of 0.4 *m*M *p*-nitrophenyl-β-D-glucuronide. All assays were run in triplicate [34]. The IC_{50} values were calculated using the EZ-Fit Enzyme Kinetics program, Perrella Scientific Inc., Amherst, U.S.A.

2.1.4 Urease enzyme inhibition

Reaction mixtures comprising 1 unit of urease enzyme (*Bacillus pasteurii or Jack bean*) solution and 55μl of buffers containing 100 *m*M urea were incubated with 5 μl of test compounds (1 *m*M concentration) at 30°C for 15 min in 96-well plates.

Urease activity was determined by measuring ammonia production using the indophenol method as described by Weatherburn [35]. Briefly, 45 μl each of phenol reagent (1% w/v phenol and 0.005% w/v sodium nitroprusside) and 70 μl of alkali reagent (0.5% w/v NaOH and 0.1% active chloride NaOCl) were added to each well. The increasing absorbance at 630 nm was measured after 50 min, using a microplate reader (Molecular Device, USA). All reactions were performed in triplicate in a final volume of 200μl. The results (change in absorbance per min) were processed by using Soft- Max Pro software (Molecular Device, USA). All reactions were performed in triplicate. All the assays were performed at pH 8.2 (0.01 M $K_2HPO_4.3H_2O$, 1 *m*M EDTA and 0.01 M $LiCl_2$). Percentage inhibitions were calculated from the formula 100-($OD_{testwell}/OD_{control}$) x 100. Thiourea was used as the standard inhibitor of urease.

2.2.5 Phosphodiesterase enzyme inhibition

Activity against the snake venom phosphodiesterase I (Sigma P 4631) (EC

3.1.4.1) was assayed by using the reported method [36] with the following modifications. 33 *m*M Tris–HCl buffer, pH 8.8, 30 *m*M Mg-acetate with 0.000742 U/well final concentrations using microtiter plate assay and 0.33 *m*M bis-(p-nitro phenyl) phosphate (Sigma N-3002) as a substrate. Cysteine and EDTA (E. Merck) were used as positive controls (IC_{50} = 748 lM±0.15, 274 lM±0.07, respectively). After 30 min of incubation, the enzyme activity was monitored spectrophotometrically at 37°C on a microtiter plate spectrophotometer (Spectra Max, Molecular Devices) by following the release of *p*-nitrophenol from *p*-nitrophenyl phosphate at 410 *n*m. All the reactions were performed in triplicate and the initial rates were measured as the rate of change in OD/min (optical density/min) and used in subsequent calculations.

1H-pyrrolo[2,3-b]pyridine (7-azaindole) (1): White crystalline solid, EIMS *m/z* =118.14 amu, M. P. = 105-107°C.

1-[3-(3,4-Dihydroxyphenyl)3-oxo-ethyl]-7H-pyrrolo[2,3-b]pyridine-1-ium;bromide (2): HREI MS: $C_{15}H_{13}N_2O_3Br$; UV: 289, 227 and 208. IR: 3100,1670,1595 and 1290, EIMS *m/z* (rel. int %): M-1($C_{15}H_{12}N_2O_3$), 269 (2), 239 (14), 193 (3), 165 (12), 137 (15), 118 (100) and 109 (25). ^{1}H-NMR (D_2O, 300 MHz) *δ*: 8.72 (1H, d, *J* = 7.85 Hz, H-6), 8.24 (1H, d, *J* = 6.80 Hz, H-4), 7.70 (1H, dd, *J* = 8.46, 2.21 Hz, H-6), 7.67 (1H, d, *J* = 3.65 Hz, H-2), 7.62 (1H, d, *J* = 7.85, 6.80 Hz, H-5), 7.55 (1H, d, *J* = 2.19 Hz, H-2), 7.06 (1H, d, *J* = 8.46 Hz, H-5'), 6.97 (1H-d, *J* =3.65 Hz, H-3), 6.48 (2H, *s*, N-CH_2), 4.83 (1H,br-*s*, OH), 4.65 (1H, br-*s*, OH).

7-[2-(3-Nitrophenyl)-2-oxo-ethyl]-1H-pyrrolo[2,3-*b*]pyridinium;bromide (3): Yield =80%, HREI MS: = $C_{15}H_{12}N_3O_3Br$, EI MS: *m/z* =362.178 amu, M. P. = 224 + 2°C, UV λ_{max} (MeOH) *nm* = 299.6, 228, 200 nm, IR v_{max} (KBr) cm^{-1} = 3232, 2916, 1965, 1797, 1697, 1529, 1475, 1356, 1224.9, 1172.8, 1091.5, 930, 883, 799, 598, 519, 481.2, EIMS = *m/z* (relative int., %): 282(M^+, $C_{15}H_{12}N_3O_3$, 9.8), 252(100), 236(41), 206(57), 193(6.1), 150(15.6), 131(53.5), 118(13.3), 90(6.2), 77(17.1), 63(6.6),^{1}H-NMR (MeOD, 300 MHz) *δ*: 8.93 (1H, s, H-6), 8.81 (1H, d, *J* = 7.79 Hz, H-4), 8.69 (1H, dd, *J* = 7.71, Hz, H-2`), 8.52 (1H, m, H-5'), 8.39 (1H, *s*, H-4`), 7.93 (1H, dd, *J* = 8.02, 3.35 Hz, H-6'), 7.77(1H, d, *J* = 3.35 Hz, H-5), 7.67 (1H, m, H-2), 7.01 (1H, d, *J* =3.50 Hz, H-3), 6.90 (2H, d, *J* = 3.57 Hz, H-2``).

7-[2-(2-Nitrophenyl)-2-oxo-ethyl]-1H-pyrrolo[2,3-*b*]pyridinium;bromide (4): Yield =80%, HREI MS: = $C_{15}H_{12}N_3O_3Br$, EIMS: *m/z* = 362.178 amu, M.P. = 224 + 2°C, UV λ_{max} (MeOH) nm = 299.6, 228, 200 nm, IR v_{max} (KBr) cm^{-1} = 3232, 2916, 1965, 1797, 1697, 1529, 1475, 1356, 1224.9, 1172.8, 1091.5, 930, 883, 799, 598, 519, 481.2, EIMS = *m/z* (relative int., %): 282(M^+, $C_{15}H_{12}N_3O_3$, 9.8), 252(100), 236(41), 206(57), 193(6.1), 150 (15.6), 131(53.5), 118 (13.3), 90(6.2), 77(17.1), 63(6.6),^{1}H-NMR (MeOD, 300 MHz) *δ*: 8.93 (1H, *s*, H-6), 8.81 (1H, d, *J* = 7.79 Hz, H-4), 8.69 (1H, dd, *J* = 7.71, Hz, H-2`), 8.52 (1H, m, H-5'), 8.39 (1H, *s*, H-4`), 7.93 (1H, dd, *J* = 8.02, 3.35Hz, H-6'), 7.77(1H, d, *J*= 3.35 Hz, H-5), 7.67 (1H, m, H-2), 7.01 (1H, d, *J*=3.50 Hz, H-3), 6.9 (2H, d, *J* = 3.57Hz, H-2``).

7-[2-(4-Nitrophenyl)-2-oxo-ethyl]-1H-pyrrolo[2,3-b]pyridinium;bromide (5): Yield = 61%, HREI MS: $C_{15}H_{12}N_3O_3Br$, EIMS: *m/z* = 362.18 amu, M.P. = 252 + 2°C, UV λ_{max} (MeOH) nm: 270, 248 and 201; IR υ_{max} (KBr) cm^{-1}: 3390, 2920, 1690, 1600 and 1340, EIMS, *m/z* (relative int., %): 282 (M^+-Br, $C_{15}H_{12}N_3O_3$, 8), 164(4), 160(12), 132 (11), 122(18), 118(100) and 77(23), ^{1}H-NMR (MeOD, 300 MHz) *δ*: 8.81 (1H, d, *J* = 7.84, 1.08 Hz, H-6), 8.84 (2H, d, *J* = 8.02, Hz, H-3', 5'), 8.26 (1H, d, *J* = 8.99 Hz, H-4), 7.78 (2H, d, 8.02 Hz, H-2', 6'), 7.71 (1H, d, *J* =3.51 Hz, H-2), 7.68 (1H, dd, *J* = 8.99, 7.84 Hz, H-5), 7.02 (1H, d, *J* =3.51 Hz, H-3), 6.02 (2H, *s*, H-2").

7-[2-(2,4-Diflouro-phenyl)-2-oxo-ethyl)-1H-pyrrolo[2,3-b]pyridine-7-ium,bromide (6): HREI MS: $C_{15}H_{12}N_2OF_2Br$, UV (MeOH) nm: 298, 261 and 200, IR (KBr): 3400, 2900,1690,1600,1580 and 1450, EIMS *m/z* (%): M-1 ($C_{15}H_{11}BrF_2N_2O$), 316 (8), 286 (100), 205 (15), 182 (25), 131 (50), ^{1}H-NMR(D_2O , 300 MHz) *δ*: 8.04 (2H, d, *J* = 8.82 Hz, H-2), 7.88 (2H, d, *J* = 8.82 Hz, H-3,5'), 6.02 (2H, *s*, H-2"), 7.67 (1H, dd, *J* = 7.81, 5.27 Hz, H-5), 7.00 (1H, d, *J* = 3.62 Hz, H-3), 5.84 (2H, *s*, N-CH_2).

7-[2-(4-Flouro-phenyl)-2-oxo-ethyl)-1H-pyrrolo[2,3-b]pyridine-7-ium,bromide (7): HREI MS: $C_{15}H_{12}N_2OFBr$, UV (MeOH) nm: 298, 261 and 200, IR (KBr): 3400, 2900,1690,1600,1580 and 1450, EIMS *m/z* (%): M-1 ($C_{15}H_{11}BrFN_2O$), 316 (8), 286 (100), 205 (15), 182 (25) and 131 (50), ^{1}H-NMR (D_2O, 300 MHz) *δ*: 8.64 (1H-dd, *J* = 7.81, 1.06 Hz, H-6'), 8.04 (2H, d, *J* = 8.82 Hz, H-2',6), 7.88 (2H, d, *J* = 8.82 Hz, H-3, 5'), 7.70 (1H, d, *J* = 3.62, H-2), 7.67 (1H, dd, *J* = 7.81, 5.27 Hz, H-5), 7.00 (1H, d, *J* = 3.62 Hz, H-3), 5.84 (2H, *s*, N-CH_2).

7-[2-(4-Chloro-phenyl)-2-oxo-ethyl)-1H-pyrrolo [2,3-b]pyridine-7-ium,bromide (8): HREI MS = $C_{15}H_{12}N_2OClBr$, UV (MeOH) nm: 298, 261 and 200: IR (KBr): 3400, 2900, 1690,1600,1580 and 1450, EIMS *m/z* (%): M-1 ($C_{15}H_{11}BrClN_2O$), 316 (8), 286 (100), 205 (15), 182 (25) and 131 (50), ^{1}H-NMR (D_2O) (300 MHz) *δ*: 8.64 (1H, dd, *J* = 7.81, 1.06 Hz, H-6'), 8.35 (1H,dd, *J* = 5.27, 1.06 Hz, H-4), 8.04 (2H, d, *J* = 8.82 Hz, H-2', 6), 7.88 (2H, d, *J* = 8.82 Hz, H-3,5'), 7.70 (1H, d, *J* = 3.62, H-2), 7.67 (1H, dd, *J* = 7.81, 5.27 Hz, H-5), 7.00 (1H, d, *J* = 3.62 Hz, H-3), 5.84 (2H, *s*, N-CH_2).

7-[2-(4-Methoxy-phenyl)-2-oxo-ethyl]-1H-pyrrolo[2,3-b]pyridine-7-ium (9): Yellow needles, HREI MS: *m/z* $C_{16}H_{15}N_2O_2Br$, UV: 389.4, 284.8, 223.0, IR: 3449.6, 3010.8, 1680.5, 1341.5 1309.5, 1236.1 and 1101.2; EIMS *m/z* (rel. int %): 266 (M $^+$-Br, $C_{16}H_{15}O_2N_2$, 31), 238(55), 149(2), 135(100), 131(10), 118(6) and 107(12),^{1}H-NMR: *δ* 8.78-8.75 (1H, d, *J* = 7.77 Hz, H-6'), 8.45-8.43 (1H, d, *J* = 6.178 Hz, H-4), 8.14-8.11 (2H, d, *J* = 8.88 Hz, H-5), 7.73-7.72 (1H,d, *J* =3.528 Hz, H-2), 7.66-7.62 (1H, m, H-6), 7.16-7.13 (2H, d, *J* = 8.925 Hz, H-3, H-5'), 6.99-6.98 (1H, d, *J* = 3.55 Hz, H-2), 6.43-6.41 (1H, d, *J* = 5.72 Hz, H-3), 3.30-3.29 (1H, *s*, H-1).

Naphthalen-2-yl-2-oxo-ethyl]-1H-pyrrolo[2,3-b]pyridinium,bromide (10): Yield = 86.2%, HREI MS: $C_{19}H_{15}N_2OBr$, EIMS = *m/z* (relative intensity; %)= 367.23 amu, M.P.= 267±2°C, UV $λ_{max}$ (MeOH) nm = 340, 293, 250, 224, 203, IR v_{max} (KBr) cm^{-1} = 3755, 2923, 1687, 1361, 987, 823.5, 783, 545, 474.5; EIMS = *m/z* (relative int, %): 286 (M$^+$ -HBr $C_{19}H_{14}N_2O$, 7.40), 259(8.2), 258(62), 257(52), 155(15), 81(18), 77(22), ^{1}H-NMR (DMSO-d6, 300 MHz), *δ* 8.86 (2H, d, *J* = 7.047, H-6, 4), 8.60(1H, d, *J* = 6.09, H-5), 8.24(1H, d, *J* = 7.71, H-10'), 8.07-8.03(2H, m, H-2', 8'), 7.96 (2H, d, *J*=3.43, H-5', 3'), 7.76(2H , m, H-6' ,7'), 6.62(2H, d , *J* =3.50, H-2', 3'), (2H, *s*, H-2").

7-(2-Adamantan-1-yl-2-oxo-ethyl]-1H-pyrrolo [2,3-b]pyridinium, bromide (11); yield = 81.5 %, HREI MS: $C_{19}H_{23}N_2OBr$, EIMS: *m/z* =375.30 amu, M. P. = 287 + 2^{0}C, UV $λ_{max}$ (MeOH) nm = 299.6, 226.0, IR v_{max} (KBr) cm^{-1}= 3409, 2918, 2850, 2773, 1712, 1620, 1458, 1357, 1164, 1097, 887, 779, 727, 661, 536, 476. EIMS = *m/z* (relative intensity; %), 295 (M$^+$, $C_{19}H_{23}N_2O$, 7.3), 294 (35), 237(2.6), 176(3), 159(15.5), 131(100), 93(18.4), 79(27),^{1}H-NMR (DMSO-d6, 300 MHz): *δ* 8.76-8.79 (1H, d, *J* = 8.68 Hz, H-6), 8.48 (1H, d, *J* = 6.15Hz, H-4), 7.95(1H, d, *J* = 3.48 Hz, H-2), 7.66 (1H, dd, *J* = 7.8, 1.5 Hz, H-5), 6.97 (1H, d, *J* =3.57, H-3), 6.06(2H, *s*, H-2``), 2.07(5H, *s*, H-4', 6, 7', 8', 9'), 1.97(5H, *s*, H-10', 5', 3', 2'), 1.78-1.68 (6H, m, H-3', 6', 6', 7', 9',10').

7-(3-Oxo-3-phenyl-propyl)-1H-pyrrolo[2,3-b]pyridine-7-ium,chloride (12): HREI MS= $C_{16}H_{15}N_2OCl$, EIMS *m/z* (%): M-1 ($C_{16}H_{14}ClN_2O$), 316 (8), 286 (100), 205 (15), 182 (25) and 131 (50). ^{1}H-NMR, D_2O (300 MHz) *δ*: 8.64 (1H-dd, *J* = 7.81, 1.06 Hz, H-3, 5), 8.35 (1H, dd, *J* = 5.27, 1.06 Hz, H-4), 8.04 (2H, d, *J* = 8.82 Hz, H-2, 6), 7.88 (2H, d, *J* = 8.82 Hz, H-3', 5'), 7.70 (1H, d, *J* =3.62, H-2'), 7.67 (1H, dd, *J* = 7.81, 5.27 Hz, H-5'), 7.00 (1H, d, *J* = 3.62 Hz, H-3'), 5.84 (2H, *s*, N-CH_2).

7-Phenethyl-1H-pyrrolo[2,3-b]pyridin-7-ium, bromide (13): UV (MeOH) nm: 390.0, 293.0, 224.0, 203.4, 198.6; IR $υ_{max}$ (KBr) cm^{-1} 51.5, 1647.6, 1495.0, 1357.4, 1307.8 and 1189.1; EIMS (relative int, %): 223 (M$^+$-Br, $C_{15}H_{14}N_2$ (10), 132(3), 118(100), 104(8), 91(7), 77(6); ^{1}H-NMR (MeOD, 300 MHz): *δ* 8.63-8.61(1H, d, *J* = 7.76 Hz, H-6), 8.20-8.18 (1H, d, *J* = 6.17 Hz, H-4), 7.74-7.73 (1H, d, *J* = 3.52 Hz, H-5), 7.45-7.40 (2H, m, H-3, H-5'), 7.2-7.18 (2H, t, *J* = 6.33 Hz, H-1, H-2), 7.02-6.99 (2H, m, H-2, m, H-2, H-3), 6.93-6.92 (1H, d, *J* = 3.51 Hz, H-1), 5.02-4.97 (1H, t, *J* = 6.92 Hz, H-1"), 4.57 (1H, *s*, H-2").

7-(4-Oxo-4-thiophen-2-yl-butyl)-1H-pyrrolo[2,3-b]pyridin-7-ium; chloride (14): UV: 389.4, 286.2, 263.8, 222.2, 204.0; IR: 3075.8, 2469, 1496.5, 1510.5, 1359.2, 1310.9, 1246.2; EIMS: 270(M$^+$-Cl, $C_{15}H_{15}ON_2$, *s*, 19), 185 (2), 159(28), 132(50), 126(48), 118(100), 111 (87), 83(9); ^{1}H-NMR: *δ* 8.68-8.65 (1H, d, *J* = 7.93 Hz, H-6), 8.55-8.53 (1H, d, *J* = 6.32 Hz, H-4), 7.87-7.80 (1H, m, H-2, H-4'), 7.58-7.53 (1H, t, *J* = 7.015 Hz, H-5), 7.20-7.17 (1H, t, *J* = 4.38 Hz, H-3), 6.97- 6.95 (2H, d, *J* = 3.48 Hz, H-2, H-3), 3.3-3.2 (1H, m, H-1), 2.47 – 2.43 (2H, m, H-2', H-3').

3. RESULTS AND DISCUSSION

3.1 Chemistry

7-azaindole (0.01 mol) and substituted phenacyl halide (0.01 mol) were dissolved separately in acetone (20 mL) and then mixed together in a round-bottommed flask (Fig. 1). Reflux reaction mixtures were monitored for about 24 h. The reaction completion was monitored by TLC. After completion of reaction, the precipitates were filtered, and washed with warm acetone. The obtained products were re-crystallized from methanol. The pure compounds were dried in vacuum over anhydrous calcium sulfate. The structures of synthetic derivatives were elucidated by different spectroscopic techniques including, UV, IR, EI-MS and ^{1}H-NMR. All compounds gave satisfactory elemental analysis. The compounds herein described were synthesized as a part of study complementary to *β*-glucuronidase, phosphodiesterase and urease inhibition investigation of fused pyrrolo pyridine system. All the 7-aza indole were prepared under the conditions recorded in experimental section (Fig. 1).

As a result of our continued interest in 7-aza indole as potential antibacterial and cytotoxic agents [13-14,30], we now wish to report the investigation for *β*-glucuronidase inhibition as well as other enzyme inhibition including of 7-aza indole derivatives and their structure activity relationship. So in our present work we designed different derivatives of aza-indole in which we replaced in different groups with phenacyl chain at different position to find a selective inhibitor.

The ^{1}H-NMR spectra were specially helpful in deciding the substitution pattern at the different positions of phenacyl part of the molecule. The NMR chemical shifts of compounds **2-14** are presented in Experimental section.

The molecular ion peaks of the compounds were obtained from EIMS spectra [31]. A series of 7-azaindole derivatives were prepared by quaternization with phenethyl bromide and substituted phenacyl halides, the side chain thus possessed certain features common to those present in sympathomimetic amines and related compounds.

The structures of compounds **1-14** were identified by spectroscopic studies previously reported in the literature [33].

Compounds **1-14** (Fig. 2) were screened for *β*-glucuronidase inhibitory activity (Table-1). Parent compound **1** showed no activity. Among the synthesized analogues of compound 1, compounds **2, 8, 9, 10, 11, 13** and **14** showed reasonable % inhibition, while compounds **3-5, 6, 7 and 12** displayed no activity. Compound **8** (IC_{50} 192.61±3.53 *μ*M), containing chloro derivative displayed maximum inhibition, while compound 14 showed 47.11% inhibition and was the second potent *β*-glucuronidase inhibitor of the series. Nevertheless, naphthyl derivative 10 (20 % inhibition), and benzyl derivative 13 (36.81% inhibition) showed almost same level of inhibition with 20 and 37.11% inhibition. The more *β*-glucuronidase activity of compound **8** than compounds **13** and **14** is due to electron withdrawing effect of chlorine as compared to having unsubstituted benzene and sulphur containing five membered ring moieties. The compounds **13** with unsubstituted benzene and **14** with sulphur containing five membered ring exhibited less activity as compared to 8 due to change in substitution group at phenacyl moiety.

7-azaindole derivatives **1-14** were screened against urease enzyme [37] according to literature protocol [38,31,32]. They showed a varying degree of urease inhibitory activity having IC_{50} values ranging between 2.19-255.11 *μ*M, when compared with standard thiourea (IC_{50} = 21.00±0.11*μ*M) (Table-1).

When parent compound **1** was screened along with its newly synthesized derivatives 2-14, it was found that parent compound showed weak urease enzyme inhibitory activity (2.91%) at 0.5 *m*M concentration. Among the synthesized analogues, compounds **2, 6, 11, 12** and **13** exhibited weak activity (44.41, 41.81, 16.10, 21.81 and 6.71%) against urease enzyme, while compounds **3, 4, 5** and **7** were found devoid of any activity. Derivatives of parent compound **1** are not making profound difference in the enzyme inhibition effect except compounds **8, 9, 10** (chloride, methoxy, naphthalene derivatives). These compounds were selected for IC_{50} value and these showed IC_{50} values 133.31±0.46, 2.19±0.37 and 255.11±6.62 *μ*M respectively. Compound **9** showed remarkable inhibition with an IC_{50} value of 2.19±0.37 *μ*M, which is about ten times more active than the standard drug thiourea (IC_{50} = 21.00±0.11 *μ*M). Compounds **8** and **10**, with IC_{50} values of 133.31±0.46 and 255.11±6.62 *μ*M showed encouraging results.

Compound **9** (IC_{50} = 2.19±0.37*μ*M) having *para* methoxy substituted phenyl ring showed potent urease inhibitory potential. However, compound 8 having *para* chloro substituted phenyl. Compound 10 having naphthalene group instead of benzene ring. The difference in activities might be due to change in nature of substituent. The more activity of Compound **9** than compounds **8** and **10** are due to more electron donating effect of methoxy as compared to chlorine group. The difference in activity might be due to numbers as well as nature of substituents.

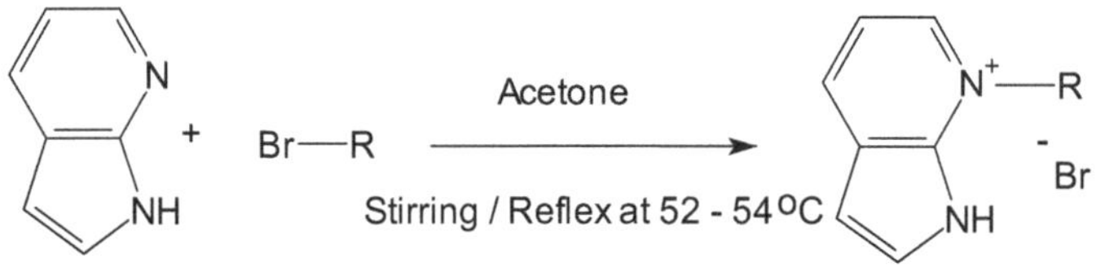

1 2a-14a 2-14

Reactants No. R

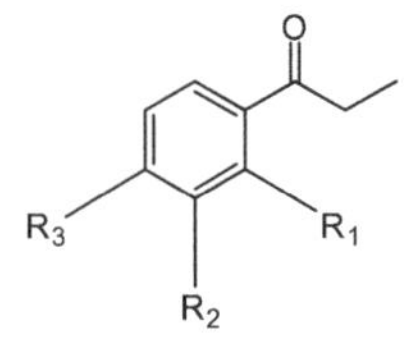

2a $R_1 = H$ $R_2 = OH$ $R_3 = OH$

3a $R_1 = H$ $R_2 = NO_2$ $R_3 = H$

4a $R_1 = NO_2$ $R_2 = H$ $R_3 = H$

5a $R_1 = H$ $R_2 = H$ $R_3 = NO_2$

6a $R_1 = F$ $R_2 = H$ $R_3 = F$

7a $R_1 = H$ $R_2 = H$ $R_3 = F$

8a $R_1 = H$ $R_2 = H$ $R_3 = Cl$

9a $R_1 = H$ $R_2 = H$ $R_3 = OCH_3$

10a 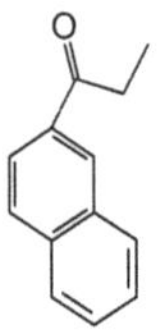

11a 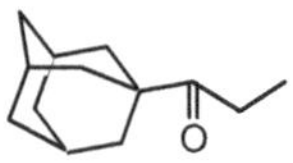

12a 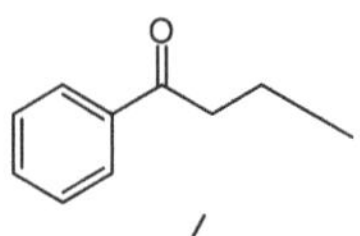

13a 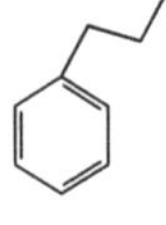

14a 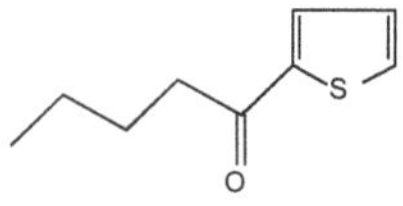

Fig. 1. Synthesis of 7-azaindole derivatives 1-14

Compound **9**, methoxy analogue is the most active in urease assay and the most active structure of the series in the urease assay suggesting that the methoxy group at para position of benzene ring provides a sterically-favored orientation for a fruitful and most significant interaction with enzymes. While compounds **8** and **10** having chlorine and naphthalene substitution instead of methoxy was unlikely to interfere with carbohydrates digestion and glycoprotein and hence found to be less active as compare to **9**.

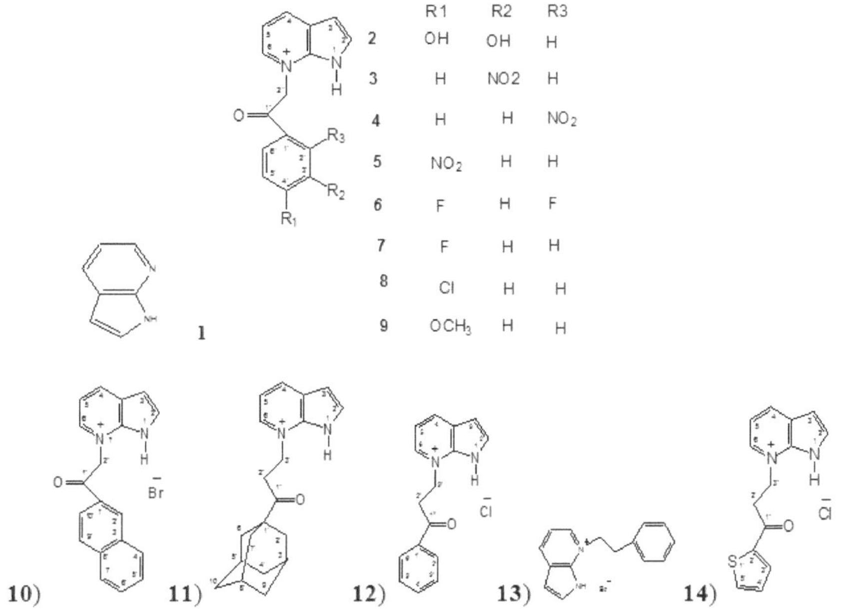

Fig. 2. Structures of 7-azaindole derivatives I -14

As we can see that attaching the different molecules at N atom of aza-indole makes the compound less active or decreases effect. An exceptional case is the compound 9 where good activity has been observed. The structural difference between all analogues except 10 is the substitution at para position may suggest that presence of methoxy group is contributing in the activity. Comparing the halide analogues, compound 8 chloro derivative showed better effect as compared to flouro derivative compound **6** (41.81% inhibition).

In phosphodiesterase-I enzyme inhibition assay the compound **2** (IC_{50} = 20.83±0.234 *u*M) showed potent phosphodiesterase inhibitory potential than standard EDTA (IC_{50} = 274.00±0.007 *u*M). caused by the presence of *meta,para* hydroxy substituted benzene ring. Among the synthesized analogues, compound **3, 4, 5, 7, 10** and **11** exhibited weak activity (5.61, 18.91, 6.71, 17.11, 20.61 and 17.81%) against phosphodiesterase enzyme, while compounds **6, 8, 9, 12, 13** and **14** were found devoid of any activity.

The more activity of compound **2** than the rest of the compounds is due to more electron withdrawing effect of *meta*, *para* hydroxy substituted benzene ring. The compound **4, 10** and **11** exhibited less activity as compared to **2** by the change in substitution group at phenacyl moiety.

After analyzing the results obtained by urease, *β*-glucosidase and phosphodiesterase-I enzymes inhibition assay, the compound **2** is the most active compound of the series in phosphodiesterase-I enzymes inhibition assay and was completely inactive in *β*-glucosidase assay. Similerly the compound **9** is the most active compound of the series in urease enzymes inhibition assay and was completely inactive in phosphodiesterase-I assay. Beside this the compound **8** is the most active compound of the series in *β*-glucosidase enzymes inhibition assay and was completely inactive in phosphodiesterase-I assay. From this study new potent inhibitors (compound **9, 2** and **8**) were identified with acceptable correlation between biological activities and binding energies. We also succeeded in identifying two potent selective inhibitors (compound **2** and **9**). Further studies will lead to a better understanding of structure and activity relationship of these compounds.

3.2 Biological Activity

Compound **1-14** were screened for phosphodiesterase inhibition activity. The parent compound (**I**) investigated for phosphodiesterase inhibition, showed insignificant activity. Compound **2** exhibited potent results as compared to standard EDTA.

In the bio-screening of the compounds for phosphodiesterase inhibition, compounds 4(ortho-nitro derivative), 7(Flouro derivative), **10** (Naphthalene derivative), 11(7-(2-Adamantan-1-yl-2-oxo-ethyl]-1H-pyrrolo[2,3-b]pyridinium bromide derivative) were found moderately active.—Compounds **3** and **5** showed week activity. During the course of synthesis of derivatives of aza-indole, compound **2** attracted our attention as potent phosphodiesterase inhibitor. Compound **2** showed remarkable phosphodiesterase inhibition with an IC_{50} value of 20.83±0.234 μM, which is about twenty times more active than the standard drug EDTA (IC_{50} = 274.00±0.007 μM). The result indicated that OH substitution in phenacyl ring at 3,4 position (meta, para benzene) is making significant change in the activity with IC_{50} value 20.83±0.234 μM. Results of phosphodiesterase activity are presented in Table 1.

Investigation for β-glucuronidase inhibition activity opposite results have been observed as compared to urease inhibition. Approximately similar inhibition of β-glucuronidase inhibition and urease inhibition by compound **8** suggested that though the presence of Cl functional group imparting inhibitory activity but the position of Cl in the phenacyl ring is not important for any significant difference in the activity Table 1.

Compound **9** showed remarkable inhibition with an IC_{50} value of (IC_{50} = 2.19±0.37 μM), which is about ten times more active than the standard drug thiourea (IC_{50} = 21.00±0.11 μM). Compounds **8** and **10**, with IC_{50} values of 133.31±0.46 and 255.11±6.62 μM showed encouraging results.

Although 7-azaindole was synthesized but the useful biological activities was unexplored for many years. There was no data available that 7-azaindole derivatives can be used as β-glucuronidase, urease and phosphodiesterase inhibitors. So we decided to do their activity. By doing so we got very promising results.

Table 1. β-Glucuronidase, phosphodiesterase and urease inhibition activity of 7- azaindole derivatives 1-14

S. no	β-Glucuronidase		Phosphodiesterase		Urease	
	% Inhibition	IC_{50} ±SEM [a] [μM]	% Inhibition	IC_{50} ±SEM [a] [μM]	% Inhibition	IC_{50} ±SEM [a] [μM]
1	-	-	17.91	-	2.91	-
2	6.31	-	86.01	20.83±0.234	44.41	-
3	-NA[b]	-	5.61	-	- NA[b]	-
4	- NA[b]	-	18.91	-	- NA[b]	-
5	- NA[b]	-	6.71	-	- NA[b]	-
6	- NA[b]	-	- NA[b]	-	41.81	-
7	- NA[b]	-	17.11	-	- NA[b]	-
8	78.51	192.61±3.53	- NA[b]	-	74.81	133.31±0.46
9	13.61	-	- NA[b]	-	75.81	2.19± 0.37
10	20.01	-	20.61	-	66.61	255.11±6.62
11	13.01	-	17.81	-	16.10	-
12	- NA[b]	-	- NA[b]	-	21.81	-
13	36.81	-	- NA[b]	-	6.71	-
14	47.11	-	- NA[b]	-	30.91	-
Standards 1,4,Lactone D saccharic acid[C]		48.41±1.24				
EDTA[d]				274.00±0.007		
Thiourea[e]						21.00± 0.11

a SEM is the standard error of the mean.
b NA–not active.
c.1,4,Lactone D saccharic acid standard inhibitor for β-Glucosidase activity
d EDTA standard inhibitor for phosphodiesterease activity.
e.Thiourea standard inhibitor for urease activity

4. CONCLUSION

There were fourteen close derivatives of 7-azaindole and out of these three i-e 7-[2-(4-chloro-phenyl)-2-oxo-ethyl)-1H-pyrrolo [2,3-b]pyridine-7-ium,bromide (**8**), 7-[2-(4-Methoxy-phenyl)-2-oxo-ethyl]-1H-pyrrolo[2,3-b]pyridine-7-ium (**9**), and Naphthalen-2-yl-2-oxo-ethyl]-1H-pyrrolo[2,3-b]pyridinium,bromide (**10**) exhibited useful anti-urease activity. Compound **9** (IC50 = 2.19±0.37 μM) showed potent urease inhibitory potential than standard thiourea (IC50 = 21.00±0.01 μM). Compound 1-[3-(3,4-dihydroxyphenyl)3-oxo-ethyl]-7H-pyrrolo[2,3-b]pyridine-1-ium;bromide (**2**) (IC50 = 20.83+0.234 μM) showed potent phosphodiesterase inhibitory potential than standard thiourea (IC50 = 21.00±0.01 μM). Compound **8** (IC50 =192.61±3.53 μM) was also found to be β-glucuronidase inhibitor, as compared to standard 1,4,lactone D saccharic acid, (IC50 = 48.41±1.24 μM). i-e 7-[2-(4-chloro-phenyl)-2-oxo-ethyl)-1H-pyrrolo [2,3-b]pyridine-7-ium,bromide (**8**). Overall results are significant and encouraging and because of their urease and phosphodiesterase inhibition potential they can be selected as a useful lead molecules. All the results of this study indicated the azaindole compound **9** as interesting as it leads for anti-urease's drug development being urease inhibitors, and compounds **2** and **8** could serve as potential lead compound with promise for further research of drug development in the field of phosphodiesterase and β-glucuronidase enzyme inhibition. The potency of these compounds, especially compounds **2, 9** and **8** suggests that this is a new class of inhibitors as it provides a solid base for using them as lead molecules, for designing effective therapeutic agent to treat these diseases.

Present study opens-up an avenue for further research on 7-azaindole derivatives for their evaluation against urease, phosphodiesterase and β-glucuronidase enzymes. Structure activity relationship studies with the standard molecules provided justification for their activity and given better understanding of their inhibiting potential depending upon their pharmacophoric regions.

COMPETING INTERESTS

Authors have declared that no competing interests exist.

REFERENCES

1. Merour JY, Joseph B. Synthesis and reactivity of 7-Azaindoles (1H-Pyrrolo[2,3-b]pyridine). Curr. Org. Chem. 2001;5:471. (b) Yakhontov LN, Prokov AA. Advances in the Chemistry of azaindoles. Russ. Chem. Rev. 1980;49:428.
2. Hugon B, Anizon F, Bailly C, Golsteyn RM, Pierré A, Léonce S, Hickman J, Pfeiffer B, Prudhomme M. Synthesis and biological activities of isogranulatimide analogues. Bioorg Med. Chem. 2007;15(17):5965-5980.
3. Marminon C, Pierre A, Pfeiffer B, Perez V, Leonce S, Joubert A, Bailly C, Renard P, Hickman J, Prudhomme M. Syntheses and antiproliferative activities of 7-arebeccamycin analogues bearing one 7-Azaindole moiety. J. Med. Chem. 2003; 46(4):609-622.
4. Kelly TA, McNeil DW, Rose JM, David E, Shih CK, Grob PM. Novel non-nucleoside inhibitors of human immunodeficiency virus type 1 reverse Transcriptase. 6. 2-Indol-3-yl- and 2-Azaindol-3- ldipyridodiaze-pinones. J. Med. Chem. 1997;40(15):2430-2433.
5. Anthony J. Verbiscar synthesis of 1-pchlorobenzyl-7-azaindole-3-.alpha.-piperidylmethanol as a potential antimalarial agent. J. Med. Chem. 1972; 15(2):149-152.
6. Song JJ, Reeves JT, Gallou F, Tan Z, Yee NK, Senanayake CH. Organometallic methods for the synthesis and functionalization of azaindoles. Chem. Soc. Rev. 2007;36(7):1120-1132.
7. Lebouvier N, Pagniez F, Duflos M, LePape P, Na YM, Le BG, Le BM. Synthesis and antifungal activities of new fluconazole analogues with azaheterocycle moiety. Bioorg. Med. Chem. Lett. 2007;17(13): 3686-3689.
8. Guillard J, Decrop M, Gallay N, Espanel C, Boissier E, Herault O, Viaud-Massuard MC. Synthesis and biological evaluation of 7-azaindole derivatives, synthetic cytokinin analogues. Bioorg. Med. Chem. Lett. 2007; 17(7):1934-1937.
9. Trejo A, Arzeno H, Browner M, Chanda S, Cheng S, Comer DD, Dalrymple SA, Dunten P, Lafargue J, Lovejoy B, Freire-Moar J, Lim J, Mcintosh J, Miller J, Papp E, Reuter D, Roberts R, Sanpablo F, Saunders J, Song K, Villasenor A, Warren SD, Welch M, Weller P, Whiteley PE, Zeng L, Goldstein DM. Design and synthesis of 4-azaindoles as inhibitors of p38 MAP kinase. J. Med. Chem. 2003;23,46(22): 4702-13.

10. Viaud MC, Jamoneau P, Baudin ML. Acylation of oxazolo[4,5-b]pyridine-2(3H)-ones, 2-phenyloxazolo[4,5-b]pyridines and pyrrolo-[2,3-b]pyridine-2(2H)-ones. Tetrahedron. 1997;53:5159.
11. Shen TY, Ellis RL, Windholz TB, Matzuk AR, Rosegay A, Lucas S, Holly FW, Wilson AN. Non steroid anti-inflammatory agents. J. Am. Chem. Soc. 1963;85:488.
12. Saify ZS, Sultana N. Role of acetylcholinesterase inhibitors and alzheimer disease. 2014;6:387-425. (39), Chapter in book "Frontiers in Drug Design and Discovery" by, Atta-ur-Rahman, Mohammad Iqbal Choudhary, DOI: 10.2174/97816080582281140601, eISBN: 978-1-60805-822-8, ISBN: 978-1-60805-823-5
13. Saify ZS. Synthesis of new carbonic anhydrase inhibitor. Pak. J. Pharm. Univ. Kar. 1982;1(1):83-88.
14. Saify ZS, Haider SM, Ahmed M, Saeed M, Khan A, Siddiqui BS. Synthesis of some 7-azaindole derivatives, their cytotoxicity and antibacterial activity. Pak. J. Sci. & Indust. Res. 1994;37(10):439-441.
15. Saify ZS, Mozzam SM, Nisa M, Khan SA, Ahmed A, Haider S, Aryne A, Khanam M, Arshad N, Ghani M. 7-Azaindole derivatives as potential antibacterial agents. Pak. J. Sci. Ind. Res. 2009;52(1):1-7.
16. Saify ZS. 7-Azaindole derivatives as potential anti-inflammatory agents. Pak. Journal of Pharmacology. 1986;2:43-46.
17. Saify ZS. Synthesis of some azaindole derivatives of potential biological interest; Significance of physical parameters in drug design. J. of Pharmacy, University of Karachi. 1984;2(2):99-103.
18. Saify ZS. The effect of substituted nicotinamide and azaindoles on lipid peroxidation and platelet aggregation. Proceedings of second International Conference on Pharmaceutical Sciences, Karachi; 1997.
19. Wang T, Zhang Z, Wallace OB, Deshpande M, Fang H, Yang Z, Zadjura L, Tweedie DL, Huang S, Zhao F, Ranadive S, Robinson B, Gong YF, Ricardi K, Spicer TP, Deminie C, Rose R, Wang H, Blair WS, Shi PY, Ling P, Colonno RJ, Meanwell NA. Discovery of 4-benzoyl-1-[(4-methoxy-1H- pyrrolo[2,3-b]pyridin-3-yl)oxoacetyl]-2- (R)-methylpiperazine (BMS-378806): a novel HIV-1 attachment inhibitor that interferes with CD4-gp120 interactions. J. Med. Chem. 2003;46:4236. (b) Curtis NR, Kulagowski JJ, Leeson PD, Ridgill MP, Emms F, Freedman SB, Patel S, Pate S. Synthesis and sar of 2- and 3-substituted 7-azaindoles as potential dopamine D4 ligands. Bioorg. Med. Chem. Lett. 1999;9:585.
20. Henry JR, Dodd JH. Synthesis of RWJ 68354: A potent inhibitor of the MAP kinase p38. Tetrahedron Lett. 1998;48: 8763. (b) Dumas, J. Exp. Opin. Ther. Patents. 2001;11:405.
21. US 7169787 B2, 7-azaindoles, use thereof as phosphodiesterase 4 inhibitors and method for producing the same.
22. Mobley, HLT. Island MD. Hausinger, RP. Molecular biology of microbial ureases, Microbiol. Rev. 1995;59:451.
23. Mobley HLT, Hausinger RP. Microbial ureases: significance, regulation, and molecular characterization. Microbiol. Rev. 1989;53:85.
24. Bremenr JM. Recent research on problems in the use of urea as a nitrogen fertilizer. Fert. Res. 1995;42:321.
25. Weatherburn MW. Phenol-hypochlorite reaction for determination of ammonia. Anal. Chem. 1967;39:971.
26. Levvy GA, Marsh CA. Preparation and properties of β-glucuronidase. Advan Carbohydrate Chem. 1959;14:381.
27. Kawasaki M, Hayashi T, Arisawa M, Morita N, Berganza HL. Flavonoids from *Physalis minima*. Phytochemistry. 1988;27:3709.
28. Bello di Cenci P, Dorling L Fellows, Winchestern B. Specific inhibition of human β-D-glucuronidase and alpha-L-iduronidase by a trihydroxy pipecolic acid of plant origin. FEBS Lett. 1984;176:61.
29. Goding JW, Grobben B, Slegers H. Physiological and pathophysiological functions of the ecto-nucleotide pyrophosphatase/phosphodiesterase family. Biochem. Biophys. 2003;1:1638.
30. Saify ZS, Haider SM, Ahmed M, Khan A. Synthesis and antibacterial activity of new quaternary ammonium-compounds. Journal of Chemical Society of Pakistan. 1998;20:141.
31. Mushtaq N, Zafer SS, Fozia N, Shamoona T, Shamim A, Muhammad A, Khalid MK. Synthesis and pharmacological activities of 7-azaindole derivatives. Pak. J. Pharm. Sci. 2008;21(1):36-39.

32. Rafiq K, Saify ZS, Ghous A, Haider S, Sultana N. The structural modification causes the enhancement of analgesic activity of 4-Hydroxy 4-Phenyl piperidine along with an excellent interaction with digestive enzyme, leading good adme. World Journal of Pharmaceutical Research. 2014;3(1):1537-1546.
33. Saify ZS, Sultana N, Walker J, Zareen T, Rafiq K. Synthesis and anti-anaphylactic activity of 7-azaindole derivatives. J. Chem. Society of Pakistan. 2013;35:3.
34. Sperker B, Backman JT, Kroemer HK. The role of β-glucuronidase in drug disposition and drug targeting in humans. Clin. Pharmacokinst. 1997;33:18.
35. Weatherburn MW. Phenol-hypochlorite reaction for determination of ammonia. Anal. Chem. 1967;39:971–974.
36. Mamillapalli R, Haimovitz R, Ohad M, Shinitzky M. Enhancement and inhibition of snake venom phosphodiesterase activity by lysophospholipids. FEBS Lett. 1998;436:256-258.
37. Horn J. The proton-pump inhibitors: Similarities and differences. Clin. Ther. 2000;22:266.
38. Cole ER, Crank G, Salam-Sheikh AJ. Antioxidant properties of benzimidazoles. Agric. Food Chem. 1974;22:918.

16

Hypolipidemic Activity of Fermented Sorghum (*Sorghum bicolor* L. Moench) Gruel (*Ogi*) in Hypercholesterolemic Rats (*Rattus nervegicus*)

Solomon Anjuwon Laleye[1], Babatunde Idowu Aderiye[1*] and Oluwole Moses David[1]

[1]Department of Microbiology, Ekiti State University, Ado-Ekiti, 360101, Nigeria.

Authors' contributions

This work was carried out in collaboration between all authors. Authors SAL and BIA designed the study, wrote the protocol and supervised the work. All Authors managed the analyses of the study. Author SAL wrote the first draft of the manuscript. Authors BIA and OMD managed the literature searches and edited the manuscript. All authors read and approved the final manuscript.

Editor(s):
(1) Rosario Gomez Garcia, Department of Biochemistry, Loyola University, USA.
(2) Luisa Di Paola, Chemical and Biochemical Engineering Fundamentals, Faculty of Engineering Università Campus Biomedico, Via Alvaro del Portillo, Roma, Italy.
Reviewers:
(1) Abdullahi m. Nuhu, Kaduna Polytechnic, Kaduna, Nigeria.
(2) Sahar Mohamed Kamal Shams El Dine, Ain Shams University, Cairo, Egypt.

ABSTRACT

Background: Fermented gruel (*Ogi*) has been reported to have high lactic acid bacteria (LAB) with proven probiotic properties. This fermented food has also has different therapeutic usages.
Objective: The activities of fermented sorghum (*Ogi*) on induced hypercholesterolemia in albino rats were investigated.
Materials and Methods: The LAB strains were isolated from traditionally fermented sorghum gruel *Ogi.* Also the ability of the isolates to tolerate bile and produced bile salt hydrolase was also determined. The hypolipidemic activity of fermented gruel (*Ogi*) with and without LAB strains in hypercholesterolemic rats was determined using standard methods.
Results: The lactic acid bacteria (LAB) count in sorghum *Ogi* ranged between 5. 0792 log_{10} CFU/g

**Corresponding author: E-mail: jadesolaaderiye@yahoo.com*

and 5.6990 $\log_{10}$ while the average values of pH and titratable acidity (TTA) were 3.8 and 0.38 respectively. There was no significant difference between the weights of the animals in the control group (Basal Diet) and the experimental groups (at $p<0.05$). The mean values of the low density lipoproteins cholesterol (LDLC) in the rats were 188.60±7.27 mg/dl in the animals fed with HLF while the least value was recorded in the Basal diet supplanted with *Ogi* (BD+*Ogi*) (110.80±26.93 mg/dl). The mean high density lipoproteins cholesterol in the serum of the rats in cholesterol free feed supplemented with BD+*Ogi* diet group was the highest (52.600 mg/dl). The amount of the total lipid in the animal fed with high lipid diet (HLF) was significantly higher than the amount in the organs of the animals feed with diets containing BD+*Ogi*, HLF+ *L. lactis* and HLF+ *L. brevis* (at $\alpha = 0.05$). In decreasing order, the amount of the glycerides in high lipid feed diets supplemented with *Ogi*, *Lactobacillus lactis* and *L. brevis*. The mean value of the Aspartate transaminase in the animals in Basal Diet group was higher at the commencement of the experiment (275.45±34.43 IU/L).while the level of the alanine transaminase was not significantly different (at $p< 0.05$). The histo-pathological micrograph of the various organs revealed the hypolipidemic potential activity of fermented sorghum.

Conclusion: The diets supplemented with the fermented food showed better activity against hypercholesterolemia in the rat groups than their microbial isolates. It would therefore be safe to recommend these foods to individuals with this disease situation and those with incidence of coronary heart disease in Nigeria.

Keywords: Hypolipidemic activity; Ogi; hypercholesterol; lactic acid bacteria; probiotes; triglyceride.

1. INTRODUCTION

Some African fermented food products have been reported to have therapeutic values [1].These food products have been used in the treatment of diarrhoea or dysentery. *Ogi* is a major indigenous traditional weaning food in West Africa. Also one of the most frequently consumed fermented foods in southwestern Nigeria [1]. The processing of *Ogi* involves either soaking the whole grain in water contained in a fermenting vat for a length of time or an initial reduction of the grain size by milling in the wet form before being allowed to ferment at room temperature for 2-3 days [2,3]. *Lactobacillus* species are the predominant bacteria isolated from *Ogi* [4,5]. The major lactic acid bacteria (LAB) responsible for the production of lactic acid are *L. plantarum.* Teniola and Odunfa [6] isolated the following species of lactic acid bacteria from fermented *Ogi; L. plantarum, L. brevis, Pediococcus* and *Pentococcus* spp.

Lactic acid bacteria (LAB) are a broad group of Gram positive, non-motile catalase negative rods and cocci that ferment carbohydrates and form lactic acid as the major end product [7]. Strains of *Lactobacillus* have been reported to be involved in the fermentation of Ogi and most other foods consumed in Africa [3,6,8]. Sufficient numbers of viable LAB alter the intestine microbiota by implantation or colonization in a compartment of the host. They are generally regarded as safe due to their long history of safe use in foods [3,5,6].

High serum cholesterol level has been linked to increased consumption of dairy and animal fats. Hypercholesterolemia is considered to be one of the major factors predisposing atherosclerotic heart disease, and has been widely accepted that diet can play a significant role [9]. The hypercholesterolemic effect of probiotes has been reported by many authors. Lin and Chen [10] reported the cholesterol lowering activity of *L. acidophilus* while Tabuchi et al. [11] showed the hypercholesterolemic effect of *Lactobacillus* in hypercholesterolemic rats.

Cholesterol found in plasma membranes can be extracted from the body by the High Density Lipoproteins (HDL). By esterification HDL carries cholesterol away from the arteries back to the liver, where it is passed out of the body. Hypercholesterolemia refers to high levels of cholesterol in the blood. A normal or desirable cholesterol level in man is described as less than 200 mg of cholesterol per deciliter of blood (mg/dl). It is considered to be at borderline when it is in the range of 200 to 239 mg/dl and elevated when it is 240 mg/dl or above in humans. De Rodas et al. [12] showed that dietary supplementation with *Lactobacillus acidophilus* reduced serum cholesterol and low density lipoprotein (LDL) cholesterol in pigs previously fed high cholesterol diet more than in pigs not receiving *L. acidophilus*. The colonization of the intestine by *Lactobacillus* was concluded to be responsible for this the cholesterol lowering effects of LAB [11,13]. Gilliland et al. [14] referred to this phenomenon as cholesterol assimilation.

Fermented sorghum gruel (*Ogi*) contains high level of LAB and has been used for the treatment of different ailments in the south western Nigeria [1]. Despite this acclaimed therapeutic usage for the treatment of microbial infections, there is dearth of scientific information on the empirical proofs on its hypolipidemic activity. This study therefore aimed at investigating the ability of fermented sorghum gruel and LAB isolated from it to treat (reverse/reduce) the level of cholesterol in hypercholesterolemic rat and to detect the enzymatic changes in the rat.

2. MATERIALS AND METHODS

2.1 Source of *Ogi*

Fermented sorghum gruel *Ogi* samples were purchased from a single producer to ensure consistency in the quality of the product. The samples were carried in the ice pack and transferred within two hours to the Microbiology Laboratory of the Adekunle Ajasin University Akungba-Akoko for analyses. The pH and the titratable acidity of the sample were determined by the method of AOAC [15].

2.2 Isolation and Identification of Lactic Acid Bacteria (LAB)

The *Ogi* samples were serially diluted and one ml of appropriate dilution was aseptically plated, using pour plate technique, on Plate Count Agar (PCA) for total viable bacterial count. The method of Harrigan and McCance [16] was used to isolate LAB. Discrete colonies on de Man Rogosa Sharpe (MRS) agar (Oxoid) plates were selected and the method of Savadogo et al. [17] was used to identify the isolates. The results were interpreted according to Holt et al. [18].

The LAB strains were harvested by centrifuging at 4000 rpm for 30 min at 15°C after washing the cells three times in physiological saline for 10 min. The washing of the bacterial cells was done to remove slime layer and nutrient from the growth medium. The cells were re-suspended in sterile normal saline solution (0.85% NaCl). The absorbance of the bacteria suspensions at 600 nm was adjusted to a reading of 0.500 with an average cell count of 1.5 x 10^9 CFU/mL. Standardized identified LAB strains were sub-cultured on MRS broth supplemented with 0.5 % taurodeoxycholic acid (TCDA) (Sigma) and incubated for 24h at 37°C. Representative tubes with growth was inoculated onto the MRS agar supplemented with 0.5 % (TCDA) and incubated for 24 h at 37°C.

2.3 Acid and Bile Tolerance Test

The ability of the isolates to tolerate bile was determined according to Liong and Shah [19]. The pH of MRS broth was adjusted by 1 M HCl (BDH) to achieve pH values of 3.0, 4.0 and 5.0 and the standardized LAB strains were inoculated. The broth was incubated for 90 min at 37°C. The optical density of the broth was read at 650nm before and after the incubation by a spectrophotometer (Jenway 6505, England). All the experiments were conducted in triplicates and repeated twice. The effects of bile on the growth of probiotic LAB strains were examined using method of Meei-Yn and Tseng-Wei [20]. Each of the standardized isolates was inoculated into MRS broth and supplemented with 0.5 ml (% v/w) sterile fresh bovine bile. The culture was incubated at 37°C for 18 h. The optical density of the broth was read at 650 nm before and after the incubation.

2.4 Bile Salt Hydrolase (BSH) Assay

According to Moser and Savage [21] was used to detect the isolates that produced bile salt hydrolase (BSH). Briefly, MRS agar was supplemented with 0.5% taurodeoxycholic acid (TDCA) and the isolates were inoculated onto the agar and incubated anaerobically at 37°C for 5 days and the plates observed for precipitation of deoxycholic acid around the colony. Isolates with pronounced activity were selected for further studies (Table 1).

2.5 Formulation of Diets for the Animal Feed Experiment

Albino rats of Wister strains were induced by feeding on a high lipid cholesterol diet (HLF) for two weeks. Following hypercholesterolemia, the feeds were supplemented with the fermented foods and later fed to the rats for another eight weeks. As shown in Table 2, the diets were composed of four test groups and one control, the basal diet (BD). The four diet groups comprised high lipid feed (HLF), high lipid feed with *Ogi* (HLF+*Ogi*), high lipid feed with different LAB strains (containing 5.6 x 10^{10} CFU/ml) (HLF+LAB) and the basal diet with Ogi (BD+Ogi). Diets HLF, HLF+*Ogi* and HLF+LAB were supplemented with 20 g/100 g lard (sourced from fatback) and 1.0 g/100 g cholesterol.

The pre-mixed vitamin and pre-mixed mineral were prepared as follows: Vitamin mixture (g/kg per diet) contained thiamine (0.02), riboflavin (0.03), pyridoxine (0.10), vitamin B12 (0.00003), niacin (0.001), calcium pantothenate (0.10), p-aminobenzoic acid (0.01) Vitamin A acetate (0.04), ergocalciferol (0.4) and choline-HCl (2.0). The mineral mixture (g/kg per diet) contained $CaCO_3$ (15.56) $CaCl_2.6H_2O$ (0.001), $CuSO_4.5H_2O$ (0.019), $FeSO_4.7H_2O$ (1.078), $MgSO_4$ (2.292), and $MnSO_4.2H_2O$ (0.025). The weight of each of the rats in the groups was taken twice a week and recorded. The difference between the initial and the final weights was taken to be the weight gained by the rat. The mean weight gained by the rats in each diet group was determined.

2.6 Source and Feeding of the Experimental Animals

Seventy post weaning healthy albino rats of Wister strains (*Rattus nervegicus*) with initial average body weight of 83.83±3.34 g were obtained from the Preclinical Animal House, University of Ibadan, Nigeria. The experimental animals were managed according to the National Research Council's Guide for the Care and Use of Laboratory Animals [22]. The animals were acclimatized for five days, fed with grower's mash (TopFeed®) and supplemented diets (Table 1) and adequately supplied with distilled water *ad libitum*. The animals were randomly assigned to each of the diet groups and they (n=5/group) were maintained at 27°C with about 13:11 h of light and dark respectively. The animals were randomly assigned (in cages) into five groups. They were fed for 8 wk and the weight taken weekly throughout the experiment. Food and water were withdrawn from the animals 18 h prior to the termination of the experiment.

Table 1. Diet composition (g/Kg) of animal feed to evaluate the hypolipidemic activity of *'ogi'*

Group	Diets					
	Basal diet	**BD+'OGI'**	**HLF**	**HLF +'OGI'**	**HLF+ *L. lactis***	**HLF+ *L. brevis***
Corn starch	64.50	--	48.5	--	48.5	48.5
'Ogi'	--	65.5	--	48.5	--	--
Soybean	21.0	21.0	21.0	21.0	21.0	21.0
Fish meal	6.50	6.50	6.50	6.50	6.50	6.50
Vegetable oil	5.0	5.0	--	--	--	--
Lard	--	--	20.0	20.0	20.0	20.0
Mineral mix*	1.5	1.5	1.5	1.5	1.5	1.5
Cholesterol	--	--	1.0	1.0	1.0	1.0
Vitamin mix*	1.0	1.0	1.0	1.0	1.0	1.0
Methionine	0.3	0.3	0.3	0.3	0.3	0.3
Choline Chl.	0.2	0.2	0.2	0.2	0.2	0.2
≈10 log_{10} CFU/ml	--	--	--	--	*L. lactis*	*L. brevis*

**Premix, Chl: chloride*

Table 2. Growth characteristics of lactic acid bacteria isolates with probiotic attributes

Organisms	Cell morphology	Growth at acidic pH (3.0)	Growth at fresh bovine bile	Ability to deconjugate TCDA
Lactobacillus bulgaricus	Rod	++	+++	++
Lactobacillus plantarium	Rod	+++	+++	+++
Lactobacillus sp. (JBI 1)	Rod	+++	+++	NG
Lactobacillus lactis	Rod	++	+++	+++
Lactobacillus sp. (JBI 4)	Rod	+	+++	++
Lactobacillus brevis	Rod	+++	+++	+++
Lactococcus sp. (JBI 9)	Coccus	+++	+++	+
Lactococcus sp. (JBI 12)	Coccus	+++	+++	NG

+++: Dense growth; ++: Moderate growth; +: Poor growth; NG: No growth

2.7 Blood Collection and Determination of Serum Cholesterol Types

Blood was collected from the rats by cardiac puncture after a deep anesthesia. One-ml syringe using 22 gauge needle was inserted 5mm from the center of the thorax towards the animal's chin, 5 to 10mm deep; the syringe was held at 25 to 30° away from the chest. Serum total cholesterol and high-density lipoprotein (HDLC) cholesterol were determined according to Grove [23] while low-density lipoprotein (LDLC) cholesterol, and triacylglycerol were measured enzymatically by the methods of Meiattini [24] and Buccolo and David [25] respectively.

2.8 Determination of Lipid and Marker Enzymes in the Organs of the Rats

The different organs (heart, kidney and liver) were removed and separately homogenized before lipid and enzyme analyses. Total lipids were determined using the method of Folch [26]. The fat contents were extracted with chloroform: methanol mixture (2:1 v/v). The aliquots of the organic phase were analyzed for total cholesterol calorimetrically using glacial acetic acid-$FeSO_4$-H_2SO_4 while the remainder of the organic phase was evaporated to dryness and weighed to give the amount of liver total lipids. Aspartate transaminase (AST) and alanine transaminase (ALT) were determined using the Randox® enzyme kits.

2.9 Histopathological Processes

The animals were sacrificed eight weeks after the commencement of the experiment and the organs (kidney, heart and liver) were extracted and processed for histopathological assessment according to the method of Gentry-Weeks *et al.* [27]. The slides were examined under the microscope and the microphotographs taken at a magnification of X 400 as described by Baker and Silverton [28].

2.10 Statistical Analysis

Statistical analysis was performed using SPSS 17.0 software. Data are expressed as mean ± SD. Data were analyzed using one-way ANOVA, followed by the Turkey Test. The result was considered significantly different at level of $p < 0.05$.

3. RESULTS

Eight LAB strains which grew well on fresh bovine bile were isolated from sorghum *Ogi*; six *Lactobacillus* species and two *Lactococcus* spp. Only two of these LAB strains ((*Lactobacillus* sp. (JBI 1) and *Lactococcus* sp. (JBI 2)) could not deconjugate TCDA. However, all the isolates are acid tolerant, with at least moderate growth at pH 3 (Table 2) but the strains of *Lactobacillus* sp. (JBI 1, JBI 4: 60%, 68.75%) and *Lactococcus* sp. (JBI 9) grew very poorly even at pH 4 (Table 3).

Table 3. The percentage survival of LAB isolates from '*ogi*' at different pH*

LAB isolates	pH 3	pH 4	pH 5
L. bulgaricus	92.59	94.74	96.08
L. plantarum	95.38	95.00	96.00
Lactobacillus sp. (JBI 1)	50.43	60.00	73.33
L. lacti	94.00	96.51	96.88
Lactobacillus sp. (JBI 4)	37.75	68.75	92.42
Lactobacillus brevis	93.33	93.75	93.82
Lactococcus sp. (JBI 9)	42.86	16.67	28.57
Lactococcus sp. (JBI 12)	44.44	80.00	89.36

** Microbial growth at 37°C for 24 h*

The total bacterial count in ogi ranged between 7.7762 log_{10} CFU/g and 7.9631 log_{10} CFU/g while the LAB count in the samples were between 5.0792 log_{10} CFU/g and 5.6990 log_{10}. LAB was dominant in the fermented gruel. The number of non-lactic acid bacteria that could survive in the sample ranged from 28.43% to 34.48% of the total bacterial load in the *ogi* sample (Table 4). The average values of pH and total titratable acidity (TTA) in all the *ogi* samples were 3.8 and 0.38 respectively (data not shown).

The weight of the animals increased with the duration of feeding. At the end of the feeding experiment animals in HLF group had the highest average weight followed by those in the Basal Diet group (Fig. 1). The animals in the HLF+*L. lactis* diet group had the least mean weight at the end of the experiment. There was no variation between the weights of the animals in the control group (Basal Diet) and the experimental groups (at 0.05 level of significance).

The amount of the total lipid in the animals fed with high lipid diet (HLF) was significantly lower than those obtained in the organs of the animals fed with diets containing BD+*Ogi*, HLF+ *L. lactis* and HLF+ *L. brevis* (at α = 0.05) (Fig. 2). At the probability of 0.001 the cholesterol concentration in the organs of the rats fed on HLF diet differs significantly from those fed with Basal diet, BD+*Ogi* and HLF+ *L. lactis*. While the levels in HLF +*Ogi* and HLF+ *L. brevis* diet groups were statistically lower than those of animals in the HLF group. The mean values of the LDLC in the rats were 188.60±7.27 mg/dl in the animals fed with HLF while the least value was recorded in the 110.80±26.93mg/dl. The values of LDLC in the animals fed Basal Diets were significantly different from those fed on HLF diet (P<0.01) while the difference in the serum LDLC of the animals fed HLF was significantly different from those on HLF+*Ogi* and BD+*Ogi* diets (at P<0.05 and P<0.001) respectively (Fig. 3).

Table 4. Lactic acid bacterial growth on taurodeoxycholic acid supplemented media

Organisms	Growth in TCDAsupplemented MRS broth	Viable count on TCDA supplemented MRS agar (CFU/ml)
Lactobacillus bulgaricus	+++	2.5 X 10^5
Lactobacillus plantarium	+++	5.0 X 10^5
Lactobacillus sp (JBI 1)	+*	1.5 X 10^4
Lactobacillus lactis	+++	2.5 X 10^5
Lactobacillus sp. (JBI 4)	++*	2.5 X 10^5
Lactobacillus brevis	+++	2.0 X 10^5
Lactococcus sp. (JBI 9)	NG	NG
Lactococcus sp. (JBI 12)	+	1.0 X 10^3

**Showed growth indicated by turbidity but there was no sign of precipitation; +: slightly turbid, ++: turbid, +++: highly turbid, NG: no growth. Data are the mean of three determinations*

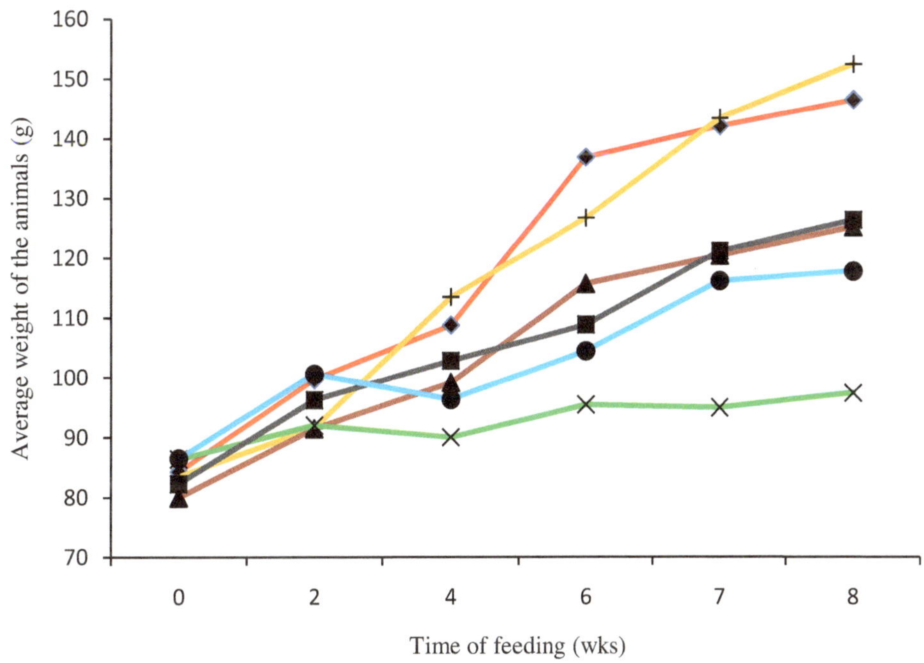

Fig. 1. Change in the body weight of rats fed high lipid cholesterol feed supplemented with *'ogi'* and *Lactobacillus* spp.

The diets include (■)HLF+L. brevis, (▲) HLF + 'OGI', (♦) Basal diet (●) BD+'OGI', (+) HLF and (x) HLF + L. lactis

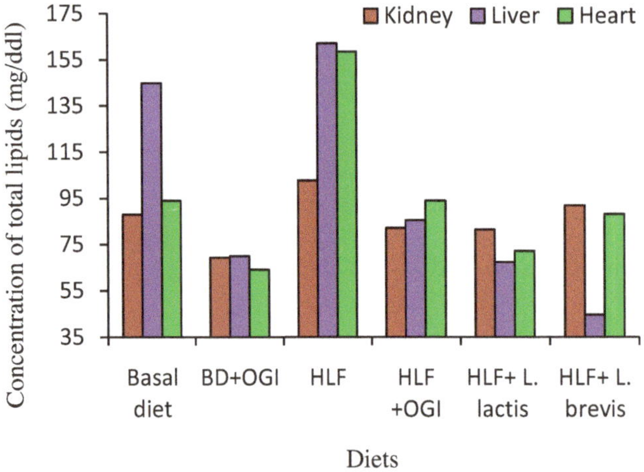

Fig. 2. The total lipid concentration in the organs of experimental rats

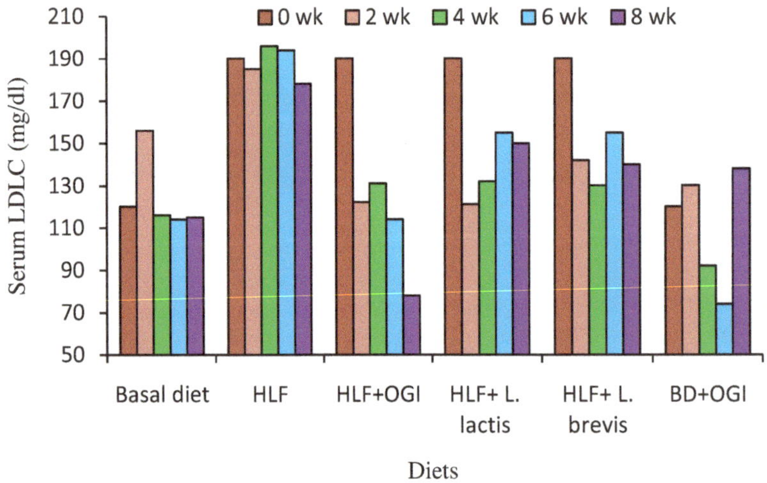

Fig. 3. The concentration of serum low density lipoproteins cholesterol in the serum of the experimental rats

The mean high density lipoprotein cholesterol (HDLC) in the serum of the rats in BD+*Ogi* diet was the highest (52.600 mg/dl) compared to the Basal Diet (49.400 mg/dl) while the HLF group had the least (24.800 mg/ml). HDLC concentration in HLF differed significantly from the basal diet, HLF+*L. brevis* and BD+Ogi at P<0.001 and was also significantly different from HLF +*Ogi* and HLF+*L. lactis* at P<0.01 (Fig. 4). As shown in Fig. 5 the highest amount of glycerides was recorded in the HLF group (100.40 ± 21.29 mg/dl) while the least amount of triglyceride was observed in the animals in the Basal Diet group (31.200±7.190 mg/dl). The levels of glycerides in the Basal and HLF diets were significantly different (P<0.001). Glyceride level in the HLF group was higher than the HLF+*L. lactis* and BD+*Ogi* groups (at P<0.05).

The mean value of alanine transaminase (ALT) in the animals in the Basal Diet group was higher at the commencement of the experiment (275.45±34.43 IU/L) compared to the value obtained after 8wks (144.27±38.63 IU/L). Meanwhile ALT activity (310.00±45.83 IU/L) was highest in the rats fed with the Basal diet supplemented with *Ogi* after 8wks of feeding. Compared to the ALT activity in rats fed on high lipid diet (HLF), the AST activity in rats fed on the Basal Diet +Ogi, HLF + Ogi, HLF+ *L. lactis* and HLF+ *L. brevis* were lower (at p<0.05). The level of AST activity in the animals in all the groups was not significantly different at 0.05 probability level (Table 5).

The histo-pathological micrographs of the various organs (heart, kidney and liver) of hypercholesterolemic and the non-induced rat groups are presented in Plates 1A to J. The heart of rats fed on the normal feed and those on HLF diets with or without *Ogi* showed no visible lesion (Plate 1). In rats fed on HLCF and *Ogi* multiple necrotic tissues of the heart muscles were observed while those on HLCF supplemented with *Ogi* showed some multiple foci of fibre necrosis and haemorrhage with mild cellular infiltration by macrophages. Plates 1E and F showed the tissues of normal and afflicted kidney. All the rats fed on the different diet regimes showed no visible lesion in the kidneys except those fed on HLF and *L. brevis* where the lumen of the tubules was distended (Plate 1F). The liver tissues of most of the rats showed no visible lesions but in some rat groups there were observable cellular changes in the central vein and portal triad of the organ while those fed on HLF were appreciably affected.

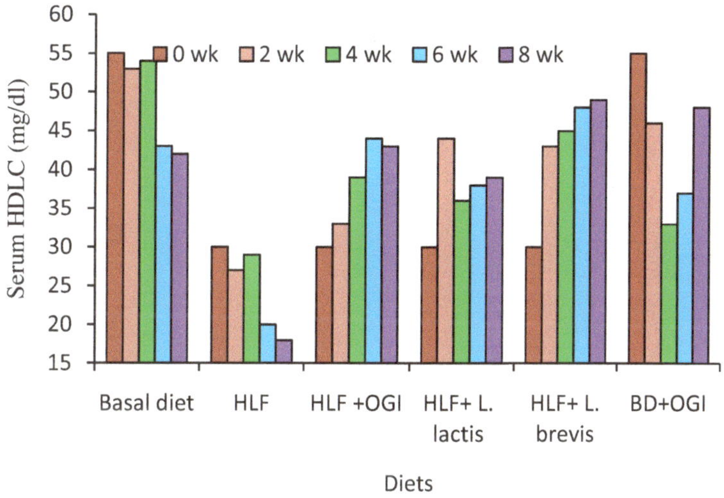

Fig. 4. The concentration of serum high density lipoproteins cholesterol in the serum of the experimental rats

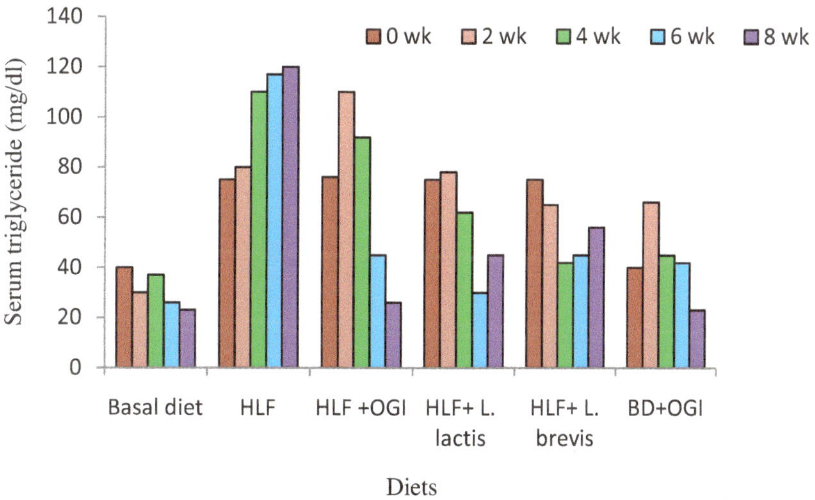

Fig. 5. The level of triglyceride in the serum of the experimental rats

Table 5. The concentration of serum Aspartate transaminase (AST) and Alanine transaminase (ALT) in the experimental rats

Weeks	Enzymes	Diets					
		Basal diet	**Basal Diet + 'OGI'**	**HLF**	**HLF + 'OGI'**	**HLF+ *L. lactis***	**HLF+ *L. brevis***
0	ALT	275.45±34.43	270.42±72.49	90.39±17.45	90.45±28.34	94.44.±19.35	90.34±23.41
	AST	80.78±12.32	90.27±18.42	95.78±14.59	90.34±16.36	93.51±28.34	105.33±41.32
2	ALT	195.27±28.39	190.44±55.27	73.52±26.49	95.22±15.43	73.23±24.31	72.34 ±22.17
	AST	60.30±11.11	250.39±59.23	150.35±48.29	152.24±33.28	103.32±32.33	242.32±34.59
4	ALT	164.49±19.15	100.35±19.46	80.69±29.43	92.34±43.14	140.12±34.22	140.42±34.11
	AST	140.49±27.38	410.34±83.42	230.00±45.81	135.46±25.34	125.23±41.12	250.39±41.39
6	ALT	100.38±28.55	176.84±17.49	121.32±39.10	120.34±25.33	120.33±31.51	110.33±51.93
	AST	150.39±51.53	180.11±45.28	250 .34±34.25	190.34±16.99	220.33±13.42	175.27±31.52
8	ALT	144.27±38.63	340.99±45.83	125.56±42.14	102.12±52.21	110.54±31.65	90.46±25.77
	AST	163.19±48.24	178.33±19.46	264.34±24.62	165.45±23.34	245.23±43.34	150.36±18.36

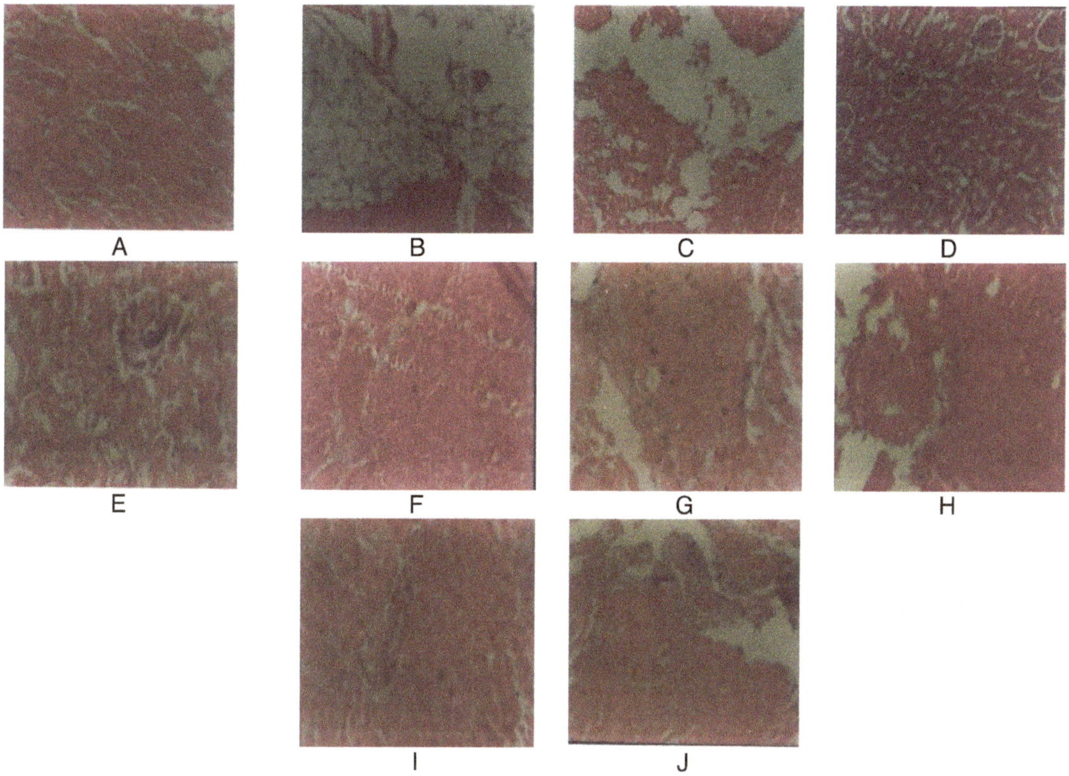

Plate 1. Histopathologic examinations of different organs of animals showing lesions due to hypercholesterolemic situation created by the HLF diet

A: Heart tissues of animals fed with Basal Diets with no visible lesion (Mag. X250), B: Heart of animal in diet group HLF+'ogi' (Mag. X250), C: Multiple necroses into the heart muscles for rats on HLF+'ogi' (Mag. X125), D: Kidney of rat on Basal Diet showing no visible lesion (Mag. X250), E: Kidney of rat in HLF+L. brevis with lumen of the tubules distended (Mag. X125), F: Liver-mild congestion of central vein of rats fed HLF+L. brevis (Mag. X250), G: Distended central vein of the liver of rats on HLF diet group (Mag. X250), H: Mild fibroplasia of portal triad in liver of rats on high lipid cholesterol (Mag. X125), I: Liver of rat on HLF+OGI with normal morphology (Mag. X125) J: Liver of rat on HLF+L. lactics having moderate cellular infiltrations of portal triad by macrophages (Mag. X125)

4. DISCUSSION

Lactic acid fermented foods have made up a significant portion of food intake by humans for a long time and are still available in the developing countries [1,2,6]. *Ogi* which is an example of lactic acid bacteria fermented food product in Nigeria was investigated for its ability to reduce the serum cholesterol level in induced hypercholesterolemic rats. The total lactic acid bacteria counts ranged between 7.7762 log_{10} CFU/g and 7.9631 log_{10} CFU/g. This high lactic acid bacteria count observed in *Ogi* may be as a result of the growth of the associated microbiota on the grains and their low pH range of between 3.5 and 4.5. This confirms the reports of Hounhouigan et al. [29], Teniola and Odunfa [6], and Adebayo and Aderiye [5], which showed the dominance of lactic acid bacteria in fermented maize for *Ogi* production. Also the viable count of LAB in *Ogi* may be due to their known complex nutritional requirements: Substrate rich in carbohydrate [30,31].

The low pH and the TTA of 3.8 and 0.38 respectively observed in *Ogi* may be attributed to the secondary metabolites (lactic acid) produced by the lactic acid bacteria involved in the fermentation of the foods. Teniola and Odunfa [6] reported that metabolites produced during the fermentation of maize (*Ogi)* are usually responsible for the changes in the physical and chemical parameters. They concluded that the changes may impact both desirable and undesirable characteristics on the products.

In all, six different lactic acid bacteria isolates were selected and used in the study. The bacteria are all catalase negative and fermented glucose with or without the production of gas. The rod-shaped isolates were identified as *Lactobacillus* spp. while the cocci were the *Lactococcus* spp. The identification of these isolates as *Lactobacillus* species is in line with Tannock [32] definition of lactobacilli as members of the Lactic acid bacteria group that are characterized by the formation of Lactic acid as sole end product and are Gram positive, catalase negative, non-spore-forming rods or coccobacilli. He described them as been strictly fermentative with complex nutritional requirement.

Bile tolerance is an important criterion for the selection of probiotes [33]. A bile concentration of 0.2% (w/v) in MRS medium has been reported to prevent growth of lactobacilli for at least 4 hrs, but the cells remained viable after 4 h incubation [34]. In this study, *L. plantarum, L. bulgaricus, L. brevis* and *L. lactis* showed considerable growth when the medium was supplemented with bovine bile. This medium and the survival of these four organisms at different pHs had been employed to simulate conditions in the upper intestinal tract of man [35]. These conditions (i.e. the acid and bile tolerance) which are required for an efficacious probiote and variability in the survival rates may help to explain some beneficial propensities of these organisms [36].

The growth of these organisms at pH 3, 4, and 5 is an indication that they are likely to survive the acidity of the stomach when ingested with foods. The proliferation of these organisms at these pH values and the ability to tolerate bile are important indices to categorize them as probiotes. Their tolerance to high acidity may be attributed to the presence of a constant gradient between extracellular and cytoplasmic pH [37].

For these organisms to be used as probiotes in humans, Moser and Savage [21] postulated that they must be able to overcome the conjugated bile acids of the duodenum. Bile acids are synthesized from cholesterol and conjugated to either glycine or taurine in the liver [38]. Isolates from the fermented foods used in this study displayed varying growth abilities in taurodeoxycholic acid (TCDA) supplemented medium. The inability of some of the isolates to precipitate TCDA may be that they cannot express bile salt hydrolase activity as suggested by Savage et al. [38] and Tanaka et al. [39].

Hypercholesterolemia refers to levels of cholesterol in the blood that are higher than the normal levels in animals. Cholesterol circulates in the blood streams and it is a major building block of cell membranes. It is essential in the formation of bile, vitamin D, other steroids and hormones [40]. All the groups of rats fed on high lipid cholesterol diet (HLCF) showed high level of serum cholesterol after 2 weeks of feeding. This may be as a result of the addition of pork fat which increased the lipid content of the feed to 20% and also the incorporation of cholesterol in the diet. Diets rich in saturated fats and contain high level of cholesterol raises serum cholesterol [41].

When compared to the control group which are fed on the normal feed and took water *ad libitum,* all the hypercholesterolemic rats were active throughout the period of the experiment. This indicates that hypercholesterolemia does not have any noticeable visible symptom(s), unless those diagnosed through blood analysis or when heart attack or stroke eventually occurred [42].

There were noticeable changes in the body of the rats in the groups in which hypercholesterolemia was induced (HLF and *Ogi*) when compared to those fed on the normal feed and *Ogi.* Rats fed on HLF supplemented with *Ogi* and lactobacilli had lower body weights compared to those fed with the Basal and HLF diets. Feeding the rats with the fermented foods did not show much difference in their body weights. It could be deduced that *Ogi* could be an antidote for the reduction of body weight. This may therefore mean that the consumption of *Ogi* had a way of enhancing the conversion/removal of the excess lipid in the body of the rats. This may also probably be attributed to the activity of the organisms contained in the fermented food which reflected on their body weight values (Fig. 1).

The cholesterol profile in the various groups showed a decline as a result of the supplementation of the diet. The animals whose diets were supplemented with the bacterial isolates also had significant reduction in their serum cholesterol concentration at the end of the feeding period. The decrease in the serum cholesterol concentration may be attributed to the ability of these organisms to take up cholesterol into their cellular membrane [13,14,36].

Rats fed on normal feed or HLF supplemented with *Ogi* showed decrease in the serum cholesterol level after the 4th week, and this continued until the 8th week. This observation can be attributed to the co-precipitation of cholesterol with deconjugated bile acids. Klaver and van der Meer [43] showed that the solubility of cholesterol is dependent on the bile salts present.

The low density lipoprotein cholesterol (LDLC) values were low in all the rat groups fed on diets supplemented with *Ogi* and *Lactobacillus* isolates. Low density lipoprotein cholesterol is a product of the liver through the packaging of triglycerides with cholesterol, phospholipids and apoprotein [44]. Very high LDLC tend to stick to the lining of the blood vessels, leading to the hardening of the arteries. The lowering effect was more pronounced in the groups fed with HLF and *Ogi.* This finding confirmed an earlier report of Fletcher et al. [45] that the principal strategy for lowering LDLC levels was to replace cholesterol raising fatty acids (FA) with dietary carbohydrates. Aderiye and David [46] reported that probiotics from African fermented food *iru* increased the concentration of the HDL and reduced the LDL.

The serum triglyceride content of the rats was also lower in all the animal groups fed with diets supplemented with the fermented foods and the organisms. This was expected since triglyceride determines LDLC composition, physical properties and cell specific binding in culture cells of organism [47]. The reduction may also be as a result of the fact that the supplements (*Ogi* and the lactobacilli) did not in any way increase the saturated fat content of the diet.

There was an increase in the high density lipoprotein cholesterol (HDLC) values in all the rat groups. High HDLC levels in the animal and human blood protect against heart attack [19]. The high values reported are in agreement with the findings of Minelli et al. [48] who observed that the administration of *Lactobacillus* sp. in animal models induced a cholagogic effect, favouring an easier bile flux and improvement in some metabolic indices of hepatic and lipidic functionality. HDLC have also been reported to increase triglyceride catabolism [49].

Alanine and Aspartate aminotransferase are normally found in diversity of tissues including the liver, heart, kidney, and brain [50]. They are found in the cells of these organs and may leak out to the blood when the cells are injured. The results from this work showed an initial increase and then a decrease in the values of the two enzymes. The reduction in activity of these enzymes over the supplementation period may indicate an improvement in the recovery of the organ injury. This may be as a result of the ability of the lactic acid bacteria present in the fermented foods to produce polyamines from amino acids [51]. Our findings agree with the report of Adawi et al. [52] that supplementation of the diet with *L. plantarum* and arginine reduced the hepatocellular necrosis and the inflammatory cell infiltration in the liver, which was reflected in the low concentrations of the enzymes released.

The lesions observed in the histopathologic plates in the different organs of the animals may probably be as a result of the hypercholesterolemic situation created by the HLF diet. The degree of the severity of the lesions could explain the impact the supplements had on arrest of the situation of the disease condition. This is further confirmed by the fact that the group fed on the normal diet only had no noticeable lesion(s) in all the organs.

5. CONCLUSION

In this study, one can attribute the reduction of serum cholesterol to the assimilation and binding of the cholesterol by the different organisms that survived the passage through the stomach and finally colonized the intestine. The organisms prevented the reabsorption of the cholesterol in the diet and thereby resulting in low values in the serum. Other ailments such as Diabetes mellitus may be challenged with these fermented foods to ascertain if there will be any effect.

ACKNOWLEDGEMENTS

The authors acknowledge the contribution of the Technical staff of the Department of Microbiology, Adekunle Ajasin University. Akungba-Akoko, Ondo State, Nigeria. The authors declared no conflict of interests.

COMPETING INTERESTS

Authors have declared that no competing interests exist.

REFERENCES

1. Aderiye BI, Laleye SA. Relevance of fermented food products in southwest Nigeria. Plant Food Hum Nutr. 2003;56: 1-16.

2. Odunfa SA, Adeyele SA. Microbiological changes during the traditional production of ogi-baba, a West African fermented Sorghum gruel. J Cer Sci. 1985,3:173-180.
3. Adebolu TT, Olodun AO, Ihunweze BC: Evaluation of *ogi* liquor from different grains for antibacterial activities against some common pathogens. Afr J Biotechn. 2007;6(9):1140-1143.
4. Fields ML, Hamad AM, Smith DK. Natural lactic acid fermentation of corn meal. J Food Sci. 1981;46:900-902.
5. Adebayo CO, Aderiye BI. Ecology and antibacterial potential of lactic acid bacteria associated with fermented cereals and cassava. Res J Microbiol. 2007;2(5): 426-435.
6. Teniola OD, Odunfa SA. Microbial assessment of quality evaluation of ogi during spoilage. World J Microbiol Biotechnol. 2003;18(8):731-736.
7. Aguirre M, Collins MD. Lactic acid bacteria and human clinical infection: A review. J Appl Bacteriol. 1993;75:95-107.
8. Odunfa SA, Oyewole OB. African fermented foods In: Microbiology of Fermented Foods, Wood, B. (Ed). Elsevier Applied Science Publishers, London. 1995;712-751.
9. American Heart Foundation (AHA). Cholesterol: AHA Scientific position; 2005. Available:http://www.americanheart.org/presenter
10. Lin M, Chen T. Reduction of cholesterol by *Lactobacillus acidophilus* in culture broth. J Food Drug Anal. 2000;8:97-102.
11. Tabuchi M, Tamura A, Yamada N, Ishida T, Hosoda M, Hosono A. Hypocholesterolemic effect of *Lactobacillus* GG in hypercholesterolemic rat. Milchwissenchaft. 2003,58(5/6): 246-249.
12. de Rodas BZ, Gilliland SE, Maxwell CV. Hypocholesterolemic action of *Lactobacillus acidophilus ATCC 43121* and calcium to swine with hyper-cholesterolemia induced by Diet. J Dairy Sci. 1996;79:2121-2128.
13. Tahri F, Grill JP, Schneider F. *Bifidobacteria* strain behaviour toward cholesterol: co-precipitation with bile salts and assimilation. Curr Microbiol. 1996;33: 187-193.
14. Gilliland SE, Nelson CR, Maxwell C. Assimilation of cholesterol by *Lactobacillus acidophilus.* Appl Environ Microbiol. 1985; 49:377-381.
15. AOAC. Official Methods ofAnalysis. 15th Edition, Association of Official Analytical Chemists Washington DC; 1990.
16. Harrigan WF, McCance ME. Laboratory methods in food Dairy Microbiology. London Academic Press; 1976.
17. Savadogo A, Quattatara CAT, Savadogo PW, Quattara AS, Barro N, Traore AS. Microorganisms involved in Fulani traditional fermented milk in Burkina Faso. Pak J Nutr. 2005;3:134-139.
18. Holt JG, Krieg NR, Sneath PHA, Staley JT, Williams ST. Bergey's Manual of Determinative Bacteriology, 9th Edn. Williams and Wilkins, Baltimore; 1994.
19. Liong MT, Shah NP. Acid and bile tolerance and the cholesterol removal ability of bifidobacteria strains. Biosci Microflora. 2005;24:1–10.
20. Meei-Yn L, Tseng–Wei C. Reduction of Cholesterol by *Lactobacillus acidophilus* in Culture. J Food Drug Analy. 2000;8: 97-102.
21. Moser SA, Savage DC. Bile salt hydrolase activity and resistance to toxicity of conjugated bile salts are unrelated properties in lactobacilli. Appl Environ Microbiol. 2001;67(8):3476-3480.
22. National Research Council (NRC). Guide for the care and use of laboratory animals. Publication No. 85-23 (rev). National Institute of Health, Bethesda, MD; 1985.
23. Grove TH. Effect of reagent pH on determination of high density lipoprotein cholesterol by precipitation with sodium phosphotungstate-magnesium. Clin Chem. 1979;25:560-564.
24. Meiattini F, Prencipe L, Bardelli F, Giannini G, Tarli P. The 4-hydroxybenzoate/ 4-aminophenazone chromogenic system used in the enzymatic determination of serum cholesterol. Clin Chem. 1978;24: 2161-2165.
25. Buccolo G, David H. Quantitative determination of serum triglycerides by use of enzymes. Clin Chem. 1973;ua19(5): 476-482.
26. Folch J, Lees M, Sloanestanley GH. A simple method for the isolation and

purification of total lipids from animal tissues. J Biolog Chem. 1957;226:497-509.

27. Gentry-Weeks C, Hutcheson HJ, Kim LM, Bolte D, Traub-Dargatz J, Morley P, Powers B, Jessen M. Identification of two phylogenetically related organisms from feces by PCR for detection of *Salmonella* spp. J Clin Microbiol. 2002;40:1487–1492.
28. Baker JF, Silverton RE. Baker and silverton's introduction to medical laboratory technology. 7th Edition. Butterworth-Heinemann. 2005;448.
29. Hounhouigan DJ, Nout MJR, Nago CM, Honben JH, Rombouts FM. Microbiological changes in maize during natural fermentation. World J Microbiol Biotechnol. 1994;10:410-413.
30. Saloff-Coste CJ. Lactic Acid Bacteria. Danone World Newsletter. 1994;5.
31. Hammes WP, Vogel RF. The genus *Lactobacillus* In: Woods BJB, Holzapfel WH (Ed). The genera lactic acid bacteria. Blackie Academic and Professional, London, United Kingdom. 1995;2:19-54.
32. Tannock GW. A special fondness for lactobacilli. Appl Environ Microbiol. 2004; 70(6):3189-3194.
33. Salminen S, Playne M, Lee YK. Successful probiotic lactobacilli: Human studies on probiotic efficacy. In: Short C. (Ed.) Functional dairy products. Technomic, Lancaster United Kingdom. 2004;13-32.
34. Lee YK, Low CS, Arvilommi H, Salminen S. Permanent colonization by *Lactobacillus casei* is hindered by the low rate of cell division in mouse gut. Appl Environ Microbiol. 2004;70(2):670-674.
35. Chou L, Weimer B. Isolation and characterization of acid and bile tolerant isolates from strains of *Lactobacillus acidophilus*. J Dairy Sci. 1999;82:23-31.
36. Pereira DIA, Gibson GB. Cholesterol assimilation by lactic acid bacteria and bifidobacteria isolated from the human gut. Appl Environ Microbiol. 2002;68(9): 4689-4693.
37. Corcoran BM, Stanton C, Fitzgerald GF, Ross RP. Survival of probiotic lactobacilli in acidic environments enhanced in the presence of metabolizable sugars. Appl Environ Microbiol. 2005;71(6):3060-3067.
38. Savage DC, Lundenn SG, O'Connor LT. Mechanism by which indigenous microorganisms colonize epithelial surfaces as a reservoir of the luminal microflora in the gastrointestinal tract. Microecol Therapy. 1995;21:27-36.
39. Tanaka H, Hashiba H, Kok J, Mierau I. Bile salt hydrolase of *Bifidobacterium longum*-biochemical and genetic characterization. Appl Environ Microbiol. 2000;66(6): 2502–2512.
40. Cardiology channel. High cholesterol; 2006.

    Available:http://www.cardiologychannel.com/hypercholestrolmia
41. American Heart Foundation (AHA). Cholesterol. AHA Scientific position; 2005. Available:http://www.americanheart.org/presenter
42. Healthatoz. Hypercholesterolemia. Encyclopedia index H; 2006. Available:http://www.healthatoz.com
43. Klaver AMF, Van der Meer R. The assumed assimilation of cholesterol by Lactobacilli and *Bifidobacterium bifidum* is due to their Bile salt-deconjugating activity. Appl Environ Microbiol. 1993;59: 1120-1124.
44. Worobetz RJ, Hilsden EA, Shaffer JB, Simon P, Para VG, Ma M, Wong F, Blandis L, Adams P, Heathcote J, Lee SS, Lilly LB, Hemming AW, Levy GA. The liver In: First principle of gastroenterology: The Basis of Diseases and Aspproach to Management. GIT Textbooks. 2005; 462-475.
45. Fletcher B, Berra K, Ades P, Burke LE, Durstine L, Fair JM, Guff D, Miller NH, Kris-Etherton P, Wilterdink J: Managing abnormal blood lipids- a collaboration approach. Circulation. 2005,112:3184-3209.
46. Aderiye BI, David OM. *In vivo* evaluation of hypolipidemic potentials of *Bacillus* species isolated from fermented locust bean (*Parkia fillicoides* Welw) Seeds (*Iru*). Br Microbiol Res J. 2013;3(4):574-584.
47. Castelli WP. Triglycerides: A risk factor for Coronary Heart Diseases. Artherosclerosis. 1996;124:557-564.
48. Minelli EB, Benini A, Marzotto M, Sbarbati A, Ruzzenente O, Ferrario R, Hendriks H, Dellaglio F. Assessment of novel probiotics *Lactobacillus casei* strain for the production of functional dairy foods. Inter Dairy J. 2004;14:723-736.

49. Tietz NM. Textbook of clinical chemistry. 3rd Ed. W.B. Saunders Company, Philadelphia, London and Toronto. 1986;510.
50. Medicine Net. Liver Blood Enzymes; 2006. Available:http://www.medicinenet.com
51. McCormack SA, Johanson LR. Role of polyamines in gastrointestinal mucosal growth, Amer J Physiol. 1991;260 (Gastrointestinal Physiology 23):G795-G806.
52. Adawi D, Kasravi FB, Molin G, Jeppsson B. Effect of *Lactobacillus* supplementation with or without arginine on liver damage and bacterial translocation in acute liver injury model in the rat. Hepatology. 1997; 25(3):642-647.

17

Investigation of Antioxidant and Cytotoxic Activities of Methanolic Extract from *Curculigo recurvata* Leaves

Mohammad Mustakim[1], Md. Mizanur Rahman[1], Prawej Ansari[1,2*], Anaytulla[1], Md. Foyez Ahmad[2] and Mohammed Abu Sayeed[1]

[1]Department of Pharmacy, International Islamic University Chittagong, 154/A, College Road, Chittagong-4203, Bangladesh.
[2]Department of Pharmaceutical Sciences, School of Health and Life Sciences, North South University, Dhaka-1229, Bangladesh.

Authors' contributions

This work was carried out in collaboration between all authors. Author MM designed the current project, performed the experiments and wrote the manuscript. Author MMR carried out the experimental process. Author PA took the responsibility for data interpretation, statistical analysis and helped in experiments and preparing the manuscript. Authors AU, MFA, MAS participated in experiments and data collection. Author PA also edited the manuscript. All authors read and approved the final version of the manuscript.

Editor(s):
(1) Carmen Lúcia de Oliveira Petkowicz, Federal University of Parana, Curitiba, Parana, Brazil.
Reviewers:
(1) Ramesh Kumar Ahirwar, Department of Botany, Govt. College Birsinghpur Pali, Madhya Pradesh, India.
(2) Anonymous, Kasdi Merbah University, Algeria.

ABSTRACT

Objective: The present study was designed to investigate antioxidant and cytotoxic effects of the methanol extract of *Circuligo recurvate* leaves, which belongs to the family Hypoxidaceae, the primary phytochemical property was also intended to screen out.
Methods: The antioxidant activity of the extracts were studied using 1,1-diphenyl-2-picryl hydroxyl (DPPH) radical scavenging activity, reducing power capacity assessment and results obtained were compared to standards. The total phenolic contents and flavonoid contents were estimated from Gallic acid and Quercetine calibration curve respectably.

**Corresponding author: E-mail: chemist89ansari@gmail.com*

Results: IC_{50} value of the extract showed 259.040±3.98 µg/mL in case of the DPPH scavenging assay, the total phenolic and the flavonoid contents of extract was 384±3.34 mg/gm and 149.23±1.12 mg/gm respectively. The reducing power capacity of the extract observed to raise as the concentration of the extract gradually increases. The mortality rate at the assessment of cytotoxicity in brine shrimp napulii model was found increasing with the increase of the concentration and the LC_{50} value of extract was 164.58±2.67 µg/ml.
Conclusion: These findings demonstrate that the leaves extract of *C. recurvata* has the presence of antioxidant compound and there is presence of molecules that can inhibit rapid growing cells.

Keywords: Antioxidant; Curculigo recurvata; DPPH; methanol extract; phytochemical.

1. INTRODUCTION

Oxidation is essential to many living organisms for the production of energy necessary for biological processes. Oxygen - centered free radicals, also known as reactive oxygen species (ROS), including superoxide, hydrogen peroxide, hydroxyl (HO), peroxyl (ROO) and alkoxyl (RO) are produced *in vivo* during oxidation [1]. Oxidative stress occurs because of an increase in oxidative metabolism, which produces a number of ROS. At a low or moderate concentration, ROS plays in important role in the host defense mechanism against microorganisms. However, by the increase in production of ROS, causes generation of free radicals and oxidants and leads to oxidative stress. Oxidative stress is a deleterious process that can damage cell structures, including lipids, proteins and DNA, which results onset of a variety of diseases including cancer, inflammation, neurodegeneration, Parkinson's disease, atherosclerosis and pre-mature aging [2]. Virtually all cellular components appear to be sensitive to oxidative damage. Antioxidants are substances known to protect the cells against the effect of free radicals through scavenging free radicals and singlet oxygen. But lipids, proteins, nucleic acids and carbohydrates are all known to undergo oxidative modification. Elimination of synthetic antioxidants in food applications has given more impetus to exploring natural sources of antioxidants [3].

The human body has several mechanisms to counteract oxidative stress by producing antioxidants, which are either naturally produced in situ, or externally supplied through foods or supplements. Ancient times, to improve flavors of food different types of spices were added which were also well known for their antioxidant capacities. Natural antioxidants constitute a broad range of compounds including phenolic compounds, nitrogen compounds and carotenoids. These antioxidants act as free radical scavengers by preventing and repairing damages caused by ROS, and therefore can enhance the immune defense and lower the risk of health complications caused by oxidative stress [4]. In recent years, there is an increasing interest in finding antioxidant activity of the natural products as oxidative stress is found to be responsible for several health complications. Cytotoxicity assays are widely used to screen for cytotoxicity of plant extract either looks for cytotoxic compounds to develop a therapeutic that targets rapidly dividing cancer cells or to screen for unwanted cytotoxic effects before investing in the development as a pharmaceutical product.

Historically, natural products have served as a source for cancer chemotherapy agents [5]. The brine shrimp lethality bioassay (BSLA) has been used routinely in the primary screening of the crude extracts to assess the toxicity towards brine shrimp. The bioassay has a good correlation with cytotoxic activity in some human solid tumors and with pesticidal activity. This *in-vivo* lethality test has been successively employed for providing a frontline screen that can be backed up by more specific and more sophisticated bioassays once the active compounds have been isolated. A number of novel antitumor and pesticidal natural products have been isolated using this bioassay [6].

C. recurvata belongs to the family of Monocotyledons. It's commonly known as Musali. *C. recurvata* recognized by several leaves, basal, sessile or petiolate; leaf blade usually lanceolate, plicate, leathery or papery. Flowering stems axillary, erect or apically bent downward, long or short. Fruit like a berry, indehiscent, apex sometimes beaked; beak (when present) formed by persistent perianth tube. Seeds are small, often striped and hilum conspicuous [7]. *C. recurvate* is used in the treatment of impotence, limb limpness, arthritis of the lumbar and knee joints, and watery diarrhea in traditional Chinese medicine. It also popular as a potent immunomodulator and aphrodisiac in the Ayurvedic practice [8].

2. MATERIALS AND METHODS

2.1 Drugs and Chemicals

DPPH, Methanol and Gallic acid were obtained from Sigma Chemical, USA. Sodium carbonates, Potassium Acetate were obtained from E. Merck, India. Aluminium chloride, Ascorbic acid (AA) and Folin-Ciocalteu reagent were obtained from Sigma-Aldrich, Bangalore, India. All other chemicals in this experiment were of analytical reagent grade.

2.2 Plant Material and Extraction Procedures

The plant, *C. recurvate*, was collected from forest of Chittagong Hill Tracts, Bangladesh and authenticated by Dr. Shaikh Bokhtear Uddin, Associate Professor, Department of Botany, University of Chittagong, Bangladesh. After cleaning, the leaves separated and sent for air-drying for 10 days. Then kept in an oven at 45°C for 72 hours. About 500 gm of the dried powder of leaves were extracted following cold extraction process with sufficient amount of solvent, methanol. Dried powder soaked in methanol for 15 days with occasional shaking and stirring. The final content was filtered by using cotton and Whatmann No. 1 filter paper to get the pure extract. Finally, we got the concentrated extract from filtrate by evaporating the solvent in water bath (at 30°C-35°C), the percent yield obtained was 5.4%.

2.3 Preliminary Phytochemical Analysis

Various tests were performed with the methanolic extracts of *C. recurvate* for identification of different phytoconsituents like alkaloids, glycosides, saponinins, tannins, terpinoids, phenolic compounds, protein, and carbohydrates using standard procedures [9].

2.4 *In vitro* Antioxidant Activity

2.4.1 DPPH radical scavenging activity

DPPH radical scavenging activity of the different *C. recurvata* extracts was determined according to the method by Braca et al. 2001. [10] 2 ml methanol solution of plant extract or reference standard ascorbic acid at different concentration was mixed with 3 ml methanol solution of DPPH in the test tube. The test tube was incubated at room temperature for 30 min in a dark place to complete the reaction. The absorbance of the solution was measured at 517 nm. DPPH free radical scavenging ability (%) was calculated with the following formula:

$$((A_{\text{absorbance of control}} - A_{\text{absorbance of sample}})/(A_{\text{absorbance of control}}) \times 100$$

The data were presented as mean values ± standard deviation (n = 3).

2.5 Reducing Power Activity

The reducing powers of *C. recurvata* extracts were carried out by using the method of Oyaizu . M, 1986 [11]. Different concentrations of the extracts or standard solutions (1 mL) were mixed with 2.5-mL potassium buffer (0.2 M) and 2.5-mL potassium ferricyanide. After heating for 20 min at 50°C, 2.5-mL trichloroacetic acid (10%) solution was added to the test tube. The total mixture was centrifuged at 3300rpm for 10 min. Next, 2.5-mL supernatant solution was withdrawn from the mixture and mixed with 2.5-mL distilled water and 0.5-mL ferric chloride (0.1%) solution. The absorbance was measured at 700 nm. AA was used for comparison.

2.6 Estimation of Total Phenolic Content

Phenolic compounds are a class of antioxidant agents, which was determined by using Shahidi and Wanasundara, 1992. [12] To measure total phenolic content 1 ml of plant extract (1000 μg/ml) of *C. recurvata* were mixed with 5ml of the Folin-Ciocalteu reagent and 4 ml of 7.5% sodium carbonate. The mixture was shaken thoroughly. Then it was allowed to stand for 20 min at 25°C. The absorbance of the solution was measured at 760 nm using a spectrophotometer against blank.

2.7 Total Flavonoid Determination

The total flavonoid content (TFC) was determined using method described by Kumaran and Karunakaran, 2007. In this test, the comparable reference was Quercetin. [10] 1.0 ml of plant extract (1000 μg/ml) of *C. recurvate* were added to 3 ml of Methanol and 200 µl of 10% aluminium chloride solution, 200 µl of 1 M potassium acetate solution and 5.6 ml of distilled water was added into the test tube. The test tube was incubated for 30 minutes at room temperature to complete the reaction. Then the absorbance of the solution was measured at 420 nm using a spectrophotometer against blank. TFC was calculated as mg Quercetin equivalent (QE) / 100 g sample dry weight.

2.8 Brine Shrimp Lethality Bioassay

The cytotoxicity of crude extract was performed at concentrations of 31.25, 62.5, 125, 250, 500 and 1000 µg/mL on brine shrimp nauplii using Meyer method. [13] Ten matured shrimps were applied to each of all experimental test tubes and the number of survived nauplii in each test tube was counted after 24 h. From this data, the percentage (%) of mortality of the brine shrimp nauplii was calculated for each concentration using the following formula:

$$\% \text{ Mortality} = {}^{Nt}/_{No} \times 100$$

Where, N_t refers to number of killed nauplii after incubation for 24 h; N_0 is number of total nauplii transferred i.e. 10. The LC_{50} was then determined using regression analysis after plotting % mortality versus log concentration of the drug.

3. RESULTS

3.1 Result of Preliminary Phytochemical Analysis

The extract of *C. recurvate* was tested for different phytoconsituents like alkaloids, glycosides, saponin, tannins, terpinoids, reducing sugars, phenol compounds, protein, carbohydrates and volatile oils using standard procedures and the results are given in Table 1.

3.2 DPPH Radical Scavenging Activity

In case DPPH radical scavenging activity of methanol extract of *C. recurvata* shown in Fig. 1 Where increasing the concentration of the extract, activity was found to increase remarkably and the inhibitory capacity of the plant extract was comparatively lower than the ascorbic acid.

Furthermore, the methanol extract of *C. recurvata* showed moderate to good DPPH scavenging activity in which IC_{50} value for the plant extracts was 259.040±3.98 µg/mL. In this test, Ascorbic acid was used as the reference antioxidant and the IC_{50} value for ascorbic acid was 27.02±4.52 µg/mL.

Table 1. Result of preliminary phytochemical analysis of *C. recurvata* extracts

SL no	Tests	Result
1.	Carbohydrate	+
2.	Steroid	+
3.	Alkaloid	+
4.	Terpenoid	-
5.	Cardiac glycoside	+
6.	Saponin	-
7.	Glycoside	+
8.	Tannin	-
9.	Resin	-
10.	Protein	-

(+) ve = Presence and (-) ve = Absence

3.3 Total Phenol and Flavonoid Content

The total phenol and total flavonoid contents of *C. recurvata* of methanol extract were expressed in gallic acid and quercetin equivalents respectively that are shown in Fig. 2. The content of phenolics of the extract under this investigation showed moderate result 384.66±3.34 mg/g GAE and the amount of flavonoid was 149.23±1.12 mg/g quercetin equivalent.

Different studies suggest that different types of polyphenolic compounds such as flavonoids, phenolic acids which are found in plants have multiple biological effects, including antioxidant activity [14].

3.4 Reducing Power

The plant extracts was identified by using the potassium ferricyanide reduction method. The reductive capabilities of this extract were compared with ascorbic acid, which demonstrated at Fig. 3. The reducing power of the extract was moderately strong while increasing dose it shows steady state condition.

3.5 Brine Shrimp Lethality Bioassay

The Methanol extract of *C. recurvata* was tested for Brine shrimp lethality bioassay by using brine shrimp nauplii and DMSO as a solvent. Control was used to see whether DMSO had any effect on brine shrimp lethality or not. The control group of brine shrimp nauplii with and without DMSO exhibited no mortality. For the extract, the number of nauplii died and percent mortality was counted. The result is shown in the Fig. 4 and Table 2. Crude extract resulting in LC_{50} value of less than 1000 µg/ml are considered as active cytotoxic herb [13], that means the *C. recurvata* crude extract, with LC_{50} value of 164.58±2.67 µg/ml has a very potent toxicity.

Table 2. Cytotoxic activity of *C. recurvata*

Concentration(µg/ml)	No. of Nauplii	Living	Death	% of mortality	$LogC_{50}$
31.25	10	6	4	40	
62.5	10	5	5	50	
125	10	4	6	60	164.583
250	10	2	8	80	
500	10	2	8	80	
1000	10	0	10	100	

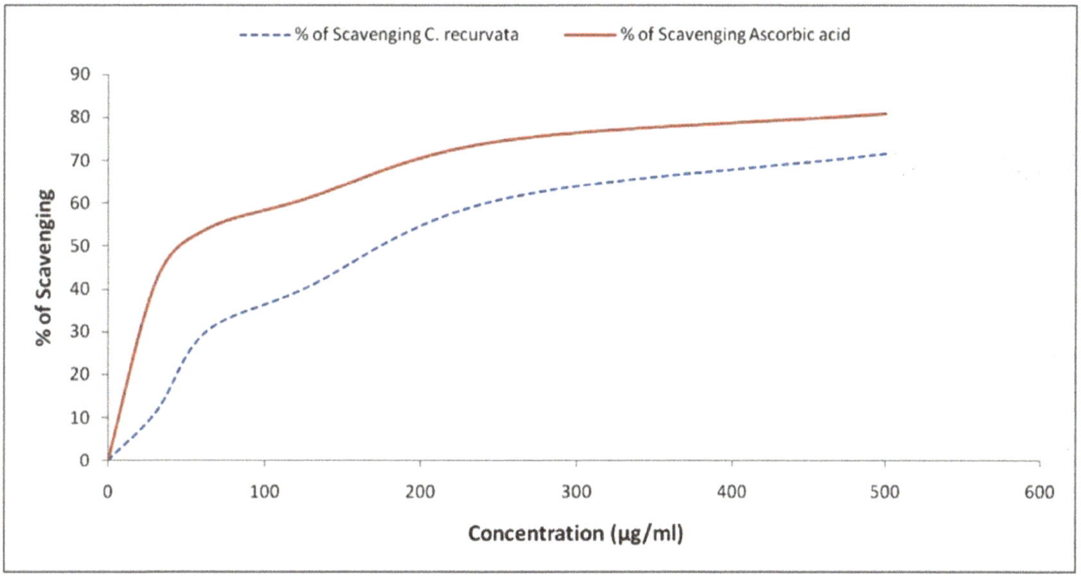

Fig. 1. Percentage of DPPH radical scavenging activity of *C. recurvata* with standard Ascorbic acid

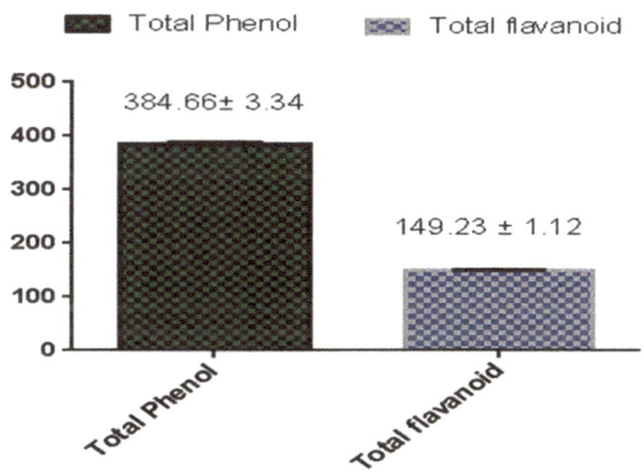

Fig. 2. Total phenol and total flavonoid content in *C. recurvata*

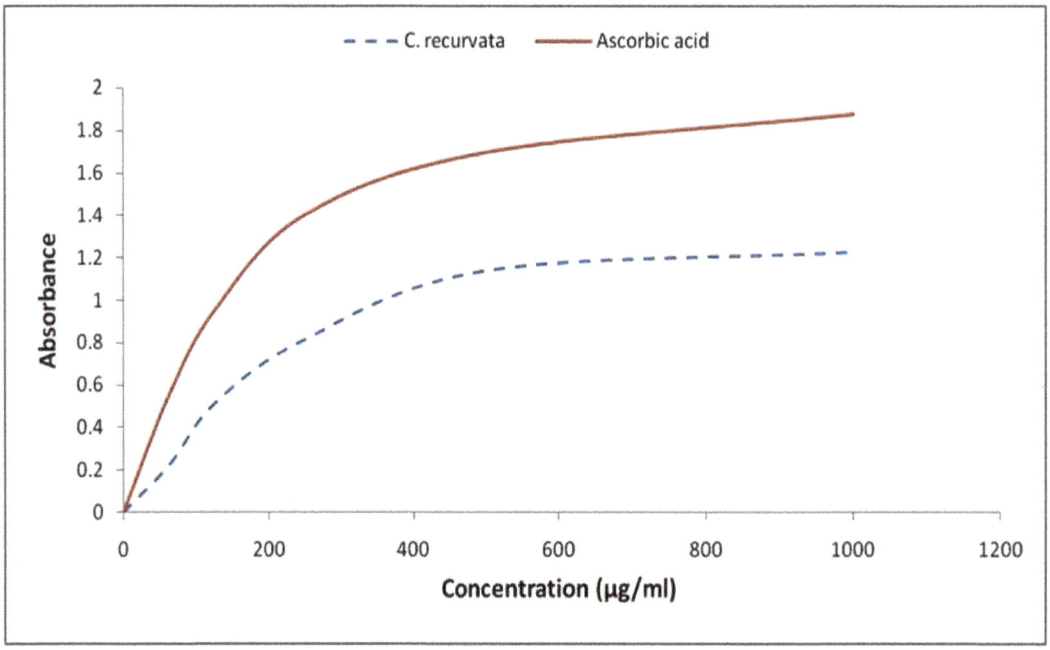

Fig. 3. Reducing power of *C. recurvata* with standard ascorbic acid

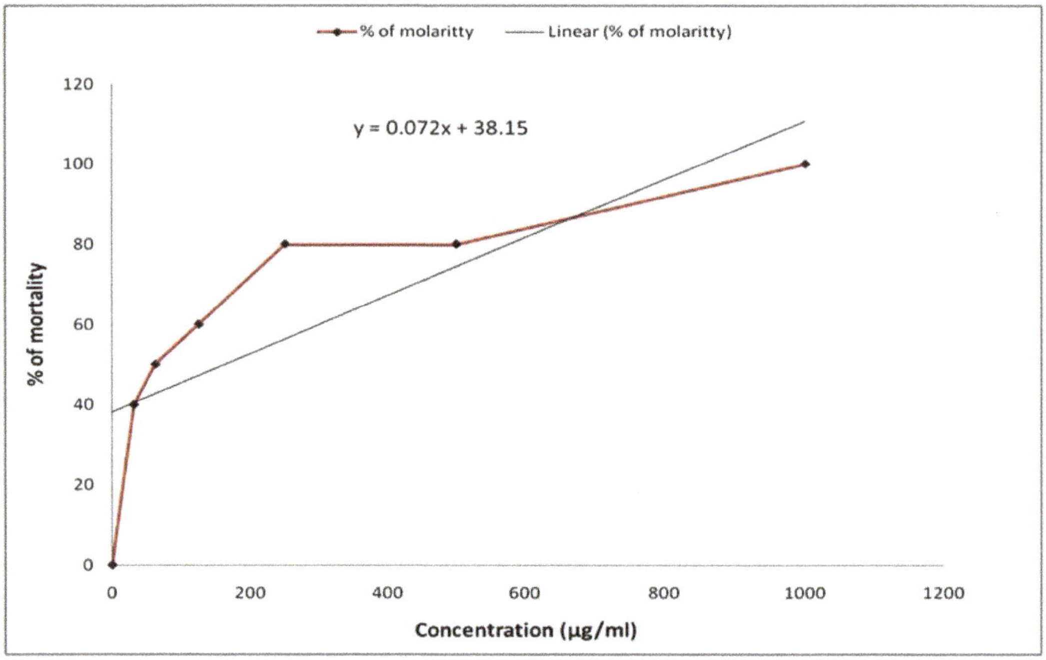

Fig. 4. Cytotoxic activity of *C. recurvata* extract

4. DISCUSSION

Many modern medicines have been discovered because of the scientific follow-up of the traditional herbal preparations. However, now a day, most of the medicines, cosmetics, chemicals etc. are synthetically prepared, however, in many cases their origin is natural more specifically of plants sources.

In addition, a high intake of vitamin C appears to protect against gastric cancer, probably through scavenging ROS formed in the gastric mucosa. Further investigations on vit-C have proved its preventive effects on inhibition of tumor promotion. [15] Flavonoid are polyphenols and they are abundantly found the fruits, vegetables, grains, bark, roots, stems, flowers and tea [16]. They possess anti-inflammatory, anti-allergic, antiviral and anti-carcinogenic properties due to their antioxidant potential [17].

This was only a preliminary study but the plant need further screening against various disease condition, occurred due to antioxidative component. This will help to find out its unexplored efficacy and can be a potential source of biologically important drug candidates. A number of scientific reports indicate certain terpenoids, steroids and phenolic compounds such as tannins, coumarins and flavonoids have a chemo preventive role in cancer, through their effects on signal transduction in cell proliferation and angiogenesis [18]. The presence of different phytochemicals in the studied methanolic extract is the best reason for its prominent antioxidant property. From the quantification of flavonoid content proved that this plant extract has scientific basis in the wide use in Ayurvedic and folk medicine.

The *in vitro* cytotoxicity activities of *C. recurvata* are probably due to the presence of tannin, terpenoids or steroids. Furthermore, flavonoids such as quercetin, kaemferol and their glycosides have been shown to possess antimutagenic and antiproliferative effect in various cancer cell line 21 [19]. So, it is suggested that for specification of the mechanism of its action against brine shrimp, next study should be carried out on human cell line or on any other rodent.

5. CONCLUSION

The methanolic leave extract of *C. recurvate* is potent against rapid proliferating cell like brine shrimp; it is due to containing various antioxidative molecules. From this present work, we got a promising molecule that could be an antioxidant supplement and will help in the prevention of different oxidation induced malfunction, for sure.

COMPETING INTERESTS

Authors have declared that no competing interests exist.

REFERENCES

1. Valko M, Izakovic M, Mazur M, Rhodes CJ, Telser J. Role of oxygen radicals in DNA damage and cancer incidence. Mole Cell Biochem. 2004;266(1-2):37-56.
2. Chen X, Guo C, Kong J. Oxidative stress in neurodegenerative diseases. Neu Regen Res. 2012;7(5):376.
3. Kumarappan CT, Thilagam E, Mandal SC. Antioxidant activity of polyphenolic extracts of Ichnocarpus frutescens. Saudi J Biol Sci. 2012;19(3):349-355.
4. Qu ZQ, Zhou Y, Zeng YS, Lin YK, Li Y, et al. Protective effects of a Rhodiola crenulata extract and salidroside on hippocampal neurogenesis against streptozotocin-induced neural injury in the rat. PLoS One. 2012;7:e296415.
5. Cragg GM, Grothaus PG and Newman DJ. Impact of natural products on developing new anti-cancer agents. Chem Rev. 2009; 109(7):3012-3043.
6. Anderson JE, Goetz CM, McLaughlin JL, Suffness M. A blind comparison of simple bench-top bioassays and human tumor cell cytotoxicities as antitumor prescreens. Phyto Ana. 1991;2:107-111.
7. Reis VS, Lipp FJ. New plant sources for drugs and foods from the New York Botanical Garden Herbarium. 1982; Harvard University Press.
8. Nie Y, Dong X, He Y, Yuana T, Han T, Rahman K, Qin L, Zhang Q. Medicinal plants of genus Curculigo: Traditional uses and a phytochemical and ethnopharmaco-logical review. J Ethnopharmacol. 2013; 147:547-563.
9. Farnsworth NR. Biological and phytochemical screening of plants. J Pharma Sci. 1966;55(3):225-276.
10. Kumaran A, Karunakaran RJ. In vitro antioxidant activities of methanol extracts of five *Phyllanthus* species from India. LWT-Food Sci Tech. 2007;40(2):344-352.
11. Duh PD, Tu YY, Yen GC. Antioxidant activity of water extract of Harng Jyur (Chrysanthemum morifolium Ramat). LWT Food Sci Tech. 1999;32(5):269-277.
12. Shahidi F, Janitha P, Wanasundara P. Phenolic antioxidants. Critical Rev Food Sci Nutr. 1992;32(1):67-103.
13. Meyer BN, Ferrigni NR, Putnam JE, Jacobsen LB, Nichols DE, McLaughlin JL. Brine shrimp: A convenient general bioassay for active plant constituents. Planta Medica. 1982;45(5):31-4.

14. Vinson JA, Jang J, Dabbagh YA. Plant flavonoids, especially tea flavonols, are powerful antioxidants using an *in vitro* oxidation model for heart disease. J Agri Food Chem. 1995;43(11):2800-2802.
15. Cemeli E, Baumgartner A, Anderson D. Antioxidants and the Comet assay. Mutation Res/ Rev Mutation Res. 2009; 681(1):51-67.
16. Lopez-Lazaro M. Distribution and biological activities of the flavonoid luteolin. Mini Rev Med Chem. 2009;9(1):31-59.
17. Benavente-Garcia O, Castillo J. Update on uses and properties of citrus flavonoids: new findings in anticancer, cardiovascular, and anti-inflammatory activity. J Agri Food Chem. 2008;56(15):6185-6205.
18. Huang WY, Cai YZ, Zhang Y. Natural phenolic compounds from medicinal herbs and dietary plants: Potential use for cancer prevention. Nutr Cancer. 2009;62(1):1-20.
19. Murakami A, Ashida H, Terao J. Multitargeted cancer prevention by quercetin. Cancer Letters. 2008;269(2): 315-325.

18

A Study on the Phytochemical Analysis, Silver Nanoparticle Synthesis and Antibacterial Activity from Seed Extract of *Areca catechu* L.

P. Rama Bhat[1*], V. H. Savitri[1], P. G. Laxmi[1] and E. P. Jenitta[1]

[1]*Department of Biotechnology, Alva's College, Moodbidri – 574 227, Karnataka, India.*

Authors' contributions

This work was carried out in collaboration between all authors. Author PRB designed the study, supervised the work, managed the analyses of the study, wrote the first draft of the manuscript and edited the manuscript. Authors VHS and PGL carried out all laboratory work. Author EPJ managed the practical work, wrote the protocol and the literature search. All authors read and approved the final manuscript.

Editor(s):
(1) Mohamed Fawzy Ramadan Hassanien, Biochemistry Department, Zagazig University, Egypt.
Reviewers:
(1) Hazem Mohammed Shaheen, Damanhour University, Egypt.
(2) Carmen Lizette Del Toro Sanchez, Universidad De Sonora, Mexico.

ABSTRACT

The present study was carried out to screen phytochemicals like carbohydrates, terpenoids, resins, saponins, tannins and alkaloids in the seed extract of *Areca catechu* by standard protocols and antibacterial activities of seed extract as well as silver nanoparticles prepared from seed extracts. The characterization of silver nanoparticles was studied by using FTIR, SEM and UV- Visible spectrophotometer. Antibacterial activity was done with aqueous, methanolic and $AgNO_3$ seed extracts against five bacterial species viz., *Salmonella typhi, Klebsiella pneumoniae, Staphylococcus aureus, Pseudomonas aeruginosa* and *Bacillus subtilis* by agar well diffusion method. The seed extract showed positive results for carbohydrate, terpenoids, tannins and saponins. The colour of the solution after treating with $AgNO_3$ was changed from light brown to dark brown confirmed the reduction of silver ion in presence of plant extract and formation of silver nanoparticles. Maximum absorption was observed at 400 nm and the size of silver nanoparticles

**Corresponding author: E-mail: bhat_pr@rediffmail.com*

produced was oval in shape with the diameter of 553-610 nm. *Pseudomonas aeruginosa* was found to be the good organism resistant against areca nut seed extracts while others were showed intermediate effects.

Keywords: Areca nut; silver nanoparticles; antibacterial activity.

1. INTRODUCTION

The medicinal properties of plants have been investigated in the recent scientific developments throughout the world, due to their potent against several disease, minor side effects and economic viability. Several compounds widely distributed in plants which have been reported to exert multiple biological effect, including antioxidant, free radical scavenging abilities, anti-inflammatory and anti-carcinogenic [1,2]. Recently, natural plants have received much attention as sources of biological active substances including antioxidants. Numerous studies have been carried out on some plants, vegetables and fruits because they are rich sources of antioxidants, such as vitamin A, vitamin C, Vitamin E, carotenoids, polyphenolic compounds and flavonoids which prevent free radical damage, and reducing risk of chronic diseases [3].

The development of antibiotic resistant bacteria and the toxicity caused by prolonged drug treatment has led to the wide use of medicinal plants by the traditional medical practitioners for curing various diseases in their day to day practice [4]. Indians have long been using many of the herbal products and formulations to counter the microbial activity on open wounds and skin infections well before the inventions of antibiotics.

Now-a-days, nanotechnology has grown to be an important research field in all areas including medicinal chemistry. It is expected to revolutionize both science and society. There is growing need to develop eco-friendly and body benign nanoparticles process without use of toxic chemicals in the synthesis to avoid adverse effects in the biochemical applications [5]. Synthesis and characterization of nanoparticles is an important area of research as selection of size and shape of nanoparticles provide an efficient control over many of the physical and chemical properties.

Nanoparticles are synthesized using plant materials. The various phytochemicals present within the plant result in effective reduction of silver salts to nanoparticles [6,7]. The synthesized silver nanoparticles were used effectively against multidrug resistant bacteria. It has reported that silver nanoparticles are nontoxic to humans and most effective against bacteria, virus and other eukaryotic micro-organisms at low concentrations and without any side effects [8,9].

Research on such plants can provide useful information for their exploitation in treating diseases. Areca nut (*Areca catechu*) is one such plant which has been used in different system of traditional medication for the treatment of diseases and aliment of human being. Areca nut is used for treatment of a mental disorder called schizophrenia and eye disorder called glaucoma; as a mild stimulant and as a digestive aid and also some people used as recreational drug because it spreads central nervous system. Areca nut contains main biochemical compounds like polyphenols, fat, starch and alkaloids [10-12]. It cleanse the mouth, impart a sweet aroma to it, enhance its beauty and cleanse and strengthen the voice, tongue and teeth, the jaws and the sense organs [13]. Areca was touted as a medicine for digestive and dental health and also used to facilitate bowel movements and reduce intestinal worms [14-16,1,17].

Areca catechu L. belongs to the family Arecaceae or Palmae. The areca nut palm is cultivated as cash crop or plantation crop (Plate 1). It is the fourth most commonly used social drug, ranking after nicotine, ethanol and caffeine. It has been mentioned in the Sanskrit manuscripts and used as food, medicine, social and religious purposes. The palm is believed to have originated in the Philippines, but is widespread in cultivation for their seeds and is considered naturalized in southern China, Taiwan, India, Sri Lanka, Cambodia, Laos, Thailand, Vietnam, Malaysia, Indonesia, Bangladesh, Maldives, New Guinea, many of the islands in the Pacific Ocean, and also in the West Indies.

The aim of the present study was to carry out the phytochemical analysis, antimicrobial activity and synthesis of silver nanoparticles from the seeds of *Areca catechu*.

Plate 1. Areca nut plant bearing fruits in bunches and dehusked dried seed – chali

2. MATERIALS AND METHODS

2.1 Preparation of Extract from the Fruit Parts of the Collected Plant

The fruits from selected plant were collected from Hebri, Udupi District, and Karnataka. They were then dried, dehusked, cleaned by washing several times with deionized water and allowed it to shade dry for 7-8 days. Then it was kept in hot air oven at 60℃ for 24- 48 hours until it was dried completely. The dried seeds were powered.

2.2 Preparation of Aqueous and Methanol Extract

Five gram of plant sample was taken in a conical flask with 50 ml of distilled water to get aqueous and 25 ml of distilled water and 25 ml of methanol to get methanol extract of the sample taken separately, mixed and covered with aluminum foil. Then it was kept in water bath at a temperature of 50℃ - 60℃ for 7-8 hours, the mixture in the flask was allowed to pass through Whatman filter paper no. 1 and filtrate was put in a clean weighed china dish. This was kept in water bath open, till all the filtrate evaporated leaving behind a film of plant residue. The china dish was weighed before and after the evaporation of the filtrate, to know that how much residue was obtained. The extract was stored in screwed tubes which is stored at -20℃ to prevent the loss of bioactive compounds until further use.

2.3 Phytochemical Screening of the Plant Extract

Phytochemical tests for carbohydrates, saponins, tannins, alkaloids, terpenoids and resins were carried out on the aqueous and methanol extracts reconstituted in respective solvents using standard protocols [18,19].

2.4 Synthesis of Silver Nanoparticles

Fruit parts were collected and air dried for 2-3 days and then kept in the hot air oven at 60℃ for 24-48 hours until it is dried completely. The dried sample was grounded to fine powder using mortar and pestle. For the synthesis of silver nanoparticles, 1 Mm aqueous extract of silver nitrate was prepared; 3 g of fruit powder was mixed with 350 ml of silver nitrate solution and centrifuged at 2000 rpm for 30 minutes. The supernatant were collected and heated at 95℃. A change in the colour of the solution was observed after heating the mixture for 10 minutes. The extracts were stored at 4℃ for further use.

2.5 Characterization

The reduction of pure Ag^+ ions was monitored by measuring the UV-Vis spectrum of the reaction medium at 5 hours after diluting a small aliquot of the sample into distilled water. UV-Vis spectral analysis was done by using UV-Vis spectrophotometer. Fourier Transform Infrared Spectroscopy (FTIR) was used to recognize the functional groups bound to the silver surface and involved in the formation of silver nanoparticles. The lyophilized powder sample was examined by Infrared (IR) spectrum at the spectral range of 1000-4000 cm^{-1}. Scanning Electron Microscopic (SEM) analysis of silver nanoparticles was done using SEM machine. Thin films of the sample was prepared on a carbon coated copper grid by dropping a very small amount of the sample on the grid, extra solution was removed using a blotting paper and then the film on the SEM grid was allowed to dry by putting it under a mercury lamp for 5 minutes.

2.6 Antimicrobial Screening of the Crude Extracts

The bacterial strains used for present investigation are *Pseudomonas aeruginosa, Klebsiella pneumoniae, Bacillus subtilis, Staphylococcus aureus* and *Salmonella typhi* were obtained from P.G Department of Biotechnology, Alva's College and Alva's MLT College, Moodbidri. A loop full of culture was inoculated into nutrient broth and incubated at 37°C for 24 hours to obtain a bacterial culture. This procedure was carried for the selected bacterial cultures to obtain inoculums of particular broth cultures. The well method was employed to assay the plant materials for antimicrobial activity. Petri dishes were plated with Mueller Hinton Agar media and allowed to solidify for 30 mins. The test organisms were then spread on the surface of the media using sterile swab stick. Cork borer was used to bore wells in media. The aqueous, methanolic and silver nanoparticle extract of different concentrations (50 μl, 100 μl and 150 μl) were separately dispensed into the wells of different plates using a micropipette. A negative control of water and a positive control of Amphicillin were kept and the extract was allowed to diffuse for 30 mins at room temperature. The plates were incubated at 37°C for 24 hours and the zones of inhibition were measured.

3. RESULTS

The phytochemical screening of the extract of the *Areca catechu* seed extract showed positive result for carbohydrates, terpenoids, tannin and saponins while negative result for resin and alkaloids as showed in Table 1.

3.1 Synthesis of Silver Nanoparticles

Formation of silver nanoparticles by the reduction of silver ions during the exposure to *Areca catechu* seed extract was recorded by change in colour of the reaction mixture from light brown to dark brown (Fig. 1) after 10 minutes of incubation which indicated the formation of silver nanoparticles.

3.2 Ultraviolet- Visible (UV- Vis) Spectra Analysis

The silver nanoparticles formed in the areca nut seed extract solution exhibited an absorption spectrum ranging from 300-800 nm with a peak at 435 nm (Fig. 2).

Table 1. Phytochemical analysis of methanol and aqueous extract of the areca nut seed

Constituents	Aqueous extract	Methanol extract
Carbohydrates	+	+
Tannins	+	+
Alkaloids	-	-
Terpenoids	+	+
Resins	-	-
Saponins	+	+

+ present, - absent

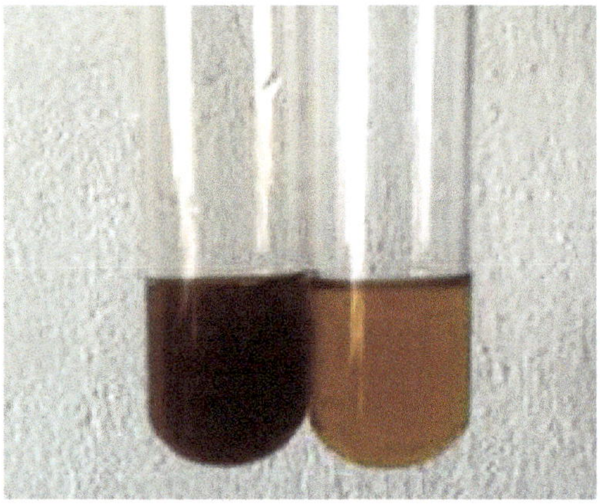

Fig. 1. Synthesis of silver nanoparticles confirmed by color change

3.3 Fourier Transform Infrared Spectroscopy (FTIR)

The spectra of nanoparticles obtained from *Areca catechu* seed extract showed absorption peaks at 3827.07 cm^{-1}, 3871.13 cm^{-1}, 1809.41 cm^{-1}, 1751.05 cm^{-1}, 2337.72 cm^{-1}, 1608.63 cm^{-1}, 1624.84 cm^{-1}, 819.76 cm^{-1}, and 867.57 cm^{-1} for different groups (Fig. 3).

3.4 Scanning Electron Microscopic (SEM) Study

The synthesized silver nanoparticles from areca nut seed extracts were oval in shape with the diameter ranged from 553 to 610 nm under 10,000X magnification (Fig. 4).

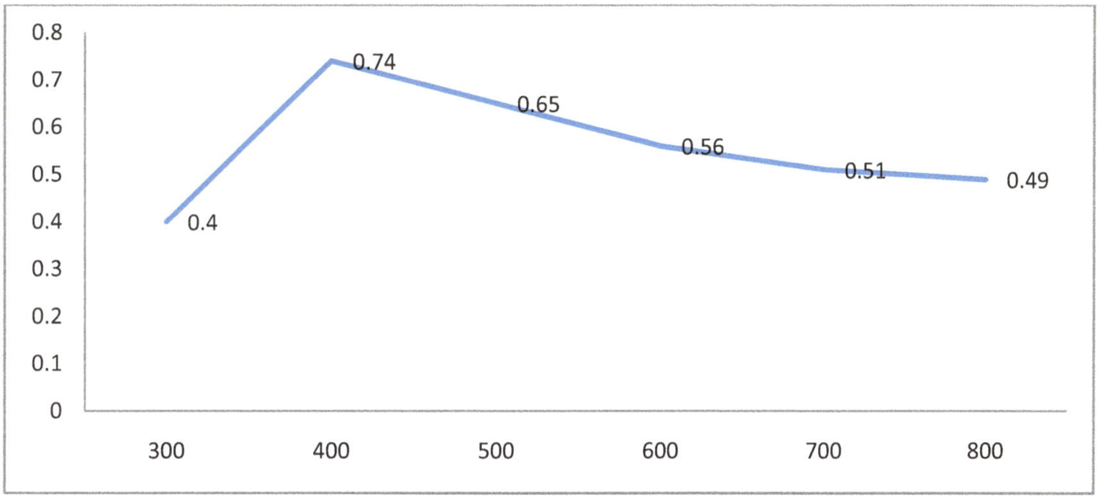

Fig. 2. UV- visible spectroscopy of areca nut silver nanoparticle extract

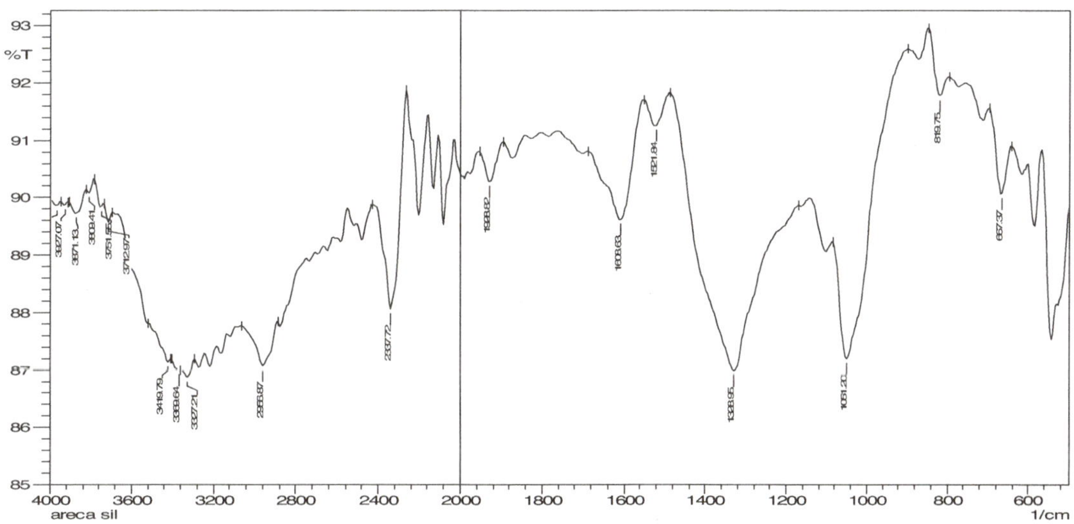

Fig. 3. FTIR analysis showing functional group spectrums

3.5 Antibacterial Activity for Different Concentrations of the Plant Extract

The antibacterial activity of aqueous, methanolic and silver nanoparticles of areca seed showed varied susceptibility and resistance. The different extract *Psuedomonas aeruginosa* showed high sensitivity for the aqueous extract of the plant i.e., the zone of inhibition for this organism was more when compared to other organisms for the aqueous extract of the plant. The *Bacillus subtilis* was less sensitive for the aqueous extract was less (Table 2). *Pseudomonas aeruginosa* was high sensitive for the methanol extract of the seed and *Bacillus subtilis* and *Klebsiella pneumoniae* showed less sensitivity compared to other organisms for the methanol extract (Table 3). *Pseudomonas aeruginosa* showed high sensitivity for the extract of silver nanoparticle synthesized from *Areca catechu. Bacillus subtilis* showed less sensitivity while *Klebsiella pneumoniae* did not exhibit any inhibition (Table 4).

4. DISCUSSION

Medicinal plants are the most productive source of new compounds and drugs of natural origin. Most of the natural products isolated from medicinal plants are the secondary metabolites, which include alkaloids, tannins, flavonoids, steroids, terpenoids, phenylpropanoids and anthraquinones Some of the products have nutritive value and antifungal and antibacterial activities. Ayurveda has related research efforts which have led to generation of enormous amount of scientific information concerning plants, crude plant extracts, and various substances from plants as medicinal agents during last 30 to 40 years. In the present investigation, the phytochemical analysis of *Areca catechu* seed extract showed presence of carbohydrates, terpenoids, tannin and saponins, and absence of resin and alkaloids by qualitative tests. In one of the earlier study the presence of phenolic constituents and antioxidant activities of areca seeds were reported [12].

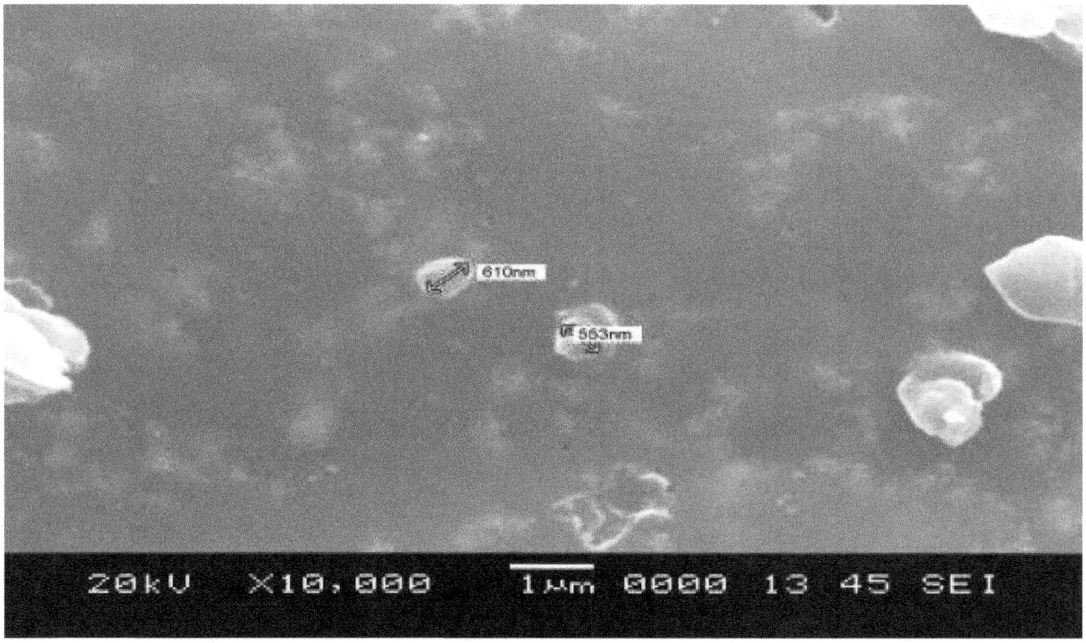

Fig. 4. Silver nanoparticles obtained from *A. catechu* seed extract under SEM

Table 2. Antibacterial activity for the aqueous extract of *A. catechu* seed

Test organisms	Zone of inhibition (mm) for different concentrations of the methanol extract		
	50 μl	100 μl	150 μl
Pseudomonas aeruginosa	13	16	17
Bacillus subtilis	10	15	15
Staphylococcus aureus	10	12	12
Klebsiella pneumoniae	09	13	13
Salmonella typhi	09	12	11

Table 3. Antibacterial activity for the methanol extracts of *A. catechu* seed

Test organisms	Zone of inhibition (mm) for different concentrations of the methanol extract		
	50 μl	100 μl	150 μl
Pseudomonas aeruginosa	5	7	8
Bacillus subtilis	3	3	5
Staphylococcus aureus	4	5	7
Klebsiella pneumoniae	2	4	5
Salmonella typhi	7	9	10

Table 4. Antibacterial activity of silver nanoparticles synthesized from *A. catechu* seed extract

Test organisms	Zone of inhibition (mm) for different concentrations of silver nanoparticles		
	50 μl	100 μl	150 μl
Pseudomonas aeruginosa	15	16	16
Bacillus subtilis	13	15	15
Staphylococcus aureus	14	15	16
Klebsiella pneumoniae	-	-	-
Salmonella typhi	12	13	14

Silver nanoparticles bear characteristic dark brown colour due to the excitation. The change in colour of the reaction mixture after 10 minutes was observed in the present study. Similarly, on other studies with the leaves of *Svensonia hyderobadensis* and stem bark of *Boswellia* [9,20] revealed the synthesis of silver nanoparticles and noted the reduction of silver ion in silver nitrate solution exposed to plant extract followed by colour. Ankanna and Savithramma [5] also reported that the stem bark of *Boswellia ovalifoliata* solution of the silver ion complex started to change the colour from yellow to dark brown colour in aqueous solution. There are few other recent works supporting the present investigation [21,22].

The synthesized nanoparticles of areca nut seed extract exhibited a maximum absorbance peak at 400 nm. Similarly, Geetha et al. [23] reported a narrow absorption band at 450 nm and Maribel et al. [24] reported that the absorption peak of silver nanoparticles by chemical reduction method was 412 nm. FTIR analysis was used for the characterization of the extract and the resulting nanoparticles. The biomolecules which are involved in the reduction of silver ions can be identified by absorption peaks in the spectral range of 1000-4000 cm^{-1}. Saifuddin et al. [25] observed the FTIR measurements to identify the possible biomolecules responsible for capping and efficient stabilization of the metal nanoparticles synthesized in leaf broth. The synthesized silver nanoparticle of areca nut seed extract was oval in shape with a diameter ranged from 553-610 nm. Christensen et al. [26] and Charusheela et al. [27] reported spherical shaped nanoparticles of 10-25 nm and Hemanth et al. [28] obtained the nanoparticles of 52-104 nm. In other studies, hexagonal shaped nanoparticles of 80-120 nm and 90-120 nm were reported [21,22]. According to Mie theory, only a single surface Plasmon resonance band is expected in the absorption spectra of spherical nanoparticles, whereas the number of peaks increases as anisotropy increases [29].

In this study on the antibacterial activity of methanol and aqueous extracts from seeds of *Areca catechu* resulted in highest inhibitory zone with *Pseudomonas aeruginosa* and minimum zone of inhibition by *Klebsiella pneumoniae*. Chin and Fernandez [16] examined the antimicrobial performance of methanolic extract of *Areca catechu* seeds against mixed oral flora from tooth scum and Gram negative laboratory isolates. Varying concentrations of *A. catechu* ethanol extract was tested for antimicrobial activity against 0.5 Mc Farl and of mixed oral flora and eight Gram negative clinical isolates by agar well diffusion method. All concentration showed to inhibit oral flora models with zone of inhibition about 7-18 mm. In the present study of silver nanoparticles synthesized using areca nut seed extract exerted a significant antibacterial activity compared with positive control and other bacterial strains which symbolized that the extract have antibacterial activity. By increasing the dose of the extract higher inhibitory zone was found. The antibacterial activity can also be compared with other silver nanoparticles synthesized from leaf extracts of *Svensonia*, which also showed activity against *E. coli, Bacillus subtilis, Staphylococcus aureus* and *Klebsiella pneumoniae* [9]. Maria et al. [15] also reported the antimicrobial effect of areca nut extract on oral pathogens which are of other bacterial strains but *Candida albicans* showed positive inhibitory effect. On the other hand, Marium and Tuslima [17] obtained inhibitory effect against Gram negative bacteria when carbon tetrachloride extract was used compared with methanol extract of areca seeds. Similarly, Kabbashi [30] tested ethanolic and methanolic fruit extract of *Balanites aegyptiaca* against two Gram positive, two Gram negative and two fungal species, among which the methanolic extract exhibited high activity only against *Aspergillus niger*, *Bacillus subtilis* and *Staphylococcus aureus*, whereas it was an intermediately active against, *E. coli, Candida albicans* and, *Pseudomonas aeruginosa*. He concluded that the ethanolic extract proved to be had high activity on all bacteria and fungi.

Senthil et al. [31] and Kumar et al. [1] reported an overview of many phytochemical characteristic of areca nut seed and therapeutic effect on various disease conditions on human beings. Areca nut seed biochemical compounds have been recently recognized as functionally active molecules, possessing antioxidant, antidiabetic, antiallergic and other useful properties, as well as exert protective effects against cardiovascular and other diseases.

5. CONCLUSION

Being an important cash and commercial crop, the areca nut has got multifacinating medicinal properties. The present study confirmed the silver nanoparticle formation using the areca nut seed extract and its antimicrobial properties against selected bacterial species, some of which are human pathogens. Further study required to use these nanoparticles for the treatment of disease using animal models.

ACKNOWELEDGEMENTS

The authors are grateful to Dr. Satyamoorthy, Director, School of Life Sciences, Manipal University, Manipal for facilities provided to lyophilizing the samples.

COMPETING INTERESTS

Authors have declared that no competing interests exist.

REFERENCES

1. Kumar AM, Kota Sravani, Veena KM, Rao JPK, Laxmikanth Chatra, Prashanth Shenai, Prabhu RV, Tashika Kush Raj, Pratima Shetty, Shaul Hameed. Areca nut and its effects on the human body. Am J Oral Medicine Radiol. 2015;2(1):10-14.
2. Rao GB, Rao PR, Murty P, Rao ES, Madhukiran P, Rao TM. Investigation on regional variation in total phenolic, alkaloid content and *in vitro* antioxidant activity *of Leucas aspera.* Internatl J Pharmace Sci Res. 2011;2(10):2699-2703.
3. Prasad K, Bisht G, Evaluation of nutritive, antioxidant and mineral composition of *Pavetta indica* L. leaves. Res J Phytochem. 2011;5:54-59.
4. Kaur R, Meena AK, Singh B, Yadan AK, Singh U, Sachan A Pal. Review on anyifungal activity of ayurvedic medicinal plants. Drug Investigation Today. 2009; 2(2):146-148.
5. Ankanna S, Savithramma N, Production of biogenic silver nanoparticles using *Boswellia ovalifoliolata* stem bark. Digest J Nanomat Biostruct. 2010;5:369-372.
6. Ahmad B, Ali J. Physiochemical, minerals, phytochemical contents, antimicrobial activities evaluation and Fourier Transform Infrared (FTIR) analysis of *Hippophae rhamnoides* L. leaves extracts. Afr J Pharmacy Pharmacol. 2008;7(7):375-388.
7. Loo Y, Buong W, Mitsuaki N, Radu S. Synthesis of silver nanoparticles by using tea leaf extract from *Camellia sinensis.* Internatl J Nanomed. 2012;7:4263-4267.
8. Kumar A, Naidu LM. Antibacterial potential of *Tridax procumbens* against human pathogens. An Internatl J Pharmace Sci. 2003;2(2):21-27.
9. Savithramma N, Linga Rao M, Rukmini K, Devi PS. Antimicrobial activity of silver nanoparticles synthesized by using medicinal plants. Internatl J Chem Tech Res. 2011;3:1394-1402.
10. Wang CK, Lee WH, Peng CH. Contents of phenolics and alkaloids in *Areca catechu* L. during maturation. J Agric Food Chem. 1997;45(4):1185-1188.
11. Winstock A. Areca nut- abuse liability, dependence and public health. Addiction Biol. 2002;7(1):133-138.
12. Wei MZ, Jing W, Wen-xue C, Hai-DZ. The chemical composition and phenolic antioxidants of Areca (*Areca catechu* L.) seeds. International Conference on Agricultural and Biosystems Engineering, Advances in Biomedical Engineering. 2011;1-2:16-22.
13. Krishnamurthy KH. Supari or betel nut (*Areca catechu* L.). Namah J New Approach Med Health. 2009;17(3):19-26.
14. Priyanka RP, Sachin UR, Burade KB. Pharmacological activities of *Areca catechu* L. – A Review. ISSN. 2009;0974-6943.
15. Maria B, Vidya P, Ipe V. Antimicrobial properties of *Areca catechu* (Areca nut) husk extract against common oral pathogens. Internatl J Res Edu Psyc. 2012;3(1).
16. Chin A, Fernandez CD. Antimicrobial performance of ethanolic extract of *Areca catechu* L. seeds against mixed oral flora

from tooth scum and gram negative laboratory isolates. Int J Res Ayurveda Pharma. 2013;4(6).

17. Marium B, Talsima B. Potential *in vivo* evaluation of analgesic investigation of *Mangifera indica* and antimicrobial activity of *Areca catechu*. Am J Bio Sci. 2015; 3(1-2): 19-22.
18. Fernandez L, Daruliza K, Sudhakaran S, Jegathambigai R. Antimicrobial activity of the crude extract of *Piper sarmentosum* against methicillin-resistant *Staphylococcus aureus* (MRSA), *Escherichia coli, Vibrio cholera* and *Streptococcus pneumoniae*. European Rev Medical Pharmacol Sci. 2010;16(3):105-111.
19. Chandrappa CP, Shilpashree CB, Karthik MR, Govindappa M, adananda TS. Antibacterial and antioxident activities of *Adiantum pedatum* L. J Phytol. 2011;3(1): 26-32.
20. Baishya D, Sharma N, Bora R. Green synthesis of silver nanoparticles using *Brophyllum pinnatum* and monitoring their antibacterial activity. Arch Appl Sci Res. 2012;4:2098-2104.
21. Lakshmi N, Jenitta Emima Packiyam, Rama Bhat P, Jayadev K. Studies on phytochemical, synthesis of silver nanoparticles from the floral extract of *Mesua ferrea* L. and their antibacterial activity. Internatl J Green and Herbal Chem. 2014;3(4):1730-1739.
22. Prathibha S, Jenitta Emima Packiyam E, Rama Bhat P, Jayadev K, Shrinidhi Shetty. Green synthesis of silver nanoparticles from fruit extracts of *Terminalia chebula* Retz. and their antibacterial activity. Internatl J Res Biosci. 2015;4(2):29-35.
23. Geetha N, Harini K, Biofabrication of nanoparticles using leaf extract of *Chromolaena odorata*. Internatl Conf Biol Sci. 2012;56-58.
24. Maribel G, Dille J, Godet S. Synthesis of silver nanoparticles by chemical reduction method and their antibacterial activity. Internatl J Chemical Biol Engg. 2009; 2:104 -110.
25. Saifuddin N, Wong CW, Yasumira NA. Rapid biosynthesis of silver nanoparticles using culture supernatant of bacteria with microwave irradiation. J Chem. 2009;6:61-70.
26. Christensen L, Vivekanandhan S, Misra M, Mohanty AK. Biosynthesis of silver nanoparticles using *Murraya koenigii* (curry leaf), An investigation on the effect of broth concentration in reduction mechanism and particle size. Adv Mat Lett. 2011;2:429-43.
27. Charusheela R, Tapan C, Sarangi. Synthesis of silver nanoparticles from the aqueous extract of leaves of *Osimum sanctum* for enhanced antibacterial activity. J Chem. 2013;1-7.
28. Hemanth NKS, Karthik L, Bhaskara RKV, Extracellular biosynthesis of silver nanoparticles using the filamentous fungus *Penicillium* sp. Arch Appl Sci Res. 2010; 2(6):161-167.
29. Rajesh W, Jaya LR, Niranjan KS, Vijay MD, Sahebrao B. Phytosynthesis of silver nanoparticles using *Gliricidia sepium*. Current Nanosci. 2009;5:117-122.
30. Kabbash AS. Antigiardial, antioxidant, antiamoebic, antimicrobial activity, cytotoxicity and phytochemical of ethanolic fruits extract of *Balanites aegyptiaca* (L.) Del. from Sudan. W J Pharmace Res. 2015;4(3):01-21.
31. Senthil AM, Hazeena BV. A review on phytochemical and pharmacological potential of *Areca catechu* L. seed. Internatl J Pharmace Sci Res. 2012;3(11): 2249-3379.

19

Biodiesel Production from Selected Used Cooking Oils and Animal Fats and Its Implementation

A. B. M. Sharif Hossain[1,2*] and Mohammed Saad AlEissa[3]

[1]Department of Biology, Program of Biotechnology, Faculty of Science, University of Hail, KSA.
[2]Program of Biotechnology, ISB, Faculty of Science, University of Malaya, Kl, Malaysia.
[3]Department of Biology, Faculty of Science, Al-imam Muhammad Ibn Saud Islamic University, KSA.

Authors' contributions

This work was carried out in collaboration between both authors. Author ABMSH designed the study, wrote the protocol and supervised the work. Authors ABMSH and MSA carried out all laboratories work and performed the statistical analysis. Author MSA managed the analyses of the study. Author ABMSH wrote the first draft of the manuscript. Author ABMSH managed the literature searches and edited the manuscript. Both authors read and approved the final manuscript.

Editor(s):
(1) Hector M. Mora Montes, Department of Biology, University of Guanajuato, Mexico.
Reviewers:
(1) Danbature Wilson Lamayi, Gombe State University Gombe, Nigeria.
(2) E. I. Bello, The Federal University of Technology, Akure, Nigeria.
(3) Anonymous, Selcuk University, Turkey.
(4) Klaus Doelle, State University of New York, USA.

ABSTRACT

Significance of the Study: Biomass derived bio-fuel is biodegradable, nontoxic, sustainable and substitute for fossil fuel as well as capable of removing emission like CO_2, CO, SOx, HC and NO_X. It is renewable and outstanding energy resource for the generation of steam and electricity, transportation fuel, manufacturing industries. Biomass (animal and plants based like, fruits, vegetable, crops, fish, chicken and other animal byproducts or waste biomass) can be used for bioenergy production like biofuel and nanocatalyst for biofuel. Agro waste like animal waste biomass can be the sources of biodiesel in the industry. The energy is an important factor for the development of social and economic development of any country. In recent years, using renewable energy as a reducing agent of environmental pollution has become an important

**Corresponding author: E-mail: abm.hossain@uoh.edu.sa*

issue in the whole world.
Aim: In this study it is highlighted that biofuel can be produced from from waste cooking oil, fish and chicken byproduct.
Results: The highest yield of fatty acid methyl esters (FAME) was found in fish oil (99.8%). However, the 2nd highest yield o biodiesel was found in chicken oil (99.2%). The kinematic viscosity at 40℃ accepted the limited value between 1.9 to 6. 0 cSt according to the ASTM D445. In addition, TAN value was followed the ASTM D664 standard. Fuel consumption, CO, HC and NOx were lower in fish and chicken biodiesel than palm and sunflower.
Conclusion: The less expensive waste feedstock can be used in the commercial and industrial purposes in future for vast energy supply.

Keywords: Bienergy; biofuel; waste cooking oil; fish; chicken byproduct.

1. INTRODUCTION

The occurrence of oil depletion, global warming and the greenhouse effect has put us on an alarming condition where we need to search for an alternative energy. Hydro, wind, solar, biomass, hydrogen and geothermal energy are the types of natural energy that are readily available and sustainable [1].

Among all the renewable energies, biofuel (biodiesel) is one of the most promising energy sources despite diesel-fuel. Biodiesel is the mono-alkyl esters of fatty acids derived from animal fats and vegetable oils. It is produced when vegetable oil or animal fat is reacted chemically with an alcohol to produce fatty acid alkyl esters (transesterification) [2,3]. Biodiesel has a low emission profile and are environmental friendly, hence not posting dangers to our vulnerable earth. The reason that biodiesel is not utilized widely around the world is due to the high cost of raw materials. To overcome this, less expensive feedstock such as waste cooking oil can be a resort of low quality oils that are produced in excess in food processing industries. By utilizing these wastes oil could help to overcome the problem of waste oils disposal [4].

The high diesel fuel price today made biodiesel an attractive alternative fuel. Biodiesel has been chosen as one of the main alternative fuel because of its better characteristics and advantages such as: a renewable source, highly biodegradable, high flash point, excellent lubricity, nontoxic, relatively low amount of polycyclic aromatics hydrocarbons as domestically available. That is why this environmental friendly biofuel can be utilized in engines without any modification [5]. It is also reported that bio-energy can be produced from starch, cellulose and lignocelluloses based vegetable and animal waste and byproducts by fermentation process using different enzyme [6].

Few researches have been carried out on the production of biodiesel from vegetable oil, animal fats and byproducts as well as fish fats. Therefore, the objectives of this present research were to produce biodiesel from waste cooking oil, fish and chicken fats and to investigate its characteristic and its application in engine emission.

2. MATERIALS AND METHODS

2.1 Preparation of Oil Samples

Fish and chicken byproducts, waste cooking sunflower and palm oil were used in this study. Waste sunflower and palm oil were collected from restaurants. Waste fish and chicken byproducts were collected from the fishes processing market and chicken slaughtered house, Kuala Lumpur, Malaysia. Firstly, the fish and chicken fats were heated at 80℃ and melted it to fish and chicken oil (Fig. 1). Secondly, filtration was done by utilizing a filter funnel with a piece of filter paper, (whatman filter paper from Sigma-Aldrich, product no. Z240249).

2.2 Biodiesel Production through Transesterification Reaction

NaOH (1%) were used as a catalyst to determine its effect on biodiesel yield. The filtered palm, sunflower, fish and chicken oil was preheated in the oven for approximately 15 minute at 60℃ before subjecting it to the transesterification process. When the transesterification process was completed, the solution was separated into two distinct layers of crude biodiesel and glycerol. distinct layers (Fig. 1). The methyl esters obtained was washed with distilled water to remove impurities such as pigments, excess alcohol, excess catalysts, glycerol, and soaps.

(A) (B) (C) (D) (E)

Fig. 1. A) Straight and waste sunflower oil, B) filtered waste sunflower cooking oil, C) fish oil and D, E) chicken fats

2.3 Drying Process

Little quantity of water was eliminated when biodiesel was dried in the oven. After drying, water was eliminated and biodiesel was produced and used for subsequent gas chromatography-mass spectrometer analysis.

3. BIODIESEL CHARACTERIZATION

3.1 Biodiesel Analysis

3.1.1 Gas-chromatography-mass spectrometry (GC-MS) analysis

The specific component of methyl esters in the biodiesel were determined using an Agilent 6890 Gas Chromatography installed with a mass spectrometry detector. A capillary column, (length: 30 m, film thickness: 0.25 m and ID: 0.25 mm) was used. Helium was used as the carrier gas. One micro-liter of the biodiesel sample was injected manually. Samples were introduced to the column via an inlet. Once in the inlet, the heated chamber vaporizes the biodiesel and a constant flow of helium moves through the inlet. A portion of the helium flow acts to transport the sample into the column. Another portion of the helium flow gets directed to purge the inlet of sample following injection. Yet another portion of the flow is directed through the split vent in a split ratio.

3.1.2 Characterization of the biodiesel

The produced biodiesel under optimum conditions was sent to the Engine Tribology Laboratory of Department of Mechanical Engineering to determine the properties of the biodiesel.

3.2 Viscosity

Samples of the biodiesel were sent to Tribology Laboratory at Faculty of Engineering, UM to determine the viscosity. Using ASTM D445, viscosity of all samples were analyzed by heating up at 40℃ and then measured using Houillon viscometer set at rpm 30 and spindle size 63 (ASTM Inc. 2005) [7].

3.3 Total Acid Number (TAN)

Total acid number was measured by using the amount of potassium hydroxide (KOH; mg) that was needed to neutralize the acids in one gram of oil.

For all samples by using ASTM D664 standard (ASTM Inc., 2006) [8].

3.4 Engine Exhaust Emission Test

The exhaust emission test was also conducted to evaluate the exhaust emission characteristics of the biodiesel. For these, the 2.50 liters of biodiesel was required to perform the engine emission test. The BOSCH gas analyzer model EET 008.36 was used to measure carbon monoxide (CO) and unburned hydrocarbon (HC). Whereas, the Bacharach model CA300NSX analyzer was used to measure NO_x concentration. The measurement of CO and NO_x were according to SAE J117 June'95 standards and HC to SAE J215 March'95 standards. Fuel consumption of the biodiesel was also measured and compared with the fuel consumption of normal petro diesel. The model of the engine was YANMAR TF120-M. This engine is horizontal, water-cooled and single cylinder. The accumulation of deposit was carried out for 8 hours at 2000 rpm constant engine speed and 15 Nm load for each test fuel.

3.5 Metal Analysis by Multielements Oil Analyzer

Samples of biodiesel were sent to the Tribology Laboratory at Faculty of Engineering, UM. By using multi element oil analyzer (MOM II) as ASTM D800 standard, lube oil analysis was conducted to all samples to determine the metal content P, K, Ca, Mg, S, Cu, Pb, and Fe in biodiesel (ASTM Inc., 2006) [9].

4. RESULTS AND DISCUSSION

It has been seen that there are a total 7, 6, 5 and 4 compounds that are present in the biodiesel samples produced from fish, chicken, palm and sun flower waste, respectively Table 1). The highest yield of fatty acid methyl esters (FAME) was found in fish oil (99.8%). However, 2nd highest was found in chicken oil (99.2%).

From Table 2, the viscosity of biodiesel produced from transesterification of waste sunflower oil was the highest, 5.7 cSt. The kinematic viscosity at 40℃ accepted the limited value between 1.9 to 6.0 cSt according to the ASTM D445. TAN value was followed the ASTM D664 standard.

The result shown in Table 3 showed that only all elements value maintained the limits of ASTM D6751 ≤5 ppm for Na, Mg, and Ca ≤10 ppm for P, Pb lead≤5, cu ≤5. This result was similar to the result found by [10].

Table 4 shows the specific fuel consumption (SFC) in ml/sec. Fuel consumption, CO, HC and NOx were lower in fish and chicken biodiesel than palm and sunflower. Carbon monoxide concentration from fish oil derived biodiesel was much lower compared to other samples because, they were possessed a lower acid and viscosity value. The amount of NOx emission is directly related to the exhaust gas temperature and engine combustion chamber temperatures.

Table 1. Identification of methyl esters in fish, chickrn, palm and sunflower based biodiesel

Total peak for different samples	Peak serial	RT (min)	Methyl ester	% relative	Total % biodiesel yield number
Fish waste	1	12.1	$C_{13}H_{24}O_2$ CH_3	9.9	99.8
	2	16.4	$C_{13}H_{30}O_2$ CH_3	6.8	
	3	20.1	$C_{12}H_{30}O_2$ CH_3	5.6	
	4	20.5	$C_{17}H_{34}O_2$ CH_3	19.7	
	5	23.7	$C_{19}H_{34}O_2$ CH_3	14.2	
	6	23.8	$C_{19}H_{36}O_2$ CH_3	40.6	
	7	24.3	$C_{19}H_{36}O_2$ CH_3	3.0	
Palm oil	1	20.1	$C_{10}H_{29}O_2CH_3$	2.1	98.4
	2	20.5	$C_{16}H_{31}O_2CH_3$	36.1	
	3	23.7	$C_{18}H_{31}O_2CH_3$	11.1	
	4	23.8	$C_{18}H_{33}O_2CH_3$	44.2	
	5	24.3	$C_{18}H_{35}O_2CH_3$	5.1	
Sunflower	1	20.5	$C_{16}H_{31}O_2CH_3$	6.1	98.6
	2	23.7	$C_{18}H_{31}O_2CH_3$	48.2	
	3	23.8	$C_{18}H_{33}O_2CH_3$	42.1	
	4	24.3	$C_{18}H_{35}O_2CH_3$	2.2	
Chicken oil	1	12.2	$C_{13}H_{24}O_2$ CH_3	9.8	99.2
	2	16.0	$C_{13}H_{30}O_2$ CH_3	6.5	
	3	20.2	$C_{12}H_{30}O_2$ CH_3	5.3	
	4	23.6	$C_{19}H_{34}O_2$ CH_3	28.4	
	5	23.1	$C_{19}H_{36}O_2$ CH_3	44.0	
	6	24.2	$C_{13}H_{36}O_2$ CH_3	5.2	

Table 2. Determination of acid value and viscosity of different waste biodiesel

Property	Unit	ASTM value	Fish waste	Chicken waste	Palm waste	Sunflower waste
Acid value	mg KHO/g oil	0.50 max	0.42	0.41	0.43	0.44
Viscosity	cSt at 40℃	1.9-6.0	5.4	5.3	4.42	5.7

Table 3. Elements present in different biodiesel compared to ASTM standard

Biodiesel sample	P (PPM)	Ca (PPM)	Mg (PPM)	Na (PPM)	Pb (PPM)	Cu (PPM)
Pure diesel	10	5	5	5	4	4
Fish biodiesel	8	2	3.5	1.5	3.5	3
Palm biodiesel	9	5	5 .5	6.0	4	4
Sunflower	9.5	5	5	5.1	4	4
Chicken waste	8.5	2.5	4.0	2.0	3.5	3

Table 4. Fuel consumption test of different biodiesel

Biodiesel sample	Load	Speed	Fuel consumption (ml/sec)	CO emission (%)	NOx emission (PPM)	HC emission (PPM)
Pure diesel	15 Nm	2000 rpm	0.588	0.030		28
Fish biodiesel	15 Nm	2000 rpm	0.581	0.025		17
Palm biodiesel	15 Nm	2000 rpm	0.613	0.029		25
Sunflower	15 Nm	2000 rpm	0.612	0.028		26
Chicken waste	15 Nm	2000rpm	0.58	0.026		20

As the exhaust gas temperature increases, NOx emission also increases. Hence, biodiesel fuelled engine had a higher potential to emit more NOx because of the exhausted gas temperature lower as the biodiesel concentration increased [11,12].

5. CONCLUSION

Fish and chicken waste oil showed the higher yield than palm and sunflower biodiesel and the quality of biodiesel produced was the best according to the engine test. Since fish and chicken waste biodiesel showed the good results, It can be the sources of less engine emission like HC, CO.

COMPETING INTERESTS

Authors have declared that no competing interests exist.

REFERENCES

1. Demribas A. Biodiesel production from vegetable oils via catalytic and non-catalytic supercritical methanol transesterification methods. Progress in Energy and Combustion Scienc. 2005;31: 466-487.
2. Hossain ABMS, Boyce AN. Biodiesel production from waste sunflower cooking oil as an environmental recycling process and renewable energy. Bulgarian Journal of Agricultural Science. 2009a;15(4):313-318.
3. Hossain ABMS, Boyce AN. Comparative study of biodiesel production from pure palm oil and waste palm oil. Arab Gulf Journal of Scientific Research. 2009; 27(1 & 2):33-38.
4. Zhang Y, Dube MA, McLean DD, Kates M. Biodiesel production from waste cooking oil: 2: Assessment. Bioresoure Technology. 2003;89:1-16.
5. Hossain ABMS, Mazen A. Effects of catalyst types and concentrations on biodiesel production from waste soybean oil biomass as renewable energy and environmental recycling process. Australian Journal of Crop Science. 2010; 4(7):550-555.
6. Hossain ABMS. Biotechnology for biodiesel fuel and bioengineering approaches. LAP Lambart Academic publishing Co. Paperback, Germany. ISBN No. 97838473349998. 2012;150.
7. ASTM International. ASTM D800 for standard test method of chemical analysis of industrial metal cleaning compositions. 100 Barr, Harbor Drive, C700, PA, USA; 2005.
8. ASTM International. ASTM D664 for standard test method of acid Number of Petroleum Products by Potentiometric Titration. 100 Barr, Harbor Drive, C700, PA, USA; 2006.
9. ASTM International. ASTM D445 for standard test method of kinemetric viscosity of transparent and opaque liquid. 100 Barr, Harbor Drive, C700, PA, USA; 2006.

10. Wisniewski AJ, Wiggers VR, Simionatto EL, Meier HF, Barros LAS, Madureira. Biofuels from waste fish oil pyrolysis: Chemical composition. Fuel. 2010;89:563–568.
11. Cherng-Yuan L, Rong-Ji L. Engine performance and emission characteristics of marine fish-oil biodiesel produced from the discarded parts of marine fish. Fuel Processing Technology. 2009a;90:883–888.
12. Cherng-Yuan L, Rong-Ji L, Fuel properties of biodiesel produced from the crude fish oil from the soapstock of marine fish. Fuel Processing Technology. 2009b;90:130–136.

20

Biochemical Assessment of the Effects of Aqueous Stem Bark Extract of *Anogeissus leiocarpus* on Wistar Strain Rats

M. I. Bello[1*], M. A. Madusolumuo[1] and I. Igbokwe[2]

[1]Department of Biochemistry, Modibbo Adama University of Technology, P.M.B. 2076 Yola, Nigeria.
[2]Department of Veterinary Medicine, University of Maiduguri, P.M.B. Maiduguri, Nigeria.

Authors' contributions

This work was carried out in collaboration between all authors. Author MIB designed the study and carry out the laboratory studies, Author MAM supervised the work and Author II carried out the histopathological analysis of the work. All authors read and approved the final manuscript.

Editor(s):
(1) Mohamed Fawzy Ramadan Hassanien, Biochemistry Department, Zagazig University, Egypt.
Reviewers:
(1) Miloslav Milichovsky, University of Pardubice, Czech Republic.
(2) Anonymous, China Medical University, Taiwan.

ABSTRACT

Aqueous stem bark extract of *Anogeissus leiocarpus* was phytochemically screened indicating the presence of alkaloids, tannins, anthraquinones, saponins and phenols. Antimicrobial activity of the aqueous extract using *Escherichia coli, Streptococcus pneumoniae, Staphylococcus aureus* and *Klebsiella pneumoniae* as test organisms showed varying zones of inhibition with 10 mm zone of inhibition exhibited by *Staphylococcus aureus* and *Escherichia coli* respectively. Minimum Inhibitory Concentration (MIC) and Minimum Bactericidal Concentration of extract were 16 µg/ml and 32 µg/ml respectively. Treatment of experimental animals with the aqueous stem bark extract for a period of six weeks and subsequent assessment of its effect biochemically revealed that alanine aminotransferase (ALT), aspartate aminotransferase (AST), acid phosphatase (ACP), and lactate dehydrogenase (LDH) levels were significantly higher than normal suggesting certain degree of liver injury. Glucose, cholesterol, total and direct bilirubin as well as total protein levels were significantly higher than normal values with respect to animals treated with 500mg/kg body weight

**Corresponding author: E-mail: modibboabubakar@gmail.com*

suggesting a dose dependent effect of the extract on the animals. Urea and creatinine levels were slightly lower than normal. Overall, the result of this study showed that the effect of the extract was more pronounced on the liver than the kidney as depicted by the photomicrographs.

Keywords: Stem bark; test organisms; zone of inhibition; dose dependent; photomicrographs.

1. INTRODUCTION

Trees, shrubs and herbs have commonly been used in traditional medicine especially in developing countries [1]. In many cultures there is this popular belief that 'every plant grown on the surface of the earth has a medicinal property or use". However, it is impossible to provide a complete list of these uses to which most of these plants may be put [2]. Despite all the progress in synthetic chemistry and biotechnology, plants are still an indispensable source of medicinals, both preventive and curative. Hundreds of species of plants are recognized as having medicinal value, and many of them are commonly used to treat or prevent ailments and diseases. The medicinal properties of plants are often associated with one or all of the parts of the plant that include bark, leaf, flower, fruit or seed [3].

1.1 *Anogeissus leiocarpus* (Guill. and Perr.)

Local names: Marke (Hausa); Kojoli (Fulfulde); Annum (Kanuri); Aina orin odon (Yoruba); Atara (Igbo). This plant belongs to the family Combretaceae.

A very graceful tree plant of up to 28 m height occurring in most of the savanna regions from the driest areas to the borders of the forest zone. It is easily recognized by the delicate pale green leaves, rather distantly spaced along slender and often drooping branchlets; the pale coloured bark often flaking off in very thin and conspicuous patches; and the terminal and axillary flower heads, about 8 mm across, develop into persistent cones about 12 mm across.

The branches of the tree are in ascending manner usually from lowdown and often drooping at the ends. The bark is grey or pale brown and is darker with age, peeling off in very thin patches which often curl at the edges. The stem exudes a brown gum.

Fruiting period of the plant is from August to January. The fruits have scale-like appearance with a diameter of about 6mm across. The fruits have 2 wings and a sharp beak, collected in a rather prickly dark brown head. The wood is greenish outside, dark brown at the heat and very hard [4].

A decoction of the leaves of *Anogeissus leiocarpus* is used locally in North Eastern Nigeria for the treatment of skin diseases, itching and anti diarrhoea. The powder stem bark is applied to wound and ulcer to quicken healing. An infusion of the stem bark is given as febrifuge [5].

2. MATERIALS AND METHODS

Stem bark sample of *Anogeissus leiocarpus,* was collected from within the premises of Modibbo Adama University of Technology Yola, Nigeria. The plant was identified at the Forestry Department of School of Agriculture and Agricultural Technology, Modibbo Adama University of Technology Yola, Nigeria.

The plant specimen was washed with tap water and then dried in an oven at 50°C over a 72 h period. The dried sample was cut into small pieces with a clean knife and then ground to powder using mortar and pestle and then put in sterile polyethylene bags for further use.

Male Wister strain rats weighing between 100–150 g were purchased from the animal house National Veterinary Research Institute Vom in Jos, Plateau State Nigeria. The animals were stabilized using standard laboratory chaw.

Bacterial and fungal specimens were obtained from the microbiology laboratory of the Pathology Department of the Federal Medical Centre Yola Adamawa State. The test organisms were *Klebsiella pneumoniae* (Gram-negative rod), *Escherichia coli* (Gram-negative rod), *Neisseria gonorrhoeae* (Gram-negative cocci), and *Staphylococcus aureus* (Gram positive cocci). The fungal specimen is *Candida albicans.* The test organisms were kept in slants of nutrient agar.

Fifty (50 g) grammes of powdered stem bark were soaked in 200 (ml) of sterile distilled water, shaken and allowed to stand for 24 hours. The mixture was filtered using whatman No.1 (115 cm) filter paper. The filtrate was concentrated over a water bath at 50°C. Dried extract was kept in sterile universal bottles and stored in a refrigerator for further use.

Twenty-five rats weighing 100 to 150 g were randomly divided into five groups of five rats each. Group 1 was the normal/control and groups 2, 3, 4, and 5 were administered daily with graded doses of 200, 300, 400 and 500 mg/kg body weight of the aqueous stem bark extract of *Anogeissus leiocarpus* by incubation. Treatment was done for the period of six weeks. Experimental animals were fed *ad libitum* with laboratory chaw throughout the period of the study.

The animals anaesthetised with diethyl-ether were sacrificed and the blood collected by cardiac puncture. The blood was allowed to clot, centrifuged for 10 minutes at 3000 rpm in a desktop centrifuge. Serum was collected carefully with Pasteur pipette and transferred into clean sample bottles. The serum was used to determine serum levels of AST, ALT, LDH, ACP, Alkaline Phosphatase, glucose, cholesterol, Bilirubin (Total and Direct bilirubin), albumin, total protein, urea and creatinine. Livers of all the rats were removed and washed in 10% formal-saline and then fixed in fresh 10 % formal saline. Livers of rats belonging to the same group were put in the same container.

2.1 Test for Alkaloids

Test for alkaloids was carried out according to the methods described by Harborne [6] and Trease and Evans [7]. Zero point five (0.5 g) grammes of each extract was stirred with 5 ml of 1% aqueous hydrochloric acid on a steam bath before being filtered and 1 ml of the filtrate was treated with a few drops of Mayer's reagent. A second 1 ml portion was treated similarly with Dragendrorff's reagent. Turbidity or precipitation with either of these reagents was an evidence of the presence of alkaloids.

2.2 Test for Saponins

Test for saponins was done according to the method described by Wall et al. [8]. Zero point five (0.5 g) grammes of the extract was shaken with water in a test tube and frothing which persisted on warming indicated the presence of saponins. This is called frothing test. To confirm the presence of saponins, 0.5 g of the extract was boiled briefly with 50ml phosphate buffer pH 7.4. The mixture was allowed to cool and after filtration about 5ml of the filtrate was passed (after 3 hours) through an asbestos disc (1.5 mm thick and 7 mm in diameter) which had been previously soaked in 3 drops of 1% cholesterol in ether and dried. The disc was washed with 0.5 ml distilled water. After drying, it was heated in 20 ml of boiling oxylol for 2 hours in order to decompose the complex formed between cholesterol and any saponins. This disc was then washed in ether and after drying, it was placed on a 7% blood nutrient agar. Complete haemolysis of red blood cells around the disc after 6 hours was evidence of the presence of saponins.

2.3 Test for Tannins

Test for tannins was carried out according to the method described by Trease and Evans [7]. Five (5 g) grammes of each portion of plant extract was stirred with 10ml of distilled water and then filtered and ferric chloride reagent added to the filtrate. A blue-black, green or blue-green precipitate indicated the presence of tannins.

2.4 Test for Phlobatannins

Test for phlobatannins was carried out according to the method described by Trease and Evans [7]. Five (5 ml) millilitres of the aqueous extract was mixed with 5 ml of 1% aqueous hydrochloric acid and boiled. The appearance of a reddish green colour indicated the presence of phlobatannins.

2.5 Test for Anthraquinones

Test for anthraquinones was carried out according to the method described by Trease and Evans [7]. Borntrager's test was used for the detection of anthraquinones. Five (5 g) grammes of each plant extract were shaken with 10 ml benzene. After filtration, 5 ml of 10% ammonia solution was added to the filtrate, the mixture was shaken and the appearance of a pink, red, or violet colour in the ammoniacal (lower) phase indicated the presence of free anthraquinones.

For combined anthraquinones, 5 g of each plant extract was boiled with 10 ml aqueous Sulphuric acid and filtered while hot. The filtrate was shaken with 5ml benzene and the benzene layer was separated and half its own volume of 10% ammonia solution added. The appearance of a pink, red, or violet colour in the ammoniacal (lower) phase indicated the presence of anthraquionone derivatives.

2.6 Test for Cardiac Glycosides

2.6.1 Legal test

The plant extract was dissolved in pyridine and a few drops of 2% sodium nitroprusside together with a few drops of 20% NaOH added. The appearance of a deep red colour which faded to brownish yellow indicated the presence of cardenolides.

2.7 Test for Flavonoids

Test for flavonoids was carried out according to the method described by Sofowora [9]. Five (5 ml) millilitres of dilute ammonia solution was added to 0.5 g of the extract and 5 ml of concentrated H_2SO_4 was added. A yellow colouration indicated the presence of flavonoids.

2.8 Test for Phenols

Test for phenols was carried out according to method described by Trease and Evans [7]. Two (2 ml) millilitres of Ferric chloride was added to 2 ml the extract. A deep blue-green colouration indicated the presence of phenols.

2.9 Quantitative Determination of Phytochemical Compounds

A total phenol was determined using spectrophotometric method. Alkaloid determination was carried out as described by Harborne [6]. Tannins was determined according to the methods described by Van-Burden and Robinson, [10]. The methods used to determine saponin was Obadoni and Ochuko, [11]. Determination of flavonoids was carried out by the method of Boham and Kocipai-Abyazan, [12].

2.10 Determination of Antimicrobial Activity of the Stem Bark Extract

Agar diffusion method described by Edward [13] was used.

2.10.1 Principle

Agar diffusion method operates on the principle that antibiotics will diffuse from a paper disc or a small cylinder into an agar medium that contains test organism. The inhibition is observed as a failure of the organism to grow in the region of the antibiotic. It may be dependent upon such factors as the diffusion rate of the antibiotic, the medium used, the sensitivity of the test organism, and incubation condition.

2.10.2 Procedure

The paper discs were put into the extracts whose concentration was 20 µg/ml and allowed to soak for some time. The discs were then removed and dried. The dried paper discs containing the absorbed extract were impregnated onto the surface of dried agar medium that has already been inoculated with the test organisms. These were left on the work bench for pre diffusion and later incubated at 37°C for 18 to 24 hours. The plates were observed and the diameter of zone of inhibition of growth was measured. Standard antibiotic discs of 20 µg/ml concentration were used in order to compare the zone of inhibition produced by both the crude extracts and standard antibiotic discs.

2.11 Determination of Minimum Inhibitory and Minimum Bactericidal Concentrations of the plant extract

Determination of Minimum Inhibitory and Minimum Bactericidal Concentrations were determined by the method described by Nester et al. [14]. Six test tubes containing serially diluted crude extracts in a nutrient broth were set. The dilutions were 64, 32, 16, 8, 4, and 0 µg/ml respectively, and the sixth test tube with 0µg/ml served as the control. 1 ml of test organism's suspension was introduced into each test tube and then incubated for 24 hours. The dilution was repeated with 32, 16, 8, 4, and 2 µg/ml respectively. The test tubes were incubated for 24 hours and then observed for growth. The test tube having the lowest concentration of the extract that do not show any growth was considered the one with minimum inhibitory concentration (MIC).

The Minimum Bactericidal Concentration was determined by using those test tubes that do not show any growth in the MIC determination. One ml suspension of each of the test tubes that showed no growth was introduced into test tubes containing extract – free nutrient broth and incubated for 24 hours. The test tubes were later observed for growth. Amongst the test tubes that do not show any growth, the one having the lowest concentration was regarded as the one with the minimum bactericidal concentration (MBC).

3. RESULT AND DISCUSSION

Alkaloids, tannins, saponins, phenols, anthraquinones, phlobatannins, and flavonoids were the phytochemicals (Table 1) that can possibly contribute to the antimicrobial and physiological effects associated with the plants extracts. The phytochemicals possibly contribute to the toxic and/or antioxidant/protective effects especially for the ones present in substantial amounts.

Some of the phytochemicals observed in this study have been implicated in the treatment of certain ailments and diseases such as diarrhoea, syphilis, and gonorrhoea [15]. The effectiveness of the use of stem bark of Anogeissus leiocarpus for the treatment of diarrhoeal disease in children by the local traditional healers could be due to the fact that they contain high concentrations of phytochemicals such as alkaloids and saponins (Table 2). This tend to agree with what was reported by Ghoshal et al. [16] that alkaloids have antidiarrhoeal effect which is probably due to their effect on transit time in the small intestine.

The effects of the crude plant extract on the test organisms as observed in this research work showed (Table 3) that the diameter of zones of inhibition produced by aqueous stem bark extract. More so, it is believed that each active principle of component parts making up the recipe complement one another in the fight against bacterial infection. These observations could be attributed to the various antimicrobial (phytochemical components) substances detected in this study which have been implicated by various authors as strong antimicrobial agents which are toxic to microorganisms. Typical amongst the phytochemical components are phenols and phenolic compounds which have been reported by Mason and Wasserman [17] as compounds whose toxicity to microorganisms include enzyme inhibition by the oxidized compounds possibly through reaction with sulphhydryl groups or through more non specific interactions with protein.

Furthermore, none of the individual crude extracts and the combined form of the extracts had any antifungal effects on *Candida albicans.* These results therefore suggest that only antibacterial substances could be present in *A. leiocarpus*. The Minimum Inhibitory Concentration (MIC) for each extract was determined against the four test organisms. A*nogeissus leiocarpus* has an MIC of 16 µg/ml each for *Staphylococcus aureus* and *Escherichia coli,* and 32 µg/ml each for *Klebsiella pneumoniae* and *Neisseria gonorrhoeae* as shown in Table 4. The effectiveness of each extract on the growth of the test organisms has manifested itself here as it did earlier in antimicrobial susceptibility test where the zones of inhibition were observed to be 10 µg/ml for Staphylococcus aureus and Escherichia coli and no zone of inhibition for Klebsiella pneumoniae and Neisseria gonorrhoeae.

It has been observed in this study that whenever there is an elevation of AST activity, a corresponding elevation of ALT activity will be followed. The AST: ALT ratio (with elevated levels of both enzymes) is approximately1:1 which is a pointer for a possible primary liver disease (Table 5). This finding was in agreement with the earlier report by the American Society of Clinical Pathologists [18]. Although AST is localized in the liver, heart, muscles, kidneys, brain, pancreas, spleen and lungs, and ALT is localized in the liver, heart, muscles, and kidneys, however, the histological studies of the liver and kidney tissues actually predicts some degree of cell damage. The elevated level of Lactate Dehydrogenase (LDH) activity indicated some degree of cell damage to those tissues where the enzyme is localized. Also high serum LDH indicates liver congestion [19]. In fact, the level of this enzyme would suggest liver damage even though the myocardium was the most probable one.

This research work has shown increased serum glucose levels in the treated groups of the experimental animals as compared to normal/ control groups, even though the elevation was not statistically significant ($p<0.05$). The rise in serum glucose levels above normal/control could be due to the effects of the crude extracts which might have affected the secretion of insulin by the islets of langerhans of the pancreas thereby reducing the rate of uptake of glucose by the body cells. Another possible cause might be due to leakage of glycogen into the blood as a result of liver cell injury.

This study has revealed that there was increase in serum level of cholesterol in the treated groups of experimental animals which was statistically significant ($p<0.05$) when compared with the normal/control groups.

Table 1. Phytochemical components detected in the stem bark extract of *Anogeissus leiocarpus*

Alkaloids	Anthraquinones	Flavonoids	Phlobatannins	Phenols	Quinones	Saponins	Glycosides	Tannins
+	+	-	-	+	-	+	+	+

+= Presence of phytochemical component; _ = Absence of phytochemical component.

Alkaloids, anthraquinones, phenols, saponins, glycosides and tannins were detected in the stem bark while flavonoids, phlobatannins, and quinones were not detected in the stem bark of Anogeissus leiocarpus

Table 2. Some phytochemical components in *Anogeissus leiocarpus*

Components	Quantity (g/100 g)
Alkaloids	9.20±0.05
Tannins	0.85±0.01
Saponins	12.35±0.03
Phenols	1.33±0.05
Flavonoids	N D

Values are mean±SEM of triplicate determinant, N D: not detected

Table 3. Zone of inhibition of the growth of the test organisms (mm)

Anogeissus leiocarpus	*S. aureus*	*K. pneumoniae*	*N. gonorrhoeae*	*E. coli*
Zone of inhibition	10	-	-	10

- = Resistance of the test organism on the antimicrobial substance

Table 4. Minimum Inhibitory concentration of test substance

Anogeissus leiocarpus	*S. aureus*	*K. pneumonia*	*N. gonorrhoeae*	*E. coli*
Min. Inhibitory Conc	16	32	32	16

Table 5. Effects of aqueous stem bark extract of *Anogeissus leiocarpus* on experimental animals

Treatment.	**Group 1 normal**	**Group 2 200 mg/kg**	**Group 3 300 mg/kg**	**Group 4 400 mg/kg**	**Group 5 500 mg/kg**
AST U/l	12.1±0.4	12.9±0.3	13.9±0.3#	14.3±0.5#	13.9±0.3#
ALT U/l	29.1±0.5	30.4±0.5	32.7±0.3#	33.5±0.1#	34.0±0.2#
ACP U/l	20.9±0.2	23.2±0.1#	24.4±0.0#	24.9±0.2#	22.4±0.3#
ALP U/l	18.6±0.2	13.1±0.1*	14.5±0.2*	14.7±0.4*	21.3±0.5#
LDH U/l	105.9±1.6	337.5±0.7#	253.7±1.8#	416.6±0.9#	347.7±1.1#
Glucose mg/dl	60.4±2.1	65.7±5.8	63.7± 4.4	64.5±2.8	79.3 ±1.6#
Cholesterol mmol/l	3.1±0.1	3.0±0.2	3.9±0.3	4.4±0.0#	4.8± 0.2#
Total bilirubin µmol/l	0.9±0.0	0.9±0.0	1.2±0.1	1.6±0.2	1.9±0.3#
Direct bilirubin µmol/l	0.6±0.0	0.9±0.0	1.0±0.0	1.0±0.1	1.4±0.2#
Albumin g/l	37.8±0.7	40.7±0.5	38.6±1.0	39.5±1.0	38.9±0.9
Total protein g/l	36.7±3.3	47.9±0.5#	15.5±0.3*	32.0±0.6	45.1±0.9#
Urea mmol/l	6.2±0.8	3.0±0.2*	5.3±0.1	4.9±0.1	3.4±0.0*
Creatinine mg/dl	0.2±0.0	0.2±0.0	0.1±0.0*	0.1±0.0*	0.2±0.0

*Values represent the mean ± SEM (n=3). P<0.05; *=Significantly lower than normal; #=Significantly higher than normal/contro*

Serum urea and creatinine levels of the experimental animals showed no significant increase as compared to normal/control groups. Serum bilirubin indicated significant increase only in group five of the experimental animals. This rise in the levels of bilirubin could be as a result of the prolonged administration of the aqueous stem bark extracts which could contain some antimicrobial agents and antimicrobials or antibiotics are among the substances that could cause increase blood levels of bilirubin. Low levels of albumin signify liver cell damage or kidney disease. This observation was in harmony with the reports on low albumin levels by Mohamadi-Nejad et al. [20].

Histological examination of the tissues (liver and kidney) of the experimental animals revealed that the cells were damaged to some extent by the extracts. The deleterious effects of the extracts were observed to be more pronounced on the liver than the kidney. The finding that the liver cells were more damaged than the kidney cells would suggest that the liver cells (plate 3 and plate 4) were more susceptible to the crude extract than the kidney (plate 1 and plate 2). It is also possible that since the crude extracts will first go the liver for detoxification, the hydrophilic conjugate may be less potent and will not have a drastic effect on the kidney, thus causing less damage to the kidney as observed in this study.

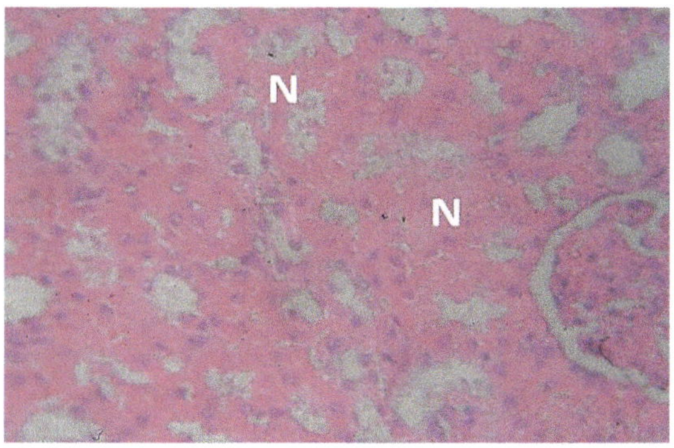

Plate 1. Kidney: Area of renal tubular necrosis (N) (x 760). Kidney section of group 4 of Anogeissus leiocarpus treated experimental animals showing areas of renal tubular necrosis as compared to normal (Plate 2 below)

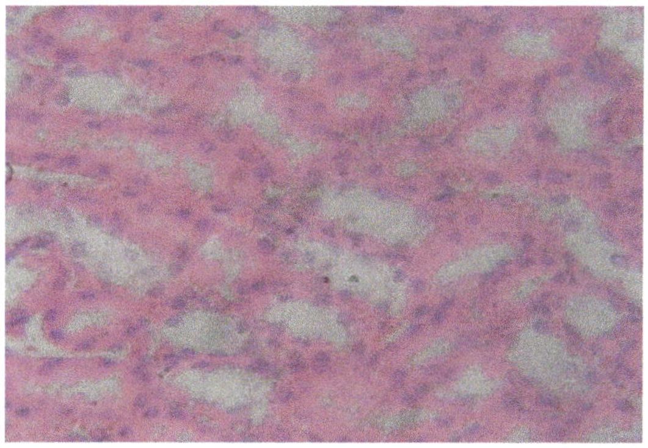

Plate 2. Kidney: Normal renal tubular epithelium (x 760)

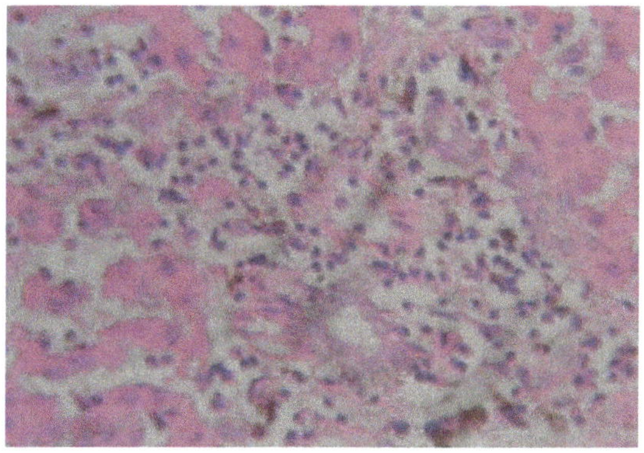

Plate 3. Liver portal triad: Infiltration of neutrophils and macrophages into connective tissue (x 760). Liver section of group 3 of Anogeissus leiocarpus showing the portal triad: Showing infiltration of neutrophils and macrophages into connective tissue when compared with the normal (Plate 4 below)

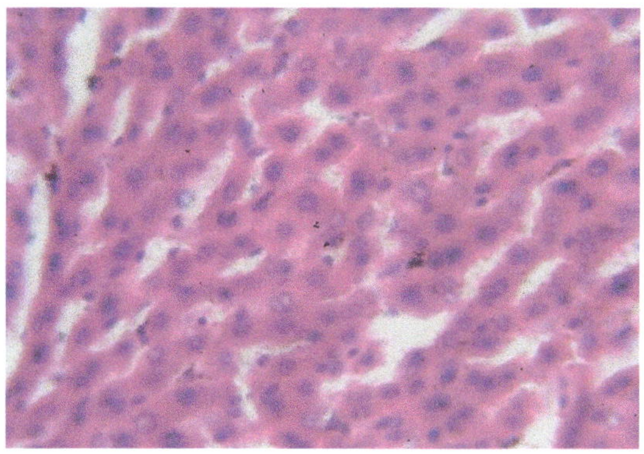

Plate 4. Liver: Normal (x 760)

4. CONCLUSION

The result of this study showed that the effect of the extract was more pronounced on the liver than the kidney as depicted by the photomicrographs.

COMPETING INTERESTS

Authors have declared that no competing interests exist.

REFERENCES

1. Sofowora A. Medicinal plants and traditional medicine in Africa. John Wiley and sons; 1982.
2. Bojang. State of Forest Genetic Resources: National Forest Inventory. 2000;1-5.
3. Srivastava J, Lambert J, Vietmeyer N. Medicinal plants: An expanding role in development world bank technical paper No.320.World Bank. Washington DC; 1996.
4. Keay RWJ. Trees of Nigeria. Clarendon press, Oxford. 1989;93-225.
5. Dalziel JM. The useful plants of West Tropical Africa. Crown Agents, London. 1937;612.
6. Harborne JB. Phytochemical methods. Chapman and Hall London. 1973;113.
7. Trease GE, Evans WC. A text book of pharmacognosy, 13th edition. Bailliere Tindall Ltd. London; 1989.
8. Wall ME, Krider MM, Krewson CF, Eddy CR, Wilaman JJ, Cordell DS, Gentry HS. Steroidal sapogenins XIII. Supplementay table of data for steroidal sapogenins VII. Agric Research Services Circ. Aic. 1954; 363:17.
9. Sofowora A. Medicinal plants and traditional medicine in Africa. Spectrum book Ltd Ibadan. 1993;289.
10. Van-Burden TP, Robinson WC. Formation of complexes between proteins and tannic acids. Journal of Agric. Food Chem. 1981;1:107.
11. Obadoni BO, Ochuko PO. Phytochemical studies and comparative efficacy of the crude extracts of homeostatic plants in Edo and Delta states of Nigeria. Global Journal of pure and applied sciences. 2001;8:203-208.
12. Boham BA, Kocipai-Abyazan C. Flavonoids and condense tannins from leaves of Hawaiian *Vaccinium vaticulum* and *Vaccinium calycimium*. Pacific Sciences. 1994;48:458-63.
13. Edward AI. Fundamental of microbiology. Addison Wesley Publishing company, Inc. 1983;691-693.
14. Nester EW, Anderson DG, Roberts CE, Pearsall NN, Nester MT. Microbiology: A human perspective. 4th Edition. 2004;518 -519.
15. Maspalma WS. Taxonomic array of medicinal plants and their local uses in and around Yola Local Government Area of Adamawa State. B. Tech. thesis Dept. of Biol. Sci. F.U.T. Yola. 1997;17-19.
16. Ghoshal S, Krishna–Prasad BN, Lakshmi V. Antiamoebic activity of *Piper longum* fruits against *Entamoeba histolytica* in vitro and in vivo. J. Ethnopharmacol. 1996;50: 167-170.
17. Mason TL, Wasserman BP. Inactivation of red beet beta –glucan synthase by native and oxidized phenolic compounds. Phytochemistry. 1987;26:2197-2202.
18. American Soceity for Clinical Pathology (ASCP 2003). American Proficiency Institute (2003) 1st Test Event.
19. Devlin TM. Text Book of Biochemistry with Clinical Correlations. 2nd Edition. A Wiley Medical Publication. John Wiley and Sons. New York, Chichester, Brisbane, Toronto, Singapore. 1986;164 –165,262-327.
20. Mohamadi-Nejad A, Moosavi-Movahedi AA, Hakimelahi GH, Sheibani N. Thermodynamic analysis of human serum albumin interactions with glucose: Insights into the diabetic range of glucose concentration. Int. J. Biochem. Cell Biol. 2002;34(9):1115 –1124.

21

Effect of Cooking Methods on Proximate and Mineral Composition of Fluted Pumpkin (*Telfairia occidentalis*) Leaves

F. G. Okibe[1*], B. Jubril[2], E. D. Paul[1], G. A. Shallangwa[1] and Y. A. Dallatu[1]

[1]Department of Chemistry, Ahmadu Bello University, Zaria, Kaduna State, Nigeria.
[2]Department of Food Science and Technology, Ibrahim Badamasi Babangida University, Lapai, Niger State, Nigeria.

Authors' contributions

This work was carried out in collaboration between all authors. Authors FGO and EDP designed the study, wrote the protocol and supervised the work. Author BJ carried out all laboratories work and performed the statistical analysis. Author FGO managed the analyses of the study. Author BJ wrote the first draft of the manuscript. Authors GAS and YAD managed the literature searches and edited the manuscript. All authors read and approved the final manuscript.

Editor(s):
(1) Luisa Di Paola, Chemical and Biochemical Engineering Fundamentals, Faculty of Engineering Università Campus Biomedico, Via Alvaro del Portillo, Roma, Italy.
Reviewers:
(1) Antonello Santini, University of Napoli Federico II, Italy.
(2) Brahmadeo Dewprashad, City University of New York, USA.
(3) Ivana Beatrice Mânica da Cruz, Federal University of Santa Maria (UFSM), Brazil.
(4) Maria Elisabeth Machado Pinto e Silva, University of Sao Paulo, Brazil.
(5) Azza A. Omran, Food Technology Research Institute (FTRI), Egypt.

ABSTRACT

The effect of boiling, steaming and microwaving on proximate and mineral compositions of fluted pumpkin (*Telfairia occidentalis*) leaves were investigated. Standard methods of analyses were applied for the determinations of selected parameters. Results obtained indicated that at $P<0.05$ confidence level no significant losses in the values of proximate composition occurred, except for protein and carbohydrate. Decrease in protein contents were 18.36, 13.54 and 20.36% for steaming, boiling and microwaving respectively. Carbohydrate content decrease as follows: steaming (15.44%), boiling (18.35%), and microwaving (2.53%). The percentage reduction in

**Corresponding author: E-mail: fgokibe@abu.edu.ng, fgokibe@gmail.com*

nitrogen content on applying the three cooking methods were 1.10, 0.72 and 0.38% for steaming, boiling and microwaving respectively. Results showed that percentage decrease in P, K, Na, Ca, Mg, and Fe was highest for boiling relative to other methods of processing. The study recommended microwaving as the most appropriate method for processing *T. occidental* leaves for optimum nutritional benefit.

Keywords: T. occidentalis; proximate composition; mineral composition.

1. INTRODUCTION

Most vegetables are commonly cooked before consumption. In general, the purpose of preparing vegetables before consumption is to improve taste rather than retention of mineral nutrients which serve as health promoting compounds [1]. It is known that cooking induces significant changes in chemical composition [1]. Cooking methods were shown to affect content of nutrients and health promoting compounds such as vitamin C, carotenoids, and polyphenols in broccoli vegetables [2,3]. However, processing makes food safe for consumption and destroys pathogenic microorganisms. The effect of processing on food depends on the sensitivity of the nutrient to the various processing conditions such as heat, oxygen, pH and light [3]. It is necessary that vegetables are prepared in such a manner that retains maximum amounts of their nutrients, that is minimizing losses through oxidation and leaching [4].

There are many methods of cooking, most of which have been known since antiquity. These include baking, roasting, frying, grilling, barbecuing, and smoking. Others include boiling, steaming and braising [4]. Various methods use differing levels of heat and moisture and vary in cooking time. The method chosen greatly affects the end result because some foods are more appropriate to some methods than others. When heat is used in the preparation of food, it can kill or inactivate harmful organisms, such as bacteria and viruses, as well as various parasites such as tapeworms and *Toxoplasma gondii* [4,5]. Food poisoning and other illness from uncooked or poorly-prepared food may be caused by bacteria such as pathogenic strains of *Escherichia coli, Salmonella typhimurium* and *Campylobacter*, viruses such as noroviruses, and protozoa such as *Entamoeba histolytica* [5]. Parasites may be introduced through salad, meat that is uncooked or rarely cooked and un-boiled water [5]. The sterilizing effect of cooking depends on temperature, cooking time, and technique used. Studies have shown that some bacteria such as *Clostridium botulinum* or *Bacillus cereus* can form spores that survive cooking, which then germinate and re-grow after the food is cooled [6]. It is therefore recommended that cooked food should not be reheated more than once to avoid repeated growths that allow the bacteria to proliferate to dangerous level [6].

Fluted pumpkin *(T. occidentalis)* is a vegetable locally known as, ugwu in Igbo and eweroko in Yoruba languages of Nigeria. This leaf is a tropical vine grown in West Africa as a leafy vegetable. It is a vegetative shrub that creeps low across the ground with large lobed leaves and long twisting tendrils. *T. occidentalis* is a vegetable which is an important food component in diets in Nigeria mostly in southern part of Nigeria. The leaf is used primarily in soups and herbal medicines, they are believe to help reduce blood pressure, improve digestion and enhance healthy metabolism [7]. Young leaves of *T. occidentalis* sliced and mixed with coconut water and salt are used to treat convulsion in ethno medicine and the roots are used as rodenticides [7]. *T. occidentalis* is traditionally used by an estimated 30 to 35 million Indigenous people in Nigeria, including the Efik, Ibibio, and Urhobo; however, it is predominantly used by the Igbo tribe, who continue to cultivate the gourd for food sources and traditional medicines [7]. The fluted gourd is noted to have healing properties and is used as a blood tonic, administered to the weak or ill [8]. *T. occidentalis* leaf contains nutrients which can be absorbed by the human system for body building, energy, regulatory and protective functions. They are also source of vitamins, minerals and phytochemicals. A study conducted by the Federal Institute of Technology, Akure, examined the chemical properties of the fluted gourd and concluded that the leaves contained high amount of antioxidants and showed antimicrobial properties [8]. Considered an "oil seed", the fluted gourd is high in oil (30%) [8]. Shoots of *T. occidentalis* contain high levels of potassium and iron, while the seeds are composed of 27% crude proteins and 53% fats [8].

A significant percentage of Nigerian population consumes this vegetable after subjecting it to one form of processing or the other. Hence, the need to evaluate the effect of the most commonly used methods of processing on the vital proximate and mineral composition of the vegetable with a view to providing useful data.

2. MATERIALS AND METHODS

2.1 Sample Collection and Treatment

Samples of *T. occidentalis* were purchased from Badeggi market, Lapai, Niger state, Nigeria. Prior to analysis, the leaves were washed with tap water then rinsed with distilled water. The residual moisture was evaporated at room temperature.

2.2 Boiling

About 200 g of homogeneous pieces of Fluted Pumpkin leaves were immersed in 400 ml of boiling water. The fluted pumpkin was drained off after boiling for 5 min.

2.3 Microwaving

A microwave oven (LOGIK Microwave appliance, model No: L20MS10, Tianjin Electronics, China) at full power (1000 W) was used for microwaving. About 200 g of Fluted Pumpkin leaves were placed in a plate and 10 ml water was added to prevent it from burning during cooking. The Fluted Pumpkin in the plate was placed in the microwave oven for 5-min, and then drained off.

2.4 Steaming

Steaming was conducted by suspending about 200 g of cut Fluted Pumpkin leaves above 200 ml of boiling water for 5 min in a steamer with a lid.

Thereafter, the leaves from the three cooking methods were separately oven dried at 60℃ until adjudged to be properly dried. The dried leaves were then ground in porcelain mortar, sieved through 2 mm mesh sieve and stored in polythene bags. The powdered samples were used for both proximate and mineral analyses.

Fresh samples of the vegetable which were not subjected to any of the cooking methods were also dried at 60℃ and grounded into powder for proximate and mineral analyses.

2.5 Proximate Analysis

Proximate compositions were estimated using standard methods of analysis. The moisture content of the leaves were determined by drying 5 g of the powdered leaves (in triplicate) in a Gallenkamp oven at 105℃ until constant weight was attained [9]. Ash content was determined according to the method described by [10] which involved dry ashing in lenton muffle furnace at 600℃ until grayish white ash was obtained. Crude protein content was determined by multiplying the value obtained from kjeldahl's nitrogen by a protein factor of 5.3, a factor recommended for vegetable analysis. Crude lipid was quantified by the method described by [9] using the soxhlet apparatus and petroleum ether (B.P. 60℃- 80℃) as solvent. Crude fiber was determined by acid-base digestion with 1.25% H_2S0_4 (W/V) and 1.25% NaOH (W/V) solutions.

The method described in [9] was used to calculate available Carbohydrates, where the total amount of carbohydrate in the sample was obtained by calculation using percentage weight difference. This involved subtracting the percentage sum of the food nutrients, % crude protein, % crude lipid, % crude fibre, and % ash from 100% dry weight. Percentage carbohydrate was calculated using equation: Carbohydrate % = 100 – (crude protein + crude lipid + crude fibre + ash).

The Energy Value (calorific Value) of a Food is given by energy Value of Food (in KJ per100 g) = [(% available Carbohydrate X 4) + (% available Protein X 4) + (% available Fat X 9)] [9].

2.6 Samples Preparation for Mineral Analysis

Six (6) grams of the powdered sample was weighed into a crucible and gently heated over a Bunsen burner until it charred. The charred sample with the crucible was transferred into a lento muffle furnace at about 600℃ and constantly heated until grayish white ash was obtained. It was cooled first at room temperature and then in a desiccator. 5 cm^3 of concentrated HCl was added and heated for 5 minutes on a hot plate in a fume cupboard. The mixture was then transferred into a beaker and the crucible washed several times with distilled water. The mixture was made up to 40 cm^3 and boiled for 10 minutes over a Bunsen burner. This mixture was then cooled, filtered and rinsed into 100 cm^3

volumetric flask and made up the volume to 100 cm^3 [10]. The solution prepared in triplicates.

2.7 Determination of Mineral Concentration

Sodium (Na) and Potassium (K) were analyzed by Flame Atomic Emission spectrophotometer. Phosphorus (P) was determined with Jenway 6100 spectrophotometer at 420 nm using vanadium phosphomolybdate (vanadate) colorimetric method with KH_2P0_4 as the standard [10]. The concentrations of calcium (Ca), magnesium (Mg), copper (Cu), Iron (Fe), Manganese (Mn) and Zinc (Zn) in the solutions were determined using Atomic Absorption Spectrophotometer AAS 969 (Bulk Scientific, MODEL VGB 210/211) [9].

2.8 Statistical Analysis

Statistical analysis was performed using the SPSS package program version 20 (SPSS Inc., Chicago, IL, USA). Data were analyzed by one-way analysis of variance (ANOVA), followed by Duncan's multiple range post-hoc test. Results are expressed as mean ± *SD* of triplicate samples. Differences were considered significant at $P<0.05$.

3. RESULTS AND DISCUSSION

3.1 Proximate Composition

The effect of different cooking methods on the proximate composition of *T. occidentalis* is presented in Tables 1 & 2. The results indicated increase in the values of crude fibre, fat, moisture and ash contents of *T. occidentalis* after processing. Protein and carbohydrate content were observed to decrease on cooking for all methods used.

Percentage decrease in crude protein values were 18.36, 13.54 and 20.36% for steaming, boiling and microwaving respectively. This means that boiling retains more of protein than the other cooking methods used in this study. This may be due to the high heating temperature involved in steaming and microwaving where there is no much water to serve as a shield as is the case with boiling. It has been reported by Rahama and Mustapha [11] that increase in temperature can cause severe protein damage in food ranging from destruction of amino acid to complete racemization. Also, Komolafe and Obayanju [12] reported that during heating, cellular protein are denatured and chlorophyll which was bound to the protein may be released. Such free chlorophyll is highly unstable and readily converted to pheophytin, which could enhance denaturation of protein when heated.

A decreasing trend in carbohydrate content was observed for the different cooking methods in the following manner, boiling > steaming > microwaving (Table 2). Microwaving in this case retains the highest amount of carbohydrate followed by steaming and then boiling, this may be as a result of some of the carbohydrate dissolving in the large volume of water in the boiling method. Increased temperature reduced the carbohydrate content of vegetable and hence limits its possible usefulness as an energy source. Carbohydrate is made up of starches and sugars. Heating starches changes crystallized starch molecules into gels during which the starch granules swell and absorb water becoming soft and easier to dissolve [13].

An increasing trend of Fat content was observed in the manner steaming<microwaving<boiling. This shows that heating frees the crude fat from the vegetable matrices. This may be due to the increase in the extraction efficiency of crude fat by water as the rate of heating increases. Some lipids in foods are in complex lipoprotein and liposaccharides, hence heat breaks the bond between them and the lipids are freed and solubilized in the extracting solvent [14].

3.2 Mineral Composition

The result of the effect of various cooking methods on the mineral compositions of *T. occidentalis* is presented in the Tables 3 & 4.

All the mineral compositions of *T. occidentalis* analyzed were observed to decrease for all the methods of cooking. The percentage changes in nitrogen content were 1.10, 0.72 and 0.38% for steaming, boiling and microwaving methods respectively. Generally, mineral contents were observed to decrease for all the cooking methods applied in this study. This general decrease in the mineral content after processing is attributed to losses arising from the solubility of these elements in water. Results showed that percentage decrease (Table 4) in P, K, Na, Ca, Mg, and Fe was highest for boiling relative to other method of processing. This observation may be related to the large volume of water used in this processing method. From the results of mineral analysis microwaving *T. occidentalis* leaves will retain more of the mineral content than steaming and boiling.

Table 1. The effect of cooking methods on the proximate compositions of *T. occidentalis*

Cooking methods	Moisture	Fat	Crude protein	Crude fibre	Ash	CHO	Energy value
	%	%	%	%	%	%	
Uncooked	4.66 ± 0.16^{a}	12.76 ± 1.26^{a}	31.74 ± 0.19^{b}	4.35 ± 0.13^{a}	13.48 ± 0.55^{c}	51.01 ± 1.34^{d}	332.62 ± 2.67^{a}
Steam	5.60 ± 0.05^{a}	13.26 ± 0.56^{a}	13.38 ± 0.44^{d}	4.50 ± 0.5^{a}	13.74 ± 0.17^{b}	35.57 ± 0.63^{b}	378.70 ± 23.51^{b}
Boil	6.50 ± 0.50^{c}	13.40 ± 0.36^{a}	18.20 ± 0.70^{c}	5.42 ± 0.15^{b}	13.86 ± 0.31^{a}	32.66 ± 0.69^{a}	378.7 ± 1.59^{b}
Microwave	5.70 ± 0.49^{b}	13.30 ± 0.35^{a}	11.38 ± 0.07^{b}	6.00 ± 0.12^{c}	15.14 ± 0.29^{d}	48.48 ± 0.51^{c}	359.14 ± 0.86^{b}

a: significant in the comparison between uncooked and cooked vegetables *b: moderately significant in the comparison between uncooked and cooked vegetables*
c:highly significant in the comparison between uncooked and cooked vegetables *d: very highly significant in the comparison between uncooked and cooked vegetables*

Table 2. Percentage change in proximate composition of *T. occidentalis*

Proximate composition	Steaming	Boiling	Microwaving	Change
	%	%	%	
Moisture	0.94	1.84	1.04	Increase
Fat	0.50	0.64	0.54	Increase
Protein	18.36	13.54	20.36	Decrease
Fibre	0.15	1.07	1.65	Increase
Ash	0.26	0.38	1.66	Increase
Carbohydrates	15.44	18.35	2.53	Decrease

Table 3. The Effects of cooking methods on the mineral compositions of *T. occidentalis*

Cooking methods	Nitrogen (N)	Phosphorus (P)	Potassium (K)	Sodium (Na)	Calcium (Ca)	Magnesium (Mg)	Iron (Fe)
	%	%	%	%	%	%	%
Uncooked	2.20 ± 0.20^{a}	0.91 ± 0.09^{a}	0.83 ± 0.23^{c}	0.180 ± 0.03^{a}	5.30 ± 0.20^{a}	0.038 ± 0.001^{a}	0.142 ± 0.004^{b}
Steam	1.10 ± 0.51^{c}	0.88 ± 0.21^{a}	0.65 ± 0.13^{a}	0.040 ± 0.01^{b}	4.20 ± 0.26^{b}	0.036 ± 0.002^{a}	0.101 ± 0.001^{a}
Boil	1.48 ± 0.51^{c}	0.77 ± 0.10^{a}	0.08 ± 0.13^{a}	0.020 ± 0.01^{a}	4.00 ± 0.50^{b}	0.030 ± 0.007^{b}	0.100 ± 0.020^{d}
Microwave	1.82 ± 0.74^{b}	0.74 ± 0.05^{b}	0.09 ± 0.51^{b}	0.055 ± 0.01^{b}	4.40 ± 0.58^{b}	0.037 ± 0.002^{c}	0.106 ± 0.010^{c}

a: significant in the comparison between uncooked and cooked vegetables *b: moderate significant in the comparison between uncooked and cooked vegetables*
c:highly significant in the comparison between uncooked and cooked vegetables *d: very highly significant in the comparison between uncooked and cooked vegetables*

Table 4. Percentage change in mineral composition of *T. occidentalis*

Minerals	Steaming %	Boiling %	Microwaving %	Change
Nitrogen	1.10	0.72	0.38	Decrease
Phosphorus	0.03	0.14	0.17	Decrease
Potassium	0.18	0.75	0.74	Decrease
Sodium	0.14	0.16	0.13	Decrease
Calcium	1.10	1.30	0.90	Decrease
Magnesium	0.002	0.008	0.001	Decrease
Iron	0.041	0.042	0.036	Decrease

4. CONCLUSION

This study has revealed that *T. occidental* leaves contain an appreciable amount of proximate and mineral nutrients. However, cooking methods of steaming, boiling and microwaving have effects on the retention of these compositions. For proximate composition, protein and carbohydrate content decreased while, crude fat, crude fibre, ash and moisture contents increased on application of the three cooking methods. A general decrease in mineral composition was observed for all the minerals evaluated. On the whole, this study recommended microwaving as the most appropriate method for processing *T. occidental* leaves for optimum nutritional benefit.

COMPETING INTERESTS

Authors have declared that no competing interests exist.

REFERENCES

1. Masrizal MA, Giraud DW, Driskell JA. Retention of vitamin C, iron, and beta-carotene in vegetables prepared using different cooking methods. J Food Qual. 1997;20(5):403–418.
   DOI: 10.1111/j.1745-4557.1997.tb00483.x
2. Lin CH, Chang CY. Textural change and antioxidant properties of broccoli under different cooking treatments. Food Chem. 2005;90(1-2):9–15.
   DOI: 10.1016/j.foodchem.2004.02.053
3. Cieslik E, Leszczynska T, Filipiak-Florkiewicz A, Sikora E, Pisulewski PM. Effects of some technological processes on glucosinolate contents in cruciferous vegetables. Food Chem. 2007;105(3): 976–981.
   DOI: 10.1016/j.foodchem.2007.04.047
4. Saidu AN, Jideobi NG. The proximate and elemental analysis of some leafy vegetables grown in Minna and environs. J. Appl. Sci. Environ. Manag. 2009;13(4): 21-22.
5. Christine Dell'Amore in Chicago (13 February 2009). "Cooking Gave Humans Edge Over Apes?" National Geographic News.
6. Haltz Christine, Rosalind S Gibson. Traditional food-processing and preparation practices to enhance the bioavailability of micronutrients in plant-based diets. The Journal of Nutrition. 2007;137(4):1097-1100.
7. Nwufo MI. Effects of water stress on the post-harvest quality of two leafy vegetables, *Telfairia occidentalis* and - Storage *Pterocarpus soyauxii* during storage. J Sci Food Agric. 1994;64:265-69.
8. Akoroda MO. Ethno botany of *Telfairia occidentalis* (Curcurbitacae) among Igbo's of Nigeria. Economic Botany. 1990;29-39. Jstor.
9. AOAC (Association of Official Analytical Chemists) Official Methods of Analysis. Washington DC: AOAC International; 1990. Official Method 985.33.
10. Ceirwyn SJ. Analytical chemistry of food. Chapman and Hall Publisher, London. 1995;76–77.
11. Rahma EH. Mustapha MM. Functional properties of peanut flour as affected by different heat treatment. J. Food Sci. Technol. 1988;25:11-15.
12. Komalafe EA, Obayanju VS. Principles of Food Processing and Preservation, 1st ed. Double Birth Publishers, USA. 2003;87.

13. Morris A, Barnet A, Burrow O. Effect of processing on nutrient content of foods. Cajaticles. 2004;37:160-164.

14. Lund DB. Effects of heat processing on nutrients. In: Nutritional Evaluation of Food Processing, Harris R, Karmas E. (Eds.). The AVI Publishing Co. Inc., Westport, CT. 1977;205-230.

22

Liver and Kidney Function Tests and Histological Study on Malaria Parasite Infected Mice Administered with Seed Extract of *Picralima nitida*

N. E. Nwankwo[1*], O. F. C. Nwodo[1], A. E. Amalunweze[1], K. U. Agbo[1] and S. C. Abugu[1]

[1]Department of Biochemistry, University of Nigeria, Nsukka, Enugu State, Nigeria.

Authors' contributions

This work was carried out in collaboration between all authors. Author NEN designed the study, wrote the protocol and the author OFCN supervised the work. Author NEN carried out all laboratories work and performed the statistical analysis. Author OFCN managed the analyses of the study. Author NEN wrote the first draft of the manuscript. All the authors managed the literature searches and edited the manuscript. All authors read and approved the final manuscript.

Editor(s):
(1) Mohamed Fawzy Ramadan Hassanien, Biochemistry Department, Zagazig University, Egypt.
Reviewers:
(1) Anonymous, International Medical University, Malaysia.
(2) Anonymous, Narsee Monjee Institute of Management Studies, Mumbai, India.
(3) Anonymous, Karpagam University, India.

ABSTRACT

The liver and kidney are very important organs of the body that play vital roles of detoxification of compounds and removal of waste products respectively. In this research, some liver and kidney function tests were carried out on albino mice infected with malaria parasite to know the effect of treatment with extract on them. Pulverised dried seeds of *Picralima nitida* were extracted using ethanol. Serum aspartate aminotransferase (AST) and alanine aminotransferase (ALT) activities, total bilirubin, total protein, urea and creatinine levels and histopathological studies on liver and kidney tissues were performed. From the results of the experiment, significant ($p < 0.05$) reductions were observed in serum levels of urea, creatinine of the treated groups compared to the positive control (group infected with malaria parasite and treated with 3% tween 80) on days 3 and 5 post

**Corresponding author: E-mail: nicemmy2000@yahoo.com*

treatment. There was non-significant ($p > 0.05$) reduction in serum AST and ALT activities of the groups treated with 20, 40 and 80 mg/kg b.w. of the extract on days 3 and 5 post treatment compared to the positive control. Groups treated with 20 and 40 mg/kg b.w. of the extract showed significant ($p < 0.05$) reduction in bilirubin level on days 3 and 5 post treatment compared to the positive control. All the treated groups showed significant ($p < 0.05$) elevation in serum total protein on day 3 post treatment while the group treated with 80 mg/kg b.w. of the extract showed significant ($p < 0.05$) elevation in total protein on day 5 treatment compared to the positive control. Histopathological studies on the hepatic tissues of the treated mice showed lesser inflammatory cells around the portal areas than the untreated. No lesions were noticed in the renal tubules of both the treated and untreated mice. From the results of this experiment, it is now obvious that *Picralima nitida* seeds possess good hepato-protective properties.

Keywords: Liver and kidney function tests; Picralima nitida; serm ALT and AST; serum total bilirubinl serum total proteinl serum urea; creatinine.

1. INTRODUCTION

Plasmodium berghei is a unicellular protozoan parasite that infects mammals other than humans. It was first described by Vincke and Lips in 1948 in the Belgian Congo. It is found in the forests of Central Africa, where its natural cyclic hosts are the thicket rat and the mosquito (*Anopheles dureni*) [1].

Like all malaria parasites of mammals, *P. berghei* infects the liver after being injected into the bloodstream by a bite of an infected female mosquito. After a short period of development and multiplication, these parasites leave the liver and invade the erythrocytes. *P. berghei* infections may also affect the braln and can cause cerebral complications in laboratory mice. The complete genome of *P. berghei* has been sequenced and it shows a high similarity both in structure and gene content with the genome of the human *Plasmodium falciparium* [2,3].

The parasite can be genetically manipulated and so, is often used for the analysis of the function of malaria genes using the technology of genetic modification [4,5]. A number of genetically modified *Plasmodium berghei* cell lines have been generated and these transgenic parasites are tools to study and visualize the parasites in the living host [6,7].

Livers function tests (LFTs of LFs) are group of clinical biochemistry laboratory blood assays designed to give information about the state of a patient's liver [8]. The parameters measured include prothrombin time (PT/INR), activated partial thromboplastin time (a PTT), albumin, bilirubin and others. Aspartate aminotransferase and alanine aminotransferase (AST and ALT) are useful biomarkers of liver injury in a patient with some degree of intact liver [9]. Some tests are associated with functionality (e.g. albumin), some with cellular integrity (e.g. transferases), and some with conditions linked to the biliary tract (e.g. gamma-glutamyl transferase and alkaline phosphate) [10]. Several biochemical tests are useful in the evaluation and management of patients with hepatic dysfunction. These tests can be used to detect the presence of liver diseases, distinguish among different types of liver disorders, guage the extent of known liver damage and follow the response to treatment.

Albumin is the main constituent of total protein made specifically by the liver and can easily be measured. Its levels are decreased in chronic liver disease, such as cirrhosis. It is also decreased in nephritic syndrome, where it is lost through the urine. The consequence of low albumin can be edema since the intravascular oncotic pressure becomes lower than the extravascular space [11].

Alanine aminotransferase (ALT), also known as serum glutamate pyruvic transaminase (SGPT), is a member of transferase family. It catalyses the reversible reaction in which an α-amino group is transferred between alanine and glutamine [12].

This enzyme is found mainly in liver and small amount in heart, muscle, and kidney. In healthy subjects, serum ALT levels are low. However, when cells are damaged, for instance, acute and chronic hepatitis, obstructive jaundice, carcinoma of liver, myocardial infarction, this enzyme may leak into the blood stream and its levels are significantly elevated [13,14].

Aspartate aminotransferase (AST) is similar to ALT in that it is another enzyme associated with parenchymal cells. It is raised in acute liver damage, but is also present in red blood cells, and cardiac and skeletal muscles, so is not specific to the liver [15]. So, elevated AST levels are not specific for liver damage. Aspartate aminotransferase, as with all aminotransferases, operates via dual substrate recognition; that is, it is able to recognize and selectively bind two amino acids (Asp and Glu) with different side-chains [16]. In either case, the aminotranferase reaction consists of two similar half-reactions that constitute what is referred to as a ping-pong mechanism. In the first half-reaction, amino acid 1 (e.g., L-Asp) reacts with the enzyme-PLP complex to generate ketoacid 1 (oxaloacetate) and the modified enzyme-PMP. In the second half-reaction, ketoacid 2 (α-ketoglutarate) reacts with enzyme-PMP to produce amino acid 2 (L-Glu), regenerating the original enzyme-PLP in the process.

Total bilirubin includes both unconjugated and conjugated bilirubin. Unconjugated bilirubin is a breakdown product of haem [17]. It is very hydrophobic and is mainly transported bound to albumin circulating in the blood. High concentrations of hydrophobic drugs and high free fatty acids can cause elevated unconjugated bilirubin. The liver is responsible for clearing the blood of unconjugated bilirubin, and about 30% of it is taken up by a normal liver on each pass of the blood through the liver by the some mechanisms.

Bilirubin can be conjugated with a molecule of glucuronic acid which makes it soluble in water. Some of the double-bonds in bilirubin isomerizes when exposed to light. This is used in the phototherapy of jaundiced newborns: the E, Z-isomers of bilirubin formed upon light exposure are more soluble than the unilluminated Z, Z-isomer, as the possibility of intramolecular hydrogen bonding is removed [18]. This allows the excretion of unconjugated bilirubin in bile.

Erythrocytes generated in the bone marrow are disposed of in the spleen when they get old or damaged. This releases haemoglobin, which is broken down to haem as the globin parts are turned into amino acids [19]. The haem is then converted into unconjugated bilirubin in the reticuloendothelial cells of the spleen. This water-insoluble unconjugated bilirubin is then bound to albumin and sent to the liver. In the liver, bilirubin is conjugated with glucuronic acid by the enzyme glucuronyltransferase making it water-soluble and afterwards goes into the bile. Conjugated bilirubin is not absorbed in the small intestine but instead is passed into the colon [19]. Colonic bacteria deconjugate and metabolize the bilirubin into colourless urobilinogen which is further oxidized to urobilin and stercobilin which give stool its characteristic brown colour. A trace amount of urobilinogen is reabsorbed into the enterohepatic circulation to be re-excreted in the bile: some of this is instead processed by the kidneys colouring the urine [20].

Kidney or renal function tests are common laboratory tests used to evaluate how well the kidneys are working. Such tests include: blood urea nitrogen (BUN), serum creatinine test, urinalysis and glomerular filtration rate (GFR) test [21]. Urea is produced as a waste product of protein digestion by the liver. Normal human adult blood should contain 6-20 mg/dL of blood. The main causes of an increase in BUN are: high protein diet, decrease in glomerular filtrate rate (GFR) and in blood volume, congestive heart failure, gastrointestinal haemorrhage, fever and increased catabolism [22]. Decreases in BUN are caused by severe liver disease, anabolic state and syndrome of inappropriate antidiuretic hormone (ADH).

Creatinine is a breakdown product of creatine phosphate in muscle, and is usually produced at a fairly constant rate by the body [23]. Serum creatinine is an important indicator of renal health because it is an easily measured by-product of muscle metabolism excreted unchanged by the kidneys. It is synthesized by the liver from the methylation of glycocyamine by S-adenosyl methionine. It is then transported through the blood to the other organs, muscles, and brain where it is phosphorylated to phosphocreatine. During this reaction, creatine and phosphocreatine are catalyzed by creatine kinase, and a spontaneous conversion to creatinine may occur [24]. Creatinine is removed from the blood by the kidneys through glomerular filtration and also by proximal tubular secretion [25]. There is little or no tubular reabsorption of creatinine. When its filtration in the kidney is impaired, blood creatinine levels rise and so its levels in blood and urine may be used to calculate the creatinine clearance (Crcl). Creatinine levels in the blood and urine correlates with the glomerular filtration rate (GFR) while blood creatinine levels alone are used to calculate the estimated GFR (eGFR).

Extracts of *Picralima nitida* parts (root, seed and stembark) have been shown to exhibit a broad spectrum of activity against bacterial strains [26]. Among the susceptible bacteria were *Staphylococcus aureus*, *Pseudomonas aeruginosa*, *Escherichia coli*, *Bacillus subtilis* and *Salmonella kintambo*. The Minimum Inhibitory Concentration (MIC) values for the ethanol extracts range from 6.25 to 50 mg/ml, while the MIC values for the cold water seed extracts (CWs) was 50 mg/ml showing that the ethanol extracts were more potent than the cold water extracts at very low concentrations. The activities of these extracts were possible because they contained a good number of phytochemicals such as alkaloids, saponins, flavonoids, steroids, tannins and terpenoids. Ubulom *et al.* [27] investigated into the antifungal activity of both aqueous and ethanolic extracts of *Picralima nitida* seeds against *Aspergillus flavus*, *Candida albicans* and *Microsporum canis* and found them to have activity against these organisms. However, the ethanol extract was found to have better antifungal activity than the aqueous extract. Okonta and Aguwa [28] compared the hypoglycemic effect of glycosides extract with that of alkaloids extract of *P. nitida* seed and were able to establish that the hypoglycemic effect of the seed lies in the glycosides extract. The alkaloids extract reduced given i.p caused increase in mean fasting blood glucose levels while the glycosides extract reduced the blood glucose levels in normoglycemic and hyperglycemic rats.

In traditional medicine, numerous plants have been used for cognitive disorders, including memory loss and antiaging. In accordance, *Picralima nitida* fruits and seeds have been found to possess memory enhancing property [29] and so could be applied in the treatment of cognitive disorders like Alzheimer's disease, amnesia, depression and schizophrenia. Soladoye et al. [30] performed ethnobotanical survey of plants used in the treatment of haemorrhoids in South-Western Nigeria and found *Picralima nitida* fruit to be very effective. From the survey, a total of 143 plant species belonging to 58 families were found to be useful for the treatment of haemorrhoids. The most prominent family among these plant families is the *Leguminosae*, and then others include *Euphorbiaceae*, *Apocynaceae*, *Meliaceae*, *Compositae* and *Poaceae*. The seed extract of *P. nitida* [31] showed inhibition of β-glucosidase activity and so the exploitation of this potential could be of immense value in the current search for new therapeutically effective drugs with inhibitory effects against β-glucosidase which is implicated in HIV infectivity of cells. The aim of this study however, is to ascertain the protective effect of the seed extract of *Picralima nitida* on hepatic and renal tissues so as to advice the users better on how to administer the drug.

2. MATERIALS AND METHODS

2.1 Materials

2.1.1 Collection of *Picralima nitida* seeds

The seeds of *Picralima nitida* were collected from Isuofia, Aguata Local Government Area of Anambra State, Nigeria and were authenticated by Mr. Alfred Ozioko of the Bioresource Development and Conservation Programme (BDCP) Research Centre, Nsukka, Nigeria.

2.1.2 Instruments/equipment

The following instruments were used for the experiment: Adjustable micropipette (PERFECT, USA), Chemical balance (Gallenkamp, England), Digital photo colorimeter (E1,312 Model, Japan), Incubator (Gallenkamp, England), Microscope slides (UNESCOPE, USA), pH meter (Pye, Unicam 293, England), Refrigerator (Kelvinator, Germany) and Spectrophotometer (NOVASPEC LKB Biochrome, model 4049, Germany

2.1.3 Chemicals/reagents

The commercial kits used for this study were products of Randox, QCA, USA and Biosystem Reagents and Instruments, Spain.

2.2 Methods

2.2.1 Extraction

Seeds of *Picralima nitida* plant were harvested and then dried at room temperature (29-35 °C) for three weeks, after which they were pulverized into powdered form with a Creston high speed milling machine. The powdered seeds (1 kg) were then macerated in 5 volume (w/v) absolute ethanol at room temperature for 48 hours. Afterwards, the extract was filtered through muslin cloth on a plug of glass wool in a glass column. The resulting ethanol extract was finely filtered using Whatman qualitative filter paper, concentrated and evaporated to dryness using rotary evaporated at an optimum temperature of between 40 and 45°C (to avoid denaturation of the active ingredients). The concentrated extract was stored in the refrigerator for subsequent studies.

2.2.2 Experimental design

A total of 180 albino mice of either sex weighing 20-34 g were housed in separate cages, acclimatized for one week and then divided into six groups of thirty mice each as follows:

- **Group 1**: Normal control (Normal mice treated with the vehicle, 3% tween 80)
- **Group 2**: Positive control (Mice inoculated with malaria parasite and treated with 3% tween 80)
- **Group 3**: Mice inoculated with malaria parasite and treated with 20 mg/kg b.w. of the extract
- **Group 4**: Mice inoculated with malaria parasite and treated with 40 mg/kg b.w. of the extract
- **Group 5**: Mice inoculated with malaria parasite and treated with 80 mg/kg b.w. of the extract
- **Group 6**: Standard control (Mice inoculated with malaria parasite and treated with the standard drug, artesunate)

The route of administration (treatment) was via oral route with the aid of an oral intubation tube.

2.2.3 Procurement of parasitaemia

Malaria parasitaemia (*Plasmodium berghei*) was obtained from malaria-infected mice at Veterinary Medicine Department of University of Nigeria, Nsukka. Ten drops of the parasitized blood obtained with the aid of a capillary tube through the ocular region of the mice, was diluted with 1 ml of normal saline. Thereafter, 0.2 ml of the diluted parasitized blood was used to passage three mice that served as the host from where subsequent ones were passaged.

2.2.4 Assay of aspartate aminotransferase (AST) activity

The activity of aspartate aminotransferase (AST) was assayed by the method of Reitman and Frankel [32] as outline in Randox kit.

Aspartate aminotransferase activity was measured by monitoring the formation of oxaloacetate hydrazone with 2, 4-dinitrophenylhydrazine.

The AST substrate phosphate buffer solutions (0.5 ml each) were pipetted into both the reagent blank and sample test tubes. Serum sample of 0.1 ml was added to the sample test tube only and mixed thoroughly. Then, 0.1 ml of distilled water was added to the reagent blank. Afterwards, the entire reaction medium was well mixed and incubated in a water bath at 37^0 C for 30 minutes.

A volume, 0.5 ml of 2, 4-dinitrophenylhydrazine was added to the reagent blank and the sample test tubes immediately after incubation. Also, 0.1 ml of the sample was added to blank only. The medium was mixed and allowed to stand for exactly 20 minutes at 25ºC. Finally, 5.0 ml of sodium hydroxide (NaOH) solution was added to both the blank and sample test tubes and mixed thoroughly. After 5 minutes, the absorbance of sample was read at 550 nm against the sample blank after 5 minutes. The activity of AST in mice serum was obtained from the already calibrated table.

2.2.5 Assay of alanine aminotransferase (ALT) activity

The activity of alanine aminotransferase was assayed by the method of Reitman and Frankel [32] as outlined in the Randox kit.

Alanine aminotransferase is measured by monitoring the concentration of pyruvate hydrazone formed with 2, 4-dinitrophenylhydrazine.

Aliquots, 0.5 ml of the ALT substrate phosphate buffer solutions were pipetted into two sets of test tubes labeled B (sample blank) and T (sample test). The serum sample (0.1 ml) was added to the sample test (T) test tubes only and mixed thoroughly and then, incubated in a water bath at temperature of 37ºC for 30 minutes.

A volume, 0.5 ml of 2, 4-dinitrophenylhydrazine was added to each of the test tubes labelled T (sample test) and B (sample blank) immediately after the incubation. Also, 0.1 ml of serum sample was added to sample blank (B) only. The entire medium was mixed thoroughly and allowed to stand at 25ºC for exactly 20 minutes. After this, 5.0 ml of sodium hydroxide (NaOH) solution was added to each test tube and mixed thoroughly. The absorbance of sample against the sample blank was read at a wavelength of 550 nm against the sample blank after 5 minutes.

The activity of ALT in the serum was obtained from a pre-calibrated table.

2.2.6 Determination of total bilirubin concentration

Total bilirubin concentration was determined using the method of Jendrassik and Grof [33] as outlined in the Randox kit.

A volume, 0.2 ml of sulphanilic acid was pipetted into the sample blank tube and sample tube. This was immediately followed by the addition of 0.05 ml of sodium nitrite to the sample tube. Caffeine (10 ml) and 0.2 ml of sample were also pipetted into each of the sample blank tube and sample tube. These mixtures were mixed and incubated for 10 minutes at 20-25°C. Finally, 1.0 ml of tartrate was pipetted into the sample blank tube and sample tube. These mixtures were once again mixed, incubated at 20-25°C for 30 minutes and their absorbances read at 578 nm against the sample blank.

2.2.7 Determination of Total Protein

Total protein concentration was determined using the method of Lowry et al. [34].

Eleven sets of test tubes were set up in a rack and filled with different concentrations (0, 10, 20, 30, 40, 50, 60, 70, 80, 90 and 100 µl) of Bovine Serum Albumin (BSA). A known volume, 2 ml of solution D (48 ml of 2 % Na_2CO_3 in 0.1 N NaOH; 1 ml of 1 % NaK Tartarate in H_2O and 1 ml of 0.5% $CuSO_4.5H_2O$ in H_2O). The mixture was incubated for 10 minutes at room temperature. An aliquot, 0.2 ml of dilute Folin-phenol as added to each test tube and vortex immediately. Afterwards, the contents were incubated for 30 minutes and the absorbance readings taken at 600 nm.

2.2.8 Determination of serum urea concentration

The concentration of serum urea was determined using the method of Tietz [35] as outlined in Randox kits, UK.

A known volume, 10 µl of the sample was pipetted into the sample tube; 10 µl of the standard was also pipetted into the standard tube while 10 µl of distilled water was added to the blank tube. A volume, 10 µl sodium nitroprusside and 10 µl urease were added to each of the three tubes. The tubes were mixed and incubated at 37°C for 15 minutes. Then, 2.50 ml of phenol was added to each of the three tubes followed by addition of 2.50 ml of sodium hypochlorite. After mixing, they were incubated at 37°C for 15 minutes and the absorbance of the sample was read against the reagent blank at 546 nm.

2.2.9 Determination of serum creatinine concentration

The concentration of serum creatinine was determined using the method of Tietz [35] as outlined in Randox kits, UK.

A volume, 100 µl, of distilled water was pipetted into the blank tube. Also 100 µl of the standard was added to the standard tube while 100 µl of the sample was pipetted into the sample tube. Then, 100 µl of the working reagent was pipetted into each of the three tubes and mixed. The absorbance of the sample was read against the blank at 492 nm.

2.2.10 Histopathological Study

The tissue specimens (liver and kidney) were excised from the sacrificed rats and fixed in 10% formaldehyde buffered saline (formal saline). After some days, the processing started with the dehydration in graded levels of alcohol (about 70-100%) in ascending order to remove the water content.

After the dehydration, the tissues were cleared in xylene impregnated with paraffin wax and sectioned at 5 microns thickness using rotary microtone. The sections were floated on a water bath maintained at a temperature of 2-3°C below melting point of the paraffin wax after which the sections were dried on a hot plate maintained at a temperature of 2-3°C above the melting point of the paraffin wax used. After proper drying, the sections were stained and mounted using haematoxylin and eosin in such a way as to avoid air bubbles.

2.3 Experimental Animals

The animals used for this study were white albino mice of either sex weighing 20-34 g. The mice were between 3-4 months old and were obtained from the Animal Unit of Faculty of Veterinary Medicine, University of Nigeria, Nsukka.

Clearance and approval for the humane use and handling of laboratory animals were given by the ethical committee of Biochemistry Department, University of Nigeria Nsukka.

3. RESULTS

In Table 1, the serum AST and ALT activities and the total protein level are presented. In normal mice, the AST activity of found to be 28.60±1.17 IU/L; the ALT activity was 44.60±2.78 IU/L and the total protein level was 6.74±0.19 g/dl. In the positive control group, AST and ALT activities was non-significantly ($p > 0.05$) elevated compared to the normal control. The total protein level in the positive control was significantly ($p < 0.05$) reduced compared to the normal control. In group 3, the AST activity was shown to be 30.20±3.58 IU/L on day 3 post treatment. This activity was non-significantly reduced to 29.40±3.44 IU/L on day 5 post treatment. In groups 4 and 5, the activity of AST was 29.60±3.01 and 29.00±3.02 IU/L on day 3 post treatment which reduced non-significantly to 29.00±3.02 and 28.80±2.85 IU/L on day 5 post treatment respectively. The total protein level was observed to be 5.78±0.10 g/dl in group 3, 5.96±0.10 g/dl in group 4 and 6.04±0.08 g/dl in group 5 on day 3 post treatment. These concentrations were non-significantly elevated to 5.86±0.10, 6.00±0.11 and 6.34±0.05 IU/L respectively on day 5 post treatment. Group 6 (standard control) showed AST activity of 30.80±2.60 IU/L on day 3 of post treatment which reduced to 29.40±1.57 IU/L on day 5 post treatment. The standard control showed ALT activity of 45.40±1.47 IU/L on day 3 post treatment which reduced further to 44.40±1.75 IU/L on day 5 post treatment. The standard control produced 5.74±0.12 g/dl on day 3 post treatment and 6.08±0.09 g/dl on day 5 post treatment. The effect of treatment with ethanol seed extract of *Picralima nitida* on AST activity and total protein level of the treated mice was dose-dependent while its effect on ALT activity of treated mice was non dose-dependent.

In Table 2 are presented the mean values of serum bilirubin (0.34±0.05 mg/dl), urea (20.00±0.32 mg/100 ml) and creatinine (0.72±0.01 mg/100 ml) of normal mice. There was significant ($p < 0.05$) elevations in serum urea and creatinine and non-significant ($p > 0.05$) elevation in serum bilirubin levels in the positive control compared to the normal control. In all experimental groups of mice, the mean values for bilirubin on day 3 post treatment were 1.46±0.10, 1.28±0.11 and 0.88±0.10 mg/dl respectively. On day 5 post treatment, the levels further decreased non-significantly to 1.40±0.11, 1.22±0.11 and 0.78±0.10 mg/dl respectively. The standard conctrol showed bilirubin levels of 1.12±0.12 mg/dl on day 3 post treatment and 1.04±0.13 mg/dl on day 5 post treatment. Compared to the positive control, groups 3 and 4 showed non-significant ($p > 0.05$) reductions in serum bilirubin level on days 3 and 5 post treatment. The treatment with ethanol seed extract of *Picralima nitida* showed a dose-dependent effect on serum bilirubin level in the treated mice. In normal mice, the serum urea level was found to be 20.00±0.32 mg/100 ml while that of the positive control was 34.80±0.37 mg/100 ml. On day 3 post treatment, groups 3, 4 and 5 produced decreases in serum urea level as 32.60±0.40, 29.20±0.37 and 31.00±0.71 mg/100 ml respectively which decreased further to 30.80±0.37, 28.40±0.51 and 24.40±0.68 mg/100 ml respectively on day 5 post treatment. The effect exerted by the treated groups on serum urea level of the treated mice followed a non dose-dependent fashion. The standard control showed serum urea level of 26.40±0.51 mg/100 ml on day 3 post treatment with a reduction of 1.00 mg/100 ml serum urea on day 5 post treatment. On days 3 and 5 post treatment, all the extract treated groups showed significant ($p < 0.05$) reductions in serum urea level compared to the positive control. The serum creatinine level of normal mice was found to be 0.72±0.01 mg/100 ml and that of positive control was 1.33±0.01 mg/100 ml. On day 3 post treatment, groups 3, 4 and 5 showed creatinine levels of 1.30±0.01, 1.14±0.01 and 1.14±0.02 which decreased to 1.22±0.01, 1.07±0.01 and 0.99±0.02 mg/100 ml respectively on day 5 post treatment. When compared with the positive control, groups 3, 4 and 5 showed significant ($p < 0.05$) reductions in serum creatinine concentration on days 3 and 5 post treatment. The effect was non dose-dependent. There was a significant ($p < 0.05$) elevation in serum creatinine level in the positive control compared to the normal control. The standard control showed the serum creatinine level of 1.07±0.02 mg/100 ml on day 3 post treatment which decreased to 1.03±0.02 mg/100 ml on day 5 post treatment.

The histomorphological changes in the liver sections of the different groups of mice (positive/untreated control, negative/normal control, standard control and treated groups) are shown in Plates 1. There is infiltration of mononuclear cells (inflammatory cells) around the portal area (PA) (portal hepatitis) (white arrow) in groups 2,5 and 6 while group 1 shows normal PA and hepatocytes (black arrow). The severity of inflammatory cells infiltration is more in 2 than 5 and 6.

The changes in the kidney sections of the different groups of mice (positive/untreated control, negative/normal control, standard control and treated groups) are presented in Plates 2. There is no infiltration of mononuclear cells around the glomeruli (G) and renal tubules (arrow) in the kidneys of all the experimental groups.

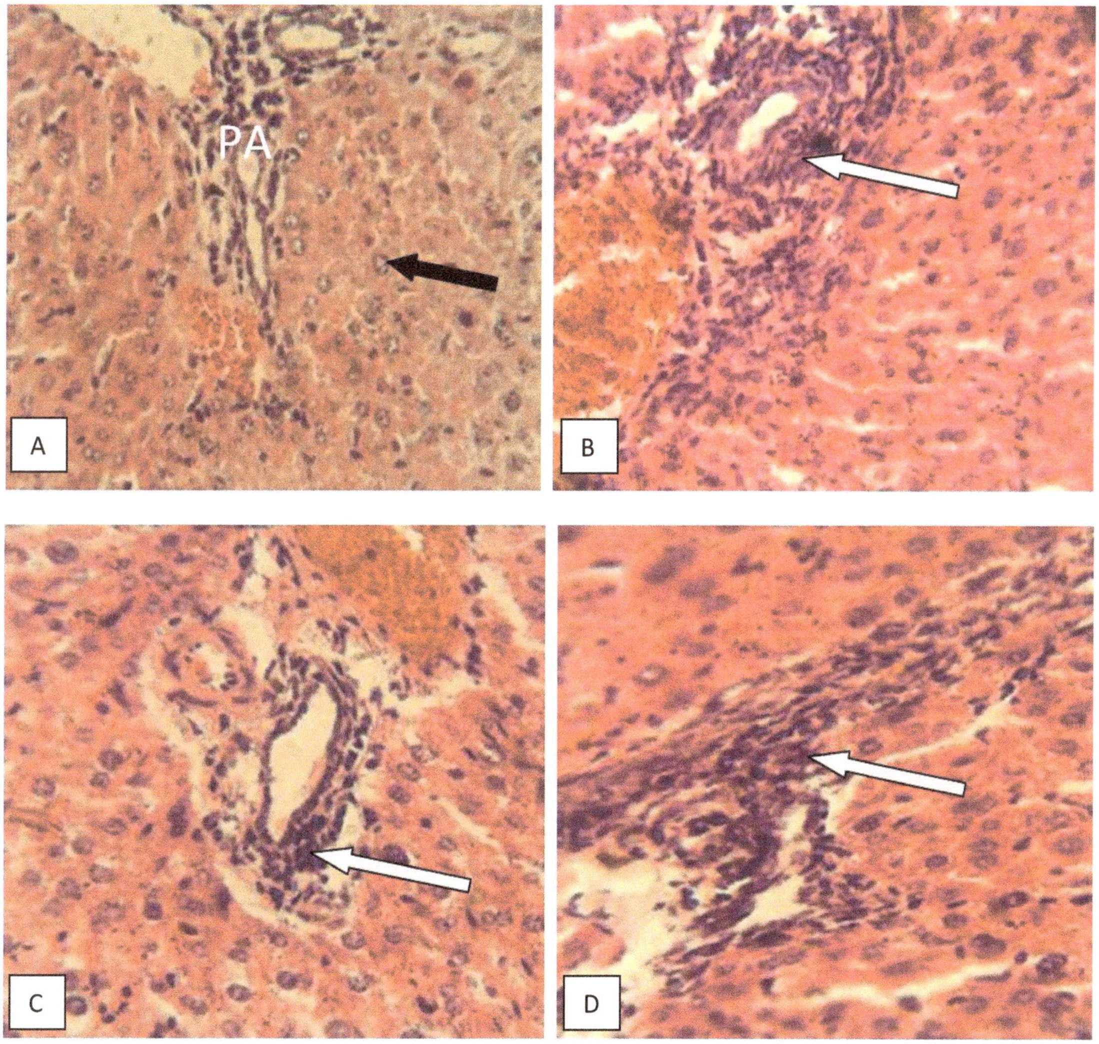

Plate 1. Photomicrograph of liver section of mice from experimental groups showing A-normal control, B-infected but untreated, C-infected and treated with artesunate, D-infected and treated with extract. The mononuclear infitration of cells around the portal area-portal hepatitis (white arrow) in groups B,C and D while the group A shows normal portal area(PA) and hepatocytes(black arrow). Note that the severity of inflammatory cells infiltration is more in B than D and C. H and E x 400.

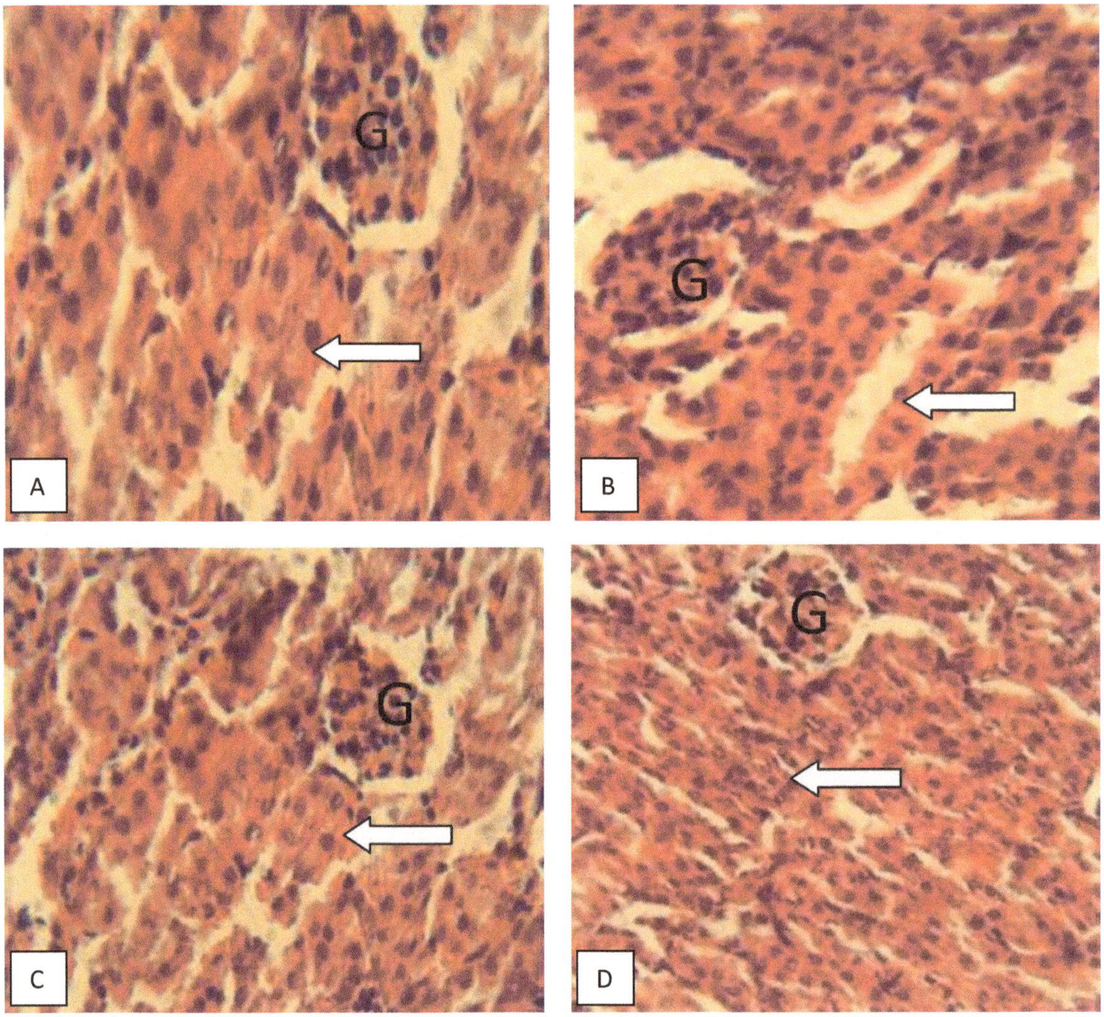

Plate 2. Photomicrograph of kidney section of mice from experimental groups showing A-normal control, B-infected but untreated, C-infected and treated with artesunate, D-infected and treated with extract. Note the apparently normal glomerulus(G) and renal tubules(arrow). H and E x 400

4. DISCUSSION

Elevated levels of AST and ALT [36] are indicative of damage to tissues or organs in the body such the parenchymal cells, red cells and other cells. However, ALT is a more specific enzyme marker for liver cells damage. The results of the effect of seed extract of *Picralima nitida* on AST activity of mice showed that group 3 had the highest mean AST activity, followed by group 4 while group 5 showed the least activity days 3 and 5 of post treatment. The extract treated groups produced better AST activity than the standard control on days 3 and 5 of post treatment. The highest ALT activity was shown by group 4 and the least activity was shown by group 5 on day 3 of post treatment. This effect of the extract on ALT of mice was non dose-dependent. However, group 3 showed the highest value; followed by group 4 and the least activity was shown by group 5 on day 5 of post treatment. The standard control showed better mean ALT activities than the extract treated groups on days 3 and 5 of post treatment. The evidence showing that the infected animals treated with the extract caused dose-dependent decreases in the activities of AST and ALT revealed that the extract protective effects on body liver and other tissues by lowering the activities of these enzymes. The time dependence of the effect of the treatment (Table 1) showed greater effect on day 5 than day 3 of post treatment. This correlates with the findings of George et al. [37] who asserted that serum ALT and AST activities were significantly higher in positive control (group inoculated with malaria but not treated) when compared with group treated with graded doses of *Aframomum sceptrum*.

Urea [22] is produced as a waste product of protein digestion by the liver which is normally removed from the body through the process of utrafiltration in the glomeruli of the kidneys. Creatinine, an important indicator of renal health, is a breakdown product of creatine phosphate in muscle produced at a fairly constant rate by the body [23]. It is an easily measured by-product of muscle metabolism excreted unchanged by the kidneys. Group 3 showed the highest mean urea and creatinine concentrations followed by group 5 while group 4 showed the least concentrations on day 3 of post treatment. Group 3 showed the highest urea and creatinine concentrations followed by group 4 and group 5 showed the least serum urea and creatinine concentrations on day 5 of post treatment. Day 3 post treatment of standard control showed more appreciable decreases in urea and creatinine concentrations when compared with the extract treated groups but group 5 produced better effects than the standard control on day 5 of post treatment. The evidence that the extracted treated groups caused non dose-dependent decreases in urea and creatinine concentrations on day 3 of post treatment and dose-dependent decreases on day 5 of post treatment revealed the ability of the extract to increase the kidney's functionality because it plays an active role in clearing urea and creatinine off the blood. The time dependence of the effects of the treatment (Table 2) produced greater effects on day 5 than day 3 of post treatment. This correlates with the findings of Zailani et al. [38] and Arise et al. [39]

Table 1. Table showing the values serum AST and ALT activities and total protein concentration

Groups	Day 3 of post treatment	Day 5 of post treatment	Parameters
			AST (IU/L)
1	28.60 ± 1.17*	N.D.	"
2	32.20 ± 3.20	N.D.	"
3	30.20 ± 3.58	29.40 ± 3.44	"
4	29.60 ± 3.01	29.00 ± 3.02	"
5	29.40 ± 3.01	28.80 ± 2.85	"
6	30.80 ± 2.60	29.40 ± 1.57	"
			ALT (IU/L)
1	44.60 ± 2.78	N.D.	"
2	49.40 ± 3.03	N.D.	"
3	48.80 ± 2.85	48.00 ± 2.88	"
4	49.20 ± 2.62	46.60 ± 2.54	"
5	46.20 ± 2.50	45.20 ± 2.87	"
6	45.40 ± 1.47	44.40 ± 1.75	
			Total Protein (g/dl)
1	6.74 ± 0.19*	N.D.	"
2	5.76 ± 0.10	N.D.	"
3	5.78 ± 0.10*	5.86 ± 0.10	"
4	5.96 ± 0.10*	6.00 ± 0.11	"
5	6.04 ± 0.08*	6.34 ± 0.05	"
6	5.74 ± 0.12	6.08 ± 0.09	"

** = p < 0.05 compared to the positive control (n=5)*

The results are expressed as mean ± standard error of mean (S.E.M)

N.D. = Not determined

***Group 1**: Normal control (Normal mice treated with the vehicle, 3% tween 80)*

***Group 2**: Positive control (Mice inoculated with malaria parasite and treated with 3% tween 80)*

***Group 3**: Mice inoculated with malaria parasite and treated with 20 mg/kg b.w. of the extract*

***Group 4**: Mice inoculated with malaria parasite and treated with 40 mg/kg b.w. of the extract*

***Group 5**: Mice inoculated with malaria parasite and treated with 80 mg/kg b.w. of the extract*

***Group 6**: Standard control (Mice inoculated with malaria parasite and treated with the standard drug, artesunate)*

Table 2. Table showing the serum total bilirubin, urea and creatinine concentrations

Groups	Day 3 of post treatment	Day 5 of post treatment	Parameters
			Bilirubin (mg/dl)
1	0.34 ± 0.05	N.D.	"
2	1.50 ± 0.11	N.D.	"
3	1.46 ± 0.10	1.40 ±0.11	"
4	1.28 ± 0.11	1.22 ± 0.11	"
5	0.88 ± 0.10*	0.78 ± 0.10*	"
6	1.12 ± 0.12	1.04 ± 0.13	
			Urea (mg/100 ml)
1	20.00 ± 0.32*	N.D.	"
2	34.80 ± 0.37	N.D.	"
3	32.60 ± 0.40*	30.80 ± 0.37	"
4	29.20 ± 0.37*	28.40 ± 0.51	"
5	31.00 ± 0.71*	24.40 ± 0.68	"
6	26.40 ± 0.51	25.40 ± 0.51	"
			Creatinine (mg/100 ml)
1	0.72 ± 0.01*	N.D.	"
2	1.33 ± 0.01	N.D.	"
3	1.30 ± 0.01*	1.22 ± 0.01	"
4	1.14 ± 0.01*	1.07 ± 0.01	"
5	1.14 ± 0.02*	0.99 ± 0.02	"
6	1.07 ± 0.02	1.03 ± 0.02	"

** = $p < 0.05$ compared to the positive control (n=5)*
The results are expressed as mean ± standard error of mean (S.E.M)
N.D. = Not determined

***Group 1**: Normal control (Normal mice treated with the vehicle, 3% tween 80)*
***Group 2**: Positive control (Mice inoculated with malaria parasite and treated with 3% tween 80)*
***Group 3**: Mice inoculated with malaria parasite and treated with 20 mg/kg b.w. of the extract*
***Group 4**: Mice inoculated with malaria parasite and treated with 40 mg/kg b.w. of the extract*
***Group 5**: Mice inoculated with malaria parasite and treated with 80 mg/kg b.w. of the extract*
***Group 6**: Standard control (Mice inoculated with malaria parasite and treated with the standard drug, artesunate)*

which stated that extracts of *Carica papaya*, *Alstonia boonei* and *Clerodendrum violaceum* Gürke have been observed to play vital roles in decreasing the serum urea and creatinine concentrations thus proving the efficacy of extracts in maintaining the integrity of the kidney function.

A rise in serum bilirubin concentration is an indication of liver damage, parenchyl or biliary [40]. Bilirubin was found to be produced in highest concentrations by group 3 and in least amounts by group 5 on days 3 and 5 of post treatment. The standard control showed better bilirubin clearance effect on mice than the extract treated groups on days 3 and 5 of post treatment. Only group 5 showed a significant decrease in serum total bilirubin concentration among the extract treated groups when compared with the positive control indicating that this extract has bilirubin clearance effect at higher doses. The evidence showing that extract treatment caused a dose-dependent decrease in total bilirubin concentration revealed that this extract enhance the liver function of clearing the blood unconjugated hydrophobic bilirubin by combining it with glucuronic acid making it water-soluble conjugated bilirubin using the enzyme UDP-glucuronide transferase. The time dependence of the effect of the treatment (Tables 1) showed greater effect on day 5 than day 3 of post treatment. This correlated with observation of Olorunnisola and Afolayan [41] who showed that leaf extract of *Sphenocentrum jollyanum* administered to malaria-infected mice significantly decreased the total bilirubin concentration. In conformation to the present study too, crude aqueous extract of the fruiting body of medicinal mushroom, *Ganoderma lucidum* has been found to significantly reduce the total bilirubin level in malaria-infected mice [42].

Decreased levels of albumin, the main constituent of total protein are noticed in chronic liver diseases such as cirrhosis and nephritic syndrome where it is lost through the urine [11]. The highest protein concentrations were produced by group 3, followed by group 4 and the least by group 3 on days 3 and 5 of post treatment. The extract treated groups produced better total protein that the standard control on days 3 and 5 of post treatment. The evidence showing that the extract treatment caused a dose-dependent increase in total protein revealed that the extract protected integrity of the liver which is involved in manufacturing of total proteins. The time dependence of the effect of the treatment (Table 1) showed greater effect on day 5 than day 3 of post treatment.

The histomorphological changes in the liver sections of the different groups of mice (Plates 1 – 4) showed inflammatory cells around the portal area (portal hepatitis) in groups 2,5 and 6 while group 1 shows normal portal area (PA) and hepatocytes. The severity of cellular infiltration (inflammation) is more in 2 than 5 and 6; an indication that the extract produced hepato-protective effect on the liver of the treated mice. The histomorphological changes in the Kidney sections of the different groups of mice (Plates 5-8) showed normal glomeruli(G) and renal tubules in the kidneys of all the experimental groups.

4. CONCLUSION

The decreased activities of the liver function enzyme markers, serum bilirubin level and increase in total protein of the treated groups are indicatives that the extract has hepato-protetective effect. The functionality of the kidney was enhance by the administration of this extract thus serum amounts of urea and creatinine were significantly reduced in the treated mice. Nevertheless, prolonged treatment and higher doses with such extract can pose a seroius danger and damage to these organs. The observed normal glomeruli and renal tubules in the untreated and treated groups shows that the kidney is not a target organ for the malaria parasite during its exo-erythrocytic stage and also indicates the apparent less toxic effect of the extract on body tissue.

COMPETING INTERESTS

Authors have declared that no competing interests exist.

REFERENCES

1. Vincke IH, Lips M. Un nouveau *Plasmodium* d'un rongeur sauvage du Congo: *Plasmodium berghei* n.sp. Ann Soc Belg Med Tr. 1948;28(1):97-104.
2. Hall N, Karras M, Raine JD, Carlton JM, Kooij TWA, Berriman M, et al. A comprehensive survey of the *Plasmodium* life cycle by genomic transcriptomic and proteomic analyses. Science. 2005;307(1): 82-86.
3. Kooij TW, Janse CJ, Waters AP. *Plasmodium* post-genomics: Better the bug you know? Nat Rev Microbiol. 2006;4(1): 344-357.
4. Janse CJ, Ramesar J, Waters AP. High-efficiency transfection and drug selection of genetically transformed blood stages of the rodent malaria parasite *Plasmodium berghei*. Nat. Protoc. 2006;1(1):346-356.
5. Janse CJ, Kroeze H, van Wigcheren A, Mededovic S, Fonager J, Franke-Fayard B, et al. A genotype and phenotype database of the genetically modified malaria parasites. Trends Parasitol. 2011;7(1):31-39.
6. Amino R, Ménard R, Frischknecht F. *In vivo* imaging of malaria parasites: Recent advances and future directions. Curr Opin Microbiol. 2005;8(1):407-414.
7. Franke-Fayard B, Waters AP, Janse CJ. Real-time *in vivo* imaging of transgenic bioluminescent blood stages or rodent malaria parasites in mice. Nature Protoc. 2006;1(1):476-485.
8. Lee, M. Basic Skills in Interpreting Laboratory Data. 5th ed. Texas: ASHP; 2009.
9. Johnston D. Special consideration in interpreting liver function tests. Am Fam Physician. 1999;59(8):2223-2230.
10. McClatchey KD. Clinical Laboratory Medicine. 2nd ed. Philadelphia: Lippincott Williams & Wilkins; 2003.
11. Menge MB, Schwiebert LP. Family Medicine: Ambulatory Care and Prevention. McGrawHill Professional; 2005.
12. Gaze DC. The role of existing and novel cardiac biomarker for cardioprotection. Curr Opin Invest Drugs. 2007;8(9):711-717.
13. Hayashi H, Mizuguchi H, Miyahara I, Nakajima Y, Hirotsu K, Kagamiyama H. Conformational change in aspartate aminotransferase substrate binding

induces strain in the catalytic group and enhances catalysis. J Biol Chem. 2003; 278(11):9481-9488.
14. Berg JM, Tymoczko JL, Stryer L. Biochemistry. New York: W.H. Freeman; 2006.
15. Anonymous. Aspartate aminotransferase. Assessed on 23rd April, 2014. Available: http/www.notebook.co.uk
16. Hirotsu K, Goto M, Okamoto A, Miyahara I. Dual substrate recognition of aminotransferases. Chemical Record. 2005; 5 (3): 160-172.
17. Pirone C, Quirke J, Martin E, Priestap HA, Lee DW. Animal pigment bilirubin discovered in plants. J Am Chem Soc. 2009; 131 (8): 2830.
18. McDonagh AF, Palma LA, Lightner DA. Blue light and bilirubin excretion. Science. 1980;208(4440):145-151.
19. Cheifetz AS. Oxford American Handbook of Gastroenterology and Hepatology. Oxford: Oxford University Press; 2010.
20. Kuntz E. Hepatology: Textbook and Atlas. 3rd ed. Germany: Springer; 2008.
21. Anonymous. National Library-National Institute of Health. 2010. Assessed 23 April, 2014. Available:https://www.kidney.org/sites/default/files/11-10-1815_HBE_PatBro_Urinalysis_v6.pdf
22. Kasper DL, Fauci AS, Longo DL, Braunwald G, Hauser SL, Jameson JI. Harrison's Principles of Internal Medicine. 18th ed. New York: McGraw-Hill; 2005.
23. Anonymous. Creatinine blood test. Assessed 22 July, 2014. Available:www.medicinenet.com/creatine_blood_test/article.htm
24. Allen PJ. Creatine metabolism and psychiatric disorders: Does creatine supplementation have therapeutic value. Neurosci Biobehav R. 2012;36(5):1442-1462.
25. Shemesh O. Golbetz H, Kriss JP, Myers RD. Limitations of creatinine as a filtration marker in glomerulopathic patients. Kidney Int. 1985;28(5):830-838.
26. Nkere CK, Iroegbu CU. Antibacterial screening of the root, seed and stembark extracts of *Picralima nitida*. Afr J Biotech. 2005;4(6):522-526.
27. Ubulom P, Akpabio E, Udobi CE, Mbon R. Antifungal activity of aqueous and ethanolic extracts of *Picralima nitida* seeds on *Aspergillu flavus*, *Candida albicans* and *Microsporium canis*. Res Pharm Biotech. 2011;3(5):57-60.
28. Okonta JM, Aguwa CN. Evaluation of hypoglemic activity of glycosides and alkaloids extracts of *Picralima nitida* Stapf (*Apocynaceae*) Seed. Int J Pharmacol. 2007;3(6):505-509.
29. Mojisola COC, Oladele AT, Elufioye TA. Ethnolobotanical survey of plants used as memory enhancer and antiaging in Ondo State, Nigeria. Int J Pharm. 2012;2(1):26-32.
30. Soladoye MO, Adetayo MO, Chukwuma EC, Adetunji AN. Ethnobotanical survey of plants used in the treatment of haemorrhoids in South-Western Nigeria. Ann Biol Res. 2010;1(4):1-15.
31. Enomfon JA, Ime BU. Inhibitory activity of seed extract from *Picralima nitida,* (Staph) on β-D-glucosidase. Biokemistri. 2004; 16(2):72-78.
32. Reitman S, Frankel S. A colorimetric method for the determination of serum glutamic oxalacetic and glutamic pyruvic transaminases. Am J Clin Path. 1957;28(1):56-63.
33. Jendrasskik J, Grof P. *In vitro* determination of total and direct bilirubin in serum or plasma. Biochemistry. 1938;297(1):81-89.
34. Lowry OH, Rosebrough NJ, Farr AL, Randall RJ. Protein measurement with the Folin-Phenol reagents. J Biol Chem. 1951;193(1):265-275.
35. Tietz NW. Textbook of Clinical Chemistry. 2nd ed. Philadelphia: W.B. Saunders Company; 1994.
36. Kew MC. Serum aminotransferase concentration as evidence of hepatocellular damage. Lancet. 2000; 355(1):591- 592.
37. George BO, Osioma E, Okpoghono J, Aina OO. Changes in liver and serum transaminases and alkaline phosphatase enzyme activities in *Plasmodium berghei* infected mice treated with aqueous extract of *Aframomum sceptrum.* Afr J Biochem Res. 2011;5(9):277-281.
38. Zailani AH, Balogun EA, Adebayo JO. Ethanolic extract of *Clerodendrum violaceum* Gürke leaves enhances kidney function in mouse model of malaria. Indian J Exp Biol. 2009;47(1):349-354.
39. Arise RO, Malomo SO, Lawal MM. Comparative antimalarial and toxicological effects of artemisinin with methanolic extract of *Carica papaya* leaves and bark

of *Alstonia boonei* in animal models. Adv Nat Appl Sci. 2012;6(2):116-123.

40. Collier J, Bassendine M. How to respond to abnormal liver function tests. Clin Med. 2002;2(5):406-409.
41. Olorunnisola OS, Afolayan AJ. *In vivo* antioxidant and biochemical evaluation of *Sphenocentrum jollyanum* leaf extract in *P. berghei* infected mice. Pakistan J Pharm Sci. 2013;26(3):445-450.
42. Oluba OM, Eidangbe GO, Adebisi KE, Odutuga AA, Onyeneke EC. *In vivo* evaluation of antimalarial and cytotoxic properties of crude aqueous extract of the fruiting body of medicinal mushroom, *Ganoderma lucidum* (W.Curt.:Fr.) P. Karst in *Plasmodium berghei*-infected mice. J Immunol Tech Infect Dis. 2013;2(4):1-6.

23

Distribution of Metabolites and Antioxidant Activity in the Berry and Shoot of *Solanum pseudocapsicum* Grown in Nigeria

A. E. Irondi[1*], E. J. Iweala[2], O. S. Agboola[3], J. K. Akintunde[1] and J. A. Ajiboye[4]

[1]Department of Biosciences and Biotechnology, Biochemistry Unit, Kwara State University, Malete, P.M.B. 1530, Ilorin, Nigeria.
[2]Department of Biological Sciences, Covenant University, Ota, Ogun State, Nigeria.
[3]Department of Veterinary Physiology, Biochemistry and Pharmacology, University of Ibadan, Nigeria.
[4]Department of Chemical Sciences, Biochemistry unit, College of Natural and Applied Sciences, Bells University of Technology, Ota, Ogun State, Nigeria.

Authors' contributions

This work was carried out in collaboration between all authors. Authors AEI and EJI designed the study, performed the statistical analysis, wrote the protocol and wrote the manuscript. Authors OSA and JKA managed the analyses of the study. Author JAA managed the literature searches. All authors read and approved the final manuscript.

Editor(s):
(1) Chunying Li, Department of Biochemistry and Molecular Biology Wayne State University School of Medicine, Detroit, USA.
Reviewers:
(1) G. Bupesh, Department of Virology, King Institute of Preventive Medicine & Research, Chennai, India.
(2) Normala Bt Halimoon, Universiti Putra Malaysia, Malaysia.

ABSTRACT

Aim: To evaluate the distribution of metabolites and *in vitro* antioxidant activity in the berry and shoot of *Solanum pseudocapsicum* (*S. pseudocapsicum*).
Study Design: Berry and shoot samples of *S. pseudocapsicum* were randomly collected from different plants of the species in the same natural habitat, and were subsequently processed and used for the spectrophotometric quantification of their metabolites and *in vitro* antioxidant activity.
Methodology: The samples were oven-dried to a constant weight, and were later milled separately into a fine particle size powder. The milled samples were extracted with methanol, and the extracts were analyzed for their metabolites content and *In vitro* antioxidant activity using appropriate

*Corresponding author: E-mail: irondi.emmanuel@yahoo.com

methods.
Results: The berry had significantly ($P < 0.05$) higher total protein, total lipid and amylose levels than the shoot. Conversely, the shoot had significantly ($P < 0.05$) higher starch, total soluble sugar, amylopectin, and vitamin C contents than the berry. Secondary metabolites (total phenol, tannin, total flavonoid, total saponin and total alkaloid) were consistently more predominant in the berry than in the shoot, except total carotenoid. Methanol extracts of both the berry and the shoot displayed antioxidant activity *in vitro*, with the berry having significantly ($P < 0.05$) higher ABTS radical scavenging ability and ferric reducing antioxidant power (FRAP) than the shoot. Both extracts also scavenged 1,1-diphenyl-2-picrylhydrazyl (DPPH) free radical in a dose-dependent manner; with the IC_{50} of the shoot (1.88 mg/mL) being lower than that of the berry (2.33 mg/mL).
Conclusion: The distribution of metabolites and antioxidant activity in the berry and the shoot of *S. pseudocapsicum* provide support for their medicinal uses. However, the preponderance of secondary metabolites and antioxidant activity in the berry than in the shoot might confer it with more potential to ameliorate oxidative stress in disease conditions, than the shoot.

Keywords: Solanum pseudocapsicum; plant metabolites; antioxidant activity; oxidative stress.

1. INTRODUCTION

Plant cells produce two types of metabolites, namely, primary and secondary metabolites. Primary metabolites including carbohydrates, lipids, proteins and organic acids are produced as a result of photosynthesis, and are directly involved in growth and metabolism. On the other hand, secondary metabolites such as alkaloids, phenolics, essential oils, terpenes, sterols, flavonoids, lignins and tannins, are considered products of primary metabolism, and are generally not involved in metabolic activity [1]; rather they protect plants against herbivores and microorganisms, attract pollinators and seed-dispersing animals, and act as signal molecules under stress conditions [2,3]. Secondary metabolites, otherwise known as phytochemicals, have been linked to many positive effects in disease conditions such as coronary heart diseases, diabetes, high blood pressure, cataracts, degenerative diseases, and obesity [4,5]. They are also the major source of pharmaceuticals, food additives, fragrances and pesticides [6]. The medicinal properties of plants have been investigated in the recent scientific developments throughout the world, due to their potent antioxidant activities, no side effects and economic viability [7]. Furthermore, novel natural antioxidants from some plants have been extensively studied in the past few years for their antioxidant and radical scavenging properties [7]. Reactive oxygen species (ROS) and reactive nitrogen species (RNS) are major sources of oxidative stress in cells, damaging proteins, lipids, and DNA [8]. Oxidative DNA damage has been implicated as a cause of cancer [9], aging and neurodegenerative diseases such as Alzheimer's and Parkinson's [10,11], cardiovascular diseases such as arteriosclerosis [12]. Therefore, prevention of oxidative stress caused by ROS and RNS has important implications for the prevention and treatment of disease.

There is increasing evidence that weeds such as *Solanum pseudocapsicum* are relatively high in bioactive secondary compounds which are likely to hold promise for drug discovery [13]. *S. pseudocapsicum* L. (*Solanaceae*) is an erect and highly branched shrub with non-spiny stem reaching a height of five meters. It is often cultivated as an indoor ornamental plant due to its beautiful berries. At maturity, the berries are attractive having red to yellow coloration. The number of seeds per berry ranges from 50 to 100 while the berries could be as many as 100 per plant [14]. The berries contain solanocapsine and other minor alkaloids [15]. The medicinal values of *S. pseudocapsicum* have been reported to include hepatoprotective [16]; antispasmodic, cytotoxic and antihypertensive [17]; antimicrobial [18]; and antitumor [19].

In view of the medicinal values of *S. pseudocapsicum*, this study was designed to evaluate the distribution of metabolites and antioxidant activity in the berry and shoot of S. *pseudocapsicum* grown in a natural habitat in Nigeria.

2. MATERIALS AND METHODS

2.1 Samples Collection and Preparation

About 20 g of berry and shoot samples of *S. pseudocapsicum* were randomly collected from different plants of the species in the same natural habitat in Akingbile area of Moniya, Ibadan, Nigeria, and were later authenticated at the

Department of Botany, University of Ibadan, Nigeria. Samples were then sorted, rinsed with distilled water and oven-dried at 50ºC to a constant moisture content of about 5%. The dried samples were subsequently milled to a fine particle size (0.5 mm), packed in air-tight plastic vials and stored at 4ºC until analysis.

2.1.1 Preparation of methanol extract

2 g of sample was extracted by steeping in 100 mL of methanol overnight, after which it was centrifuged at 3,000 rpm for 10 min. The supernatant, subsequently referred to as methanol extract, was collected and stored at -4ºC for further analysis.

2.2 Determination of Primary Metabolites

The primary metabolites (protein, total lipid, starch, total soluble sugar, vitamin C, amylose and amylopectin) levels were quantified using appropriate methods as follows: Protein and lipid [20]; starch and total soluble sugar [21]; and Vitamin C [22]. The amylose content was determined following the method of AACC [23], while the amylopectin was calculated by difference [24] using the following formula:

Amylopectin (%) = 100% – amylose (%).

2.3 Determination of Secondary Metabolites

2.3.1 Total phenol

The total phenol content of the extracts was determined according to the Folin–Ciocalteu method reported by Chan et al. [25], and was expressed as gallic acid equivalent (GAE) in mg/g sample.

2.3.2 Tannin

The tannin content of the extracts was determined according to the method reported by Amorim et al. [26], and was expressed as tannic acid equivalent (TAE) in mg/g sample.

2.3.3 Total flavonoid

The total flavonoid content, expressed as quercetin equivalent (QE) in mg/g sample, was determined using aluminum chloride method as reported by Kale et al. [27].

2.3.4 Total carotenoid

The total carotenoid content of the samples was determined according the method of Rodriguez-Amaya [28]. The total carotenoid was separated from other pigments in the ether extract by passing it through an alumina (Activity III) column, and eluting the total carotenoid with ether to a final volume of 25 mL. To avoid the degradation of the total carotenoid, analysis was carried out under subdued light and the extract was collected into a vial wrapped with aluminum foil.

2.3.5 Total saponin

The total saponin, expressed as diosgenin equivalent in mg/g of sample, was determined by the method described by Makkar et al. [29].

2.3.6 Total alkaloid

The total alkaloid content of the samples was measured using 1,10-phenanthroline method described by Singh et al. [30].

2.4 Determination of Antioxidant Activity

2.4.1 ABTS radical-scavenging ability

The ABTS* scavenging ability of both extracts were determined according to the method described by Sellappan and Akoh [31]. The ABTS* was generated by incubating equal volume of a 7 mM ABTS aqueous solution with $K_2S_2O_8$ (2.45 mM) in the dark for 16 h at room temperature and adjusting the absorbance at 734 nm to 0.7±0.02 with 95% ethanol. Then 0.2 mL appropriate dilution of the extract was added to 2.0 mL ABTS* solution and the absorbance was measured at 734 nm after 15 min. The trolox equivalent antioxidant capacity (TEAC) was subsequently calculated.

2.4.2 DPPH free radical scavenging ability

The DPPH free radical scavenging ability of the extracts was determined as described by Cervato et al. [32]. Briefly, appropriate dilution of the extracts (1 mL) was mixed with 3 mL of 60µM methanol solution of DPPH radicals; the mixture was left in the dark for 30 min before the absorbance was taken at 517 nm. The decrease in absorbance of DPPH˙ on addition of test samples in relation to the control was used to calculate the percentage scavenging ability, following the equation:

% scavenging ability = ($A517_{control}$ – $A517_{sample}$) ÷ $A517_{control}$ x 100.

The IC_{50} was calculated from the % scavenging ability versus extract concentration non-linear regression curve.

2.4.3 Ferric reducing antioxidant power

Reducing property of the extracts was determined by assessing the ability of the extract to reduce $FeCl_3$ solution as described by Oyaizu [33]. 2.5 mL aliquot was mixed with 2.5 mL of 200 mM sodium phosphate buffer (pH 6.6) and 2.5 mL of 1% potassium ferricyanide. The mixture was incubated at 50°C for 20 min. and then 2.5 mL of 10% trichloroacetic acid was added. This mixture was centrifuged at 650 rpm for 10 min. Then 5 mL supernatant was mixed with an equal volume of water and 1 mL of 0.1% ferric chloride. The absorbance was measured at 700 nm. The ferric reducing antioxidant property was subsequently calculated.

2.5 Statistical Analysis

Experimental results were reported as mean ± standard deviation of parallel triplicate measurements. Analysis of variance (ANOVA) was carried out on the result data at 95% confidence level using SPSS statistical software package (version 17).

3. RESULTS

The result of the primary metabolites distribution in the berry and the shoot of *S. pseudocapsicum* is presented in Table 1. The berry had significantly ($P < 0.05$) higher total protein, lipids and amylose levels than the shoot. Conversely, the shoot had significantly ($P < 0.05$) higher starch, total soluble sugar, amylopectin and vitamin C contents than the berry. The result further revealed that amylopectin was more abundant than amylose in both the berry and the shoot.

The result of the secondary metabolites distribution in the berry and the shoot of *S. pseudocapsicum*, presented in Table 2, showed that the berry had significantly ($P < 0.05$) higher total phenol, tannin, total flavonoid, total saponin and total alkaloid than the shoot. On the other hand, the total carotenoid content of the shoot was significantly ($P < 0.05$) higher than that of the berry.

The antioxidant activity of the berry and the shoot extracts, determined by their ability to scavenge DPPH and ABTS free radicals, and reduce Fe^{3+} to Fe^{2+}, is presented in Table 3. Both extracts exhibited DPPH free radical-scavenging activity in a dose-dependent manner, as depicted by their % scavenging ability-extract concentration curves (Fig. 1). However, their IC_{50} values (2.33 mg/mL and 1.88 mg/mL, for the berry and the shoot respectively), were higher than that of vitamin C (0.04 mg/mL), that was used as a reference antioxidant. Furthermore, the ABTS* scavenging ability of the extracts, reported as trolox equivalent antioxidant capacity (TEAC), revealed that the ability of the berry to scavenge ABTS* was significantly ($P < 0.05$) higher than that of shoot. The FRAP of the extracts, reported in gallic acid equivalent (GAE/g), revealed that the berry had a significantly ($P < 0.05$) higher reducing power than the shoot.

Table 1. Distribution of primary metabolites in the berry and shoot of *S. pseudocapsicum* on dry weight basis

Metabolite	Berry	Shoot
Protein (%)	18.23 ± 0.25^{a}	10.64 ± 0.25^{b}
Total lipid (%)	8.20 ± 0.03^{a}	1.16 ± 0.01^{b}
Starch (%)	15.20 ± 0.16^{b}	25.49 ± 0.61^{a}
Total soluble sugar (%)	9.63 ± 0.14^{b}	11.33 ± 0.10^{a}
Amylose (%)	1.02 ± 0.04^{a}	0.11 ± 0.01^{b}
Amylopectin (%)	99.01 ± 0.07^{b}	99.90 ± 0.01^{a}
Vitamin C (mg/g)	1.55 ± 0.03^{b}	1.91 ± 0.02^{a}

Data represent the mean ± standard deviation of triplicate readings.
Values with different lowercase superscript letter along the same row are significantly different (P<0.05)

Table 2. Distribution of secondary metabolites in the berry and shoot *S. pseudocapsicum* on dry weight basis

Metabolite	Berry	Shoot
Total phenol (mg/g)	7.02±0.07[a]	5.01±0.05[b]
Tannin (mg/g)	7.97±0.13[a]	5.87±0.13[b]
Total flavonoid (mg/g)	0.45±0.03[a]	0.31±0.01[b]
Total carotenoid (μg/g)	9.92±0.14[b]	21.39±0.18[a]
Total saponin (mg/g)	44.73±0.07[a]	20.11±0.27[b]
Total alkaloid (mg/g)	29.03±0.92[a]	18.12±0.11[b]

Data represent the mean ± standard deviation of triplicate readings. values with the different lowercase superscript letter along the same row are significantly different (P<0.05)

Table 3. DPPH $IC_{50,}$ trolox equivalent antioxidant capacity (TEAC) and ferric reducing antioxidant power (FRAP) of the methanol extracts of the berry and shoot of *S. pseudocapsicum*

Antioxidant activity	Berry	Shoot	Vitamin C
DPPH IC_{50} (mg/mL)	2.33±0.02[a]	1.88±0.01[b]	0.04±0.01[c]
ABTS* scavenging ability (μM TEAC/g)	451.12±0.67[a]	422.88±1.34[b]	NA
FRAP(mg GAE/g)	8.74±0.02[a]	6.25±0.07[b]	NA

Data represent the mean ± standard deviation of triplicate readings. TE is trolox equivalent. Values with different lowercase superscript letter along the same row are significantly different (P<0.05). NA means "not applicable"

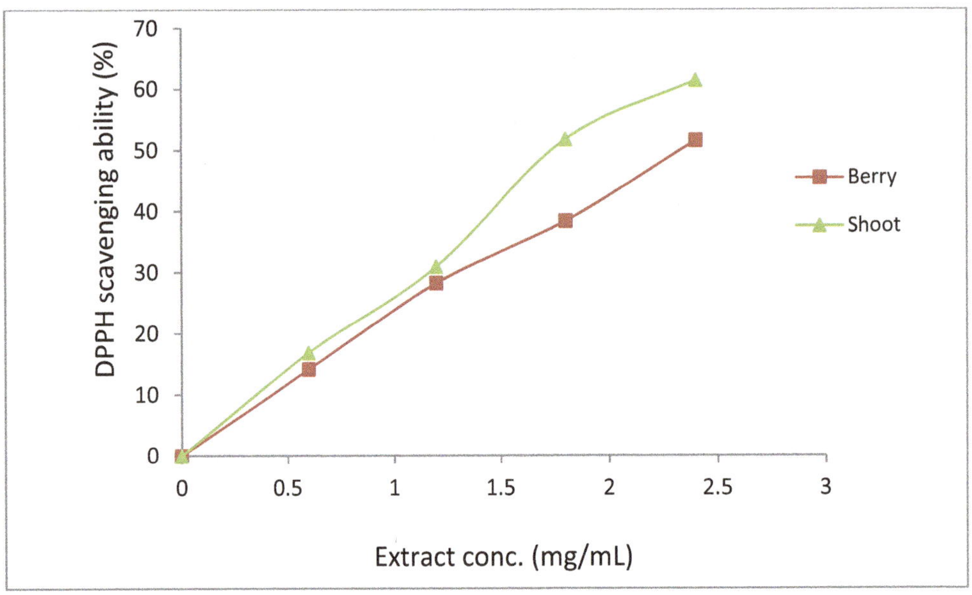

Fig. 1. DPPH % scavenging ability-extract concentration curve of the methanol extracts of the berry and shoot of *S. pseudocapsicum*

4. DISCUSSION

The distribution of the primary metabolites indicated that both the berry and the shoot of *S. pseudocapsicum* are good sources of protein. Proteins provide structural material for the human body; and they also function as enzymes, hormones, and antibodies. Dietary proteins are the major source of the essential amino acids for humans [34]. Plant storage proteins such as lectins are known to be involved in several biological processes, such as antimicrobial and antiviral action, and inhibitory effect on the growth of tumor cells [35]. Lipids, in addition to providing fuel for metabolism, are major components of cell membranes. Some plant lipids contain some bioactive polyunsaturated fatty acids (omega 3 and omega 6) that have been shown to be beneficial in preventing cardiovascular diseases, and decreasing the incorporation of cholesterol into the walls of arteries [36].

Intake of the berry and shoot of *S. pseudocapsicum* may not pose any glycemic problem, due to their low level of digestible carbohydrates (sugars and starches). Sugars and starches function primarily to provide energy to body cells, including the brain, which is entirely carbohydrate-dependent for its energy need. Sugars are involved in control of blood glucose and insulin metabolism, intestinal microflora activity, and food fermentation. Monosaccharides bound to protein and lipid molecules (glycoproteins and glycolipids) are involved in cell signaling [34]. The predominance of amylopectin over amylose in both the berry and the shoot of *S. pseudocapsicum* observed in this study is in agreement with the reports of Yotsawimonwat et al. [37] and Irondi et al. [38], who observed that amylopectin is the major component in most starch. Starch is composed of amylose (straight chains of glucose) and amylopectin (highly branched chains of glucose). Hydrolysis of amylose results in the release of fewer glucose molecules at once than the hydrolysis of the highly branched amylopectin chains. Thus, unlike amylopectin, high amylose content implies a lower glycemic index [39] due to its lower glucose response [40].

Vitamin C is a water-soluble antioxidant, acting as a leading natural antioxidant that can scavenge ROS, and it has anticarcinogenic effects [41,42]. It promotes iron absorption, increases antibody concentration, and improves human immunity. In addition, vitamin C is beneficial in strengthening blood vessel and decreasing cholesterol concentration; it also helps to prevent arteriosclerosis-related cardiovascular diseases, as well as hypertension and apoplexy [43].

Phenolic compounds are considered as being a major group of plant secondary metabolites that contribute to the antioxidant activity of plants. In addition, they possess biological activities, such as anti-histamine [44], anti-inflammatory and anticarcinogenic [7], antibacterial [45], and antiviral activities [46]. Several recent studies have also reported that phenolic compounds present in extracts from different plants foods and their products, including soybean and *Mangifera indica*, inhibited carbohydrate-digesting gastric enzymes [47,48], that have been implicated in the pathology of type 2 diabetes.

Saponins have a potential as pharmaceutical synthons, and they have been used in hormone synthesis [49]. They also have other biological activities including, hypoglycemic, antioxidant and hypolipidaemic activities; lowering of cancer risks, virucidal activity, reduction of protein digestibility and antimicrobial activity [50]. Alkaloids, as the most efficient therapeutically significant plant substances, have such pharmacological uses as anticancer agents, gout suppressant, muscle relaxant, antiarrythmic, antibiotic, and sedative agents [51]. Sharma et al. [52] also reported that alkaloids are preventive or therapeutic agents against human free radical associated diseases.

The higher level of total carotenoid in the shoot than in the berry may be related to the presence of chlorophyll in shoot. Gitelson et al. [53] has earlier established a linear relationship between chlorophyll and carotenoids contents in higher plant leaves. Carotenoids are lipid-soluble antioxidant, which have been reported to have health benefits such as pro-vitamin A activity [54], antioxidant [55], anticancer [56], and antiobesity effects [57].

Increase in the free radicals and reactive oxygen species burden of the body cells, without a corresponding robust antioxidant defense system leads to oxidative stress, and this plays an important role in the pathogenesis of degenerative diseases [58]. Antioxidants react with DPPH, which is a stable free radical, and convert it to α,α-diphenyl-β-picryl hydrazine. As antioxidants donate protons to this radical, its absorption decreases. So the degree of discoloration indicates the scavenging potentials of the antioxidant extract [59]. The lower DPPH IC_{50} of vitamin C, a reference antioxidant, in comparison with the berry and the shoot of *S. pseudocapsicum*, indicated that vitamin C was more potent in scavenging the DPPH free radical than the berry and the shoot. However, the scavenging ability of the shoot was stronger than that of the berry. Irondi et al. [60] noted that since IC_{50} is a measure of inhibitory concentration, a lower IC_{50} value indicated a greater antioxidant activity.

Extracts of both the berry and the shoot of *S. pseudocapsicum* also scavenged ABTS radical.

The ABTS assay is based on the inhibition of the absorbance of the radical cation ABTS* by antioxidants in a reaction that involves electron transfer process [7]. The ferric reducing antioxidant power (FRAP) of the extracts indicated their ability to reduce Fe^{3+} to Fe^{2+}. In FRAP assay, potassium ferricyanide is used as an oxidant, in a reaction that involves electron transfer, during which there is the reduction of ferric 2,4,6-tripyridyl-*s*-triazine (TPTZ) to the coloured ferrous form [61].

5. CONCLUSION

The results of this study revealed that the berry and the shoot of *S. pseudocapsicum* contain appreciable levels primary and secondary metabolites, including protein, lipid, starch, sugar, amylose, amylopectin, vitamin C, total phenol, tannin, total flavonoid, total saponin, total alkaloid, and total carotenoid; and possess antioxidant activity *In vitro*. Comparatively, the berry might have more pharmacological benefits in alleviating oxidative stress in disease conditions than the shoot, due to its higher content of secondary metabolites and *In vitro* antioxidant activity.

COMPETING INTERESTS

Authors have declared that no competing interests exist.

REFERENCES

1. Ramawat KG, Dass S, Mathur M. The chemical diversity of bioactive molecules and therapeutic potential of medicinal plants. In: Ramawat KG. editor. Herbal Drugs: Ethnomedicine to Modern Medicine. 2009;7-32.
2. Seiger DS. Plant secondary metabolism. Dordrecht, the Netherlands: Kluwer Academic Publishers. 1998;592.
3. Crozier A, Clifford M, Ashihara H. Plant secondary metabolites: Occurrence structure and role in the human diet. Oxford, U.K. Blackwell Publishing Ltd. 2006;384.
4. Liu S, Manson JE, Lee IM, et al. Fruit and vegetable intake and risk of cardiovascular disease: The women's health study. Am J Clin Nutr. 2000;72:922-8.
5. Djouss´e L, Arnett DK, Coon H, et al. Fruit and vegetable consumption and LDL cholesterol: The National Heart Lung and Blood Institute Family Heart Study. Am J Clin Nutr. 2004;79:213-7.
6. Ramawat KG. Secondary metabolites in nature. In: Ramawat KG, Merillon JM. Editors. Biotechnology: Secondary Metabolites. Science Publishers, Enfield, CT. 2007;21.
7. Srikanth G, Babu SM, Kavitha CHN, et al. Studies on *In-vitro* antioxidant activities of *Carica papaya* aqueous leaf extract. Res J Pharm Biol Chem Sci. 2010;1(2):59-65.
8. Orrenius S, Gogvadze V, Zhivotovsky B. Mitochondrial oxidative stress: Implications for cell death. Annu Rev Pharmacol Toxicol. 2007;47:143-183.
9. Huang X. Iron overload and its association with cancer risk in humans: Evidence for iron as a carcinogenic metal. Mutat Res. 2003;533:153-171.
10. Markesbery WR, Lovell MA. DNA oxidation in Alzheimer's disease. Antioxid Redox Signaling. 2006;8:2039-2045.
11. Halliwell B. Role of free radicals in the neurodegenerative diseases. Drugs and Aging. 2001;18:685-716.
12. Vokurkova M, Xu S, Touyz RM. Reactive oxygen species, cell growth, cell cycle progression and vascular remodeling in hypertension. Future Cardiol. 2007;3:53-63.
13. Stepp JR. The role of weeds as sources of Pharmaceuticals. J Ethnopharmacol. 2004;92:25-31.
14. Bassett IJ, Munro DB. The biology of Canadian weeds: *Solanum ptycanthum* Dun, *S. nigrum* L. and *S. sarrachoides* Sendt. Can J Plant Sci. 1985;65:401-414.
15. Chakravarty AK, Das B, Ali E, et al. Studies on Indian Medicinal Plants. Part 77. Structure and stereochemistry of some new steroidal alkaloids from *Solanum pseudocapsicum* and Solanum giganteum by nuclear magnetic resonance spectroscopy. J Chem Soc Pekin Trans. 1984;1:467-74.
16. Vijayan P, Prashanth HC, Preethy Vijayaraj, et al. Hepatoprotective Effect of the Total Alkaloid Fraction of *Solanum pseudocapsicum* Leaves. Pharm Biol. 2003;41(6):443-48.
17. Dhar ML, Dhar MM, Dhawan BN, et al. Screening of Indian plants for biological activity. Part IV. Indian J Exp Biol. 1973;11:43-54.
18. Mitscher LA, Juvarkar JV, Beal JL. Solacasine, a new steroidal alkaloid from *Solanum pseudocapsicum* possessing

antimicrobial activity. Experientia. 1976;32:415-16.
19. Vijayan P, Kumar SV, Dhanaraj SA, et al. In vitro citotoxity and anti-tumor properties of the total alkaloid fraction of unripe fruits of *Solanum pseudocapsicum*. Pharm Biol. 2002;40(6):456-60.
20. Association of Official Analytical Chemists (AOAC). Official Methods of Analysis of Association of Analytical Chemists International, Horwitz, W. (ed). AOAC International Publs, Maryland USA. Ch. Edition 17. 2000;1,2:12-20.
21. Onitilo MO, Sanni LO, Daniel I, Maziya-Dixon B, Dixon A. Physicochemical and Functional Properties of Native Starches from Cassava Varieties in South-west Nigeria. J Food Agric Environ. 2007;5(3&4):108-114.
22. Benderitter M, Maupoil V, Vergely C, et al. Studies by electron paramagnetic resonance of the importance of iron in the hydroxyl scavenging properties of ascorbic acid in plasma: Effects of iron chelators. Fundam Clin Pharmacol. 1998;12:510-16.
23. American Association of Cereal Chemists, "AACC method 61-03.01: Amylose content of milled rice", in "Approved Methods of the American Association of Cereal Chemists" (Ed. K Cummings), AACC International, St. Paul MN. 2000;1-4.
24. Juan G, Luis A, David B. Isolation and molecular characterization of Makal (*Xanthosoma yucatanensis*) starch. Starch. 2006;58:300-307.
25. Chan EWC, Lim YY, Chew YL. Antioxidant activity of *Camellia sinensis* leaves and tea from a lowland plantation in Malaysia. J food chem. 2007;102:1214-1222.
26. Amorim ELC, Nascimento JE, Monteiro JM, et al. A simple and accurate procedure for the determination of tannin and flavonoid levels and some applications in ethnobotany and ethnopharmacology. Funct Ecosyst Commun. 2008;2:88-94.
27. Kale A, Gaikwad S, Mundhe K, et al. Quantification of Phenolics and Flavonoids by Spectrophotometer from -*Juglans regia*. Int J Pharma Bio Sci. 2010;1:1-4.
28. Rodriguez-Amaya DB. A guide to carotenoid analysis in foods. ILSI Press, Washington DC; 1999.
29. Makkar HPS, Siddhuraju P, Becker K. Plant secondary metabolites. Humana Press Inc., Totowa, NJ, USA; 2007.
30. Singh DK, Srivastva B, Sahu A. Spectrophotometric determination of Rauwolfia alkaloids, estimation of reserpine in pharmaceuticals. Anal Sci. 2004;20:571-573.
31. Sellappan S, Akoh CC. Flavonoids and antioxidant capacity of Georgia-grown vidalia onions. J Agric Food Chem. 2002;50:5338–5342.
32. Cervato G, Carabelli M, Gervasio S, et al. Antioxidant properties of oregano (*Origanum vulgare*) leaf extracts. J Food Biochem. 2000;24:453-65.
33. Oyaizu M. Studies on products of browning reaction: Antioxidative activity of products of browning reaction prepared from glucosamine. Jpn J Nutr. 1986;44:307-15.
34. Hounsome N, Hounsome B, Tomos D, et al. Plant Metabolites and Nutritional Quality of Vegetables. J Food Sci. 2008;73(4):R48-R65.
35. De Souza Candido E, Pinto MFS, Pelegrini PB, et al. Plant storage proteins with antimicrobial activity: Novel insights into plant defense mechanisms. FASEB J. 2011;25:3290–3305.
36. Vrablík M, Prusíková M, Šnejdrlová M, et al. Omega-3 Fatty Acids and Cardiovascular Disease Risk: Do We Understand the Relationship? Physiol Res. 2009;58(Suppl. 1):S19-S26.
37. Yotsawimonwat S, Sriroth K, Kaewvichit S, et al. Effect of pH on complex formation between debranched waxy rice starch and fatty acids. Int J Biol Macromol. 2008;43:94-99.
38. Irondi AE, Anokam KK, Chukwuma PC, Akintunde JK, Nurain IO. Variation in nutrients composition of *Tetrapleura tetraptera* fruit at two maturity stages. Int J Biosci. 2013;3(9):304-312.
39. Shanita SN, Hasnah H, Khoo CW. Amylose and amylopectin in selected Malaysian foods and its relationship to glycemic index. Sains Malaysiana. 2011;40(8):865–870.
40. Hallfrisch J, Behall KM. Mechanisms of the effects of grains on insulin and glucose responses. J Am College Nutri. 2000;19(3):320S–325S.
41. Kim DO, Lee CY. Comprehensive study on vitamin C equivalent antioxidant capacity (VCEAC) of various polyphenolics in scavenging a free radical and its structural relationship. Crit Rev Food Sci Nutr. 2004;44:253-73.
42. Lee SH, Oe T, Bliar IA. Vitamin C induced decomposition of lipid hydroperoxides to

endogenous genotoxins. Sci. 2001;292:2083-2086.
43. Iqbal Y, Ihsanullah I, Shaheen N, et al. Significance of vitamin C in plants. J Chem Soc Pak. 2009;31:169–170.
44. Nitta Y, Kikuzaki H, Ueno H. Food components inhibiting recombinant human histidine decarboxylase activity. J Agric Food Chem. 2007;55:299–304.
45. Romero C, Medina E, Vargas J, et al. In vitro activity of olive oil polyphenols against Helicobacter pylori. J Agric Food Chem. 2007;55:680–686.
46. Song JM, Lee KH, Seong BL. Antiviral effect of catechins in green tea on influenza virus. Antivir Res. 2005;68:66–74.
47. Ademiluyi AO, Oboh G. Soybean phenolic-rich extracts inhibit key-enzymes linked to type 2 diabetes (α-amylase and α-glucosidase) and hypertension (angiotensin I converting enzyme) *In vitro*. Exp Toxicol Pathol. 2013;65:305–309.
48. Irondi EA, Anokam KK, Chukwuma PC. Phenological Variation in the *In-vitro* Antioxidant Properties and α-Amylase Inhibitory Activity of *Tetrapleura tetraptera* Pod. Int J Pharm Sci Drug Res. 2013;5(3):108-112.
49. Yücekutlu AN, Bildacı I. Determination of Plant Saponins and Some of *Gypsophila* Species: A review of the literature. Hacet J Biol Chem. 2008;36(2):129-135.
50. Desai SD, Desai DG, Kaur H. Saponins and their Biological Activities. Pharm Times. 2009;41(3):13-16.
51. Facchini PJ. Alkaloid biosynthesis in plants: Biochemistry, cell biology, molecular regulation, and metabolic engineering applications. Annu Rev Plant Physiol Plant Mol Biol. 2001;52:29–66.
52. Sharma GC, Sharma A, Singh SK, et al. Ethanomedicinal trends in Indian tradition for treatment of vitiligo. J Med Plants Res. 2012;6(10):1827-1833.
53. Gitelson, AA, Gritz Y, Merzlyak MN. Relationships between leaf chlorophyll content and spectral reflectance and algorithms for non-destructive chlorophyll assessment in higher plant leaves. J Plant Physiol. 2003;160:271–282.
54. Rodriguez-Amaya DB. Carotenoids and food preparation: The retention of Provitamin A carotenoids in prepared, processed, and stored foods. Department of Food Science Faculty of Food Engineering, State University of Campinas. Brazil. John Snow, Inc./OMNI Project. 1997;93.
55. Tian B, Xu Z, Sun Z, et al. Evaluation of the antioxidant effects of carotenoids from *Deinococcus radiodurans* through targeted mutagenesis, chemiluminescence, and DNA damage analyses. Biochem Biophys Acta. 2007;1770:902-911.
56. Nishino H. Cancer prevention by carotenoids. Mutat. Res. 1998;402:159-163.
57. Jaswir I, Noviendri D, Hasrini RF, et al. Carotenoids: Sources, medicinal properties and their application in food and nutraceutical industry. J Med Plants Res. 2011;5(33):7119-7131.
58. Halliwell B, Gutteridge JMC. Free Radicals in Biology and Medicine, 3rd edn. Oxford University Press, Inc., New York. 1999;105-245.
59. Huang D, Ou B, Prior RL. The chemistry behind antioxidant capacity assays. J Agric Food Chem. 2005;53(6):1841-1856.
60. Irondi AE, Oboh G, Akintunde JK. Comparative and Synergistic Antioxidant properties of *Carica papaya* and *Azadarichta indica* Leaves. Int J Pharm Sci Res. 2012;3(12):4773-4779.
61. Ndhlala AR, Moyo M, Staden JV. Natural antioxidants: Fascinating or mythical biomolecules? Molecules. 2010;15:6905-6930.
DOI: 10.3390/molecules15106905.

24

Some Indices of Carbohydrate Metabolism in Rats Maintained on Heat-treated *Vigna unguiculata* Husk

A. S. Idoko[1*], N. Lawal[1] and A. Balkisu[1]

[1]Department of Biochemistry, Federal University Dutsin-Ma, Katsina State, Nigeria.

Authors' contributions

This work was carried out in collaboration between all authors. Author ASI designed the experiment, performed the statistical analysis and interpreted the results. Author NL was in charge of the literature search and writing of the first draft. Author AB administered the diets and typed the manuscript. All authors read and approved the final manuscript.

Editor(s):
(1) Mohamed Fawzy Ramadan Hassanien, Biochemistry Department, Zagazig University, Egypt.
Reviewers:
(1) Barnabé Lucien Nkono Ya Nkono, University of Yaounde, Cameroon.
(2) Ji Yeon Kim, Seoul National University of Science and Technology, South Korea.

ABSTRACT

The high cost of feed ingredients such as corn is one of the most important factors in livestock production. If agricultural by-products such as *Vigna unguiculata* husk could replace corn starch in animal feed, the cost of raising livestock would be cheaper. Some indices of carbohydrate metabolism in rats maintained on heat-treated *Vigna unguiculata* husk were investigated to ascertain the extent of carbohydrate metabolism in rats maintained on the husk. Forty-six weanling albino rats (31.11 g±3.32 g) were divided into seven groups of equal average weight. The groups were then assigned to seven experimental diets. The rats were fed on their respective experimental diets and water *ad libitum.* Blood pyruvate and lactate concentrations, glucose 6-phosphatase activities in selected tissues, and serum lipid profiles were determined. The results revealed that the concentrations of pytuvate (0.86 mg per 100 ml blood) and lactate (12.96 mg per 100 ml blood) in rats fed 50% inclusion level of 90-minute boiled husk were not significantly different (p =.05) from the respective values (0.85 mg per 100 ml blood and 12.53 mg per 100 ml blood) in rats fed the control diet. Also, 50% inclusion level of 90-minute boiled husk caused no significant changes in

**Corresponding author: E-mail: asidoko1@gmail.com*

the lipid profiles. With the exception of rats maintained on the control diet, the liver (37.46 nmol/hr/mg Protein) and kidney (19.02 nmol/hr/mg Protein) concentrations of glucose 6-phosphatase activities in rats fed 50% inclusion level of 90-minute boiled husk were significantly lower ($p = 0.05$) than the concentrations in the liver and kidney of rats maintained on other diets. The findings show that diets based on 50% inclusion level of 90-minute boiled husk are well metabolized. Therefore, 50% replacement of corn starch by 90-minute boiled *Vigna unguiculata* pod husk may be considered as a way of reducing the cost of livestock production.

Keywords: Heat-treated; Vigna unguiculata; pod husk; carbohydrate; pyruvae; lactate.

1. INTRODUCTION

The need for increased production of live stock had long been established. This in part is due to the rapidly growing demand for animal products. Animal products such as meat, milk and eggs provide much required nutrition, but developing country consumption levels are still low due to the high cost of these products. Apart from the importance of animal production to national economies in contributing to national income, improved human nutrition and foreign exchange, livestock play an important role in contributing to rural livelihoods, employment and poverty relief [1]. Since high cost of feed ingredients such as corn is one of the most important factors in livestock production, looking closely at locally available feed resources becomes necessary. If agricultural by-products such as *Vigna unguiculata* pod husk could replace corn starch in animal feed, the cost of raising livestock would be much cheaper. Most *Vigna unguiculata* are grown on the African continent, particularly in Nigeria and Niger which account for 72% of world *Vigna unguiculata* production [2]. Commonly called cowpea beans, *Vigna unguiculata* is an annual herbaceous legume cultivated for its edible seeds or for fodder. It is one of several species of the widely cultivated genus *Vigna* [3]. *Vigna unguiculata* seeds are contained in cylindrical pods, 6 to 2 cm long and 3-12 mm broad and each pod may contain 8 to 20 seeds. The seeds are usually separated from the pods by threshing. The pod husks are used as they are in many places to feed livestock [4]. However, information on the metabolizability of carbohydrate in rats maintained on *Vigna unguiculata* seed pod husk remained scanty. This research was therefore designed to ascertain the extent of carbohydrate metabolism in rats maintained on heat-treated *Vigna unguiculata* pod husk.

2. MATERIALS AND METHODS

Fresh and mature *Vigna unguiculata* pods were bought from a farm in Katagum Local Government of Bauchi State of Nigeria. Corn, soybeans, soybean oil and sucrose were bought from Ilorin central market, while rice bran was collected from a rice milling factory in Ilorin.

Forty-six (46) weanling albino rats (31.11 g±3.32 g) were obtained from the animal holding unit of the Department of Biochemistry, University of Ilorin, Ilorin, Nigeria.

All the assay kits used in this study were from Randox Laboratories Ltd. (United Kingdom).

2.1 Sample Preparation

The pods were sorted out to remove bad ones. The pod husks were manually removed from the pods, washed with water, sun-dried and then divided into 3 portions: 1, 2 and 3. Portion 1 was left unheated, while portions 2 and 3 were subjected to 60-minute and 90-minute boiling respectively. The three portions were separately pulverised and stored in clean polyethylene bags.

2.2 Feed Formulation

The pulverised unheated, 60-minute boiled and 90-minute boiled *Vigna unguiculata* husk were separately mixed with other ingredients to formulate seven experimental diets to replace dietary corn starch at 0%, 50% and 70%. Diet C (Control) contained 100% inclusion level of corn starch. Diets R, B60 and B90 contained 70% inclusion levels of unheated, 60-minute boiled and 90-minute boiled *Vigna unguiculata* husk respectively, while diets CR, CB60 and CB90 contained 50% inclusion levels of unheated, 60-minute boiled and 90-minute boiled *Vigna unguiculata* husk. The food items were thoroughly mixed together and manually made into pellets to feed albino rats.

2.3 Assigning of Experimental Animals

The forty-six weanling albino rats were assigned into seven groups. The six rats in each group were housed in standard plastic laboratory cages

Table 1. Components of the formulated diets (gkg^{-1})

	C	R	B60	B90	CR	CB60	CB90
B60	-	-	361.2	-	-	-	-
B90	-	-	-	361.2	-	-	-
R	-	361.2		-	-	-	-
C	516	154.8	154.8	154.8	258	258	258
CA	-	-	-	-	258	-	-
CB	-	-	-	-	-	258	-
CR	-	-	-	-	-	-	258
Soy meal	250	250	250	250	250	250	250
Soybean oil	40	40	40	40	40	40	40
Rice bran	40	40	40	40	40	40	40
D-L methionine	04	04	04	04	04	04	04
sucrose	100	100	100	100	100	100	100
Vit/min mix	50	50	50	50	50	50	50

with stainless steel covers. The groups were then maintained on their respective experimental diets and water *ad libitum* for six weeks. The rats were kept in accordance with the recommendation of ARRP [5].

2.4 Animal Sacrifice, Collection of Blood Samples and Preparation of Tissue Homogenate

After 42 days of feeding trial, the animals were weighed and sacrificed by anaesthetizing them in a jar containing cotton wool soaked in diethylether. Blood was obtained through the jugular veins in tubes containing EDTA as anticoagulant and in plain tubes. Serum was separated from the blood in plain tubes by allowing the blood to clot for 3 hours. The clotted blood was spun in a bench top centrifuge at 1500 rpm for 15 minutes to obtain serum. The serum samples were thereafter separated into another set of plain sample tubes and kept in the refrigerator [6]. The sacrificed rats were dissected to remove the liver and kidney. The isolated tissues were weighed and a portion cut out, cut into very small pieces and then homogenized in ice-cold 0.25 M sucrose solution to make a 1 in 6 dilution. The tissue homogenates were kept frozen while being used [7].

2.5 Determination of Pyruvate and Lactate Concentration in the Blood

The method of Gloster and Harris [8] was employed in the determination of pyruvate and lactate. The method is based on the reversible conversion of lactate to pyruvate catalyzed by lactic dehydrogenase and is driven to completion by excess of oxidized form of nicotinamide adenine dinucleotide (NAD) or reduced form of nicotinamide adenine dinucleotide (NADH).

2.6 Determination of Glucose-6-Phosphatase Activity

The method described by Swanson [9] was used for the determination of glucose 6-phosphatase activity in the liver and kidney. This method is based on the incubation of the specific substrate with the enzyme source and the determination of liberated inorganic orthophosphate.

$$\text{Glucose 6-phosphate} \xrightarrow{\text{glucose-6-phosphatase}} \text{Glucose} + P_i$$

2.7 Determination of Serum Lipid Profile and Atherogenic Indices

The total cholesterol (TC) was estimated according to the methods of Abell et al. [10] while triglycerides (TG) and high density lipoprotein cholesterol (HDL-C) were estimated according to the methods of Tietz [11]. Low density lipoprotein cholesterol (LDL-C) was computed from the expression: LDL-C (mmol/l) = TC-TG/ (2.2-HDL-C) [12]. Both the Cardiac Risk Ratio (CRR) and Atherogenic Coefficient (AC) were determined using the method of Ikewuchi and Ikewuchi [13], while the atherogenic index was by the method of Tan et al. [14].

Cardiac Risk Ratio (CRR) = TC/HDLC.
Atherogenic Coefficient (AC) = (TC– HDL-C)/HDL-C
Atherogenic Index (AI) = log (triglyceride/HDL-C).

2.8 Statistical Analysis

Results were expressed as means ± SEM. Using statistical package program (SPSS 16.0 version), one-way ANOVA was used to assess statistical significance where p value less than 0.05 was considered significant.

3. RESULTS AND DISCUSSION

3.1 Concentrations of Lactate, Pyruvate and Lactate/Pyruvate Ratio in the Serum of Rats Fed *Vigna unguiculata* Husk Based Diets

The concentrations of pyruvate and lactate were not significantly affected (p>0 .05) by 50% inclusion level of 90-minute boiled *Vigna unguiculata* pod husk (Table 2). On the other hand, 70% inclusion levels of unheated, 90-minute boiled and 60-minute boiled, and 50% inclusion levels of unheated and 60-minute boiled *Vigna unguiculata* pod husk caused significant increase (p<0.05) in both the levels of pyruvate and lactate. The rats fed on 70% inclusion level of unheated *Vigna unguiculata* husk had significantly highest concentrations of pyruvate and lactate. However, lactate:pyruvate ratio did not significantly vary among the groups.

The observed increase in the blood pyruvate and lactate levels of rats maintained on 70% inclusion levels of unheated, 90-minute, and 60-minute boiled, and 50% inclusion levels of unheated and 60-minute boiled *Vigna unguiculata* pod husk could have been due to low supply of exogenous energy by these diets. This could have resulted from the high feed conversion ratio coupled with the low intake of the diets (Idoko and Ilouno, Federal University Dutsinma, Katsina State, Nigeria). Low supply of exogenous energy necessitates the mobilization of triacylglycerols in adipose tissue and gluconeogenesis by the liver. The oxidation of triacylglycerols causes the concentrations of acetyl CoA and citrate to increase [15]. Acetyl CoA is an allosteric activator of pyruvate carboxylase and an allosteric inhibitor of pyruvate dehydrogenase, and thus pushes pyruvate to gluconeogenesis [16]. As the pyruvate accumulates, its metabolism shifts towards lactate formation, the excess of which is excreted into the bloodstream. The result also shows that 50% inclusion level of 90-minute boiled *Vigna unguiculata* pod husk has the potential to supply adequate energy. Under such metabolic condition, pyruvates are channelled towards acetyl CoA formation and triacylglycerols are synthesized rather than broken down. This finding is consistent with the pattern of observed weight gains by the rats on these diets (Idoko and Ilouno, Federal University Dutsinma, Katsina State, Nigeria). Idoko et al. [17] reported similar increase in circulating levels of lactate and pyruvate in rats fed 100% inclusion levels of raw, boiled, autoclaved *Citrullus colocynthis* seed coat and 50% inclusion level of raw *Citrullus colocynthis* seed coat, which they attributed to altered carbohydrate metabolism.

3.2 Concentrations of Glucose 6-Phosphatase Activities in Selected Tissues of Rats Fed *Vigna unguiculata* Husk Based Diets

The concentrations of glucose 6-phosphatase activities in the liver and kidney of rats maintained on the control diet were significantly lower (p<0.05) than the concentrations in the

Table 2. Concentrations of lactate, pyruvate and lactate/pyruvate ratio in the serum of rats fed *Vigna unguiculata* pod husk based diets

	Pyruvate (*)	Lactate (*)	Lac:Pyr
C	0.85±0.01^{a}	12.53±0.23^{a}	14.68±0.27^{b}
R	2.08±0.03^{f}	29.09±0.46^{f}	13.99±0.25^{a}
B60	1.91±0.01^{e}	25.18±0.02^{e}	13.16±0.08^{ab}
B90	1.43±0.03^{c}	21.29±0.15^{c}	14.93±0.21^{b}
CR	1.50±0.04^{d}	22.95±0.64^{d}	15.33±0.79^{b}
CB60	1.16±0.01^{b}	17.43±0.21^{b}	15.03±0.26^{b}
CB90	0.86±0.00^{a}	12.96±0.02^{a}	15.02±0.08^{b}

** (mg per 100 ml blood), Results are means of 3 determinations ± S. E. M .Values along the same row with the same superscript are NOT significantly different (P = .05), and are significantly different if the superscripts are different, C: (control): rats fed on diet based on100% inclusion level of corn starch, R: rats fed on diet based on 70% inclusion level of unheated Vigna unguiculata pod husk, B60: rats fed on diet based on 70% inclusion level of 60-minute boiled Vigna unguiculata pod husk, B90: rats fed on diet based on 70% inclusion level of 90-minute boiled Vigna unguiculata pod husk, CR: rats fed on diet based on 50% inclusion level of unheated Vigna unguiculata pod husk, CB60: rats fed on diet based on 50% inclusion level of 60-minute boiled Vigna unguiculata pod husk, CB90: rats fed on diet based on 50% inclusion level of 90-minute boiled Vigna unguiculata pod husk* liver and kidney of rats fed other diets. With the exception of rats maintained on the control diet, the liver and kidney concentrations of glucose 6-phosphatase activities in rats fed 50% inclusion level of 90-minute boiled *Vigna unguiculata* husk were significantly lower ($p<0.05$) than the concentrations in the liver and kidney of rats maintained on other diets (Table 3).

Table 3. Concentrations of glucose 6-phosphatase activities (nmol/hr/mg protein) in selected tissues of experimental rats fed *Vigna unguiculata* pod husk based diets

	Liver	Kidney
C	30.11±0.30[a]	18.13±0.27[a]
R	59.67±0.73[e]	28.67±0.04[e]
B60	51.16±0.51[d]	23.12±0.04[d]
B90	50.47±0.30[d]	22.82±0.24[d]
CR	50.04±0.02[d]	22.18±0.02[c]
CB60	40.78±0.46[c]	21.99±0.03[c]
CB90	37.46±0.36[b]	19.02±0.00[b]

Results are means of 3 determinations ± S. E. M Values along the same row with the same superscript are NOT significantly different (P = .05), and are significantly different if the superscripts are different. C: (control): rats fed on diet based on100% inclusion level of corn starch, R: rats fed on diet based on 70% inclusion level of unheated Vigna unguiculata pod husk, B60: rats fed on diet based on 70% inclusion level of 60-minute boiled Vigna unguiculata pod husk B90: rats fed on diet based on 70% inclusion level of 90-minute boiled Vigna unguiculata pod husk, CR: rats fed on diet based on 50% inclusion level of unheated Vigna unguiculata pod husk, CB60: rats fed on diet based on 50% inclusion level of 60-minute boiled Vigna unguiculata pod husk, CB90: rats fed on diet based on 50% inclusion level of 90-minute boiled Vigna unguiculata pod husk

Glucose 6-phosphatase is an enzyme that hydrolyzes glucose 6-phosphate, resulting in the creation of a phosphate group and free glucose. Glucose is then exported from the cell via glucose transporter membrane proteins. This catalysis completes the final step in gluconeogenesis and glycogenolysis and therefore plays a key role in the homeostatic regulation of blood glucose levels [18]. The increase in glucose 6-phosphatase activities could be due to increased need of gluconeogenesis. With the exception of rats on the control diet, the result also shows that rats fed 50% inclusion level of 90-minute boiled *Vigna unguiculata* husk had the least requirement for the endogenous supply of glucose, indicating that 50% inclusion level of 90-minute boiled *Vigna unguiculata* husk supplied more exogenous glucose. Consequently, rats fed the diet had less dependence on the activity of glucose 6- phosphatase for glucose. This is consistent with the trend in pyruvate and lactate levels. The regulation of hepatic glucose 6-phosphatase by changes in hormones and nutritional status has been reported by a number of authors [19,20]. In particular, the expression has been shown to be regulated by dietary carbohydrates in mammals [21].

3.3 Serum Lipid Profile of Rats Fed Heat-treated *Vigna unguiculata* Husk Based Diets

The serum lipid profiles of rats fed the control diet and rats fed 50% inclusion levels of 90-minute boiled *Vigna unguiculata* pod husk did not differ significantly ($p>0.05$) from each other. Similarly, the concentrations of total cholesterol and high density lipoprotein cholesterol did not vary significantly among all the experimental groups. However, triglyceride concentrations were significantly decreased ($p<0.05$) in rats fed 70% inclusion levels of unheated, 90-minute and 60-minute boiled, and 50% inclusion levels of unheated and 60-minute boiled *Vigna unguiculata* pod husk (Table 4). Although, the calculated values of Cardiac Risk Ratio (CRR) and Atherogenic Coefficient (AC) did not significantly differ from one group to another, rats maintained on 70% inclusion levels of unheated and 60-minute boiled *Vigna unguiculata* pod husk had significantly ($p<0.05$) lowest values of Atherogenic Index (AI) (Table 5).

The lower level of TG in rats fed 70% inclusion levels of unheated, 90-minute and 60-minute boiled, and 50% inclusion levels of unheated and 60-minute boiled *Vigna unguiculata* pod husk is understandable; low supply of exogenous energy usually involves oxidation of fatty acids obtained from TG hydrolysis in adipose tissue. Because rats fed 50% inclusion levels of 90-minute boiled *Vigna unguiculata* pod husk had adequate supply of dietary energy (carbohydrate), the TG level was unaffected when compared to the level in rats fed the control diet. The lower atherogenic index (AI) of rats fed 70% inclusion levels of unheated, 90-minute and 60-minute boiled, and 50% inclusion levels of unheated *Vigna unguiculata* pod husk is understandable sincerats fed on these diets had significantly lower triglyceride. The atherognic index is calculated as log (triglyceride/HDL-C). Takasaki [22] and Dobiásová [23] reported atherogenic indices to be powerful indicators of the risk of atherosclerosis with increase in their values

Table 4. Serum lipid profile of experimental rats fed heat-treated *Vigna unguiculata* husk based diets

	TC	TG	HDL-C	LDL-C
C	1.90±0.09[a]	0.72±0.02[a]	0.84±0.02[a]	0.87±0.07[a]
R	2.07±0.07[a]	0.43±0.01[c]	0.83±0.01[a]	1.20±0.06[dbc]
B60	1.99±0.01[a]	0.44±0.01[c]	0.84±0.00[a]	1.14±0.01[c]
B90	2.00±0.05[a]	0.54±0.01[b]	0.84±0.01[a]	1.07±0.07[bc]
CR	2.06±0.03[a]	0.56±0.03[b]	0.84±0.02[a]	1.10±0.04[c]
CB60	1.99±0.02[a]	0.60±0.01[b]	0.81±0.01[a]	1.00±0.01[abc]
CB90	1.98±0.06[a]	0.70±0.01[a]	0.82±0.01[a]	0.93±0.05[ab]

indicating increased risk of developing cardiovascular diseases. The amount of triglycerides in blood is one important barometer of metabolic health; high levels are associated with coronary heart disease [24]. According to Woodward et al. [25], elevated triglycerides levels are strongly associated with an increased mortality of coronary heart disease. It is evident from this finding that all the studied inclusion levels of *Vigna unguiculata* pod husk may not be implicated in the development of atherosclerosis and cardiovascular disease. The lower triglyceride and consequent lower AI is consistent with the low feed intake by rats fed on these diets since raised triglycerides is usually caused by regular overeating [26].

Table 5. Atherogenic indices of rats fed heat-treated *Vigna unguiculata* husk based diets

	CRR	AC	AI
C	2.27[a]	1.27[a]	-0.07[d]
R	2.49[b]	1.49[b]	-0.28[a]
B60	2.37[ab]	1.37[ab]	-0.28[a]
B90	2.39[ab]	1.39[ab]	-0.19[b]
CR	2.44[ab]	1.44[ab]	-0.18[bc]
CB60	2.47[ab]	1.47[ab]	-0.13[cd]
CB90	2.43[ab]	1.43[ab]	-0.07[d]

Results are means of 3 determinations ± S. E. M. Values along the same row with the same superscript are NOT significantly different (P = .05), and are significantly different if the superscripts are different

4. CONCLUSION

Having investigated some indices of carbohydrate metabolism in rats maintained on *Vigna unguiculata* pod husk, it is our conclusion that 50% inclusion level of 90-minute boiled pod husk are well metabolized in rats. Therefore, 50% replacement of corn starch by 90-minute boiled *Vigna unguiculata* pod husk may be considered as a way of reducing the cost of livestock production.

COMPETING INTERESTS

Authors have declared that no competing interests exist.

REFERENCES

1. Martin U. The role of livestock in economic development and poverty reduction. A Technical Report of Food and Agriculture Organization - Animal Production and Health Division Viale delle Terme di Caracalla 00100 Rome, Italy; 2004.
2. FAO. Cowpea; 2012. Accessed 10 August 2015.
   Available:https://en.wikipedia.org/wiki/Cowpea
3. Cowpea; 2012. Accessed 10 August 2015. Available:https://en.wikipedia.org/wiki/Cowpea
4. Oluokun JA. Intake, digestion and nitrogen balance of diets blended with urea treated and untreated cowpea husk by growing rabbit. Afr. J. Biotechnol. 2005;4(10): 1203-1208.
5. Animal Research Review Panel. Guideline 20: Guidelines for the Housing of Rats in Scientific Institutions; 2007. Accessed 15 July 2013.
   Available: www.animalethics.org.au
6. Akanji MA, Ngaha EO. Effect of repeated administration of berenil on urinary excretion with corresponding tissue pattern in rats. Pharmacol. Toxicol. 1989;64:272-275.
7. Akanji MA, Yakubu MT. α-Tocopherol protects against metabiosulphate- induced tissue damage in rats. Nig. J. Biochem. Mol. Biol. 2000;15(2):179-183.
8. Gloster JA, Harris P. Observations on anenzymatic method for the estimation of pyruvate inblood. Clinica Chimica Acta. 1962;7:206-211.

9. Swanson MA. Phosphatases of liver. I. Glucose-6phosphatase. J. Biological Chemistry. 1950;184:647-660.
10. Abell LL, Levey BB, Brodie BB. Determination of cholesterol in serum and plasma. J. BioL Chem. 1952;195:357.
11. Tietz NW. Clinical Guide to Laboratory Tests. 2nd ed. Philadelphia, USA W.B. Sauders Company; 1990.
12. Friedewald WT, Levy RI, Fredrickson DS. Estimation of low-density lipoprotein cholesterol without the use of the preparative ultracentrifuge. Clin. Chem. 1972;18(6):499-502.
13. Ikewuchi CJ, Ikewuchi CC. Alteration of Plasma Lipid Profiles and Atherogenic Indices by *Stachytarpheta jamaicensis* L. (Vahl). Biokemistry. 2009;21(2):71-77.
14. Tan MH, Johns D, Glazer NB. Pioglitazone reduces atherogenic index of plasma in patients with type 2 diabetes. Clin. Chem. 2004;50:1184-1188.
15. Jeremy B, Tymoczko JL, Stryer L. Biochemistry 5th ed. San Francisco: W.H. Freeman; 2002.
16. Richard AH, Denise RF. Lippincott's Illustrated Reviews: Biochemistry. 5th ed. Lippincott Williams & Wilkins, a Wolters Kluwer business 351 West Camden Street Baltimore; 2008.
17. Idoko AS, Oladiji AT, Iouno LE. Growth performance of rats maintained on *Citrullus colocynthis* seed coat-based diet. IOSR-JBB. 2015;1(4):09-14.
18. Nordlie R. The Enzymes of biological membranes. 2nd e d. New York: Plenum Press; 1985.
19. Argaud D, Zhang Q, Pan W, Maitra S, Pilkis SJ, Lange AJ. Regulation of rat liver glucose-6-phosphatase gene expression in different nutritional and hormonal states. Diabetes. 1996;45:1563-1571.
20. Streeper RS, Svitek CA, Chapman S, Greenbaum LE, Taub R, O'Brien RM. A multicomponent insulin response sequence mediates a strong repression of mouse glucose-6phosphatase gene transcription by insulin. J Biol Chem. 1997;272(18):11698-11701.
21. Rencurel F. Girard J. Regulation of liver gene expression by glucose. Proc. Nutr. Soc. 1998;57:265-275.
22. Takasaki Y. Serum lipid levels and factors affecting atherogenic index in japanese children. J Physiol Anthropol Appl Human Sci. 2005;24(4):511-515.
23. Dobiásová, M. Atherogenic index of plasma [log(triglyceride/HDLCholesterol)]: Theoretical and Practical Implications. Clin Chem. 2004;50 (7):1113-1115.
24. American Heart Association (AHA) and American Stroke Association. Triglycerides: Frequently Asked Questions; 2005. Accessed July 14 2014.
Available:https://my.americanheart.org/idc/groups/ahamah.../ucm_425988.pdf
25. Woodward M, Huxley H, Lam TH, Barzi F, Lawes CM, Ueshima H. A comparison of the associations between risk factors and cardiovascular disease in Asia and Australasia. Eur J Cardiovasc Prev Rehabil. 2005;12(5):484-491.
26. Karen S. Common cause of High triglyceride. Retrieved July 10 2014. Available:cholesterol.answers.com/risks...high.../common-causes-of-high-triglyceride

25

Acute Toxicity of Aqueous and Ethanolic Extracts of *Strophanthus hispidus* Stem Bark

M. Osibemhe[1*], B. O. Abdulrahman[1] and I. O. Onoagbe[2]

[1]Department of Biochemistry and Molecular Biology, Faculty of Science and Education, Federal University Dutsin-Ma, Katsina State, Nigeria.
[2]Department of Biochemistry, Faculty of Life Sciences, University of Benin, Benin City, Edo State, Nigeria.

Authors' contributions

This work was carried out in collaboration between all authors. Author IOO designed the study, wrote the protocol and supervised the work. Authors MO and BOA carried out all laboratories work and performed the statistical analysis. Author MO wrote the first draft of the manuscript, managed the literature searches and edited the manuscript. All authors read and approved the final manuscript.

Editor(s):
(1) Dileep G. Nair, Ministry of Higher Education, Sultanate of Oman.
Reviewers:
(1) Abdullahi M. Nuhu, Kaduna Polytechnic, Kaduna, Nigeria.
(2) Begum Rokeya, Bangladesh University of Health Sciences, Bangladesh.

ABSTRACT

The aim of this study was to determine the median lethal dose (LD_{50}) of the extracts (aqueous and ethanol) of *S. hispidus* stem bark. A modified Lorke's (1983) method was used in this study. Varying doses of the extracts were administered orally to male albino rats. Treatment-related signs of toxicity and mortality were monitored for 24 hrs and continued for 72 hrs. Treatment-related mortality was observed at the dose of 1000 mg/kg body weight and above in the ethanolic extract and 1600 mg/kg body weight in the aqueous extract. Signs of sedation, exophthalmos, decreased locomotion and appetite in the aqueous or ethanolic treated animals were observed in the early stages of experimentation. However, these signs were not sustained in surviving animals. The LD_{50} was 2154 and 2039 mg/kg body weight for aqueous and ethanolic extracts of *S. hispidus* stem bark respectively. The results showed that 1000 mg/kg body weight and above of *S. hispidus* stem bark may be slightly toxic. Therefore, moderate dosage is advised for purposes of medication.

**Corresponding author: E-mail: mosibemhe@fudutsinma.edu.ng*

Keywords: Medication; exophthalmos; mortality; experimentation.

1. INTRODUCTION

Plants produce a great diversity of substances that can have therapeutic significance for maintaining human health and improving the quality of human life, thus justifying their use in traditional medicine [1]. According to their traditional use, natural compounds are often assumed to be safe. However, several studies have reported that a great number of plant species used as food ingredients or in traditional medicine present mutagenic, carcinogenic or toxic properties [2-4]. Drug toxicity can occur on many different time-scales. Acute toxicity results from a single exposure to a drug, with adverse effects resulting in minutes to hours [5]. A fundamental goal of toxicology is to determine safe levels of exposure to potentially poisonous substances for human and the environment [5].

The scientific exploitation of herbs used ethno-medically for pain relief, wound healing and abolishing fevers has resulted in the identification of a wide range of compounds that have been developed as new therapies for several ailments including diabetes [6].

A few of these herbal products have scientific data on their toxic levels in literatures. Acute toxicity testing can provide a guide to selection of doses of herbal medicines, in order to avoid potential harmful effects when used. *Strophanthus hispidus* is an African shrub that belongs to the genus; *Strophanthus*, the family of Apocynaceae. Its popularity in folklore medicine for the treatment of diabetes, ulcer, gonorrhea as well as its bitter taste may not be unconnected with the presence of important phytochemicals such as saponins, alkaloids, phenols, flavonoids and tannins [7]. This work is therefore aimed at providing scientific records as to the amount of *S. hispidus* that may constitute potential risk in animal model.

2. MATERIALS AND METHODS

2.1 Animals

Male rats (Wistar strain) obtained from the Department of Anatomy, University of Benin, Benin City, Nigeria were used. The rats weighing between 100-232 g were maintained under standard animal house condition and were allowed free access to food (growers mash produced by Bendel Feed and flour Mill, Ewu Edo State) and water for two weeks to acclimatize to the new environment. All animals were handled with proper care and humanely treated according to our Institutional Animal Ethics Committee guidelines as well as the internationally accepted practices for use and care of laboratory animals as contained in US guidelines [8].

2.2 Medicinal Plant

The stems of *S. hispidus* were collected from Galadimawa, in Giwa Local Government, Kaduna State, Nigeria. They were identified by Mr. U. S. Gallah of the Department of Biological Science, Ahmadu Bello University, Zaria, Nigeria where a specimen with voucher number (no: 2714) was deposited.

2.3 Preparation and Extraction of Plant Materials

A modified method of Onoagbe et al. [9] was used. The stems of *S. hispidus* were thoroughly washed with clean water and the barks were peeled off by incision. They were then dried under shade for two weeks and then pulverized into fine powder with the aid of a mechanical pulverizer. Measured quantities of the powdered sample were extracted separately in aqueous and 99% ethanol for 72 hrs followed by periodic stirring and they were kept in a refrigerator to avoid any microbial growth. The extracts were filtered using cheese-cloth and the filtrate re-filtered using Whatman No. 42 (125 mm) filter paper. The filtrates collected were lyophilized using a freeze-dryer and stored in an airtight container for further analysis.

2.4 Acute Toxicity Study

The method of Lorke, (1983) [10] for the determination of lethal dose (LD_{50}) with little modification was used in this experiment. A total of 26 male rats were used. The rats were randomly selected into two major groups of 13 rats each and were kept in standard cage. The groups were labeled 1 and 2. The experiment was carried out in two phases. In the phase one, rats in group 1 were again selected randomly into three sub-groups of three rats each and were administered orally with the aid of a gavage, doses of 10, 100, and 1000 mg/kg body weight aqueous extract respectively. Similarly, the second group (2) which was also randomly selected into three sub-groups of three rats each was administered respective doses of ethanolic extract. Both groups were monitored frequently during working hours for signs of toxicity and mortality and subsequently for 24 hours. Observations were continued for 72 hours for any late sign of toxicity. In the second phase, four rats in group 1 were made into four groups of one rat each and were administered doses of 1000, 1600, 2900 and 5000 mg/kg body weight aqueous extract respectively. In a similar manner, four rats were also selected from group 2 into four groups of one rat each. They were administered ethanolic extract at the doses of 800, 1600, 2600 and 3600 mg/kg body weight respectively. Again, both groups were monitored for signs of toxicity and mortality as in the phase one. The second phase was used in the calculation of the median lethal dose (LD_{50}).

3. RESULTS

The results for the acute toxicity study of both extracts of *S. hispidus* stem bark are presented in Tables 1-2. 1000 mg/kg body weight and above in the ethanolic extract (Table 2) and 1600 mg/kg body weight and above in the aqueous extract (Table 1) resulted in mortality. The LD_{50} was beyond 2000 mg/kg for both extracts. (Tables 1-2).

4. DISCUSSION

Diabetes mellitus is a disease that is increasingly affecting millions of people all over the world. And this disease has no known cure in spite of its age long existence. Currently there are over 150 millions diabetics worldwide and this is likely to increase to 300 million or more by 2025 [11]. Research to provide scientific data for plants with acclaimed hypoglycemic and anti-diabetic properties are on the increase. This became more apparent following WHO (1994) [12] recommendations regarding the need to develop and evaluate better pharmacological agents for improving insulin secretion, enhancing insulin sensitivity, preventing beta-cell destruction, promoting beta-cell regeneration or repair and interrupting pathways leading to the various

Table 1. LD_{50} determination on aqueous extract

Phase one		
Doses (mg/kg)	**Number of rats/Group**	**Ratio of dead rats**
10	3	0/3*
100	3	0/3
1000	3	0/3
Phase two		
Doses (mg/kg)	**Number of rats/Group**	**Ratio of dead rats**
1000	1	0/1
1600	1	0/1
2900	1	1/1
5000	1	1/1

* *Number of dead animals/ number of animals used,* $LD_{50}=\sqrt{a} \times b$. *(Where a= highest non-lethal dose, b= least lethal dose) within 24 hours of extract administration,* $LD_{50}=\sqrt{1600} \times 2900 \ =\sqrt{4640000}$ *= 2154 mg/kg (oral)*

Table 2. LD_{50} determination on ethanolic extract

Phase one		
Doses (mg/kg)	**Number of rats/Group**	**Ratio of dead rats**
10	3	0/3*
100	3	0/3
1000	3	1/3
Phase two		
Doses (mg/kg)	**Number of rats/Group**	**Ratio of dead rats**
800	1	0/1
1600	1	0/1
2600	1	1/1
3600	1	1/1

* *Number of dead animals/ number of animals used,* $LD_{50}=\sqrt{a} \times b$. *(Where a= highest non-lethal dose, b= least lethal dose). Within 24 hours of extract administration,* $LD_{50}=\sqrt{1600} \times 2600 \ =\sqrt{4160000}$ *= 2039 mg/kg (oral)*

complications of diabetes. The major contributory factors to this growing interest include these recommendations, the cost and side effects of most orthodox hypoglycaemic agents, low therapeutic index of synthetic compounds and growing incidences of drug resistance [13,14]. Ojiako and Igwe, [15] has reported the use of *S. hispidus* for the treatment of diabetes by the ethnic tribal people of Africa. The major hindrance to the use of traditional herbal preparations is the lack of scientific and clinical data in support of better understanding of the efficacy and safety of the drugs [16]. In screening drugs, determination of LD_{50} is usually an initial step in the assessment and evaluation of the toxic characteristics of a substance. The oral LD_{50} of aqueous and ethanolic extracts of *S. hispidus* stem bark were found to be 2154 and 2039 mg/kg body weight (Tables 1 and 2) respectively in the present study. Treatment-related mortality (Tables 1 and 2) and signs of sedation, exophthalmos, decreased locomotion and appetite in the aqueous or ethanolic treated animals were observed in the early stages of experimentation at the dose of 1000 mg/kg body weight and above in the ethanolic extract and 1600 mg/kg body weight and above in the aqueous extract. According to [17], any substance with oral LD_{50} above 1000 mg/kg body weight in rats is regarded as being of low toxicity or relatively safe. Based on Hodge and Sterner scale [18], a test drug administered orally is considered extremely toxic at≤ 1 mg kg^{-1}, highly toxic at 1-50 mg kg^{-1}, moderately toxic at 50-500 mg kg^{-1}, slightly toxic at 500-5000 mg kg^{-1}, practically non toxic at 5000-15,000 mg kg^{-1} and relatively harmless at≥15,00 mg kg^{-1}. Agbaje et al. [19] has reported LD_{50} for both intraperitoneal and oral routes of *Syzigium aromaticum* (L.) in rodents as 265 and 2500 mg/kg body weight. The oral (rat) LD50 of ethanol extract of *Vitex leucoxylon* leaf (>3000 mg/kg), cold water infusion extract of the same plant (1050 mg/kg), ethanolic extracts of *Ailanthus excelsa* (1000 mg/kg), *Toddalia asiatica* (350 mg/kg) and *Araucaria bidwilli* (250 mg/kg) have been reported [20]. The scarcity of scientific data to validate local claims on the use of *S. hispidus* for the treatment of diabetes as well as lack of data about its tolerable dosage hinders meaningful comparison.

5. CONCLUSION

The findings from this study indicated that *S. hispidus* stem bark extracts may be slightly toxic at high doses. *S. hispidus* may also be considered to be relatively safe since the qualitative and quantitative acute toxicity in animals only has little relevance to human acute toxicity in special cases and after many years of usage. The results also showed that ethanolic extract has lower acute toxicity than aqueous extract of *S. hispidus*. Despite these justifications that *S. hispidus* stem bark extracts may be relatively safe, moderate dosage is advised for purposes of medication. Also, studies to establish safer and therapeutic doses of *S hispidus* stem bark for specific ailments as well as research on its cytotoxic effect on animal model are recommended in order to ascertain its true toxic level.

COMPETING INTERESTS

Authors have declared that no competing interests exist.

REFERENCES

1. Ravikumar YS, Mahadevan KM, Kumaraswamy MN, Vaidya VP, Manjunatha H, Kumar V, Satyanarayana ND. Antioxidant, cytotoxic and genotoxic evaluation of alcoholic extract of polyalthia cerasoides(roxb.) bedd. Environmental Toxicology and Pharmacology. 2008;26(2) 142-146.
2. Deciga-Campos M, Rivero-Cruz I, Arriaga-Alba M, Castaneda-Corral G, Angeles-Lopez GE, Navarrete A, Mata R. Acute toxicity and mutagenic activity of Mexican plants used in traditional medicine. Journal of Ethnopharmacology. 2007;110(2): 334-342.
3. Ferreira ICFS, Vargas VMF. Mutagenicity of medicinal plant extracts in *Salmonella*/ microsome assay. Phytotherapy Research. 1999;13(5):397-400.
4. Mohd-Fuat AR, Kofi EA, Allan GG. Mutagenic and cytotoxic properties of three herbal plants from Southeast Asia. Tropical Biomedicine. 2007;389(1):3-122.
5. David EG, Armen HT Jr, Ehrin JA, April WA. Principles of pharmacology: The pathophysiologic basis of drug therapy (3rd ed.). Philadelphia, PA: Lippincott Williams & Wilkins; 2011.
6. Harvey AL. Natural products in drug discovery. Drug discovery Today. 2008; 13(19/20):894-901.

7. Osibemhe M, Onoagbe IO. Qualitative and quantitative phytochemical evaluations of *Strophanthus hispidus* stem bark. IOSR Journal of Pharmacy and Biological Sciences (IOSR-JPBS). 2015;10(2): 124-128.
8. National Institute of Health. Institutional Animal Care and use Committee Guidebook, NIH Publication no. 92-345. Washington, D. C. U.S. Government Printing Office; 1992.
9. Onoagbe IO, Ebhota AO, Udegbe HC, Omondia M, Edeni D, Ebengho SO. Assessment of some medicinal plants for hypoglycemic activities in rats and rabbits. Biosci. Res. Commun. 1999a;11:159-163.
10. Lorke D. A new approach to practical acute toxicity testing. Arch. Toxicol. 1983;54: 275-287.
11. Tanko Y, Yaro AH, Isa AI, Yerima M, Saleh MIA, Mohammed A. Toxicological and hypoglycemic studies on the leaves of *Cissampelos mucronata (Menispermaceae)* on blood glucose levels of streptozotocin–induced diabetic wistar rats. J. Med. Plant Res. 2007;1(5): 113-116.
12. WHO Study Group Report on Prevention of Diabetes Mellitus. WHO, Geneva; 1994.
13. Onyeyilli PA, Egwu GO. Chemotherapy of African Trypanosomiasis: A historical perspective. Protozoological Abstracts. 1995;19(5):229-241.
14. Seed JR. Current status of African Trypanosomiasis. ASM News. 2000;66(7): 395-402.
15. Ojiako OA, Igwe CU. A time-trend hypoglycemic study of ethanol and chloroform extracts of *Strophanthus hispidus*. Journal of Herbs, Spices & Medicinal Plants. 2009;15(1):1-8.
16. Saidu Y, Bilbis LS, Lawal M, Isezuo SA, Hassan SW, Abbas AY. Acute and sub-chronic toxicity studies of crude aqueous extract of *Albizzia chevalieri* Harms (Leguminosae). Asian Journal of Biochemistry. 2007;2:224-236.
17. Clarke EGC, Clarke ML. Veterinary Toxicology (1st ed.) London: Bailliere Tindall; 1977.
18. CCOHS. What is an LD50 and LC50; Canadian's National Occupational Health and Safety Resource: Canadian Centre for Occupational Health and Safety. 2005. Available:http://www.ccohs.ca/oshanswers/chemicals/ld50.html
19. Agbaje EO, Adeneye AA, Daramola AO. Biochemical and toxicological studies of aqueous extract of *Syzigium aromaticum* (L) Merry & Perry (Myrtaceae) in rodents. Afr J Tradit Compement Altern Med. 2009; 6(3):241-254.
20. Dahanukar SA, Kulkarni RA, Rege NN. Pharmacology of medicinal plants and natural products. Indian Journal of Pharmacology. 2000;32:S81–S118.

26

Evaluation of Macro Minerals in Patients with Type II Diabetes Mellitus in Southern Nigeria

A. Z. Ogunleye[1*] and M. F. Asaolu[1]

[1]Department of Biochemistry, Ekiti State University, Ado Ekiti, Ekiti State, Nigeria.

Authors' contributions

This work was carried out in collaboration between both authors. Author MFA designed the study, wrote the protocol, supervised the work and edited the manuscript. Author AZO carried out all laboratories work, performed the statistical analysis, managed the analyses of the study, wrote the first draft of the manuscript and managed the literature searches. Both authors read and approved the final manuscript.

Editor(s):
(1) Yi-Ren Hong, College of Medicine , Kaohsiung Medical University, Taiwan.
(2) Dileep G. Nair, Ministry of Higher Education, Sultanate of Oman.
Reviewers:
(1) Larry A. Weinrauch, Harvard Medical School, Boston, USA.
(2) Mohamed Hamdy Ibrahim, Ain Shams University, Cairo, Egypt.
(3) Aida Abdeen Mahmoud, Sohag University, Egypt.
(4) Kumar M. R. Bhat, Manipal University, India.
(5) Khalid S. Aljabri, King Fahad Armed Forces Hospital, Jeddah, Saudi Arabia.
(6) Anonymous, Keio University, Japan.

ABSTRACT

Aims: To evaluate serum sodium, potassium, calcium, chloride, magnesium, phosphorus and sulphur in patients with type II DM and compare with non-diabetic subjects.
Study Design: A cross sectional study.
Place and Duration of Study: Department of Chemical Pathology, Ekiti State University Teaching Hospital, Ado Ekiti and Federal Medical Centre, Ido Ekiti, Ekiti State, between April 2013 and February 2014.
Methodology: This study was conducted on 150 subjects, out of which 100 were type II diabetes mellitus patients and 50 were non diabetic (control) subjects. Glucose level was determined by Glucose oxidase –Peroxidase method, serum calcium, potassium, sodium and chloride

**Corresponding author: E-mail: ogunleyeadeoluz@gmail.com*

concentrations were measured using Biolyte *Spin 6,* Full Automated Electrolyte Analyzer while magnesium was analyzed by Atomic Absorption Spectrophotometry. Phosphorus concentrations was determined by spectrophotometer using BioSystem reagent kit specific for phosphorus and Sulphur concentrations were measured using auto analyzer.

Results: The results showed that sodium, potassium, chloride, calcium, phosphorous and magnesium were significantly lower (*P<.001*) in diabetic subjects when compared with the control subjects whereas the mean sulphur concentration was significantly higher *(P<.001)* in diabetic subjects when compared with the non-diabetic (control) subjects.

Conclusion: This study showed Impair metabolism of macro minerals in diabetic group which result in variations in the levels of these minerals in diabetic male and female as well as in different diabetic age groups when compare with non-diabetic group.

Keywords: Macro minerals; diabetes mellitus; full automated electrolyte analyzer; Atomic Absorption Spectrophotometry (AAS); auto analyzer.

1. INTRODUCTION

Type II diabetes mellitus (DM2) is a chronic and progressive metabolic disorder characterized by insulin resistance and pancreatic beta islet cell failure. Diabetes mellitus is usually characterized by hyperglycaemia resulting from defects in insulin secretion, insulin action, or both [1]. Three specific abnormalities that contribute to hyperglycaemia in DM2 are: impaired insulin secretion, increased hepatic glucose production, and decreased insulin-stimulated uptake of glucose in peripheral tissues. Macro minerals (sodium, potassium, calcium, magnesium, chloride, sulphur and phosphorus) play an important role in intermediary metabolism and cellular function, including enzyme activities and electrical gradients [2]. Serum concentrations of these minerals have been shown to change with plasma glucose levels [3]. Disturbances in the levels of macro minerals were found to be associated with diabetes mellitus [4-6]. Potassium influences the release of insulin from the pancreas so it could be said that low potassium contributes to pancreatic stress and future insulin resistance leading to type II diabetes. Even slight decrease in potassium level in the blood reflects a larger decrease of potassium in the cells and this apparently impaired glucose metabolism and function [7]. Sodium depletion is another common feature of type II diabetes, reduction in serum sodium in diabetic subjects might be a result of electrolyte loss which arises due to dehydration or a result of kidney dysfunction and diabetic nephropathy. Also, several alterations in the renin-angiotensin system have been described in diabetes mellitus. In patients with established autonomic neuropathy, decreased plasma renin activity responses have been found, which suggests that neural control of renin release is altered in diabetes which may in turn alter sodium concentration in diabetes mellitus [8].

Glycosuria which is very common in diabetes might produce chloride depletion, in the absence of adequate fluid intake, the obligatory excretion of water necessary for the urinary excretion of glucose may result in a deficit of body water; under some circumstances this may be followed by a secondary depletion of the body stores of chloride. Insulin secretion which is a calcium-dependent process [9] may alter calcium flux and thus have adverse effects on β-cell secretory function. It may also be speculated that inadequate calcium intake may alter the balance between the extracellular and intra-cellular β-cell calcium pools, which may interfere with normal insulin release, especially in response to a glucose load. Sulphur containing compounds have also been established as risk factor for cardiovascular diseases and it occurs with high prevalence in patients with type II diabetes [10]. Diabetes is also known to be associated with various degrees of intracellular phosphate depletion, partially because of a shift of phosphorus from the intracellular to the extracellular compartments and because of prolonged and excessive hyperphosphaturia [11]. The occurrence of a paradoxical imbalance in phosphate metabolism from the early onset of diabetes mellitus have indicated that the imbalance may lead to a reduction of high energy phosphates and tissue hypoxia [11,12]. Mild changes in minerals such as low magnesium levels can predict mortality in type II DM [13], and supplementation of some of this mineral have been found to reduce fasting plasma glucose levels in DM patients [14]. Therefore, Several Studies have been carried out on the potential alteration of minerals in diabetes mellitus. However, results in these study have been inconsistent and contradictory [15-18] in view of these conflicting reports, the present study was undertaken in order to investigate these parameters in diabetic patients and as well compare the result with age and sex matched non-diabetic patients.

2. MATERIALS AND METHODS

2.1 Study Subjects

The subjects used for the present study were made up of two groups. Group 1 comprised of 100 diagnosed type II diabetic patients and group 2 was made up of 50 non-diabetic aged matched control subjects. Among the total diabetic subjects, 43 were male and 57 were female while that of non-diabetic group were 26 males and 24 females. Both groups were of age range 35 – 80 years.

2.2 Inclusion and Exclusion Criteria

The study included diabetic patients and non-diabetic subjects.

Lactating mothers, patients with serious co-morbid diseases (stroke, major surgery, mal-absorption, myocardial infarction), Smoking and alcoholic individuals, history of using drugs that significantly affect glucose metabolism (glucocorticoids, oral contraceptives, high-dose thiazide diuretics), pregnant women, patients with other chronic illnesses or taking any other medications that could potentially affect levels of macro minerals were all excluded.

2.3 Sample Collection

5 ml of fasting blood sample was collected from antecubital vein into plain bottles from each of the subjects. The whole blood samples were allowed to clot and were centrifuged at 4000 rpm for 15 minutes. The supernatant serum was taken and delivered into plastic tubes with screw caps and was used for the analysis of macro-minerals (sodium, potassium, calcium, magnesium, chloride, sulphur and phosphorus).

2.4 Laboratory Analysis

Glucose level was determined by Glucose oxidase –Peroxidase method [19], calcium, potassium, sodium and chlorine concentration were measured using Biolyte *Spin 6,* Full Automated Electrolyte Analyzer, magnesium was analyzed by flame atomic absorption spectrophotometry, Serum phosphorus concentrations was determined by spectrophotometer using BioSystem reagent kit specific for phosphorus [20] and Serum Sulphur concentrations were measured using auto analyzer.

2.5 Statistical Analysis

Data analysis was performed using Statistical Package for Social Science (SPSS) for Windows version 16. The statistical significance was determined by one-way analysis of variance (ANOVA). The p-value less than .001 *(P<.001)* was considered as significant.

3. RESULTS AND DISCUSSION

3.1 Results

One hundred diabetic patients (43 Males and 57 Females) and fifty non-diabetic (control) subjects (26 Males and 24 Females) defined by clinical examination and with no history of any disease were compared for glucose levels and macro-minerals. Values were expressed as mean ±SD.

Table 1. Sex distribution

Parameter	Diabetic group	Control group
Male	43 (43%)	26 (52%)
Female	57 (57%)	24 (48%)
Total	100 (100%)	50 (100%)

Table 2. Age distribution

Age code (years)	Diabetic subjects	Control subjects
30-39	16 (16%)	16 (32.0%)
40-49	17 (17%)	11 (22.0%)
50-59	21 (21%)	10 (20.0%)
60-69	23 (23%	9 (18.0%)
70 and above	23 (23%)	4 (8.0%)
Total	100 (100%)	50 (100%)

3.2 Discussion

Type 2 Diabetes mellitus is one of the most serious public health problems being faced globally. It may therefore be prudent in medical practice to periodically monitor the macro minerals status of diabetics, because evaluation of these minerals may help in suggesting adequate management for type II diabetes mellitus patients. In the present study, serum magnesium levels were found to be significantly reduced in the diabetic patients *(P<.001),* when compared with the control group this was in correlation with the findings of Berhane et al. [21], Chetan et al. [22], Diwan et al. [23], Tosiello [24], Walter et al. [25], Tripathy et al. [16], McNair et al. [26], Anetor et al. [27], Supriya et al. [28] and Sharma et al. [29]. It was also observed in this study that Magnesium levels of diabetic group differ between the different age groups but the difference was not statistically significant *(p>.05).* Diabetic male has a slightly higher Magnesium levels than their female counterpart but the difference was not statistically significant *(p>.05).* Magnesium is an essential element which is involved in glucose homeostasis at multiple levels; it is an integral part of the activated MgATP complex regulating protein kinases which is directly involved in the control of glucose metabolism. Low levels of magnesium can reduce secretion of insulin by the pancreas (Durlach et al. [30]. The cause of diabetic hypomagnesaemia is multifactorial; Osmotic actions of glycosuria are known to depress the net tubular reabsorption of magnesium in normal humans (Elaine et al. [31]; Ishrat et al. [32]; Nsonwu et al. [33].

Sodium depletion is another common feature of type II diabetes. In this study, sodium levels of diabetic group differ between the different age groups but the difference was not statistically significant *(p>.05).* Sodium level in male diabetic group was slightly higher *(P>.05)* than their female counterpart. Despite the variation in sodium level among different ages and sexes of diabetic patients, the serum sodium levels were found to be significantly lower in diabetic group *(P<.001)* when compared with control subjects, this result was consistent with the findings of Hasan et al. [34], Al-Rubeaan et al. [5]. The sodium depletion might be due to inhibition of the renin-angiotensin-aldosterone system, which plays a key role in the regulation of fluid and electrolyte balance. This enzyme system has been reported to be affected in many endocrine and cardiovascular diseases particularly diabetes [35].

Potassium, the main intracellular cation in the human body is required for vital cellular processes. Recent research has led to renewed interest in low potassium as a possible risk factor for diabetes. In this study, the serum potassium levels were found to be significantly lower in diabetic group *(P<.001)* when compared with control subjects, which was in accordance with findings of Hasan et al. [34], (Ranee et al. [36], Al-Rubeaan et al. [5]. Potassium levels of diabetic group differ between the different age groups but the difference was not statistically significant *(p>.05).* Also, potassium level in male diabetic group was significantly *(P<.05)* lower than their female counterpart. The observed depression in serum potassium in the diabetic cohort might be a result of electrolyte loss which arises due to dehydration or a result of kidney dysfunction and diabetic nephropathy. Potassium depletion was associated with a decrease in pancreatic β-cell sensitivity to hyperglycemia with a reduction in insulin release; it is a well-established correlate of disturbances in glucose metabolism. Potassium depletion might also occur due to activation of the renin-angiotensin-aldosterone system which plays a key role in the regulation of fluid and electrolyte balance in the body in type II diabetes mellitus patients. In type II diabetes low renin, low aldosterone responsiveness is associated with increased level of serum potassium [37]. In this study, serum chloride levels were also found to be significantly lowered in diabetic group *(P<.001)* when compared with control subjects, this result was in correlation with findings of Hasan et al. [34]. Chloride levels of diabetic group differ between the different age groups but the difference was not statistically significant *(p>.05),* chloride level in male diabetic group was slightly higher than their female counterpart but the difference was not statistically significant *(p>.05).* The observed lower serum chloride in the diabetic cohort might be a result of electrolyte loss which arises due to dehydration or a result of kidney dysfunction or diabetic nephropathy.

Sulphur containing compounds have been established as risk factor for cardiovascular diseases and occur with high prevalence in patients with type 2 diabetes. A large amount of evidence supports increased plasma sulphur containing compounds in type 2 diabetes, this can be found in the findings of Emoto et al. [10], Fiorina et al. [38], Hultberg et al. [39], Chico et al. [40], Stabler et al. [41], Buysschaert et al. [42]. In this study, serum sulphur concentrations were significantly higher *(P<.001)* in diabetic subject when compare with controls. Sulphur level among the diabetic group differs between the different age groups but the difference was not statistically significant *(p>.05)* whereas the level of Sulphur in both male and female diabetic group remain the same.

Table 3. Comparison between the glucose levels of diabetic group and control group

Parameter	Diabetic group N = 100 (MEAN±SD)	Control group N = 50 (MEAN±SD)	P-value
Glucose (Mg/dL)	160.94±61.81	67.41±7.95	<.001

The mean and SD of glucose in control group was (67.41±7.95) which was significantly lower (P<.001) than that of the diabetic group (160.94±61.82)

Table 4. Comparison between the ages of diabetic group and control group

Parameter	Diabetic group N = 100 (MEAN±SD)	Control group N = 50 (MEAN±SD)	P-value
Age(years)	58.11±13.51	47.65±9.03	0.000

The age range of these subjects was between 35 and 80years. The mean age and standard deviation of the control group was 47.65±9.03 years, while that of the diabetic group was (58.11±13.51) years. The mean age of the diabetic group is greater than that of the control group; the mean age difference was significant (p < 0.001)

Table 5. Association between serum macro-minerals, blood sugar and age of diabetic mellitus patients

Age range (year)	30-39	40-49	50-59	60-69	≥70	P-value
FBS (Mg/dL)	123.00±9.81	198.49±82.89	151.20±28.34	159.14±61.76	167.31±73.65	0.083
K (Mg/dL)	15.64±0.86	15.74±0.84	15.30±1.59	15.39±1.68	14.93±1.59	0.589
Na (Mg/dL)	302.58±6.20	302.55±8.44	297.78±8.80	301.61±9.34	303.27±6.52	0.306
Cl (Mg/dL)	349.48±8.89	348.22±4.62	344.35±9.80	346.29±8.95	343.17±9.21	0.314
Ca (Mg/dL)	12.32±0.84	12.24±1.30	12.00±1.47	11.55±1.40	12.25±1.43	0.427
Mg (Mg/dL)	0.73±0.14	0.89±0.16	0.87±0.29	0.83±0.21	0.91±0.20	0.337
P (Mg/dL)	1.73±0.71	2.35±0.65	2.19±0.93	1.96±1.14	2.20±0.99	0.569
S (μg/dL)	3.80±0.32	3.49±0.74	3.70±0.65	3.94±1.07	3.83±0.70	0.630

There was no significant difference (P>0.05) in FBS levels among the different age groups but the age range 30 39 has the lowest level of FBS whereas the age range 40-49 has the highest level.
There were no significant differences (P>0.05) in Potassium, Sodium, Chloride, Calcium, Magnesium, phosphorous and sulphur level among the different age groups, although the levels of these minerals vary between the different age ranges.

Table 6. Comparism between serum macro-minerals and blood sugar in Male and Female diabetic subjects

Parameters	Male N=43	Female N=57	P-value
Age (year)	63.64±11.30	54.02±14.06	0.002
FBS (Mg/dL)	177.11±74.45	149.59±48.87	0.049
K (Mg/dL)	14.94±1.80	15.60±1.12	0.046
Na (Mg/dL)	302.76±9.94	300.57±6.61	0.238
Cl (Mg/dL)	347.04±8.88	344.73±8.73	0.251
Ca (Mg/dL)	12.28±1.27	11.82±1.41	0.135
Mg (Mg/dL)	0.94±0.22	0.80±0.20	0.003
P (Mg/dL)	2.41±1.13	1.88±0.74	0.013
S (μg/dL)	3.78±0.74	3.78±0.82	0.972

The levels of Na, Cl, Ca, and S were slightly higher in diabetic male than in their female counterpart but the differences were not statistically significant. The levels of FBS, Mg and P were significantly higher (p<0.05) in diabetic male when compare with diabetic female whereas the K level was significantly higher (p<0.05) in diabetic female than in male

This study also showed that the mean levels of serum calcium were significantly lower *(p<0.01)* in the serum of types II diabetic patients when compared with the control group, similar results were recorded by Djalali et al. [43] and Anastassios et al. [44]. The loss of these elements might be attributed to impaired absorption and/or the excessive excretion of

Table 7. Comparison between serum macro-elements of diabetic group with control

Parameter (Mg/dL)	Diabetic group N = 100 (MEAN±SD)	Control group N = 50 (MEAN±SD)	P-value
Na	301.47±8.16	318.44±6.22	<.001
K	15.33±1.47	17.57±1.82	<.001
Cl	345.68±8.81	355.00±6.81	<.001
Ca	12.01±1.36	13.43±0.41	<.001
P	2.10±0.95	3.46±0.58	<.001
Mg	0.85±0.22	1.11±0.24	<.001
S	3.78±0.78	2.84±1.10	<.001

The results showed that, Sodium (301.47±8.16 mg/dL, P<.001), Potassium (15.33±1.47 mg/dL, P<.001), Chloride (15.33±1.47 mg/dL, P<.001), Calcium (12.01±1.36mg/dL, P<.001), Phosphorous (2.10±0.95 mg/dL P<.001) and Magnesium (0.85±0.22 mg/dL, P<.001) were significantly lower in diabetic subjects when compared with the control subjects (318.44±6.22 mg/dL), (17.57±1.82 mg/dL), (355.00±6.81mg/dL), (13.43±0.41 mg/dL), (3.46±0.58 Mg/dL) and (1.11±0.24 mg/dL)respectively. Conversely, the mean Sulphur (3.78±0.78 mg/dL, P<.001) was significantly higher in diabetic subjects when compared with the control group (2.84±1.10 mg/dL)

these elements in urine (glycosuria) in these patients, which may induce a deficiency or marginal state of these minerals in blood of diabetic patients. Brown et al. [45], Isbir et al. [46]. Calcium levels in diabetic group differ between the different age groups in both male and female but the differences was not statistically significant *(p>0.005)*. Moreover, calcium level in male diabetic group was slightly higher than their female counterpart but the difference was not statistically significant *(p>0.05)*.

Analysis of serum phosphorus in this study revealed a significantly lower level *(P<.001)* in diabetic patients when compared with control group, similar result has also been recorded by Djalali et al. [43]. Phosphorous level of diabetic group differs between the different age groups but the difference was not statistically significant *(p>0.05)*, and also male diabetic group has a slightly higher phosphorous level than their female counterpart but the difference was not statistically significant *(p>0.05)*. This reduction in phosphorus level may be due to loss as a result of increase glucosuria. The evidence of the occurrence of a paradoxical imbalance in phosphate metabolism from the early onset of diabetes mellitus has indicated that this imbalance may lead to a reduction of high energy phosphates. Cellular phosphorus depletion and hypophosphatemia have played a role in the development of acute, occasionally life-threatening complications in diabetic mellitus. When insufficient phosphate and oxygen are available for adenosine triphosphate (ATP) synthesis, cell homeostasis cannot be maintained and may result in cell lyses; this has been recognized as a cause of morbidity and mortality in Diabetic ketoacidosis [47].

4. CONCLUSION

I. Significantly reduced concentration of sodium, potassium, Chloride calcium, magnesium, phosphorus and sulphur were observed in diabetic group when compared with the non-diabetic group. The macro-minerals level also vary among the different age group and in both sexes but the differences were not statistically significant except calcium which was significantly reduced in female when compare with their male counterparts.

II. Alterations in Macro minerals levels may underlie many of the pathophysiologic and clinical characteristics of diabetes and as such must be routinely assayed in Diabetic patients so as to prescribe appropriate management which could prevent complications.

III. The impaired mineral elements metabolism of the present work may have a role in the pathogenesis and progression of diabetic mellitus type II. Imbalances in mineral elements level in the body may disturb antioxidants levels, hormone secretion, enzymatic activities and secretion of some enzymes e.g. the secretion of pancreatic amylase and it may also enhance lipid peroxidation.

IV. Sex related differences in serum minerals and mostly lipid fractions were observed in the diabetic population of this study. The differences observed in the diabetic males and female may be attributed to increased urinary excretion of these minerals, bone mineralization, physical activity, life style and hormonal imbalances may associate with the diabetic state in both sexes.

ETHICAL APPROVAL

This study was approved by the Department of Biochemistry of Ekiti State University, Ado Ekiti, Department of Chemical Pathology of Ekiti State University Teaching Hospital, Ado Ekiti and Federal Medical Centre of Ido Ekiti, Ekiti State, Nigeria. All procedures followed were in accordance with the ethical standards Ministry of Health, Nigeria.

COMPETING INTERESTS

Authors have declared that no competing interests exist.

REFERENCES

1. Odewabi AO, Akinola EG, Ogundahunsi OA, Oyegunle VA, Amballi AA, Raimi TH, Adeniyi FA. Liver enzymes and its correlates in treated and newly diagnosed type 2 diabetes mellitus patients in Osogbo, South West, Nigeria. Asian J. Med. Sci. 2013;5(5):108-112.
2. Lobo DN. Fluid, electrolytes and nutrition: Physiological and clinical aspects. Proc Nutr Soc. 2004;63(3):453–466.
3. DeFronzo RA, Goldberg M, Agus ZS. The effects of glucose and insulin on renal electrolyte transport. J Clin Invest. 1976; 58(1):83–90.
4. Katz MA. Hyperglycemia-induced hyponatremia–calculation of expected serum sodium depression. N Engl J Med. 1973;289(16):843–844.
5. Al-Rubeaan K, Siddiqui K, Abu Risheh K, Hamsirani R, Alzekri A, Alaseem A, Saleh SM, Al-Yami Z, Al-Ghamdi A, Alayed K. Correlation between serum electrolytes and fasting glucose and Hb1Ac in Saudi diabetic patients. Biol Trace Elem Res. 2011;144(1–3):463–468.
6. Levy J, Stern Z, Gutman A, Naparstek Y, Gavin JR, Avioli LV. Plasma calcium and phosphate levels in an adult noninsulin-dependent diabetic population. Calcif Tissue Int. 1986;39(5):316–318.
7. Chatterjee R, Yeh HC, Shafi T, et al. Serum and dietary potassium and risk of incident Type 2 diabetes mellitus: The Atherosclerosis Risk in Communities (ARIC) study. Arch Intern Med. 2010;170: 1745–1751.
8. Tuck ML, Sambhi MP, Levin L. Selective hyporeninism and hypoaldosteronism in diabetes mellitus: Studies of the autonomic nervous system control of renin release. Diabetes. 1979;28:237-246.
9. Milner RD, Hales CN. The role of calcium and magnesium in insulin secretion from rabbit pancreas studied *in vitro*. Diabetologia. 1967;3:47–49.
10. Emoto M, Kanda H, Shoji T, Kawagishi T, Komatsu M, Mori K, Tahara H, Ishimura E, Inaba M, Okuno Y, Nishizawa Y. Impact of insulin resistance and nephropathy on homocysteine in type 2 diabetes. Diabetes Care. 2001;24:533-538.
11. Ditzel J, Lervang HH. Disturbance of inorganic phosphate metabolism in diabetes mellitus: Temporary therapeutic intervention trials. Diabetes Metab Syndr Obes. 2009;2:173–177.
12. Ditzel J, Lervang HH. Disturbance of inorganic phosphate metabolism in diabetes mellitus: Its impact on the development of diabetic late complications. Curr Diabetes Rev. 2010;6(5):323–333.
13. Haglin L, Tornkvist B, Backman L. Prediction of all-cause mortality in a patient population with hypertension and type 2 DM by using traditional risk factors and serum-phosphate, -calcium and-magnesium. Acta Diabetol. 2007;44(3): 138–143.
14. Song Y, He K, Levitan E, Manson J, Liu S. Effects of oral magnesium supplementation on glycaemic control in Type 2 diabetes: A meta-analysis of randomized double-blind controlled trials. Diabetic Med. 2006;23(10):1050–1056.
15. Mumayun M, Khalid A, Ali A, Ahmed S, Javed A. To study levels of serum Cr, Cu, Mg and Zn in patients with diabetes mellitus type 2. Pak. J. Med. Health Sci. 2011;5:34-38.
16. Tripathy S, Sumathi S, Raj GB. Minerals nutritional status of type-2 diabetic subjects. Int J Diab Dev Countries. 2004; 24:27-8.
17. Adewumi MT, Njoku CH, Saidu Y, Abubakar MK, Shehu RA, Bilbis LS, Serum CC, Mn C. Levels of diabetic subjects in Katsina, Nigeria. Asian J. Biochem. 2007;2:284-288.
18. Chinyere NA, Opara UCA, Henrieta EM, Nathanie UI. Serum and urine levels of chromium and magnesium in type 2 diabetics in Calabar, Nigeria. Malays. J. Nutr. 2005;11:133-242.

19. Barham D, Trinder P. An improved color reagent for the determination of blood glucose by the oxidase system. Analyst. 1972;97(151):142-5.
20. Munoz MA, Balon M, Fernandez C. Direct determination of inorganic phosphorous in serum with with a single reagent. Clin Chem. 1983;29:372-374.
21. Berhane Seyoum, Elias S Siraj, Christopher Saenz, Jemal Abdulkadir. Hypo- magnesemia in Ethiopians with Diabetes Mellitus. Ethnicity and Disease. 2008;18:4-8.
22. Chetan P Hans, Sialy R, Devi D Bansal. Magnesium deficiency and diabetes mellitus. Current Science. 2000;83:12.
23. Diwan AG, Pradhan AB, Lingojwar D, Krishna KK, Singh P, Almelkar SI. Serum zinc, chromium and magnesium levels in type-2 diabetes. Int J Diabet Dev Countries. 2006;26:122-3.
24. Tosiello L. Hypomagnesaemia and diabetes mellitus. A review of clinical implications. Arch Intern Med. 1996;156: 1143-8.
25. Walter RM jr, Uriu-Hare JY, Olin KL, Oster MH, Anawalt BD, Critchfield JW. Copper, zinc, manganese and magnesium status and complications of diabetes mellitus. Diabetes Care. 1991;14:1050-6.
26. McNair P, Christensen MS, Christensen C, Madsbad S, Transbol I. Renal hypomagnesaemia in human diabetes mellitus: its relation to glucose homeostasis. Euro. J. Clin. Invest. 1982; 12:81-85.
27. Anetor JI, Senjobi A, Ajose OA, Agbedana EO. Decreased serum magnesium and zinc levels: Artherogenic implications in type-2 diabetes mellitus in Nigerians. Nutr Health. 2002;16:291-300.
28. Supriya, Shrabani Mohanty, Venkata Bharatkumar Pinnelli, Roopa Murgod, Raghavendra Ds. Evaluation Of serum copper, magnesium and glycated haemoglobin in type 2 diabetes mellitus. Asian Journal of Pharmaceutical and Clinical Research. 2013;6(2).
29. Sharma A, Dabla S, Agrawal RP, Barjatya H, Kothari RP, Kochar DK. Serum magnesium: An early predictor of course complications of diabetes mellitus. J Indian Med Assoc. 2007;105:16-20.
30. Durlach J, Altura B, Altura BM. Highlights and summary of the 10th Annual French Colloquium on Magnesium. Magnesium. 1983;2:330-36.
31. Elaine M Worcester, Fredric L Coe. New insights into the pathogenesis of idiopathic hypercalciuria. 2008;28(2):120-32.
32. Ishrat Kareem, Jaweed SA, Bardapurkar JS, Patil VP. Study of magnesium, glycosylated hemoglobin and lipid profile in diabetic retinopathy. Indian Journal of Clinical Biochemistry. 2004;19(2):124-27.
33. Nsonwu AC, Usoro CAO, Etukudo MH, Usoro IN. Influence of age, gender and duration of diabetes on serum and urine levels of zinc, magnesium, selenium and chromium in type-2 diabetics in Calabar, Nigeria. Turk J Biochem. 2006;31:107-14.
34. Hasan CMM, Parial R, Islam MM, Ahmad MNU, Rakhing HC. Evaluation of biochemical parameters in controlled and uncontrolled type- 2 diabetic patients of Bangladesh. International Research Journal of Biological Sciences. 2014;3(3): 19-22.
35. Cowie CC, Harris MI. Physical and metabolic characteristics of persons with diabetes, Diabetes in America, 2nd ed. National Institutes of Health. 1995;117-164.
36. Ranee Chatterjee, Hsin-Chieh Yeh, David Edelman, Frederick Brancati. Potassium and risk of type 2 diabetes. Expert Rev Endocrinol Metab. 2011;6(5):665–672.
37. Christlieb AR, Kaldany A, D'Elia J, Williams G. Aldosterone responsiveness in patients with diabetes mellitus. Diabetes. 1978;27: 732-737.
38. Fiorina P, Lanfredini M, Montanari A, Peca MG, Verpnelli A, Mello A, Astorri E, Craveri A. Plasma homocysteine and folate are related to arterial blood pressure in type 2 diabetes mellitus. Am J Hypertens. 1998;11:1100-1107.
39. Hultberg B, Agardh E, Andersson A, Brattström L, Isaksson A, Israelsson B, Agardh CD. Increased levels of plasma homocysteine are associated with nephropathy, but not severe retinopathy in type 1 diabetes mellitus. Scand J Clin Lab Invest. 1991;51:277-282.
40. Chico A, Perez A, Cordoba A, Arcelus R, Carreras G, de Leiva A, Gonzales-Sastre F, Blanco-Vaca F. Plasma homocysteine is related to albumin excretion rate in patients with diabetes mellitus: A new link between diabetic nephropathy and

cardiovascular disease? Diabetologia. 1998;41:684-693.

41. Stabler SP, Estacio R, Jeffers BW, Cohen JA, Allen RH, Schrier RW. Total homocysteine is associated with nephropathy in non-insulin-dependent diabetes mellitus. Metabolism. 1999;48: 1096-1101. 21.
42. Buysschaert M, Dramaıs AS, Wallemaco PE, Hermansa MP. Homocysteinemia in type 2 diabetes. Diabetes Care. 2000;23: 1816-1822.
43. Djalali M, Taheri E, Saedisomeolia A, Djazayeri A, Rahemi A, Hashemi M, Larijani B. Vitamin D status of type 2 diabetic patients compared with healthy subjects in the Islamic Republic of Iran. Eastern Mediterranean Health Journal. 2013;19(Supplement 3).
44. Anastassios GP, Joseph L, Frank BH, Bess Dawson-Hughes. REVIEW: The role of vitamin D and calcium in type 2 diabetes. A systematic review and meta-analysis. The Journal of Clinical Endocrinology & Metabolism. 2007;92(6): 2017–2029.
45. Brown IR, McBain AM, Chalmers J, Campbell IW, Brown ER, Lewis MJ. Sex difference in the relationship of calcium and magnesium excretion to glycaemic control in type-1 diabetes mellitus. Clin Chim Acta.1999;283:119–28.
46. Isbir T, Tamer L, Taylor A, Isbir M. Zinc, copper and magnesium in insulin-dependent diabetes. Diabetes Res. 1994;26(1):41– 5.
47. Farber E. ATP and cell integrity. Fed Proc. 1973;32(4):1534–1539.

27

Toxicological Evaluation of Some Artemisinin Combination Therapies (ACTs) on the Kidney and Liver of Albino Wistar Rats

O. E. Etim[1*], U. E. Bassey[1], G. E. Charles[2], E. E. Sambo[1] and E. J. Akpan[1]

[1]Department of Biochemistry, Faculty of Basic Medical Sciences, University of Uyo, Nigeria.
[2]Department of Pharmaceutical/Medicinal Chemistry, Faculty of Pharmacy, University of Uyo, Nigeria.

Authors' contributions

This work was carried out in collaboration between all authors. Author OEE designed the study, wrote the protocol and supervised the work. Authors EES and GEC carried out all laboratories work and performed the statistical analysis. Author UEB managed the analyses of the study and wrote the first draft of the manuscript. Author EJA managed the literature searches and edited the manuscript. All authors read and approved the final manuscript.

Editor(s):
(1) Dileep G. Nair, Ministry of Higher Education, Sultanate of Oman.
Reviewers:
(1) Aina, Oluwagbemiga, Olanrewaju, Nigerian Institute of Medical Research, Nigeria.
(2) Ioana Stanciu, University of Bucharest, Romania.

ABSTRACT

World Health Organisation (WHO) recommends the use of artemisinin-based combination therapies (ACTs) as drug of choice for the treatment of malaria in endemic regions of the world. This study was designed to evaluate the effects of therapeutic doses of ACTs: Artesunate (Artesunat®), Artesunate-Mefloquine (Artequin®), Artemether-Lumefantrin (Coartem®) and Dihydroartesiminin-Piperaquine (P-Alaxin®) on the integrity of the liver and kidneys of albino Wistar rats. Thirty (30) albino Wistar rats weighing between 200 g – 280 g were randomly divided into 5 groups with 6 animals per group. Group 1 served as control (CTR) while Group 2 received artesunate (AS) for 5 days. Groups 3, 4 and 5 received therapeutic doses of artequine (AQ), coartem (CT) and p-alaxin (PA) respectively for 3 days. The animals were sacrificed under chloroform anaesthesia and blood samples obtained through cardiac puncture for biochemical investigations. Serum ALT activity of

**Corresponding author: E-mail: utibeevans@yahoo.com, yireobong_123@yahoo.com*

Groups 2 and 3 were significantly elevated ($p<0.05$) while Groups 4 and 5 experienced marginal increased. Group 4 also showed significant increase in total bilirubin though it was marginally increased in all other groups when compared to the control. Creatinine levels were marginally increased in all groups while the urea levels were relatively stable across all groups except in Group 3 where significant increase was observed. The results therefore indicate possible hepatic injury and renal toxicity in albino Wistar rats hence the need for caution and proper attention while undergoing malaria treatment with ACTs.

Keywords: Malaria; Artemisinin-based Combination Therapies (ACTs); liver; kidney and toxicity.

1. INTRODUCTION

Malaria remains one of the greatest causes of mortality and morbidity in the world. Globally, there are between 124 – 283 million cases of clinical malaria every year. Currently, estimated 367,000 to 755,000 deaths are attributed to malaria annually, 90% of them in Africa [1]. In Nigeria, malaria is holoendemic hence clinical cases of the disease are seen throughout the year. It is the commonest cause of out-patient hospital attendance in all age groups in the country. Studies have revealed that the readily available and cheap antimalarial drugs: chloroquine and sulphadoxine-pyrimethamine are no longer efficacious in the treatment of malaria in Nigeria due to the development of resistant to these agents [2,3]. Prompt and adequate treatment with an effective antimalarial drug is an essential strategy to reduce the morbidity and mortality resulting from malaria infections. Presently, artemisinin-based combination therapy is the drug of choice for the treatment of malaria in the country and this has proven to be effective in combating the disease. The rationale for the use of ACTs is to reduce the probability of resistant developing simultaneously to the two drugs with independent mechanism of action [3].

The artemisinins are developed from the Chinese wormwood (*Artemisia annua*) and the derivatives namely; artemether, artesunate and dihydroartemisinin have now gained popularity as short acting drugs which could be used in combination with drugs which have longer half-life [4]. Artemether-lumefantrine, artesunate-mefloquin, dihydroartemisinin-piperaquin alongside artesunate are among the commercially available artemisinin-based antimalaria therapies. ACTs have been reported to be safe and efficacious in the treatment of uncomplicated malaria in Nigeria and other countries [1,5-7].

Artemisinin and its derivatives have a short half life while the other drugs in ACTs are often with longer half life hence the risk of drug resistant is substantially reduced. Artemisinins are considered to have high safety margins, however, they may be toxic under certain conditions [8] and when used in combination as ACTs on renal function [9]. This study was designed to evaluate and compare the toxicities of ACTs; artemether-lumefantrine, artesunate-mefloquin, dihydroartemisinin-piperaquin and artesunate on the liver and kidney of albino Wistar rats. These organs are involved in the metabolism and elimination of drugs as such they are susceptible to drug induced injuries.

2. MATERIALS AND METHODS

2.1 Materials

Artemether-lumefantrin (Coartem 80 mg/480 mg), Artesunate-Mefloquin (Atequin 600 mg/750 mg), Dihydroartemisinin-piperaquin (P-Alaxin 40 mg/ 320 mg) and Artesunate (Artesunat 50 mg) were purchased from Tonay Pharmacy, Itu Road, Uyo, Akwa Ibom State, Nigeria. Reagents used for analysis were gotten from Randox Laboratories Ltd, UK. All reagents and chemicals were of analytical grades.

2.2 Experimental Design and Treatment of Animals

Thirty (30) albino Wistar rats weighing between 200g – 280 g were randomly divided into 5 groups with 6 animals per group. The rats were kept in standard cages and laboratory conditions, and were fed with normal rat pellets and water *ad libitum*. Group 1 served as control while Groups 2 – 5 were administered therapeutic doses of ACTs as shown in Table 1.

24 hours after the last dosage, the animals were sacrificed under chloroform anaesthesia. Blood samples were removed through cardiac puncture into a plane sample bottle. Serum was gotten from the blood by centrifugation and then preserved for biochemical investigations.

Table 1. Administration of ACTs to groups 2–5

Groups	ACT brand	Dosage	Duration
2. AS	Artesunat®	1.43 mg of artesunate per kg b.w of rat.	Twice on day 1 followed by once daily from day 2 to 5
3. AQ	Artequin®	2.86 mg of artesunate and 3.57 mg of mefloquine per kg b.w of rat.	Daily for 3days
4. CT	Coartem®	1.14 mg of artemether and 6.86 mg of lumefantrine per kg b.w of rat.	Twice daily for 3days
5. PA	P-Alaxin®	1.71 mg of dihydroartemisinin and 13.71mg of piperaquine per b.w of rat.	Once daily for 2 days followed by two third of the dosage on day 3

2.3 Estimation of Biochemical Parameters

Serum ALT activity was assayed based on the method developed by Wroblewski and Ladue [10], while the method of Doumas and Briggs [11] was used to assay AST activity. The method of Bowers and McComb [12] was used to estimate ALP activity in the serum. Determination of Total Bilirubin (TB) concentration was carried out according to the method of Jendrassik and Grof, [13]. Serum urea and creatinine concentration were estimated based on the methods of Tobacco et al. [14] and Narayanan and Appleton [15] respectively.

2.4 Statistical Analysis

The data from biochemical analysis were subject to one way analysis of variance (ANOVA) and Student's T-test using SPSS package. Values ($P<0.05$) were considered significant against the control.

3. RESULTS AND DISCUSSION

3.1 Results

The results of liver enzyme activities and evaluation of kidney function following administration of some ACTs are presented in Tables 2 and 3 respectively.

Table 2. Serum enzyme activities of albino wistar rats administered with different ACTS

Groups	ALT (U/L)	AST (U/L)	ALP (U/L)
1. CTR	16.04±3.20	16.36±4.31	36.99±1.68
2. AS	20.29±3.54*	13.93±4.66	24.67±3.03*
3. AQ	20.32±3.09*	12.40±1.87*	35.32±2.31
4. CT	18.90±4.70	12.00±3.68*	37.37±3.84
5. PA	18.90±5.57	15.26±3.99	28.86±5.77*

*Values are presented as mean ± standard deviation. * = significantly different from control (p<0.05). Number of rats (n) = 6*

Table 3. Effect of administration of ACTs on urea, creatinine and bilirubin concentration in albino wistar rats

Groups	Creatinine (µmol/l)	Urea (mmol/l)	Total bilirubin (µmol/l)	CONJ. bilirubin (µmol/l)
1. CTR	84.21±20.39	3.01±0.36	22.90±4.30	6.10±2.10
2. AS	87.42±18.05	2.51±0.77	23.38±4.00	5.17±1.77
3. AQ	89.84±28.92	3.76±0.40*	23.60±4.57	3.53±1.75*
4. CT	106.82±30.75	2.75±2.31	29.40±3.21*	6.52±2.70
5. PA	101.82±32.25	3.03±0.74	24.79±4.00	5.12±2.09

*Values are presented as mean ± standard deviation. * = significantly different from control (p<0.05). Number of rats (n) = 6*

3.2 Discussion

Hepatocyte membrane distortion is associated with membrane leakage of the hepatocyte cytosolic contents which is manifested by significant elevation of serum/plasma enzymes namely; ALT, AST and ALP [16]. These enzymes serve as biomarkers for evaluation of hepatocellular damage, however, ALT is the most reliable. AST is known to be abundant in the cardiac muscle, skeletal muscle, kidney and testes thus any disease affecting any of these extra hepatic tissues significantly elevates the serum level of the enzyme [17]. The observed significant increase in ALT activity in Groups 2 and 3 indicates possible drug induced hepatic damage or hepatotoxicity in the albino Wistar rats. This is in collaboration with other studies where the effects of P-Alaxin and Coartem on pregnant albino Wistar rats were evaluated [18,19,20].

Furthermore, a study by Ofem et al. [21] reported that activities of serum ALT, AST and ALP increased significantly upon the administration of Coartem and P-Alaxin on albino Wistar rats. Elevation of serum enzyme activities can also be attributed to causes other than liver injury such as muscle damage. These enzymes exist as isoenzymes which originates from other tissues hence the need for evaluation of other biomarkers of liver function. Total bilirubin was found to be significantly elevated in coartem treated group, suggestive of hepatobillary injury. There were non-significant increases in total bilirubin in other groups. Intravascular haemolysis of red blood cells due to drug treatment has been reported. In particular, quinine based drug, when used for chemoprophylaxis has been shown to stimulate the production of drug dependent complement fixing capable of causing intravascular red cell lyses [22].

Elevated levels of creatinine and urea in serum are an indication of poorly functioning kidneys and malfunctioning such as impairment of glomerular filtration. A rise in blood creatinine is observed only with marked damage to functioning nephrons [23]. This study revealed a non-significant increase in the levels of creatinine in all groups while urea was significantly increased in artequin treated group. However, a better evaluation of kidney function would be through a creatinine clearance test as suggested by American Diabetes Association [24].

4. CONCLUSION

Summarily, it can be concluded from this study that ACT therapies can have deleterious effect on the kidney and liver. Furthermore, dihydroartemisinin-piperaquin has the best safety profile among the ACTs studied while artemether-lumefantrin and artesunate-mefloquin are more likely to induce significant toxicological effects.

COMPETING INTERESTS

Authors have declared that no competing interests exist.

REFERENCES

1. World Health Organisation. World malaria report 2014. WHO press, Geneva; 2014.
2. Ekanem OJ, Weisfeld LA, Salako Nahlen MU. Sensitivity of *Plasmodium falciparum* to chloroquine and sulphadoxine-pyrimethamine in Nigerian children. Bulletin of World Health Organisation. 1990;68:45-52.
3. White NJ. Delaying antimalarial drug resistance with combination chemotherapy. Parasitology. 1999;41:301-308.
4. White NJ, Olliaro P. Artemisinin and derivatives in the treatment of uncomplicated Malaria. Tropical Medicine. 1998;58:54–56.
5. Mayxay M, Thongpraseuth V, Khanthavong M, Lindergardh N, Barends M, Keola S, et al. An open randomised comparison of artesunate plus mefloquin vs dihydroartemisinin-piperaquin for the treatment of uncomplicated *Plasmodium falciparum* malaria in the Lao people's democratic republic (Laos). Tropical Medicine and International Health. 2006;11(8):1157-1165.
6. Frey SG, Chelo D, Kinkela MN, Djouknone F, Tietche F, Hatz C. et al. Artesunate-mefloquin combination therapy in acute *Plasmodium falciparum* malaria in young children: A field study regarding neurological and neuropsychiatric safety. Malaria Journal. 2010;9:291-299.
7. Olayemi SO, Anikawe AP, Akinyede A, Oreagba AI, Awodele O. Effect of malarial treatment on biochemical parameters and plasma pH of mice infected with

Plasmodium berghei. International Journal of Pharmacology. 2012;8:549-553.
8. Udobre A, Edoho EJ, Eseyin O, Etim EJ. Effect of artemisinin with folic acid on the activities of aspartate aminotransferase, alanine aminotransferase and alkaline phosphatase in rats. Asian Journal of Biochemistry. 2009;4:55-59.
9. Etim EI, Udoh IE, Udoh AE. Evaluation of plasma creatinine and urea after oral administration of artemisinin and some artemisinin combination therapy regimen on healthy Nigerian males. World Journal of Applied Sciences and Technology. 2012;4:1-6.
10. Wroblewski F, Ladue JS. Serum glutamic pyruvic transaminase in cardiac and hepatic disease. Proceedings of the Society for Experimental Biology and Medicine. 1956;91:569.
11. Doumas B, Briggs HG. Standard Methods of Clinical Chemistry. London Academic Press, New York; 1972.
12. Bowers GN, McComb RB. A Continous spectrophotometric method for measuring the activity of serum alkaline phosphatase. Clinical Chemistry. 1966;12:70
13. Jendrassik L, Grof P. Colorimetric Method of determination of bilirubin. Biochemistry. 1938;297:81-82.
14. Tobacco A, Meiatini F, Moda E, Tarli P. Simplified enzymatic/colorimetric serum urea nitrogen determination. Clinical Chemistry. 1979;25:336.
15. Narayanan S, Appleton HD. Creatinine: A review. Clinical Chemistry. 1980;26:1119–1126.
16. Bhaltacharya M, Western V, Pan K. Elevation of serum/plasma enzymes of acute hepatocellular damage. Clinical Pharmacology and Therapeutics. 2003;41(7):16-19.
17. Olayinka ET, Ore A. Alteration in antioxidant status and biochemical indices following administration of dihydroartemisinin-piperaquin phosphate (P-Alaxin). IOSR Journal of Pharmacy and Biological Sciences. 2013;5(4):43-53.
18. Farombi A, Ademola B, Olatunde M. Effect of coartem and p-alaxin on pregnant albino wistar rats. Asian Journal of Biochemical Science. 2000;16(7):66-69.
19. Obi E, Orisakwe OE, Asongha LA, Udemezue OO, Orish VN. The hepatotoxic effect of halofrantrine in guinea pigs. Indian Journal Pharmacology. 2014;26:303-305.
20. Akam U, Okon E, Edem E. Effect of chloroquine, amodiaquine, quinine and halofantrine on serum enzymes. Journal of Pharmacy and Biological Sciences. 2013;15(8):33-35.
21. Ofem OE, Nna VU, Archibong AN. Comparative effects of two antimalarial drugs (P-alaxin and Coartem) on serum electrolytes and serum enzymes in albino wistar rats. Research Journal of Pharmaceutical. Biological and Chemical Sciences 2014;5(1):54-63.
22. Weatherall DJ, Miller LH, Baruch DI, Marsh K, Doumbo OK, Gasals-Pascual C, et al. Malaria and the Red Cell. In: Haematology. American Society of Haematology Education Program Book; 2002.
23. Etim OE, Akpan EJ, Akpanabiatu MI, Iba U, Bassey UE. Effect of co-administration of chloroquine and cefuroxime on kidney function in albino wistar rats. Journal of Physiology and Pharmacological Advances. 2014;4(8):407-412.
24. Gross JL, de Azevedo MJ, Silveiro SP, Canani LH, Caramori ML, Zelmanovitz T. Diabetic nephropathy: diagnosis, prevention and treatment. Diabetes Care. 2005;28(1):164-76.

28

Some Biochemical Changes in Serum of Female Albino Rats Administered Aqueous Extract of *Piper guineense* Schumach Seeds

Friday O. Uhegbu[1*], Chinedu Imo[1] and Amadike E. Ugbogu[1]

[1]*Department of Biochemistry, Abia State University, PMB 2000, Uturu, Nigeria.*

Authors' contributions

This work was carried out in collaboration of all authors. Author FOU conceived and designed the study. Authors FOU and CI wrote the protocol and first draft of the manuscript. Author AEU collected processed and prepared the plant seeds extract and also collected the albino rats. Authors FOU and CI managed the literature search, animal experiments and administration of plant extract and collection of biochemical samples from the sacrificed animals. Authors CI and AEU performed the statistical analysis, proofread and formatted the final manuscript. All authors read, corrected and approved the final manuscript.

Editor(s):
(1) Richard A. Manderville, Departments of Chemistry and Toxicology University of Guelph, Canada.
Reviewers:
(1) Anonymous, Medical Univeristy of Lublin, Poland.
(2) Atef Mahmoud Mahmoud Attia, Biochemistry department, National Research Centre, Egypt.
(3) Anonymous, Benghazi University, Benghazi, Libya.

ABSTRACT

This study examined the biochemical changes in serum of female albino rats administered aqueous extract of *Piper guineense* Schumach seeds extract. The test animals (groups 2, 3 and 4) received 25 mg/kg, 50 mg/kg and 75 mg/kg of the extract respectively for 21 days, while group 1 served as control and received a placebo of 0.9% physiological saline which is the vehicle for administration. On the 22nd day the test animals which have been fasted overnight were sacrificed with chloroform anesthesia. Blood was collected through cardiac puncture. Results show that serum protein, albumin and packed cell volume concentration significantly ($p \leq 0.05$) increased. Creatinine and urea levels decreased ($p \leq 0.05$) significantly, 95% confidence level, while cholesterol profile decreased

*Corresponding author: E-mail: fouhegbu@yahoo.com

($p \leq 0.05$) significantly, except HDL. Superoxide dismutase, glutathione-s-transferase, increased ($p \leq 0.05$) significantly, while lipid peroxidation decreased. This study has shown that nursing mothers in South East of Nigeria who use the *Piper guineense* Schumach seeds may derive immense benefit from its effect on the biochemical parameters assayed in this study.

Keywords: Spices; lipid; antioxidants; protein; creatinine; urea; Piper guineense.

1. INTRODUCTION

Spices are aromatic vegetable substances that are used in small quantities for seasoning of foods and also as preservatives or flavoring [1]. Spices enhance the taste of food, beverage and drug [2]. Spices contain phytochemicals which give them their characteristic odor, flavor and other properties [3].

Piper guineense (Schumach.et Thonn / Piperaceae) (Uziza or black pepper) is a spice, a West African tropical plant that is widely consumed and is considered medicinal by some people [4]. The aroma, flavor and preservative properties of *Piper guineense* Schumach are probably due to the nutritional and non-nutritional factors of the plant. It is rich in protein, fat, fiber and carbohydrate. The vitamin content is also high, especially vitamin C, which is an antioxidant. Peroxide value and free fatty acid level are generally low. The essential oils content is fairly high and range from 0.1 to 5.2%. The fruits of *Piper guineense* Schumach (which is the part of the plant that is traditionally used) are rich in a wide range of natural products, including volatile oils, lignans, amides, alkaloids, flavonoids and polyphenols [5]. Juliani et al. [6] reported that *Piper guineense* Schumach is an important plant used in traditional medicine and as a spice. As a medicinal plant, *Piper guineense* Schumach seeds are used to relief discomfort in the stomach caused by excess gas, while the leaves are used to treat respiratory infections, rheumatism and syphilis [7]. The leaves and seeds of the plant are used to flavor soup, stew and food generally [8]. The presence of an alkaloid – piperine has also been reported. The piperine contains strong anticonvulsant called anti- epileparine [N- 3-4- methylene dioxycinamyl piperidine] which has been synthesized for the treatment of epilepsy (Abila et al. [9], Oleifa et al. [10], Amosan et al. [11]. Udoh et al. [12] and Echo et al. [5] have also reported that the plant has therapeutic effect for the management of convulsion, leprosy, inflammation and rheumatoid pains. Bacteriostatic and bactericidal properties of *Piper guineense* Schumach has also been reported by Zollo et al. [13], Okigbo and Igwe [14], while Echo et al. [5] reported the molluscidal properties of the methanolic seed extract of the plant, and suggested it could be used for the treatment of Bilharzia. Mbongwe et al. [15] reported that the leaves of *Piper guineense* Schumach seeds are used for treating female infertility, while the seeds are used as an aphrodisiac. Hence the Igbos of the South Eastern part of Nigeria use the seeds to prepare soups for nursing mothers from the first day of delivery to prevent post partum contractions and to aid the fast return of the uterine muscles of the uterus to their original shape. It is also believed that it aids the increase in breast milk production and flow.

The present study is therefore aimed at investigating the biochemical changes that may occur in female albino rats due the administration of aqueous extract of *Piper guineense* Schumach seeds.

2. MATERIALS AND METHODS

2.1 *Piper guineense* Schumach Seeds

The *Piper guineense* Schumach seeds were purchased from the Eke Okigwe market as sold. The seeds were identified and authenticated at the Department of Plant Science and Biotechnology, Abia State University, Uturu Nigeria and voucher specimens deposited at the departmental herbarium.

2.2 Preparation of Extract

The *Piper guineense* Schumach seeds were cleaned and sorted for healthy seeds. The seeds were sun dried for five days, and milled into fine powder using a milling machine. 40 g of the powder was dissolved in 300 ml of distilled water and left to stand for 48 hrs with occasional shaking. The mixture was then filtered using a muslin bag. 263 ml of the filtrate was recovered. The concentration of the extract was determined to be 10.15 mg/ml.

2.3 Experimental Animals and Design

Forty female albino rats, 7 weeks old and weighing between 128 g-142 g were used for the study. The animals were randomly placed into four groups of 10 animals each. Group 1 served as control, Group 2, 3 and 4 served as test animals. The animals were allowed feed and water *ad libitum*. Standard laboratory protocols for animal studies were maintained. Approval for animal studies was obtained from the Animal Ethics Committee of the College of Medicine and Health Sciences, Abia State University, Uturu, Nigeria.

2.4 Treatment of Animals

The animals were starved overnight prior to the commencement of the administration of the *Piper guineense* Schumach seeds extract. The administration of the *Piper guineense* Schumach seeds extract was by oral intubation once daily at 25 mg/kg body weight, 50 mg/kg body weight and 75 mg/kg body weight for group 2, 3 and 4 animals respectively for 21 days.

2.5 Collection of Blood Samples and Serum Preparation

At the 22nd day the animals were sacrificed after overnight starving. Incisions were made into their thoracic cavity. Blood samples were collected by heart aorta puncture using a 10 mL hypodermic syringe and allowed to clot in sample vials. The samples were centrifuged at 3000 rpm for 5 min. using the Bran Scientific and Instrument Company England centrifuge. The supernatant (serum) was harvested by simple aspiration with Pasteur pipette and stored in clean tubes at -4°C until analysis.

2.6 Estimation of Parameters

2.6.1 Determination of total cholesterol

Total cholesterol was determined by the enzymatic colorimetric cho-PAP method, High density lipoprotein (HDL) Very low density lipoprotein (VLDL), and Low density lipoprotein (LDL) were determined as described by Burstein et al. [16]. Triacylglycerol and cholesterol were assayed as described by Fletcher [17], Richmond [18] and Barlett [19], using Biosystem kits. The measurements were performed according to the manufacturers' instruction.

2.6.2 Determination of total protein

Total protein test kits produced by Randox laboratories were used. This test involves formation of colored complex between cupric ions in alkaline medium with peptide bonds. The intensity of the color formed is proportional to the concentration of the protein.

2.6.3 Determination of albumin

Albumin test kit produced by Biosystem kits were used to estimate albumin. The principle of this test is based on albumin quantitatively binding to bromocresol green (BCG). The complex of albumin-BCG so formed absorbs maximally at 578 nm, and the absorbance is directly proportional to the concentration of albumin in the sample.

2.6.4 Determination of lipid peroxidation

Thiobarbituric acid reactive substances (TBARS) method as described by Buege and Aust [20] was used to determine lipid peroxidation,

2.6.5 Determination of antioxidant enzymes

Glutathione-s- transferase (GST) activity was determined according to the method described by Habig et al. [21], while Superoxide Dismutase was measured as described by Kakkar et al. [22].

2.6.6 Hematological parameters

Commercial kits (Biosystem kits) were used to estimate packed cell volume, creatinine and urea. Packed cell volume and creatinine were assayed as described by Dacie and Lewis [23].

2.7 Statistical Analysis

Values were represented as Mean ± SD. Data obtained were subjected to one way Analysis of Variance (ANOVA) and group means were compared using Duncan's new multiple range tests. Differences were considered to be significant at ($p<0.05$).

3. RESULTS

The results of the analysis is presented in Tables 1-4 below.

4. DISCUSSION

Body weight of experimental animal increased ($p \leq 0.05$) significantly ($p \leq 0.05$) compared to control (Table 1). Animals treated with 75 mg/kg body weight of extract showed significant increase in body weight. The effect of the extract on body weight seems to be dose dependent, as shown on Table 1. The increase in weight may be due to the high content of protein in the extract and the phytochemicals which could elicit increase in protein synthesis.

Triacylglycerol, total cholesterol, Low density and very low density cholesterol decreased ($p \leq 0.05$) non-significantly compared to control; while high density cholesterol increased ($p \leq 0.05$) significantly (Table 2). Cholesterol is a precursor of other steroids such as sex hormones. It is also an important constituent of cell membranes and bile acids [24,25]. Low density and high density cholesterol are involved in the transport of cholesterol to and from the liver. The reduction of these fatty acids may show that the *Piper guineense* seed extract may have hypocholesterolmic effect. The increase in High density cholesterol could be of immense benefit to the animals, since HDL is also known as the 'good' cholesterol and is involved in transport of other lipids to the liver for disposal in bile [26].

Lipid peroxidation decreased ($p \leq 0.05$) significantly (Table 3) compared to control. The significant decrease in lipid peroxidation values in the experimental animals may be as a result of the extract aiding in the reduction of lipid oxidation since the *Piper guineense* seeds are rich in vitamin C which is an antioxidant. Lipid peroxidation is due to the oxidation of lipids especially polyunsaturated lipids to generate peroxides and aldehydes which have been associated with the development of diseases [27,28]. The aldehydes and peroxides are long-lived and can easily spread to distant sites through blood circulation to cause further peroxidation of cells, hence they have been reported to be more dangerous than reactive oxygen species (ROS) [27]. Lipid peroxidation causes cytotoxicity and injury that leads to cell damage due to its ability to form malondialdehyde (MDA) [29].

The antioxidant enzymes superoxide dismutase (SOD) and glutathione s-transferase (GST) increased ($p \leq 0.05$) significantly (Table 3) compared to control. The increase in SOD and GST could be attributed to the presence of phytochemicals in the *Piper guineense* seeds extract especially flavonoids. The seeds of *Piper guineense* are rich in flavonoids, which may have induced detoxification enzymes through up-regulation of their genes by interacting with antioxidant response elements [30-32]. Prestera et al. [33] reported that substances like flavonoids that posses electrophilic centers are able to react with sulfhydryl groups through oxido-reduction or alkylation. Pantuck et al. [34] also reported that an increase in GST activity translates to increased capacity to conjugate and excrete toxic intermediates that can cause diseases such as diabetes. The significant increase in antioxidant enzymes further confirms the high antioxidant capacities of the seeds of *Piper guineense*.

In humans the damaging effects of oxidants and lipid peroxidation can be checked by an organized antioxidant defense mechanism. Several enzymatic and non-enzymatic processes are involved that protect cells against oxidative damage. Farombi et al. [35] reported that the major enzymatic anti-oxidant systems include superoxide dismutase (SOD), glutathione peroxidase (GSH), glutathione – s- transferase (GST) and catalase, while the non – enzymatic antioxidants include glutathione and dietary constituents such as vitamins, flavonoids and carotenoids. The antioxidants protect cells from the damaging effects of free radicals and lipid peroxidation. Opara et al. [36] recorded significant ($p \leq 0.05$) increase in total protein, albumin and globulin, while cholesterol and glucose levels decreased significantly ($p \leq 0.05$) in biochemical responses of adult rabbits to aqueous extract of *Ocimum gratissium* leaves. Aziza et al. [37] reported a decrease in triglycerides, total cholesterol and malondialdehyde level, while GSH and HDL-cholesterol levels increased in rats administered *Gymnema sylvestre* R.Br. leaves extract.

Serum protein concentration increased ($p \leq 0.05$) significantly, while albumin increased non-significantly (Table 4). This could be attributed to the high protein content of the *Piper guineense* seeds [5]. However, the non-significant increase in albumin content could infer a possible preservation of the stability of the liver. Packed cell volume increased ($p \leq 0.05$) significantly, while creatinine and urea decreased ($p \leq 0.05$) significantly (Table 4). The increase in PCV may be contributed by the rich phytochemical content of *Piper guineense* seeds which increased the rate of hemoglobin synthesis in the experimental animals, while increasing the excretion of creatinine and urea [38].

Table 1. Effect of *Piper guineense* seed extract on animal weight (g)

	Group 1 (Control)	Group 2	Group 3	Group 4
Body weight	92.56±0.65*	98.74±0.35	101.46±0.86*	102.15±0.63*

**Values are mean±SD of triplicate determinations (n=10)*
*Values with * are statistically significant ($p<0.05$) compared with the control*

Table 2. Effect of *Piper guineense* seed extract on fatty acid levels (mg/dL)

Parameter	Group 1 (Control)	Group 2	Group 3	Group 4
Triacylglycerol	73.46±3.28*	72.95±4.72	70.54±2.35*	68.42±3.82*
Total cholesterol	1.43 ±0.18	1.37±0.09	1.26±0.10	1.20±0.12*
HDL	35.56±0.86*	36.43±0.16*	37.76±0.24*	38.57±0.11*
LDL	32.68±2.14	31.82±2.06	30.32±2.21*	28.54±2.16
VLDL	11.56±0.85*	11.14±0.14*	10.76±0.56*	10.25±0.43*

**Values are mean±SD of triplicate determinations (n=10)*
*Values with * are statistically significant ($p<0.05$) compared with the control*

Table 3. Effect of *Piper guineense* seed extract on lipid peroxidation levels and antioxidant enzymes

Parameter	Group 1 (Control)	Group 2	Group 3	Group 4
Lipid peroxidation (Mol/L)	4.69±0.65*	3.28±0.16	3.11±0.72	3.01±0.52*
SOD (IU/L)	68.76±6.74	104.57±5.64*	108.26±5.96*	112.39±6.86*
GST (IU/L)	6.86±1.64	7.56±1.22	8.73±1.45*	9.28±1.32*

**Values are mean±SD of triplicate determinations (n=10)*
*Values with * are statistically significant ($p<0.05$) compared with the control*

Table 4. Effect of *Piper guineense* seed extract on some macromolecules

Parameter	Group 1 (Control)	Group 2	Group 3	Group 4
Serum protein(mg/dl)	246.45±12.25	252.25±10.24	265.53±9.65*	278.48±9.82*
Albumin (mg/dl)	2.16±0.13*	2.37±0.17	2.52±0.15*	2.58±0.22*
PCV (L/L)%	30.25±0.56*	33.14±1.05	34.68±1.27*	34.94±1.21*
Creatinine (mg/dl)	0.65±0.03	0.52±0.02	0.46±0.06	0.41±0.04
Urea (mg/dl)	38.74±1.43*	38.14±1.69	36.45±1.36	36.16±1.76*

**Values are mean±SD of triplicate determinations (n=10)*
*Values with * are statistically significant ($p<0.05$) compared with the control*

The kidney is a major organ for excretion and its stability is estimated by the creatinine clearance, which is a function of the glomerular filtration rate.

5. CONCLUSION

This study has been able to show that the *Piper guineense* seeds increased the body weight of the experimental animals, hence may elicit and increase protein synthesis. Total cholesterol and triacylglycerol were significantly reduced, hence the seeds may also have the ability to control cholesterol levels. Lipid peroxidation levels were reduced while antioxidant enzymes increased; albumin non-significantly increased, pack cell volume significantly increased, while creatinine and urea were reduced. The reduction in urea, creatinine and cholesterol levels as observed in this study, shows that *Piper guineense* seeds can be used to improve kidney function and for the management of hypercholesterolemia. These observations have lend credence to the claims of alternative medicine practitioners that nursing mothers who use these spices may benefit from its effect on some biochemical indices as reported in this study.

CONSENT

It is not applicable.

ETHICAL APPROVAL

It is not applicable.

ACKNOWLEGEMENTS

The authors wish to acknowledge the financial assistance of the Vice Chancellor, Abia State University, Uturu- Nigeria. Prof. Chibuzo Ogbuagu. We also thank Mr. Uche Arukwe who helped in carrying out some of the analysis.

COMPETING INTEREST

Authors have declared that no competing interest exists.

REFERENCES

1. Hafner UC, John DA, Mickey CB. Nutrition for Developing countries. Journal of Nutrition. 2003;6:18-25
2. Marcel KK, Akhanovna MJ, Yves-Alain B, Gabin DB, Jérôme ZB. *In vitro* antioxidant activities of total flavonoids extracts from leaves and stems of *Adenia lobata* (Jacq.) Engl. (Passifloraceae). Journal Pharmacog Phytother. 2010;3(1):8-12.
3. Olusegun Ekundayo, Intolaakso Rita Marie, Babajide Oguntimei, Abayomi Sofowora Essential oil constituents of Ashanti pepper (*Piper guineense*) fruits. J. Agric. Food Chem. 2000;26(5):880-882.
4. Iwu M, Hand Book of African medicinal plants. CRC Press, Boca Raton, Flip. 1993;183-184.
5. Echo IA, Osuagwu AN, Agbor RB, Okpako EC, Ekanem BE. Phytochemical Composition of *Aframomum melegueta* and *Piper guineense* Seeds World Journal of Applied Environmental Chemistry. 2012;2(1):17-21.
6. Juliani Rodolfo H, Koroch AR, Giordano L, Amekuse L, Koffa SJ, Asante-Dartey SE, Simon JE. *Piper guineense* (Piperaceae): Chemistry, Traditional Uses and Functional Properties of West African Black Pepper. African Natural Plant Products. Discoveries and Challenges in Chemistry, Health and Nutrition. ACS Symposium Series. 2013;2:1127:33-48.
7. Bep EO. Medicinal Plants in Nigeria. J. of Agric and Food Sci. 1999;7:218-225.
8. Agoha AE. Uses of *Piper guineense* fruits. Journal of Food Sci. 2002;10(3):231-238.
9. Abila B, Richens A, Davies JA. Anticonvulsant effects of extracts of the West African black pepper, *Piper guineense*. Journal of Ethnopharmacology. 1993;39(2):113-117.
10. Olaifa JI, Eehum WO, Akingbohungbe EA. Insecticidal activity of some Nigerian Plants. Insect Science Application. 2002;8: 221-224.
11. Amosan CA, Okerie TG, Mike BC. The use of *Piper guineense* fruit oil as a protectant of dried fish against dermatis and maculates. Global J. Pure Sci. 2002;8(2): 197-201.
12. Udo FVI, Lot TY, Braide VB. Effects of extracts of seed and leaf of *Piper guineense* on skeletal muscle activity in rat and frog. Phytother Res. 1999;13(2):106-10.
13. Zollo PHA, Biyiti L, Tchoumbougnang F, Menut C, Lamaty G, Bouchet P. Aromatic plants of Tropical Central. Part 32. Chemical composition and antifungal activity of thirteen essential oils from aromatic plants of Cameroon. Flav. Fragr. J. 1998;13:107-114.
14. Okigbo RN, Igwe DI. Antimicrobial effects of *Piper guineense* 'Uziza' and *Phyllantus amarus* 'Ebe-Benizo' on *Candida albicans* and *Streptococcus faecalis*. Acta Microbiologica et Immunologica Hungarica. 2007;54(4):353-366.
15. Mbongue FG, Kamtchouing P, Essame O J, Yewah P M, Dimo T, Lontsi D. Effect of the aqueous extract of dry fruits of *Piper guineense* on the reproductive function of adult male rats. Indian Journal Pharmacol. 2005;37:30-32.
16. Burstein M, Scholnick HR, Morfin R. Rapid method for isolation of lipoproteins from human serum by precipitation with polyanions. J. Lipid Res. 1970;11:583-595.
17. Fletcher MJ. A colorimetric method for estimation of Serum triglycerides. Clin. Chem. Acta, 1968;22:393-397.
18. Richmond N. Determination of Cholesterol in Serum. Clin. Chem. 1973;19:1350-1396.
19. Barlett GR. Colorimetric assay methods for free and phosphorylated glyceric acids. Journal Biol. Chem, 1959;234:469-471.
20. Buege J, Aust S. Microsomal lipid peroxidation. In: Methods in Ezymology, Biomembranes. Part C, (Packer L and Fleischer S. eds) Academic Press Inc. 1978;52:302-310.
21. Habig WH, Pabst MJ, Jakoby WB. Glutathione S-transferases: The first enzymatic step in mercapturic acid formation. J. Biol. Chem. 1974;n 249:7130-7139.
22. Kakkar P, Das B, Viswanathan PN. A modified spectrophotometric assay of

Superoxide dismutase. Indian J. Biochem. Biol. 1984;21:130-132.
23. Dacie JV, Lewis SM. Practical Hematology. Churchill Livingstone. New York. 1990;34.
24. Murrray IK, Mayer PA, Granner OK, Rodwel V. Cholesterol biosynthesis, egradation and transport. In: Harper's Biochemistry. 2nd Edn. Prentice Hall International Inc. 1990;240-254.
25. Boles SJ, Bradley D, Force RA, Brown BR. The new dietary Fats in Health and Diseases. J. Am. Diabetic Assoc. 1994; 97:280-286.
26. Vance IS. Metabolism of cholesterol. In: Biochemistry of Lipids and Lipoprotein and Membranes. Amsterdam Elsevier Science Publishers, Amsterdam; 1990.
27. Esterbaur H, Gebicki J, Puhl H, Jurgens G. The role of Lipid peroxidation and antioxidants in oxidation modification of LDL. Free Radic. Biol. Med. 1992;13:341-390.
28. Steinberg D. Thematic review series. The pathogenesis of atherosclerosis. An interpretive history of cholesterol controversy, part 5. The discovery of the statins and the end of the controversy. J. Lipid Res. 2006;47:1339-1351.
29. Del RD, Stewart AJ, Pellegrini N. A review of recent studies on malondialdehyde as toxic molecule and biological marker of oxidative stress. Nutr. Metab Cardiovasc. Dis. 2005;15(4):316-328.
30. Ferguson LR. Role of plant polyphenols in genomic stability. Mutat. Res. 2001; 475:89-111.
31. Ross JA, Kasum CM. Dietary flavonoids: Bioavailability, metabolic effects, and safety. Annu. Rev. Nutr. 2002;22:19–34.
32. Ren W, Qiao Z, Wang H, Zhu L, Zhang L. Flavonoids: Promising anticancer agents. Med. Res. Rev. 2003;23(4):519-534.
33. Prestera T, Holtzclaw WD, Zhang Y, Talalay P. Chemical and molecular regulation of enzymes that detoxify carcinogens. Proc. Natl. Acad. Sci. USA. 1993;90:2965-2969.
34. Pantuck EJ, Pantuck CB, Anderson KE, Wattenberg LW, Conney AH, Kappas A. Effect of Brussels sprouts and Cabbage on drug conjugation. Clin. Pharmacol. Ther. 1984;35:161-169.
35. Farombi EO. African Indigenous plants with chemotherapeutic potentials and biotechnological approach to the production of bioactive prophylactic agents. Afr. J. Biotechnol. 2003;2(12):662-671.
36. Opara MN, Iwuji TC, Igwe IN, Etuk IF, Maxwell JA. Haematological and Biochemical Responses of Adult Rabbits to Aqueous Extract of *Ocimum gratissimum* Leaves. J. Phys. Pharm. Adv. 2012;2(9):301-306.
37. Aziza AME, Magda ME, Moshira MES, Hannen HMO, Doaa SI. Effect of *Gymnema sylvestre* R. Br. Leaves extract on certain physiological parameters of diabetic rats. Journal of King Saud University-Science. 2013;25(2):135-141.
38. Adepoju GKA, Odubena OO. Effects of Mucuna pruriens on some hematological and biochemical parameters in Albino rats. Journal of Medicinal Plants Research. 2009;3(2):73-76.

29

Physicochemical Properties and Organoleptic Quality of *M. oleifera* Seed Oil

A. N. Tsado[1*], B. Lawal[2], I. F. Ossamulu[2], E. S. Santali[1], E. N. Gana[1], M. I. Umar[3] and A. Y. Gimba[1]

[1]*Department of Basic and applied sciences, Niger State Polythechnic, Zungeru, Nigeria.*
[2]*Department of Biochemistry, Federal University of Technology, PMB 65, Minna, Niger State, Nigeria.*
[3]*National Cereals Research institute (NCRI) Baddegi P.M.B 8, Niger State, Nigeria.*

Authors' contributions

This work was carried out in collaboration between all authors. Authors ANT, BL and IFO designed the study, wrote the protocol and carried out all laboratories work and performed the statistical analysis. Author ENG managed the analyses of the study. Author BL wrote the first draft of the manuscript. Authors MIU and AYG managed the literature searches and edited the manuscript. All authors read and approved the final manuscript.

Editor(s):
(1) Carmen Lúcia de Oliveira Petkowicz, Federal University of Parana, Curitiba, Parana, Brazil.
Reviewers:
(1) Anonymous, Italy.
(2) Atef Mahmoud Mahmoud Attia, Biochemistry department, Biophysical laboratory, National Research Centre, Egypt.
(3) Bhaskar Sharma, Gyan Vihar University, Jaipur, India.
(4) Anonymous, Universidad Autónoma de Tamaulipas, México.

ABSTRACT

The study were undertaken in order to determine the physico-chemical properties and sensory attribute of oils extracted from the seeds of *Moringa oleifera.* The oils were extracted from the seeds and analyzed for moisture content, specific gravity, saponification value, refractive index, peroxide value, acid number, iodine value and sensory quality. The fresh oil was also characterized for its sensory attribute using three different foods. The seed oil had the following physical and chemical parameters. The color (yellow), moisture content (2.0%), specific gravity 1.0598 kg/dm$^{3)}$, refractive index (1.465±0.04), acid values (0.67 mg/KOH/g), iodine value (110g/100g), peroxide value (7.0 meq/kg), saponification value (178.10 mg/KOHg), melting point (36.7), Flash point (0c) (162) and rancidity (14.0). Organoleptic scoring show that yam fried with *Moringa* seed oil had

**Corresponding author: E-mail: Amos.infonet@gmail.com*

more taste, color, aroma than yam fried with vegetable oil while chicken and egg fried with vegetable oil had higher taste, color, aroma and overall acceptance scoring than chicken and egg fried with *Moringa* seed oil. The study showed that *Moringa* seed oil can be used as nutritional oil, industrial raw material and nutraceuticals.

Keywords: Moringa oleifera seed oil; physicochemical parameters; organoleptic assesment; oil extraction and characterization.

1. INTRODUCTION

Fats and oil whether the source is of animal or vegetable in origin represent the highest source of energy per unit weight that man can consume. Apart from being a source of reserved energy, fats deposits insulate the body against loss of heat and protect the vital organs against mechanical injury. They are an important food source for man, and are also extensively used for nutritional, cosmetic and industrial purposes. They are used for supplying essential fatty acids such as linolenic and arachidonic acids. Fats and oils are also used for producing drug dispersants in therapeutics [1].

The high world demands for oils and fats to meet the multiplex human consumption and the multitudinous industrial needs are the reasons for the increase in the importance of oil seeds and make them play an important role in the national economy of the producing countries [2].

Nigeria has over 150 million people and estimated to be 170,123,740 as at July 2012 which particularly use palm oil, ground nut oil and soya beans oil as edible oil [3]. To ensure the increase of oils and fats, it is necessary to continue not only with the development of new varieties with improved oil yields, but also to search for new sources of oil [4]. In identifying new source of edible oil, there is need to determine the physicochemical parameters of the oil so as to determine its edibility in terms of properties and composition.

Moringa oleifera Lam. (Moringaceae) is one of the 14 species of family Moringaceae, native to India, Africa, Arabia, Southeast Asia, South America, and the Pacific and Caribbean Islands [5]. Because *M. oleifera* has been naturalized in many tropic and sub-tropic regions worldwide, the plant is referred to by a number of names such as horseradish tree, drumstick tree, ben oil tree, miracle tree, *Moringa* plant and "Mother's Best Friend".

The *Moringa* plant has been consumed by humans throughout the century in diverse culinary ways [5]. Almost all parts of the plant are used culturally for its nutritional value, purported medicinal properties and for taste and flavor as a vegetable and seed. The leaves of *M. oleifera* can be eaten fresh, cooked, or stored as a dried powder for many months reportedly without any major loss of its nutritional value [6]. Epidemiological studies have indicated that *M. oleifera* leaves are a good source of nutrition and exhibit anti-tumor, anti-inflammatory, anti-ulcer, anti-atherosclerotic, anti-convulsant activities [7], and antianaemic effect [8]. Seeds of *M. oleifera* are also known as a great source of vitamins and minerals including calcium, copper, sulphur, vitamin A and B-vitamins [9]. In this study, attempt was made to evaluate the physicochemical properties and organoleptic quality of *M. oleifera* seed oil.

2. MATERIALS AND METHODS

2.1 Plant Collection

Moringa oleifera seeds were collected from Baddegi, Niger State Nigeria. Taxonomic identification of the plant was conducted by a Botanist in the Department of Biological Science, Federal University of Technology Minna, Niger State. The seeds were then dried to constant weight in an oven at 60°C for 24 h to remove moisture content and then ground using mechanical grinder, put in air tight container and stored in a desiccators for further analysis [10].

2.2 Oil Extraction

Oil from the Moringa seeds was extracted by continuous extraction in Soxhlet apparatus for 6 h using petroleum ether (40-60°C) as solvent according to the method described [11]. At the end of the extraction, the extraction solvent was evaporated in a rotary evaporator. The extracted oil was assayed using standard methods.

2.3 Physicochemical Characterization

Standard procedures of American Oil Chemists Society were used for indices values [11]. The procedures were also applied for acid value, iodine value, peroxide value and saponification value. Refractive index, color, melting point and specific gravity were determined using the recommended methods. (B+S) Bellingham + Stanley Refractometer (England) and tintometer were used to determine refractive index and color, respectively. The saponifiable matter in the oil was determined using standard methods [11], while the ester value was obtained by subtracting the acid value from the saponification value. Each of the analysis was performed in triplicate.

2.4 Organoleptic Evaluation

The *M. oleifera* seed oil was use to fry 3 different foods (yam, egg and chicken). The foods were evaluated for organoleptic properties (in comparison with the same kind of food fried with vegetable oil) according to the method as described by Stone [12]. Their organoleptic evaluations were carried out by 8 judges. All the judges forming the panel were conversant with the factor governing the quality of the sample. The fried food was evaluated organoleptically for color, taste, aroma and overall acceptability. The taste panelists were asked to rate the samples for color, taste, aroma and overall acceptability on 1-7 point hedonic scale,

where,

7=like extremely; 6=like well; 5=like
4=neither like nor dislike; 3=dislike
2=dislike well; 1=dislike extremely.

2.5 Statistical Analysis

Values were analyzed by one way analysis of variance (ANOVA) using statistical package for social science (SPSS) version 16 and presented as means±SE of the mean.

3. RESULTS

3.1 Physicochemical Parameters

The results of the physicochemical parameters of *M. oleifera* seed oil are presented in Table 1. The seed oil had the following physical and chemical parameters. The color (yellow), moisture content (2.0 %), specific gravity 1.0598 kg/dm$^{3)}$, refractive index (1.465±0.04), acid values (0.67 mg/KOH/g), iodine value (110 g/100 g), peroxide value (7.0meq/kg), saponification value (178.10 mg/KOHg), melting point (36.7), flash point(0c) (162) and rancidity (14.0).

Table 1. Physicochemical characteristics of *Moringa oleifera* seed oil

Physicochemical properties	*Moringa oleifera* seed oil
Specific gravity at 200C	1.0598±0.12
Refractive index at 290C	1.465±0.04
Acid value (mg/KOH/g)	0.67±1.93
Iodine value mg/100g	110.89±4.13
Peroxide value (mEq/kg)	7.00±0.12
Saponification value (mg-KOH/g)	178.10±2.35
Free fatty acid (as oleic acid)	0.99±0.02
Melting point	36.7±3.04
Flash point (0c)	162±7.32
Rancidity	14.0±1.09
Color	Yellow

Data are mean±SEM of triplicate determination

3.2 Orgnaoleptic Scoring

The results of the organoleptic scoring of foods fried with *Moringa* seed oil in comparison with foods fried with vegetable oil are shown in Table 2. Yam fried with *Moringa* seed oil had more taste, color, aroma and overall acceptance than plantain fried with vegetable oil while egg and chicken fried with vegetable oil had higher taste, color aroma and overall acceptance scoring than egg and chicken fried with *Moringa* seed oil (Table 2).

Table 2. Organoleptic assessment of *Moringa* seeds

Sensory attribute	Yam		Egg		Chicken	
	M.S.O	V.O	M.S.O	V.O	M.S.O	V.O
Taste	6.12±0.64	5.87±0.99	5.50±0.53	5.87±1.12	5.25±1.03	5.50±0.92
Color	5.7±1.03,	5.12±1.35	5.50±0.92	5.62±1.30	5.25±1.28	5.37±0.51
Aroma	5.75±1.03	5.00±0.53	5.50±1.30	5.87±1.12	5.00±1.06	5.12±1.12
Overall acceptance	6.12±0.83	5.69±0.74	5.50±.75	5.87±0.64	5.00±1.06	5.3±0.51

Data are Mean±SEM (N= 8); M.S.O = Moringa seed oil; V.O= Vegetable oil

4. DISCUSSION

The oil extracted from Moringa seed is yellow in color. The color of the oil is used preliminarily in judging the quality and in determining the degree of bleaching of the oil. The darker the color, the poorer the quality. Therefore the yellow color shows that the quality of the oil is good and confirms to Encyclopedia of Chemical Technology [13].

The moisture content of the oil was 2.0%; low moisture content of this seed oil is an indication of low perishability of the oil and this responsible for it relatively long shelf life [14]. In addition, the low moisture content shows the presence of a lesser amount of dirt and impurities in the oil [15]. Thus low moisture content observed in this study is an indication that the dirt and impurities in *Moringa* seed oil are very low. However this value is lower than the moisture contents (10.9%) of melon seed oil [13].

Acid value gives an idea of the free fatty acid composition due to enzymatic activities. The acid value obtained for *Moringa* seed oil in this study (0.67 mg/KOH/g) is very low as compared to the acid value (7.09 mgKOH/g) reported for melon seed oil [15], 1.68 mgKOH/g for almond seed oil [16] and 2.37 mgKOH/g for coconut oil [17]. This implies that Moringa seed oils contain low level of free fatty acid.

Oil with iodine value in the range of 100-150mg/100 g have higher affinity for oxygen when expose to atmosphere and cannot be classify as drying oil, thus the result obtained in this study (110g/100 g) compare with iodine value obtained for water melon (114.94 g/100 g) by [15]. The oil can form thicken, sticky but cannot develop into hard dry film. However this property makes them suitable for the production of soap [18]. These findings indicate that *Moringa* seed oil can be used as a raw material for the production of soap.

The value of saponification reflects the molecular weight of oil [19]. In this study saponification value obtained for Moringa seed oil was found to be high (178.10 mg/KOHg) and is comparable with the saponification value reported for palm kernel (190-206 mg/KOHg) and oil obtained from groundnut (188-196 mg/KOHg) [20]. High saponification value implies greater proportion of fatty acids of low molecular weight. Thus the high saponification value of Moringa seed oil indicates that the oil contained higher proportion of low molecular weight fatty acids. High saponification is an additional property that makes it suitable as a raw material for soap and lather industry [21].

The peroxide value reflects the degree of oil oxidative rancidity [22]. The peroxide values of fresh and stored *Moringa seed* oils were obtained to be 7.0. The peroxide values obtained for oil in this study is lower compared to the value (20.0 mEq/kg) reported for melon seed oil [15]. Oxidative rancidity is the addition of oxygen across the double bonds in unsaturated fatty acids in the presence of enzyme or certain chemical compounds [23]. The odour and flavor associated with rancidity are due to liberation of short chain carboxylic acids. High peroxide values are associated with higher rate of rancidity. The low peroxide values of *Moringa* seed oils obtained in this study is an indication that they are less liable to oxidative rancidity at room temperature [24]. These oils are fresh because the content of peroxide lower than 10 mEq O_2/kg and oil grow rancid when the content peroxide lies between 22 and 40.0 mEq O_2/kg [25].

The specific gravity of *Moringa* seed oil obtained in this study (1.0598 kg/dm^3) is higher than the value (0.956 kg/dm^3) for *Blighia sapida* oil [16]. However the refractive index (which reflects the purity the oil) obtained in this study (1.465) is similar to the reported value (1.462) for *Blighia sapida* oil [16]. The results indicate that the *Moringa* seed oils are of high purity.

The panelist's assessment indicated that yam fried with *Moringa* seed oil have higher acceptability for color, taste and flavor than the yam fried with commercial vegetable oil. However chicken and egg fried with commercial vegetable oil have more acceptable taste, color and aroma than those fried with *Moringa* seed oil. This finding is an indication that the sensory attribute of oil depends on the kinds of food in which the oil is used to prepare and thus *Moringa* seed oil is suitable for frying yam.

5. CONCLUSIONS

The study showed that *M. oleifera* may be considered as Nigeria's potential asset in the seed oil for industrial application. The oil has a considerable high quality and can be used as nutritional oil, industrial raw material and source of nutraceuticals.

COMPETING INTERESTS

Authors have declared that no competing interests exist.

REFERENCES

1. Ranften CO, Billy EH. Post-harvest Biotechnology of oil seeds. CRC Press. 1993;161-170
2. Charley H. Food science, 2nd Ed. New York: John Wiley and Sons Publishers. 2002;91-95.
3. Ottong, JG. Population of Nigeria. Online Nigeria.com, Available:www.onlinenigeria.com/Population/?blurb=137. Retrieved Jan 19, 2013.
4. Fadlalla E. Review of oil seeds production in the Sudan and it is importance to the Sudan economy. Second Sudanese oilseed symposium. 2000;7.
5. Iqbal S, Bhanger MI. Effect of season and production location on antioxidant activity of *Moringa oleifera* leaves grown in Pakistan. J. of Food Comp. and Anal. 2006;19:544-551.
6. Arabshahi DS, Devi DV, Urooj A. Evaluation of antioxidant activity of some plant extracts and their heat, pH and storage stability. Food Chemistry. 2007; 100:1100-1105.
7. Dahiru D, Obnubiyi JA, Umar HA. Phytochemical screening and anti-ulcerogenic effect of *Moringa*. Afri J. Trad. CAM. 2006;3(3):70-75.
8. Hisham MO, Mohamed ES, Elsiddig MB, Bashier O, Ali ME. Effect of ethanolic leaf extract of *Moringa oleifera* on aluminum-induced anemia in white albino rats. Jordan Journal of Biological Sciences, 2012;5(4):255–260.
9. Lowell JF. The miracle tree. ACP – EU, Dakar. 2002;137-139.
10. Ekpa OD. Bioinorganic constituent and possible use of female influorescence of oil palm W. Afri. J. Bio and Appl. Chem. 1995;40:13-18.
11. A.O.A.C. Association of official analytical chemist. Method of analysis (ised). Pub by association of official analytical chemist washinton D.C; 1990.
12. Stone CV. Sensory evaluation prattices. Florida: Academic Press. 1985;311.
13. Power WC. Kirk-othmer encyclopedia of chemistry technology, 3rd ed. New York: Wiley-Interscience Publishers. 2008;41.
14. Kester DE, Kader AA. Almonds: In: Encyclopedia of food science, food technology and nutrition, edited academic press limited (London and San Dicgo, C.A). Edited by Macrae R, Robinson RK, Sudler MJ. 1993;1:124-726.
15. Edidiong AE, Ubong M. Chemical analysis of *Citrullus lanatus* seed oil obtained from Southern Nigeria. Elixir Org. Chem. 2013; 54:12700-12703.
16. Akpabio UD. Evaluation of proximate composition, mineral element and anti-nutrient in almond (*Terminalia catappa)* seeds. Advances in Applied Science Research. 2012;3(4):2247-2252.
17. Peter AO. Chemical properties of coconut oil. West African Journal of Biological and Applied Chemistry. 1956;14:120-130.
18. Kinkela MI. Physicochemical properties and fatty acid composition of lipids extracted from some Nigerian fruits and seeds. Global Journal of Pure and Applied Sciences.2006;6:23-29.
19. Booth FEM, Wickens GE. Nontimber uses of selected and zone trees and shrubs in Africa. FAO Conservation Guide, Rome. 1988;92-101.
20. Cocks LV, Van Reed C. Laboratory handbook for oil and fat analyst. Academic Press, London. 1996;113-117.
21. Nzikou JM, Mvoula-Tsieri M, Matos L, Matouba E, Ngakegni AC, Linder M, Desobry SC. *Solanum nigrum L. seeds* as an Alternative sources of edible lipids and nutrition in Congo Brazzarille. Journal Applied Science. 2007;7:1107-1115.
22. Ekpa OD, Ekpa UJ. Comparison of the characteristics parameters and deterioration properties of oils from the *Fenera and dura* Variety of the oil palm. Nigerian Journal of Chemical Research. 2006;1:26-33.
23. Oyenuga VA, Fetuga BL. Some aspects of biochemical and nutritive value of thewater melon seeds (*Citrullus colocynthis*). J Sci Food Agric. 1975;26:843–854.

24. Odoemelam SA. Proximate composition and selected physicochemical properties of the seeds of African oil bean (*Pentaclethra marcrophylla*). Pak. J. Nutr. 2005;4:382-383.

25. Ojeh O. Effects of refining in the physical and chemical properties of cashew kernel oil. Journal Fats and Oils Technology. 1981;16:513-517.

Permissions

All chapters in this book were first published in IJBRR, by SCIENCE DOMAIN international; hereby published with permission under the Creative Commons Attribution License or equivalent. Every chapter published in this book has been scrutinized by our experts. Their significance has been extensively debated. The topics covered herein carry significant findings which will fuel the growth of the discipline. They may even be implemented as practical applications or may be referred to as a beginning point for another development.

The contributors of this book come from diverse backgrounds, making this book a truly international effort. This book will bring forth new frontiers with its revolutionizing research information and detailed analysis of the nascent developments around the world.

We would like to thank all the contributing authors for lending their expertise to make the book truly unique. They have played a crucial role in the development of this book. Without their invaluable contributions this book wouldn't have been possible. They have made vital efforts to compile up to date information on the varied aspects of this subject to make this book a valuable addition to the collection of many professionals and students.

This book was conceptualized with the vision of imparting up-to-date information and advanced data in this field. To ensure the same, a matchless editorial board was set up. Every individual on the board went through rigorous rounds of assessment to prove their worth. After which they invested a large part of their time researching and compiling the most relevant data for our readers.

The editorial board has been involved in producing this book since its inception. They have spent rigorous hours researching and exploring the diverse topics which have resulted in the successful publishing of this book. They have passed on their knowledge of decades through this book. To expedite this challenging task, the publisher supported the team at every step. A small team of assistant editors was also appointed to further simplify the editing procedure and attain best results for the readers.

Apart from the editorial board, the designing team has also invested a significant amount of their time in understanding the subject and creating the most relevant covers. They scrutinized every image to scout for the most suitable representation of the subject and create an appropriate cover for the book.

The publishing team has been an ardent support to the editorial, designing and production team. Their endless efforts to recruit the best for this project, has resulted in the accomplishment of this book. They are a veteran in the field of academics and their pool of knowledge is as vast as their experience in printing. Their expertise and guidance has proved useful at every step. Their uncompromising quality standards have made this book an exceptional effort. Their encouragement from time to time has been an inspiration for everyone.

The publisher and the editorial board hope that this book will prove to be a valuable piece of knowledge for researchers, students, practitioners and scholars across the globe.

List of Contributors

Hanaa M. Abdeen
Department of Medical Biochemistry and Molecular Biology, Faculty of Medicine, Mansoura University, Mansoura, Egypt

Nadia El-Menshawy and Ziad Tawheed
Department of Clinical Pathology, Faculty of Medicine, Mansoura University, Mansoura, Egypt

Nader El-Malky
Department of Internal Medicine, Faculty of Medicine, Mansoura University, Mansoura, Egypt

Abeer M. Khattab
Department of Ophthalmology, Faculty of Medicine, Mansoura University, Mansoura, Egypt

Ebenezer Idowu O. Ajayi, Emmanuel U. Modo, Adetayo O. Adebamowo, and Olufunso Olabode Olorunsogo
Department of Biochemistry, Laboratories for Biomembrane Research and Biotechnology, University of Ibadan, Ibadan, Nigeria

Ebenezer Idowu O. Ajayi
Department of Chemical Science, Biochemistry Unit, Faculty of Basic and Applied Sciences, College of Science, Engineering and Technology, Osun State University, Osogbo, Nigeria

Uttam Chand Banerjee and Ebenezer Idowu O. Ajayi
Department of Pharmaceutical Technology/ Biotechnology, National Institute of Pharmaceutical Education and Research, SAS Nagar, Punjab, 160 062, India

Olumide O. Tewe
Department of Animal Science, Agricultural Biochemistry and Nutrition Division, University of Ibadan, Nigeria

Parisa Sistani, Mina Adampourezare, Gholamreza Dehghan, Amin Adami and Leila Mehdizadeh Fanid
Department of Biology, Faculty of Natural Science, University of Tabriz, Tabriz, Iran

Kouangbé Mani Adrien, Boga Gogo Lucien and Djaman Allico Joseph and N'Guessan Jean David
Biosciences Department, Biochemical Pharmacodynamy Laboratory, University of Félix Houphouët- Boigny, Côte d'Ivoire

Kouangbé Mani Adrien, Bahi Calixte, Tia Honoré and Edoh Vincent
Department of Bacteriology-Virology, Central Laboratory of University Hospital of Treichville, Abidjan, Côte d'Ivoire

Djaman Allico Joseph
Department of Clinical and Basic Biochemistry, Pasteur Institute of Côte d'Ivoire, Côte d'Ivoire

Hifsa Mobeen, Hamdan Hamid, Almina Shafiq, Muhammad Adil and Muhammad Kashif
Department of Immunology, University of Health Sciences Lahore, Pakistan

Noshin Wasim Yousuf
Department of Pathology, Allama Iqbal Medical College Lahore, Pakistan

Shalini Gupta, Lasanthi Jayathilaka, Jin-Sheng Huang, and Bao-Shiang Lee
Protein Research Laboratory, Research Resources Center, University of Illinois at Chicago, 835 S. Wolcott Avenue, Chicago, IL 60612, USA

Jenny Lee
Department of Chemical Engineering, Columbia University, 500 W. 120th Street, New York, NY 10027, USA

Ayman Zaky Elsamanoudy, Moustafa Ahmed Mohamed Neamat-Allah, Dina Abd el Daim Ahmed Ibrahim and Raymonde Hanna Assaf
Medical Biochemistry and Molecular Biology, Faculty of Medicine, Mansoura University, Egypt

Ayman Zaky Elsamanoudy
Clinical Biochemistry, Faculty of Medicine, King Abdulaziz University (KAU), KSA

Hatem A. El-Alfy
Tropical Medicine, Faculty of Medicine, Mansoura University, Egypt

Abousree Taha Ellethy
Department of Oral and Medical Basic Sciences, Biochemistry Division, College of Dentistry, Qassim University, AL-Qassim, Saudi Arabia

J. Y. Jr. Hena, H. N. Nyambaka and M. P. Nawiri
Kenyatta University, Nairobi, Kenya

J. Y. Jr. Hena
University of Liberia, Monrovia, Liberia

D. M. Andala
Multimedia University of Kenya, Nairobi, Kenya

Sadiq Mareai and Javaraiah Rajesh
Department of Biochemistry, Yuvaraja's College, University of Mysore, Mysore, Karnataka, India

Sadiq Mareai
Department of Biochemistry, Applied Science College, University of Thamar, Thamar, Yemen

Shaukath Khanum
Department of Chemistry, Yuvaraja's College, University of Mysore, Mysore, Karnataka, India

A. C. C. Argolo and L. C. B. B Coelho
Departamento de Bioquímica, Centro de Ciências Biológicas, Universidade Federal de Pernambuco, UFPE, Recife, PE, Brasil

V. C. R. Dantas, A. C. R. D. Saturnino and J. Brandão-Neto
Programa de Pós-Graduação em Ciências da Saúde, RN, Brasil

A. F. S. Santos
CEB-Centre of Biological Engineering, University of Minho, Braga, Portugal

A. D. O. Paixão
Departamento de Fisiologia e Farmacologia, Centro de Ciências Biológicas, Universidade Federal de Pernambuco, UFPE, Recife, PE, Brasil

Ruby Singh and C. G. Suresh
Division of Biochemical Sciences, CSIR-National Chemical Laboratory, Pune, 411008, India

Shaik Rahiman and Tarek H El-Metwally
Departments of Medical Biochemistry, Al Jouf University, Saudi Arabia

Shaik Rahiman and Divya Shrivastava
Department of Life and Basic Sciences, Jaipur National University, Rajasthan, India

Tarek H El-Metwally
Departments of Medical Biochemistry, Assiut University, Egypt

Parisa Sadighara and Rooholla Ghanbari
Department of Environmental Health Engineer, Food Safety Division, Faculty of Public Health, Tehran University of Medical Sciences, Tehran, Iran

Zafar S. Saify and Ajmal Khan
HEJ Research Institute of Chemistry, University of Karachi, Karachi-75270, Pakistan

Nighat Sultana
Pharmaceutical Research Center, PCSIR Laboratories Complex, Karachi-75280, Pakistan

Shazia Haider
Department of Pharmaceutical Chemistry, Faculty of Pharmacy, University of Karachi, Karachi-75270, Pakistan

Solomon Anjuwon Laleye, Babatunde Idowu Aderiye and Oluwole Moses David
Department of Microbiology, Ekiti State University, Ado-Ekiti, 360101, Nigeria

Mohammad Mustakim, Md. Mizanur Rahman, Anaytulla and Mohammed Abu Sayeed
Department of Pharmacy, International Islamic University Chittagong, 154/A, College Road,
Chittagong-4203, Bangladesh

Prawej Ansari and Md. Foyez Ahmad
Department of Pharmaceutical Sciences, School of Health and Life Sciences, North South University, Dhaka-1229, Bangladesh

P. Rama Bhat, V. H. Savitri, P. G. Laxmi and E. P. Jenitta
Department of Biotechnology, Alva's College, Moodbidri - 574 227, Karnataka, India

A. B. M. Sharif Hossain
Department of Biology, Program of Biotechnology, Faculty of Science, University of Hail, KSA

Program of Biotechnology, ISB, Faculty of Science, University of Malaya, Kl, Malaysia

Mohammed Saad AlEissa
Department of Biology, Faculty of Science, AI-imam Muhammad Ibn Saud Islamic University, KSA

M. I. Bello and M. A. Madusolumuo
Department of Biochemistry, Modibbo Adama University of Technology, P.M.B. 2076 Yola, Nigeria

I. Igbokwe
Department of Veterinary Medicine, University of Maiduguri, P.M.B. Maiduguri, Nigeria

F. G. Okibe, E. D. Paul, G. A. Shallangwa and Y. A. Dallatu
Department of Chemistry, Ahmadu Bello University, Zaria, Kaduna State, Nigeria

B. Jubril
Department of Food Science and Technology, Ibrahim Badamasi Babangida University, Lapai,
Niger State, Nigeria

N. E. Nwankwo, O. F. C. Nwodo, A. E. Amalunweze, K. U. Agbo and S. C. Abugu
Department of Biochemistry, University of Nigeria, Nsukka, Enugu State, Nigeria

A. E. Irondi and J. K. Akintunde
Department of Biosciences and Biotechnology, Biochemistry Unit, Kwara State University, Malete, P.M.B. 1530, Ilorin, Nigeria

E. J. Iweala
Department of Biological Sciences, Covenant University, Ota, Ogun State, Nigeria

O. S. Agboola
Department of Veterinary Physiology, Biochemistry and Pharmacology, University of Ibadan, Nigeria

J. A. Ajiboye
Department of Chemical Sciences, Biochemistry unit, College of Natural and Applied Sciences,
Bells University of Technology, Ota, Ogun State, Nigeria

A. S. Idoko, N. Lawal and A. Balkisu
Department of Biochemistry, Federal University Dutsin-Ma, Katsina State, Nigeria

M. Osibemhe and B. O. Abdulrahman
Department of Biochemistry and Molecular Biology, Faculty of Science and Education, Federal
University Dutsin-Ma, Katsina State, Nigeria

I. O. Onoagbe
Department of Biochemistry, Faculty of Life Sciences, University of Benin, Benin City, Edo State,
Nigeria

A. Z. Ogunleye and M. F. Asaolu
Department of Biochemistry, Ekiti State University, Ado Ekiti, Ekiti State, Nigeria

O. E. Etim, U. E. Bassey, E. E. Sambo and E. J. Akpan
Department of Biochemistry, Faculty of Basic Medical Sciences, University of Uyo, Nigeria

G. E. Charles
Department of Pharmaceutical/Medicinal Chemistry, Faculty of Pharmacy, University of Uyo,
Nigeria

Friday O. Uhegbu, Chinedu Imo and Amadike E. Ugbogu
Department of Biochemistry, Abia State University, PMB 2000, Uturu, Nigeria

A. N. Tsado, E. S. Santali, E. N. Gana and A. Y. Gimba
Department of Basic and applied sciences, Niger State Polythechnic, Zungeru, Nigeria

B. Lawal and I. F. Ossamulu
Department of Biochemistry, Federal University of Technology, PMB 65, Minna, Niger State, Nigeria

M. I. Umar
National Cereals Research institute (NCRI) Baddegi P.M.B 8, Niger State, Nigeria